T0192145

Lecture Notes in Bioinformatics 13760

Subseries of Lecture Notes in Computer Science

More information about this subseries at https://link.springer.com/bookseries/5381

Mukul S. Bansal · Zhipeng Cai ·
Serghei Mangul (Eds.)

Bioinformatics Research and Applications

18th International Symposium, ISBRA 2022
Haifa, Israel, November 14–17, 2022
Proceedings

Editors
Mukul S. Bansal
University of Connecticut
Storrs, CT, USA

Zhipeng Cai
Georgia State University
Atlanta, GA, USA

Serghei Mangul
University of Southern California
Los Angeles, CA, USA

ISSN 0302-9743 ISSN 1611-3349 (electronic)
Lecture Notes in Bioinformatics
ISBN 978-3-031-23197-1 ISBN 978-3-031-23198-8 (eBook)
https://doi.org/10.1007/978-3-031-23198-8

LNCS Sublibrary: SL8 – Bioinformatics

This Springer imprint is published by the registered company Springer Nature Switzerland AG
The registered company address is: Gewerbestrasse 11, 6330 Cham, Switzerland

Preface

The 18th International Symposium on Bioinformatics Research and Applications (ISBRA 2022) was held at the University of Haifa, Israel, during November 14–17, 2022. The symposium provides a forum for the exchange of ideas and results among researchers, developers, and practitioners working on all aspects of bioinformatics and computational biology and their applications. This proceedings volume includes 30 regular papers and 4 short papers selected by the Program Committee from among 72 submissions received in response to the call for papers. Each submission received at least three reviews in an single-blind process.

We would like to thank the Program Committee members and external reviewers for volunteering their time to review and discuss symposium papers. We would like to extend special thanks to the Steering and General Chairs of the symposium for their leadership, and to the Publication, Publicity, and Local Organization Chairs for their hard work in making ISBRA 2022 a successful event. Last but not least, we would like to thank all authors for presenting their work at the symposium.

October 2022

Mukul S. Bansal
Zhipeng Cai
Serghei Mangul

Organization

Steering Committee

Dan Gusfield	UC Davis, USA
Ion Mandoiu	UConn, USA
Yi Pan (Chair)	SIAT, China and GSU, USA
Marie-France Sagot	Inria, France
Zhirong Sun	Tsinghua University, China
Ying Xu	UGA, USA
Aidong Zhang	SUNY, USA

General Co-chairs

Sagi Snir	University of Haifa, Israel
Alexander Zelikovsky	GSU, USA

Vice General Chair

Tianwei Yu	The Chinese University of Hong Kong – Shenzhen, China

Program Co-chairs

Zhipeng Cai	GSU, USA
Mukul Bansal	UConn, USA
Serghei Mangul	University of Southern California, USA

Publicity Chairs

Olga Glebova	GSU, USA
Yanjie Wei	SIAT, China

Web Chair

Jaden Moore	University of Southern California, USA
Filipp Rondel	GSU, USA

Program Committee

Derek Aguiar	University of Connecticut, USA
Tatsuya Akutsu	Kyoto University, Japan
Kamal Al Nasr	Tennessee State University, USA
Mukul S. Bansal	University of Connecticut, USA
Paola Bonizzoni	Università di Milano-Bicocca, Italy
Dan Brown	University of Waterloo, Canada
Zhipeng Cai	Georgia State University, USA
Hongmin Cai	South China University of Technology, China
Weiyang Chen	Qufu Normal University, China
A. Ercument Cicek	Bilkent University, Turkey
Ovidiu Daescu	The University of Texas at Dallas, USA
Alexandre G. De Brevern	DSIMB, INSERM and Paris Cité University, France
Pufeng Du	Tianjin University, China
Nadia El-Mabrouk	University of Montreal, Canada
Oliver Eulenstein	Iowa State University, USA
David Fernández-Baca	Iowa State University, USA
Xin Gao	King Abdullah University of Science and Technology, Saudi Arabia
Pawel Gorecki	University of Warsaw, Poland
Xuan Guo	University of North Texas, USA
Nurit Haspel	University of Massachusetts Boston, USA
Matthew Hayes	Xavier University of Louisiana, USA
Zengyou He	Dalian University of Technology, China
Steffen Heber	North Carolina State University, USA
Wooyoung Kim	University of Washington, USA
Sergey Knyazev	University of California, Los Angeles, USA
Danny Krizanc	Wesleyan University, USA
Soumya Kundu	Stanford University, USA
Manuel Lafond	Université de Sherbrooke, Canada
Yaohang Li	Old Dominion University, USA
Ran Libeskind-Hadas	Claremont McKenna College, USA
Kevin Liu	Michigan State University, USA
Ion Mandoiu	University of Connecticut, USA
Erin Molloy	University of Maryland, College Park, USA
Marmar Moussa	University of Connecticut, USA
Sheida Nabavi	University of Connecticut, USA
Murray Patterson	Georgia State University, USA
Andrei Paun	University of Bucharest, Romania
Nadia Pisanti	University of Pisa, Italy

Yuri Porozov	I.M. Sechenov First Moscow State Medical University, Russia
Russell Schwartz	Carnegie Mellon University, USA
Joao Setubal	University of São Paulo, Brazil
Jian-Yu Shi	Northwestern Polytechnical University, China
Xinghua Shi	Temple University, USA
Yi Shi	Shanghai Jiao Tong University, China
Pavel Skums	Georgia State University, USA
Ileana Streinu	Smith College, USA
Emily Chia-Yu Su	Taipei Medical University, Taiwan
Sing-Hoi Sze	Texas A&M University, USA
Weitian Tong	Georgia Southern University, USA
Valentina Tozzini	Istituto Nanoscienze - CNR, Italy
Jianrong Wang	Michigan State University, USA
Jianxin Wang	Central South University, China
Seth Weinberg	The Ohio State University, USA
Fangxiang Wu	University of Saskatchewan, Canada
Yi-Chieh Wu	Harvey Mudd College, USA
Yufeng Wu	University of Connecticut, USA
Alex Zelikovsky	Georgia State University, USA
Le Zhang	Sichuan University, China
Fa Zhang	Institute of Computing Technology, Chinese Academy of Sciences, China
Lu Zhang	Hong Kong Baptist University, Hong Kong, China
Louxin Zhang	National University of Singapore, Singapore
Wei Zhang	University of Central Florida, USA
Quan Zou	Tianjin University, China

Additional Reviewers

Agnieszka Mykowiecka
Alina Nemira
Arjun Srivatsa
Changsheng Ma
Chinmayee Rayguru
Fatemeh Mohebbi
Guihua Tao
Gun Kaynar
Hanmin Li
Herdiantri Sufriyana
Jiao Sun
Julia Zheng
Jun Gao
Ke Xu
Khandakar Tanvir Ahmed
Kiril Kuzmin
Lei Liu
Liu Liangjie
Manal Almaeen
Mattéo Delabre
Meijun Gao
Mofan Feng
Pavel Avdeyev
Qiang Yang
Runqiu Chi
Samson Weiner

Contents

MLMVFE: A Machine Learning Approach Based on Muli-view Features Extraction for Drug-Disease Associations Prediction

Ying Wang[1], Ying-Lian Gao[2], Juan Wang[1], Junliang Shang[1], and Jin-Xing Liu[1(✉)]

[1] School of Computer Science, Qufu Normal University, Rizhao 276826, Shandong, China
sdcavell@126.com

[2] Qufu Normal University Library, Qufu Normal University, Rizhao 276826, Shandong, China

Abstract. Determining the associations between drugs and diseases plays an important role in the drugs development processes. However, current drug-disease associations (DDAs) prediction methods are too homogeneous for features extraction, so a machine learning approach based on multi-view features extraction (MLMVFE) is proposed for DDAs prediction. Firstly, proteins are introduced to form a new heterogeneous network, which enriches the associations information. Then, nodes features are extracted from two perspectives: network topology and biological knowledge. Finally, the Light Gradient Boosting Machine classifier is utilized to predict DDAs. The MLMVFE achieves satisfactory results on both B-dataset and F-dataset through 10-fold cross-validation. In addition, to further demonstrate the reliability of the MLMVFE, case study is done where clozapine is used as a case. The result suggests that the MLMVFE has the potential to tap into novel DDAs.

Keywords: Biological knowledge · Network topology · Multi-view features extraction · Drug-disease associations prediction

1 Introduction

In recent years, a growing number of diseases are endangering human health. There-fore, it has become urgent to discover drugs to treat these diseases. Nevertheless, the discovery and development of drugs cost a lot of time and money. Previous studies have shown that the research for a novel drug costs $2.6 billion and takes an average of 12 years [1]. Therefore, it is critical to find an effective way to identify drug-disease associations (DDAs), which can increase the speed of drug development.

Previous studies have demonstrated that network-based approaches outperform other approaches, which can extract useful information from different kinds of bio-logical networks to improve the performance of drug repositioning [2] providing a new direction for DDAs prediction. However, the common way of network-based approaches is to construct heterogeneous network, which ignores the inherent features of various molecules. Therefore, it is arduous to utilize the potential knowledge in the network. In addition, it has been shown that considering nodes attributes can contribute to the accuracy analysis

M. S. Bansal et al. (Eds.): ISBRA 2022, LNBI 13760, pp. 1–8, 2022.
https://doi.org/10.1007/978-3-031-23198-8_1

of complex networks [3–5]. However, it is uncommon to consider both network topology and biological knowledge (BK) of the heterogeneous networks in DDAs. One of the main reasons for this phenomenon is that there are few approaches that can correctly handle the two types of information. Moreover, the key role of proteins is ignored in most prediction methods for DDAs. The expression of proteins affects the symptom of diseases and the effect of drugs. To be specific, drug can alleviate disease effect by affecting enzymes in the organism. For example, valproic acid can reduce the lifecycle of breast cancer cells by affecting the expression of histones [6]. Hence, the introduction of proteins is crucial in DDAs prediction. In addition, heterogeneous network in DDAs predictions is usually sparse, therefore, proteins can be integrated into DDAs prediction network to enhance connectivity.

In this paper, a novel approach named MLMVFE is proposed in which proteins are introduced to form a heterogeneous network. Then, network topology and BK are considered simultaneously for better learning drugs and diseases feature. Especially, the BK of drugs and diseases can be attained in two different ways, and then autoencoder is utilized to attain the feature vectors more simply. In addition, the DeepWalk [7] is utilized to obtain the network representations (NRs) of the drugs and diseases from a network topology perspective. The biological representations and the feature vectors obtained from the network topology are stitched together and then fed into the Light Gradient Boosting Machine (LGB) [8] classifier to obtain the final prediction result. In addition, 10-fold cross-validation (10-CV) is conducted on two datasets, and the consequences show that the MLMVFE outperforms the extant approaches.

2 Materials and Methods

2.1 Datasets

In this paper, the B-dataset is utilized as a benchmark dataset, which consists of drug-disease, drug-protein, and protein-protein associations. The DDAs network containing 269 drugs, 598 diseases, and 18416 known DDAs is obtained from the CTD database [9]. Besides, drug-protein associations network containing 969 drugs, 613 proteins, and 11107 drug-protein associations can be obtained from the Drugbank database [10]. Protein-disease associations network can be attained from the DisGeNET database [11], which consists of 832 proteins, 692 diseases, and 25087 protein-disease associations.

In addition, the F-dataset is utilized to further validate the generalizability of the approach. The DDAs network includes 593 drugs, 313 diseases, and 1933 DDAs. The drug-protein and protein-disease associations in F-dataset are also downloaded via Drugbank and DisGeNET, which includes 3243 drug-protein associations and 71840 protein-disease associations. The flow chart of MLMVFE can be shown in Fig. 1.

2.2 Heterogeneous Network

To describe a heterogeneous network (HN), a three-element tuple is proposed. HN = {N, B, I} in which $N = \{N^{dr}, N^{di}, N^{pr}\}$ represents the nodes of drugs, diseases, and proteins, respectively. The $B = \{B^{dr}, B^{di}\}$ stands for the BK of the drugs and diseases. I

= {DDAs, DPAs, PDAs} is drug-disease, drug-protein, and protein-disease associations, respectively. In addition, it is assumed that the n and k are the number of drugs and the biological attributes of drugs, respectively, and m represents the number of diseases. The $\mathbf{B}^{dr} \in R^{n*k}$ and $\mathbf{B}^{di} \in R^{m*m}$ are n*k matrix and m*m matrix, respectively.

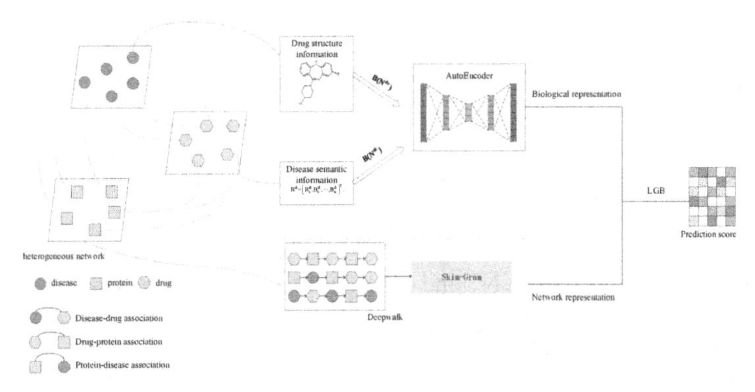

Fig. 1. The overall flow chart of MLMVFE.

2.3 Biological Knowledge Extraction

It is known from previous studies that if drug molecules have akin chemical structures, they are usually participated in the same biological activities [1]. Thus, chemical information can be utilized. To obtain the chemical structures of drugs, chemical descriptors can be attained from the Simplified Molecular Input Line Entry System (SMILES) [13], which is available in the DrugBank database (https://go.drugbank.com/). Then, the RDKit tool is used to check if a specific chemical structure is present in drug molecules. Then, the B^{dr} can be attained by applying the above process to all drugs. If the corresponding chemical structure is present, it is set to 1, otherwise it is set as 0 in which k chemical structures exist.

Based on the fact that if the disease-related drugs are similar then diseases are similar [14], medical subject descriptors obtained from the Medical Subject Headings (MeSH) thesaurus can obtain biological information about diseases. Especially, the Mesh tree structure represents the correlations among diseases and is calculated as representation vectors. Thus, a directed acyclic graph (DAG) is applied to describe each disease. The generalized Jaccard formula is then applied to compute the similarity of any two diseases. $\mathbf{DAG}_{N_a^{di}} = (N_a^{di}, A(N_a^{di}), C(N_a^{di}))$ in which the $A(N_a^{di})$ and $C(N_a^{di})$ represent the ancestor nodes and a collection of all links of the N_a^{di}. The $V(N_a^{di})$, the contribution of N_k^{di} to N_a^{di}, is calculated as follow:

$$\begin{cases} V_{N_a^{di}}(N_k^{di}) = 1, \text{ if } N_a^{di} = N_k^{di}, \\ V_{N_a^{di}}(N_k^{di}) = \max\left\{\lambda * V_{N_a^{di}}(N_k^{di'}) | N_k^{di'} \in \text{children of } N_k^{di}\right\}, \text{ if } N_a^{di} \neq N_k^{di}, \end{cases} \tag{1}$$

where λ represents the semantic contribution. The contribution of N_k^{di} is related to the distance between diseases and the semantic value, which can be calculated by the following formula:

$$VD(N_a^{di}) = \sum_{N_k^{di} \in A(N_a^{di})} V_{N_a^{di}}(N_k^{di}). \tag{2}$$

According to the above two formulas, the semantic similarity of N_c^{di} and N_a^{di} can be obtained as follows:

$$\mathbf{Lik}(N_a^{di}, N_c^{di}) = \frac{\sum_{N_k^{di} \in A(N_a^{di}) \cap A(N_c^{di})} \left(V_{N_a^{di}}(N_k^{di}) + V_{N_c^{di}}(N_k^{di})\right)}{VD(N_a^{di}) + VD(N_c^{di})}. \tag{3}$$

The semantic similarity between N_a^{di} and other diseases can be expressed as the attribute information of N_a^{di}. $B_a^{di} = [\mathbf{Lik}\,(N_a^{di}, N_c^{di})](1 \le c \le M)$ in the assumption in which B_a^{di} is the corresponding row of N_a^{di} in B^{di}. The B^{di} is described as follows:

$$B^{di} = \left[B_a^{di}, B_b^{di}, \cdots, B_m^{di}\right]^T. \tag{4}$$

In addition, since the diseases identifiers in the F-dataset are different from the Mesh, the Mesh is not available. Therefore, when performing this procedure in the F-dataset, each element in B^{di} should be set to 0. In addition, to handle redundancy and sparsity of raw data, the autoencoder, a neural network model, can be used to reduce the dimensionality of B^{di} and B^{dr}.

2.4 Network Representations for Diseases and Drugs

Network topology information denotes the links between pairs of nodes and is more complex than biological information involving only single disease and drug, which means that the incorporation of the information is necessary. DeepWalk, a graph representation learning algorithm, is utilized to extract NRs of drugs and diseases, which inputs pairs of nodes and learns each node by random walk, and the output is obtained by skip-gram. The sequence of random walk from y_a to y_{i-1} $(1 \le i \le |Y|)$ can be represents as $\{y_a, y_b, \cdots y_{i-1}\}$. Then the possibility of the next node arriving at y_i can be expressed as:

$$Pr(y_i | (y_a, \cdots, y_{i-1})). \tag{5}$$

The mapping function $\Phi : y \in Y \to R^{|Y| \times n_2}$ is imported to attain vector representation of each node in which n_2 is set as 64. The above formula can be reworded as:

$$Pr(y_i | (\Phi(y_a), \cdots, \Phi(y_{i-1}))). \tag{6}$$

Then, the skip-gram model is utilized to compute the formula (9). The formula is as follows:

$$\underset{\Phi}{\text{minimize}} - \log \Pr(\{y_{i-p} \cdots, y_{i-1}, y_{i+1}, \cdots, y_{i+p}\} \\ |\Phi(y_i)) = \prod_{\substack{o=i-p \\ o \neq i}}^{i+p} \Pr(y_j|\Phi(i)), \tag{7}$$

where w is the range of neighboring nodes of y_i. The $\Phi(N) \in R^{|N| \times n_2}$ of all nodes of Y can be attained by solving the formula (10). In addition, a $(J + M) \times n_2$ matrix $\mathbf{T}$ represents the representation vectors of drugs and diseases where $\mathbf{T} = \Phi(\{N^{dr}, N^{di}\})$.

2.5 Light Gradient Boosting Machine Classifier

To accomplish the DDAs prediction, the MLMVFE employs LGB classifier. Pairs of drugs and diseases are trained as training sets in which drugs and diseases representation vectors are concatenated as input to the LGB. Then, the **F** is introduced to store the prediction results with 1 or 0 in which the 1 represents the presence of associations, and the 0 represents the absence of associations.

3 Results and Discussion

3.1 Effects of Parameters

In the process of analyzing the parameters, the 10-CV is used to repeat ten times, and then is averaged to obtain the prediction results. The depth and learning rate are used as examples to demonstrate the tuning process where depth belongs to {2, 4, 6, 8}, and learning rate belongs to {0.01, 0.03, 0.06, 0.09, 0.1}. Firstly, the depth is fixed in turn, and then the 10-CV results corresponding to the learning rate are obtained separately, so that until all the 10-CV results are obtained, the parameter corresponding to the highest prediction result is the optimal parameter. The tuning diagram of the two indicators can be shown in Fig. 2. It can be seen that the optimal result is obtained when the learning rate is 0.09, and depth is 6.

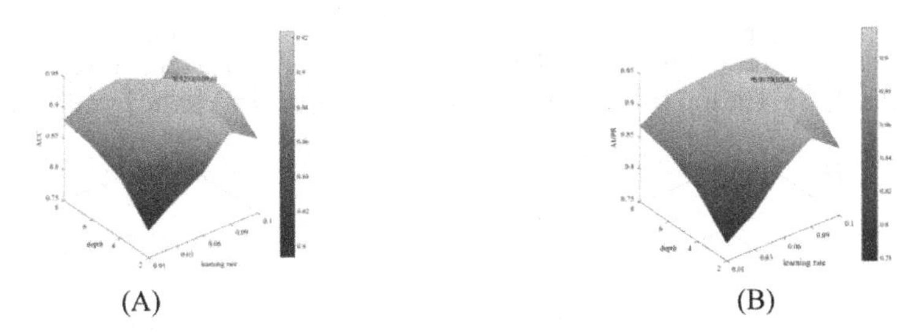

Fig. 2. The 10-CV analysis of depth and learning rate on B-dataset: (A) AUC; (B) AUPR.

3.2 Ablation Experiments

To study the effect of different features extraction of the MLMVFE, ablation experiments are done. The model using only BK is denoted as BK1, and the model using only NRs is denoted as NR. Since BK and NRs yield similar performance in both datasets, the B-dataset is chosen for analysis. In addition, the comparison graphs of AUPR and AUC are shown in Fig. 3. The BK1 has the worst performance, which means the BK of drugs and diseases alone cannot be relied on for better prediction of DDAs. However, the importance of BK cannot be denied, especially in cases where new diseases are not associated with existing drugs, which may be a strong indicator. In the model using only NRs, the performance is less different from our approach on each metric, which suggests that NRs can better attain the features of drugs and diseases.

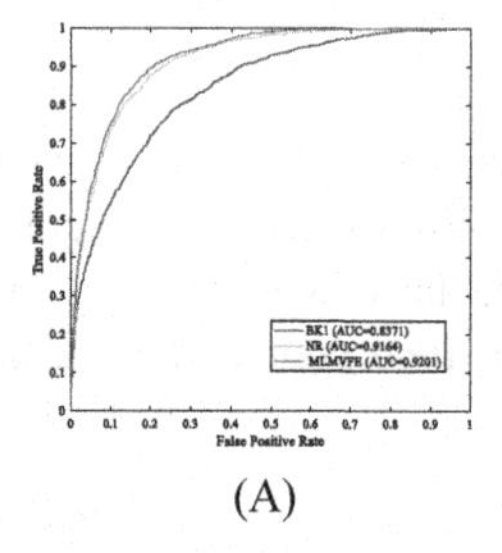

(A)

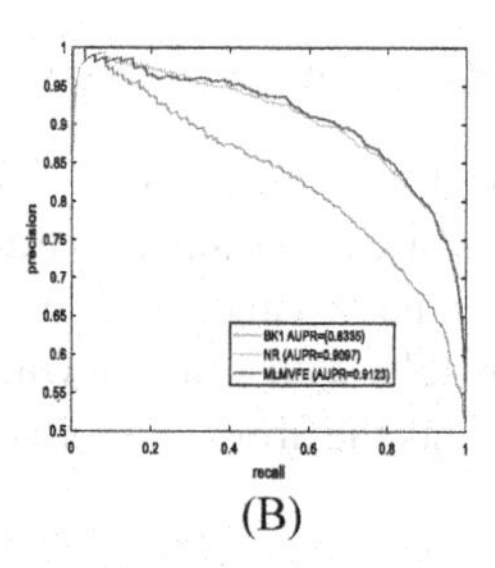

(B)

Fig. 3. The AUC and AUPR results for MLMVFE, MLMVFE without BK, and MLMVFE without NRs at 10-CV on B-dataset.

3.3 Comparison with the Latest Methods

To further appraise the performance of the MLMVFE, some existing prediction methods for DDAs are used for comparison with it, namely LAGCN, DTINet, deepDR, and HINGRL. The Fig. 4 shows more graphically the comparative results of these methods on B-dataset. The consequences indicate that the MLMVFE is excellent in the prediction of DDAs.

3.4 Case Study

To further validate the ability of the MLMVFE to discover novel associations, clozapine is utilized as a case. The top 10 diseases associated with clozapine can be listed in Table 1.

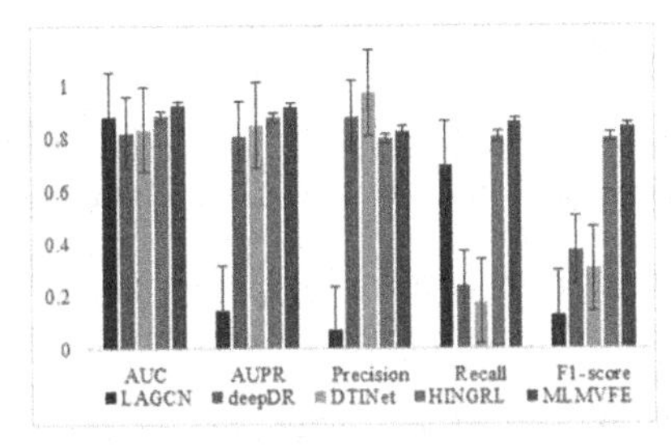

Fig. 4. The 10-CV comparison consequences of AUC, AUPR, Precision, Recall, and F1-score of five methods on B-dataset.

Table 1. The top 10 diseases related to clozapine obtained by MLMVFE.

Drug	Disease	Evidence
Clozapine	Anxiety disorders	CTD
	Somnambulism	NA
	Stress disorders, post-traumatic	NA
	Headache	CTD
	Parkinson disease, secondary	CTD
	Sleep initiation and maintenance disorders	CTD
	Torsades de pointes	[15]
	Memory disorders	CTD
	Hyperalgesia	CTD
	Inappropriate adh syndrome	[16]

4 Conclusion

In this paper, a novel approach is proposed, which takes a more comprehensive perspective on features extraction. Firstly, proteins are integrated into the original network constituting a complex HN, which can further enrich the associations information. Then, MLMVFE captures drugs and diseases features from the perspective of network topology and BK, respectively, which facilitates a more comprehensive access to diseases and drugs features. Finally, the LGB classifier is used to predict DDAs. Experimental consequences on two datasets show that MLMVFE has superior accuracy as well as generalization. In addition, case study further demonstrates that MLMVFE can be a powerful tool for discovering DDAs, especially for novel diseases or drugs.

Acknowledgment. This work was supported by the National Natural Science Foundation of China (Grant No. 62172254).

References

1. Chan, H.C.S., Shan, H.B., Dahoun, T., et al.: Advancing drug discovery via artificial intelligence. Trends. Pharmacol. Sci. **40**, 801 (2019). https://doi.org/10.1016/j.tips.2019.07.013
2. Yang, M.Y., Wu, G.Y., Zhao, Q.C., et al.: Computational drug repositioning based on multi-similarities bilinear matrix factorization. Brief. Bioinform. **22**, bbaa267 (2021). https://doi.org/10.1093/bib/bbaa267
3. Chu, Y.Y., Kaushik, A.C., Wang, X.G., et al.: DTI-CDF: a cascade deep forest model towards the prediction of drug-target interactions based on hybrid features. Brief. Bioinform. **22**, 451–462 (2021). https://doi.org/10.1093/bib/bbz15
4. Hu, L., Chan, K.C.C., Yuan, X.H., et al.: A variational bayesian framework for cluster analysis in a complex network. IEEE Trans. Knowl. Data. Eng. **32**, 2115–2128 (2020). https://doi.org/10.1109/TKDE.2019.2914200
5. Hu, L., Zhang, J., Pan, X.Y., et al.: HiSCF: leveraging higher-order structures for clustering analysis in biological networks. Bioinformatics **37**, 542–550 (2021). https://doi.org/10.1093/bioinformatics/btaa775
6. Aztopal, N., Erkisa, M., Erturk, E., et al.: Valproic acid, a histone deacetylase inhibitor, induces apoptosis in breast cancer stem cells. Chem. -Biol. Interact. **280**, 51–58 (2018). https://doi.org/10.1016/j.cbi.2017.12.003
7. Li, G.H., Luo, J.W., Wang, D.C., et al.: Potential circRNA-disease association prediction using DeepWalk and network consistency projection. J. Biomed. Inform. **112**, 103624 (2020). https://doi.org/10.1016/j.jbi.2020.103624
8. Liang, J.C., Bu, Y.D., Tan, K.F., et al.: Estimation of stellar atmospheric parameters with light gradient boosting machine algorithm and principal component analysis. Astron. J. **163**, 153 (2022). https://doi.org/10.3847/1538-3881/ac4d97
9. Davis, A.P., Grondin, C.J., Johnson, R.J., et al.: The comparative toxicogenomics database: update 2019. Nucleic. Acids. Res. **47**, D948–D954 (2019). https://doi.org/10.1093/nar/gky868
10. DrugBank 5.0: a major update to the DrugBank database for 2018. Nucleic. Acids. Res. 46, D1074-D1082 (2017). https://doi.org/10.1093/nar/gkx1037
11. Pinero, J., Bravo, A., Queralt-Rosinach, N., et al.: DisGeNET: a comprehensive platform integrating information on human disease-associated genes and variants. Nucleic. Acids. Res. **45**, D833–D839 (2017). https://doi.org/10.1093/nar/gkw943
12. Huang, L., Luo, H.M., Li, S.N., et al.: Drug-drug similarity measure and its applications. Brief. Bioinform. **22**, bbaa265 (2021). https://doi.org/10.1093/bib/bbaa265
13. Nesmerak, K., Toropov, A.A., Toropova, A.P., et al.: SMILES-based quantitative structure-property relationships for half-wave potential of N-benzylsalicylthioamides. Eur. J. Med. Chem. **67**, 111–114 (2013). https://doi.org/10.1016/j.ejmech.2013.05.031
14. Yan, S.H., Yang, A.M., Kong, S.S. et al.: Predictive intelligence powered attentional stacking matrix factorization algorithm for the computational drug repositioning. Appl. Soft. Comput. **110**, 107633 (2021).
15. Warner, B., Hoffmann, P.: Investigation of the potential of clozapine to cause torsade de pointes. Adverse. Drug. React. Toxicol. Rev. **21**, 189–203 (2002)
16. Fujimoto, M., Hashimoto, R., Yamamori, H., et al.: Clozapine improved the syndrome of inappropriate antidiuretic hormone secretion in a patient with treatment-resistant schizophrenia. Psychiat. Clin. Neuros. **70**, 469 (2016). https://doi.org/10.1111/pcn.12435

STgcor: A Distribution-Based Correlation Measurement Method for Spatial Transcriptome Data

Xiaoshu Zhu[1,2](✉), Liyuan Pang[1], Wei Lan[1], Shuang Meng[4], and Xiaoqing Peng[3]

[1] School of Computer, Electronics, and Information Science and Engineering, Guangxi University, Nanning 530004, Guangxi, China
xszhu@csu.edu.cn

[2] School of Computer Science and Engineering, Yulin Normal University, Yulin 537000, Guangxi, China

[3] Center for Medical Genetics School of Life Sciences, Central South University, Changsha 400083, Hunan, China

[4] School of Computer Science and Engineering, Guangxi Normal University, Guilin 541006, Guangxi, China

Abstract. Spatial transcriptome technology can provide the transcript profiles of different regions in a tissue sample while preserving the positional information of spots. Based on spatial transcriptome data, the construction of gene regulatory networks can help researchers to identify gene modules and understand the biological process. The correlation measurement is the basis of the construction of gene regulatory networks. Typical correlation coefficients, such as the Pearson correlation coefficient and Spearman correlation coefficient, are hard to be applied to spatial transcriptome data, due to the outliers and sparsity in spatial transcriptome data. In this study, by observing the distribution of gene-pair expression values, we propose a novel gene correlation measurement method for spatial transcriptome data, named STgcor. In STgcor, a gene pair (X, Y) expressed on spots is represented as a vertex in a two-dimensional plane consisting of the gene pair vectors X and Y as coordinate axes. We calculate the joint probability density of Gaussian distributions of the gene pair (X, Y) to exclude outliers. Then, to overcome the sparsity of spatial transcriptome data, the correlation between the gene pair (X, Y) is measured based on the degree, trend, and location of aggregation of the distribution of gene-pair expression values. To validate the effectiveness of the STgcor, STgcor and two other correlation coefficients were applied to a weighted co-expression network analysis method on two spatial transcriptome datasets published by 10x genomics including the breast cancer dataset and prostate cancer dataset. They were evaluated based on the gene modules identified from the corresponding gene co-expression networks. The results showed that STgcor combined with the weighted gene co-expression network analysis method can discover special gene modules and some pathways related to cancer.

Keywords: Spatial transcriptome data · Correlation measurement · Gene module identification · Gene co-expression network · Distribution of gene-pair expression · Probability density

M. S. Bansal et al. (Eds.): ISBRA 2022, LNBI 13760, pp. 9–18, 2022.
https://doi.org/10.1007/978-3-031-23198-8_2

1 Introduction

With the development of sequencing technology, many technologies had been developed to sequence gene expressions [1–4]. Recently, spatial transcriptome (ST) sequencing technology emerged, in which tissue was divided into many spots that contained a dozen to dozens of cells, and the average gene expression of the spot was sequenced [5, 6]. Therefore, the location information of spots was preserved [7–11]. Due to solving the problem of single cell location loss, the ST technology had been further investigated in cell heterogeneity during cell differentiation [12–17].

In ST data analysis, spatial domain partitioning and gene module identification were focused on by researchers. The former discovered the cell types to explore cell differentiation, and the latter helped researchers understand gene function and construct gene regulatory networks at the system level [18–21]. Gene correlation measurement was an essential step in gene module identification. PCC and SPCC were the typical gene correlation measurement methods for ST datasets [22–24]. However, PCC was easily disturbed by outliers, and difficult to accurately measure the correlation between high-sparseness vectors. SPCC improved the anti-interference against outliers, but it also suffered difficulty in high-sparseness vectors.

Therefore, for the high-noise and high-sparseness in ST data, a novel gene correlation measurement method, named STgcor, was designed based on considering the distribution of gene-pair expression vertexes, such as the degree, trend, and location. To verify the validity of STgcor, the weighted gene co-expression network analysis (WGCNA) method [25–28] was combined to construct a gene co-expression network and identify gene modules [29–32]. STgcor was compared to PCC and SPCC on the breast cancer dataset and prostate cancer dataset of the spatial transcriptome.

2 Methods and Materials

Inspired by the gene co-expression and the distribution of gene expression, it was found that the degree, trend, and location of aggregation of distribution of gene expression had an important impact on gene correlation measures. PCC and SPCC had not considered this and treated each gene equally, which would suffer a poor accuracy in ST data with high sparseness. On this basis, a novel STgcor method was proposed to measure the correlation between gene pairs in ST data.

ST data included tissue image, gene expression matrix, and spots coordinate matrix. In this work, the gene expression matrix $\boldsymbol{M}$ with n genes and m spots was as input, in which m_{ij} of matrix $\boldsymbol{M}$ denoted the expression level of the i-th gene in the j-th spot. For each gene, m expression values in m spots were represented as an m-dimensional vector X. To characterize and observe the distribution of gene pairs expression in spot, a two-dimensional plane was constructed based on the gene pair vectors X and Y as the coordinate axes, and the spot expressed on a gene pair was represented as a gene-pair expression vertex. So, for the gene pair (X, Y), there were m gene-pair expression vertexes in the two-dimensional plane. The flowchart of the STgcor method was shown in Fig. 1.

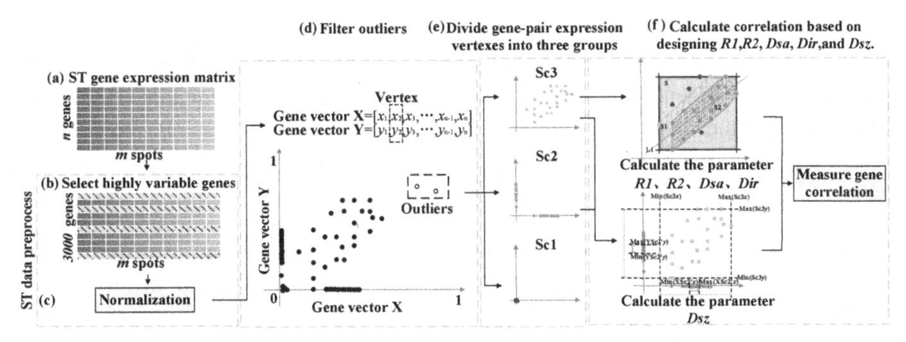

Fig. 1. The flowchart of STgcor. (a) An ST gene expression matrix ***M*** as input; (b) 3000 high variable genes were selected to construct a new gene expression matrix; (c) Max-min normalization was performed; (d) Based on a two-dimensional coordinate plane formed by gene pair vectors *X* and *Y*, outliers were filtered; (e) Gene-pair expression vertexes were divided to three groups; (f) Correlation based on designing *R1*, *R2*, *Dsa*, *Dir*, and *Dsz* was measured.

2.1 Preprocessing

Gene Selection. The increasing research showed that the top 2000 or 3000 highly variable genes hold more effective information to cluster cells, so the seurat_v3 package in the scanpy method was performed to select 3000 highly variable genes [33].

Normalization. To convert the range of gene expression values to the interval [0,1], the gene expression matrix was normalized using the max-min normalization.

2.2 Gene Correlation Measurement Based on the Distribution of Vertexes

Filter Outliers. To filter the outliers of gene-pair expression vertexes, the joint probability density of the Gaussian distribution of gene-pair vectors *X* and *Y* were calculated. P_i was the product of Px_i and Py_i, denoting the joint probability density of the *i*-th vertex, shown in formula (1). Px_i and Py_i represented the probability density of the *i*-th vertex in vectors *X* and *Y*, respectively. The probability density was commonly 0.03. So if $P_i \leq 0.001$, the *i*-th vertex were considered outliers and removed.

$$P_i = P_{x_i} \times P_{y_i} = \frac{e^{\frac{-(x_i-\mu_x)^2}{2\sigma_x{}^2}}}{\sqrt{2\pi\sigma_x{}^2}} \times \frac{e^{\frac{-(y_i-\mu_y)^2}{2\sigma_y{}^2}}}{\sqrt{2\pi\sigma_y{}^2}} \tag{1}$$

where μ_x was the mathematical expectation of vector *X*, and σ_x was the standard deviation of vector *X*.

Divide the Gene-Pair Expression Vertexes into Three Groups. To resolve the high-sparseness in ST data, a gene-pair expression vertexes division strategy was designed. According to the degree of gene pair expressed on a spot, the gene-pair expression vertexes were divided into three groups. In the first group *Sc1*, the *x* coordinate and *y* coordinate of gene-pair expression vertexes were less than 0.03, that is, gene pairs were both zero-expressed at the spots; In the second group *Sc2,* only one of the x coordinate and y coordinate of gene-pair expression vertexes was less than or equal to 0.03; In the third group *Sc3,* the x coordinate and y coordinate of gene-pair expression vertexes were greater than 0.03. *Sc3* was the main group for that gene pairs was both expressed at the spots.

Estimate the Trend of Co-expression based on Linear Regression. For *Sc3,* based on the assumption that the gene pair vectors X and Y were related when the distribution of gene-pair expression vertexes was aggregated, an area coverage ratio was designed to measure the correlation.

First, the area S was calculated by constructing the minimum rectangle covering the four gene-pair expression vertexes with the maximum or minimum x-coordinate, and maximum or minimum y-coordinate, respectively.

Second, based on the unary linear regression line, the trend of co-expression of gene pair vectors X and Y was estimated according to the location of gene-pair expression vertexes. The unary linear regression lines L_1 with the gene pair vectors X and Y were calculated as x- and y-axes, respectively. L_2 was just the opposite. Then, the angle between L_1 and 45 angles was denoted as θ_{L1}, and the angle between L_2 and 45 angles as θ_{L2}. When the angle of the unary linear regression line was close to 0 angle or 90 angles, meaning that the unary linear regression was close to x- or y-axes, there was little correlation between vectors X and Y.

Third, the sum of the distances from each vertex to L_1 and L_2 was calculated, and the top 10% of vertexes were removed. Then, the area S_1 was calculated, which was formed by connecting the fewest vertexes using polylines to enclose all vertexes. The top 10% of vertexes in S_1 were removed in the same way, forming the area S_2. So, the ratio of S_1 and S_2 to S was calculated respectively, as $R1$ and $R2$. When R_1 and R_2 were close to 1, the gene pair vectors X and Y were independent. Otherwise, the gene pair vectors X and Y may be related.

Measure Gene Correlation based on the Location of the Distribution. For the remaining 80% of vertexes in *Sc3, Dsa* was defined to further describe the location of the distribution of vertexes according to *L1* and *L2,* shown in formula (2).

$$Dsa = \begin{cases} \begin{matrix} R_1 > 0.5 \text{ and} \\ R_2 > 0.4 \end{matrix} & : \begin{cases} 0, & \theta_L \leq 20 \\ (\theta_L - 450)^{3.5}, & \theta_L > 20 \end{cases} \\ \begin{matrix} R_1 > 0.5 \text{ or} \\ R_2 > 0.4 \end{matrix} & : \begin{cases} 0, & \theta_L \leq 25 \\ (\theta_L - 45)^5, & \theta_L \geq 35 \\ (\frac{\theta_L}{45})^{10-\frac{\theta_L - 25}{2}}, & 35 > \theta_L > 25 \end{cases} \end{cases} \tag{2}$$

where θ_L was equal to θ_{L2}, when the angle between θ_{L1} and 45° was less than the angle between θ_{L2} and 45°. Otherwise, it was equal to θ_{L1}.

Then, the ratio of the number of vertexes in *Sc2* and *Sc3* was calculated, as N. The larger the N was, the less correlated the vertexes were. Meanwhile, the negative correlation between gene pair vectors X and Y was defined when the slope of the linear regression line corresponding to θ_L was less than 0, denoted as $Dir = -1$.

For $Sc2$, the gene-pair expression vertexes close to the X- or Y-axis were defined as $XSc2$ or $YSc2$. Based on the idea of reducing the influence of extreme vertexes, half of the vertexes located at the boundary of Gaussian distribution were removed while the central vertexes were retained, as $XSc2'$ and $YSc2'$. So, $XSc2'_x$ and $YSc2'_y$ correspondingly represented the x- and y-coordinate. Dsz and Dsr were defined as formula (3).

$$
\begin{aligned}
T_x &= \begin{cases} \frac{\max(XSc2'_x)-\min(XSc2'_x, Sc3_x)}{\max(Sc3_x)-\min(Sc3_x)}, & Dir = 1 \\ \frac{|\min(XSc2'_x)-\max(XSc2'_x, Sc3_x)|}{\max(Sc3_x)-\min(Sc3_x)}, & Dir = -1 \end{cases} \\
T_y &= \begin{cases} \frac{\max(XSc2'_y)-\min\left(XSc2'_y, Sc3_y\right)}{\max(Sc3_y)-\min(Sc3_y)}, & Dir = 1 \\ \frac{|\min(XSc2'_y)-\max(XSc2'_y, Sc3_y)|}{\max(Sc3_y)-\min(Sc3_y)}, & Dir = -1 \end{cases} \\
Dsz &= \begin{cases} ,0 & N \leq 0.1 \\ \max(T_x, T_y \times 1.5^{\lfloor N \rfloor} \times N), & N > 0.1 \end{cases} \\
Dsr &= 1 - R1^{2.55-\frac{\lfloor 10\times R_1 \rfloor}{10}} - R2^{2.35-\frac{\lfloor 10\times R_2 \rfloor}{10}} - (R1 - R2)^{2-(R_1-R_2)\times 3}
\end{aligned}
\tag{3}
$$

Finally, $STgcor$ was defined to measure the correlation between gene pair vectors X and Y, shown in formula (4).

$$
STgcor = relu(Dsr - Dsa - Dsz) \times Dir \tag{4}
$$

3 Result

To test the validity of STgcor, STgcor was compared to two classical gene correlation measurement methods PCC and SPCC by embedding in WGCNA. The workflow of our experiment was shown in Fig. 2.

3.1 Datasets

Two ST datasets published by 10x genomics from the websites (https://www.10xgenomics.com/resources/datasets) were downloaded, including the prostate cancer dataset and the Breast Cancer dataset. The former was Human Prostate Cancer, Adenocarcinoma with Invasive Carcinoma (HPCAIC), containing 4371 spots and 5391 median genes per spot. The latter was Human Breast Cancer Ductal Carcinoma In Situ, Invasive Carcinoma (HBCDC), containing 2518 spots and 5244 median genes per spot. The histological images of the two ST datasets were shown in Fig. 3.

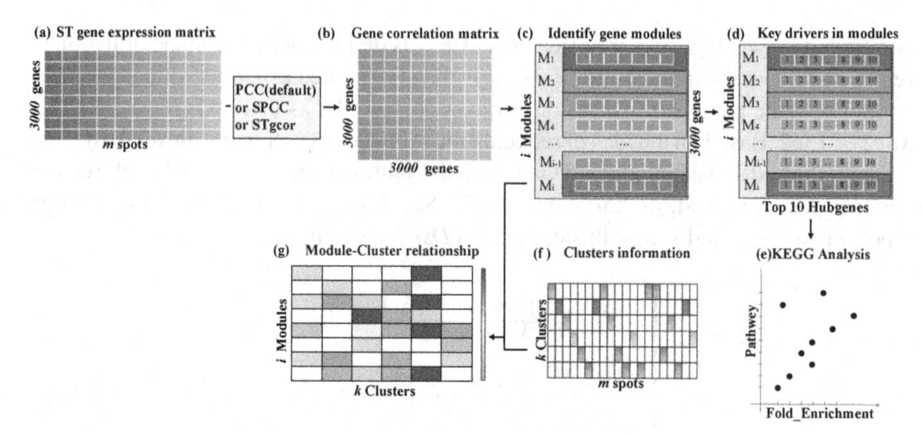

Fig. 2. The workflow of analyzing the ST dataset combining WGCNA with STgcor, PCC, and SPCC. (a)An ST gene expression matrix (3000 × m) was input; (b) A gene relation matrix was calculated by PCC, SPCC, of STgcor, respectively; (c) Gene modules were identified; (d) Key drivers in modules were identified; (g) By combining the clusters information (f), the module-cluster relationship was calculated; (e) KEGG analysis was performed.

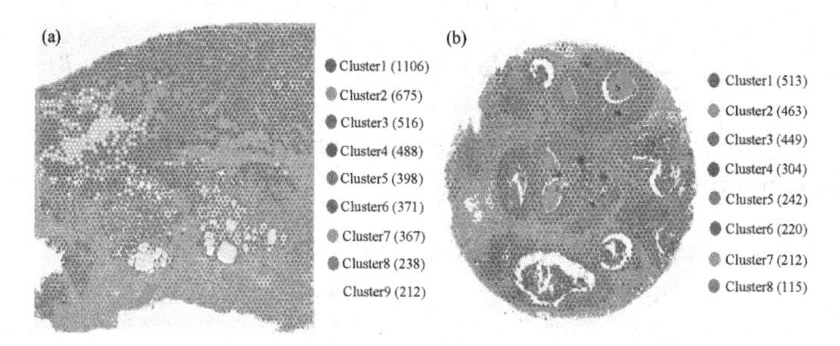

Fig. 3. The clustering of spots in histological images. (a) HPCAIC dataset; (b) HBCDC dataset.

3.2 Comparison of the Performance of Gene Module Identification

STgcor, SPCC, and PCC were applied in WGCNA to compare their performance in gene module identification. The heat map of the relationship between gene modules and clusters from STgcor, PCC, and SPCC were provided, respectively, shown in Fig. 4.

In Fig. 4, each color block on the left column represented a gene module. For example, In Fig. 4 (a), there were five gene modules including the brown, yellow, blue, turquoise, and grey gene modules. MEbrown represented the important features extracted by PCA in the brown gene module containing 50 genes. The value at the intersection of the MEbrown row and Cluster1 column was the STgcor correlation between the MEbrown gene module and Cluster1.

From Fig. 4, STgcor showed better performance in measuring gene correlation and identifying gene modules than PCC and SPCC. Such as, in the HPCAIC dataset, STgcor discovered a brown gene module strongly positively correlated with Cluster1 and a blue gene module that only strongly negatively correlated with Cluster1. In the same way, in the HBCDC dataset, STgcor discovered a turquoise gene module strongly positively

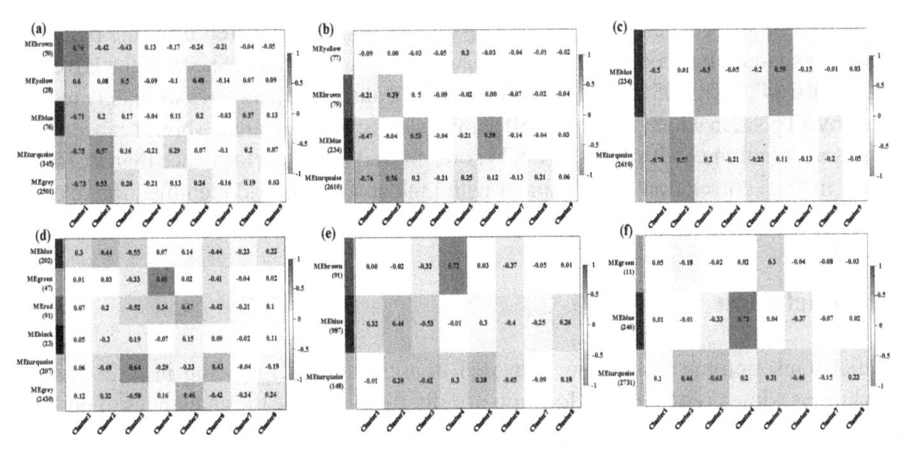

Fig. 4. The heat map of the relationship between gene modules and clusters. (a) STgcor was performed on the HPCAIC datasets; (b) PCC was performed on the HPCAIC datasets; (c) SPCC was performed on the HPCAIC datasets; (d) STgcor was performed on the HBCDC datasets; (e) PCC was performed on the HBCDC datasets; (f) SPCC was performed on the HBCDC datasets.

correlated with Cluster3. However, PCC and SPCC could not find them. The two novel strong positive correlations may help researchers discover something.

3.3 Comparison of the KEGG Enrichment Analysis

Based on the database of the Kyoto Encyclopedia of Genes and Genomes (KEGG), the top 10 hub genes in every gene module were analyzed. Dozens or hundreds of KEGG pathways were discovered by PCC, SPCC, and STgcor, and the KEGG pathways found only by one method were displayed. All pathways discovered by SPCC were also found in PCC and STgcor, so the pathways discovered by SPCC were not shown. The results of the KEGG analysis were shown in Fig. 5.

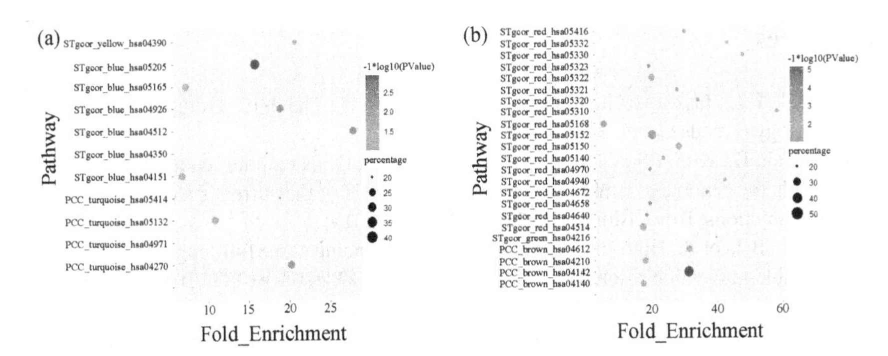

Fig. 5. The pathways discover from KEGG enrichment analysis. (a) HPCAIC dataset; (b) HBCDC dataset.

From Fig. 5(a), seven special KEGG pathways were discovered from the blue and yellow gene modules identified by STgcor, which were related to cancer. In the HBCDC dataset, hsa04216, hsa05150, hsa05168, hsa04970, and other pathways were only recognized by STgcor, in which the hsa05150 pathway was closely related to breast cancer[34]. Therefore, combined with WGCNA, STgcor achieved better performance than PCC and SPCC in gene correlation measurement with high sparseness and identified the relative genes into a module.

4 Conclusion

Spatial transcriptome technology provided transcriptome sequencing in a dozen cell resolutions while preserving the location of cells. To overcome the challenges of outliers and high sparseness in ST data, a novel gene correlation measurement STgcor was proposed by considering the degree, trend, and location of the distribution of gene-pair expression vertexes. After highly variance gene selection and normalization, the joint probability density of Gaussian distributions of gene pairs (X, Y) was calculated to filter outliers. By observing the distribution of the gene-pair expression vertexes, they were divided into three categories according to the degree of gene expression level. For *Sc3* with both expressed on gene pairs, the trend of gene pair co-expression was estimated based on the unary linear regression. The correlation of gene pairs (X, Y) was measured by combining the number and distribution of vertexes in *Sc2*. In the breast cancer ST dataset and prostate cancer ST dataset, STgcor combined with WGCNA discovered some pathways related to cancer from several discovered special gene modules.

Authors' Contributions. Conceptualization, X.Z., L.P., and X.P.; Methodology, X.Z., L.P., and W.L.; Software, L.P., and S.M.; Writing-Original Draft Preparation, X.Z., L.P., and X.P.; Visualization, L.P.; Funding Acquisition, X.Z.

Funding Statement. This research was supported by the National Natural Science Foundation of China (No. 62141207).

References

1. Van Dijk, E.L., Jaszczyszyn, Y., Naquin, D., Thermes, C.: The third revolution in sequencing technology. Trends Genet. **34**, 666–681 (2018)
2. Senol Cali, D., Kim, J.S., Ghose, S., Alkan, C., Mutlu, O.: Nanopore sequencing technology and tools for genome assembly: computational analysis of the current state, bottlenecks and future directions. Brief. Bioinform. **20**, 1542–1559 (2019)
3. Callahan, B.J., et al.: High-throughput amplicon sequencing of the full-length 16S rRNA gene with single-nucleotide resolution. Nucleic Acids Res. **47**, e103–e103 (2019)
4. Usoskin, D., et al.: Unbiased classification of sensory neuron types by large-scale single-cell RNA sequencing. Nat. Neurosci. **18**, 145–153 (2015)
5. Lun, A., Bach, K., Marioni, J.C.: Pooling across cells to normalize single-cell RNA sequencing data with many zero counts. Genome Biol. **17**, 75 (2016)
6. Wei, Y., Zhang, X.-D., Hu, M.-M., Wu, Z.-Q., Cheng, M., Guo, Y.: Advances in spatial transcriptome technologies. Progress Biochem. Biophys. **49**, 561–571 (2022)

7. Li, Z., Peng, G.: Spatial transcriptomics: new dimension of understanding biological complexity. Biophys. Rep. **7**, 1–17 (2021)
8. Xu, C., et al.: Human-specific features of spatial gene expression and regulation in eight brain regions. Genome Res. **28**, 1097–1110 (2018)
9. Moncada, R., et al.: Integrating microarray-based spatial transcriptomics and single-cell RNA-seq reveals tissue architecture in pancreatic ductal adenocarcinomas. Nat. Biotechnol. **38**, 333–342 (2020)
10. Hildebrandt, F., et al.: Spatial Transcriptomics to define transcriptional patterns of zonation and structural components in the mouse liver. Nat. Commun. **12**, 1–14 (2021)
11. Friedrich, S., Sonnhammer, E.L.: Fusion transcript detection using spatial transcriptomics. BMC Med. Genomics **13**, 1–11 (2020)
12. Crosetto, N., Bienko, M., Van Oudenaarden, A.: Spatially resolved transcriptomics and beyond. Nat. Rev. Genet. **16**, 57–66 (2015)
13. Ståhl, P.L., et al.: Visualization and analysis of gene expression in tissue sections by spatial transcriptomics. Science **353**, 78–82 (2016)
14. Vickovic, S., et al.: High-definition spatial transcriptomics for in situ tissue profiling. Nat. Methods **16**, 987–990 (2019)
15. Cho, C.-S., et al.: Microscopic examination of spatial transcriptome using Seq-Scope. Cell **184**, 3559–3572, e3522 (2021)
16. Xia, C., Fan, J., Emanuel, G., Hao, J., Zhuang, X.: Spatial transcriptome profiling by MERFISH reveals subcellular RNA compartmentalization and cell cycle-dependent gene expression. Proc. Natl. Acad. Sci. **116**, 19490–19499 (2019)
17. Yamazaki, M., et al.: Effective microtissue RNA extraction coupled with Smart-seq2 for reproducible and robust spatial transcriptome analysis. Sci. Rep. **10**, 1–8 (2020)
18. Chen, C., et al.: Two gene co-expression modules differentiate psychotics and controls. Mol. Psychiatry **18**, 1308–1314 (2013)
19. Deng, S.-P., Zhu, L., Huang, D.-S.: Predicting hub genes associated with cervical cancer through gene co-expression networks. IEEE/ACM Trans. Comput. Biol. Bioinf. **13**, 27–35 (2015)
20. Gaiteri, C., Ding, Y., French, B., Tseng, G.C., Sibille, E.: Beyond modules and hubs: the potential of gene coexpression networks for investigating molecular mechanisms of complex brain disorders. Genes Brain Behav. **13**, 13–24 (2014)
21. Wang, D.-W., et al.: Identification of prognostic genes for colon cancer through gene co-expression network analysis. Current Med. Sci. **41**(5), 1012–1022 (2021). https://doi.org/10.1007/s11596-021-2386-2
22. Zhu, X.-S., Li, H.-D., Guo, L.-L., Wu, F.-X., Wang, J.-X.: Analysis of single-cell RNA-seq data by clustering approaches. Curr. Bioinform. **14**, 314–322 (2019)
23. Van den Heuvel, E., Zhan, Z.: Myths about linear and monotonic associations: pearson'sr, Spearman's ρ, and Kendall's τ. Am. Stat. **76**, 44–52 (2022)
24. Fraidouni, N., Záruba, G.V.: Computational techniques to recover missing gene expression data. Adv. Sci. Tecnol. Eng. Syst. J. **3**, 233–242 (2018)
25. Langfelder, P., Horvath, S.: WGCNA: an R package for weighted correlation network analysis. BMC Bioinform. **9**, 1–13 (2008)
26. Liu, W., Li, L., Ye, H., Tu, W.: Weighted gene co-expression network analysis in biomedicine research. Sheng wu gong cheng xue bao Chinese J. Biotechnol. **33**, 1791–1801 (2017)
27. Tian, H., Guan, D., Li, J.: Identifying osteosarcoma metastasis associated genes by weighted gene co-expression network analysis (WGCNA). Medicine **97** (2018)
28. Jia, R., Zhao, H., Jia, M.: Identification of co-expression modules and potential biomarkers of breast cancer by WGCNA. Gene **750**, 144757 (2020)
29. Chen, W.-T., et al.: Spatial transcriptomics and in situ sequencing to study Alzheimer's disease. Cell **182**, 976–991. e919 (2020)

30. Fawkner-Corbett, D., et al.: Spatiotemporal analysis of human intestinal development at single-cell resolution. Cell **184**, 810–826, e823 (2021)
31. Hou, X., et al.: Integrating spatial transcriptomics and single-cell RNA-seq reveals the gene expression profling of the human embryonic liver. Front. Cell Dev. Biol. **9**, 652408 (2021)
32. Guo, X., Liang, J., Lin, R., Zhang, L., Wu, J., Wang, X.: Series-spatial transcriptome profiling of leafy head reveals the key transition leaves for head formation in Chinese cabbage. Front. Plant Sci. **12** (2021)
33. Wolf, F.A., Angerer, P., Theis, F.J.: SCANPY: large-scale single-cell gene expression data analysis. Genome Biol. **19**, 1–5 (2018)
34. Zhang, Y.-X., Du, Z.-G., Li, H.-J.: Weighted gene co-expression network analysis for excavation of Hub genes related tothe development of breast cancer. West China Med. J. **35**, 1074–1081 (2020)

Automatic ICD Coding Based on Multi-granularity Feature Fusion

Ying Yu[1,2,3], Junwen Duan[1,2(✉)], Han Jiang[1,2], and Jianxin Wang[1,2]

[1] School of Computer Science and Engineering, Central South University, Changsha, China
{yuying,jwduan,jh-better,jxwang}@mail.csu.edu.cn
[2] Hunan Provincial Key Lab on Bioinformatics, Central South University, Changsha, China
[3] School of Computer Science, University of South China, Hengyang, China

Abstract. International Classification of Disease (ICD) coding is to assign standard codes, which describe the state of a patient, to a clinical note. It is challenging given the complexity and the number of codes. The ICD taxonomy is hierarchically organized with several level codes (chapter, category, subcategory and its subdivision). However, most existing studies focus on the prediction of the fine-grained subcategory codes, neglecting the hierarchical relations of ICD codes. Those models pay less attention to common features related to sibling subcategories. The common features could be helpful for rare sample prediction and could be captured in the task of coarse-grained code prediction. In this paper, we propose a multi-task learning model, which explicitly trains multiple classifiers for different code levels. Simultaneously, we capture the relations between finer-grained and coarser-grained labels through a reinforcement mechanism. Extensive experiments on an English and a Chinese dataset show that our approach achieves competitive performance compared with baseline models, especially on Macro-F1 results.

Keywords: Multi-granularity · ICD code · Clinical notes · Automatic assignment

1 Introduction

International Classification of Disease (ICD)[1] proposed by the World Health Organization (WHO), has become the international standard diagnostic classification. It is convenient for storage, retrieval and analysis of the clinical data. ICD coding is generally carried out by professional and experienced coders according to a patient admission data, such as clinical notes or electronic medical record (EMR). Usually, several ICD codes, representing clinician diagnoses or hospital procedures, are assigned to a patient visit. However, the high dimensional coding space brings great challenge even for experienced coders. For example, the most widely used tenth version (ICD-10) contains over 20,000 codes. Meanwhile, manually ICD coding is inclined to be time-consuming, laborious and error-prone [2,3]. Accordingly, the quality and efficiency of coding is a matter of concern in practice.

Computer-assisted automatic ICD coding has drawn much attention recently. Automatic ICD coding can be essentially regarded as a multi-label text classification task.

M. S. Bansal et al. (Eds.): ISBRA 2022, LNBI 13760, pp. 19–29, 2022.
https://doi.org/10.1007/978-3-031-23198-8_3

Most approaches [4–8] focus on mining distinctive feature representation of input text in order to receive better classification performances. They treat the labels in one flat label space disregarding the nature of the hierarchical structure. Those models pay more attention to capturing distinctive features of text during training, while the common feature related to sibling subcategories, which could give a clue for zero-shot or few-shot prediction, are neglected. Actually, the classification is divided into several chapters (e.g. 22 chapters in ICD-10) at first. Each chapter are subdivided into homogeneous blocks of three-character categories. The fourth digit following a dot indicates additional details of three-character categories, with some optional five-digit or six-digit subdivisions based on local rules of countries or hospitals. Most of the time, four to six characters ICD-10 codes are used to label an EMR. Considering the nature of the hierarchical structure of ICD, we wonder whether we can extract different level features related to different level codes from the input text, and obtain a better representation of the clinical text by fusing multi-grained features.

In this paper, we propose a multitask model to learn label-specific features of different granularity and assign ICD codes to a discharge summary automatically by fusing multi-grained features. It aims to train multiple classifiers for different level ICD codes separately and predict candidate codes by combining features related to labels of different granularity. By using label-wise attention mechanism, label-specific feature of each label granularity is obtained independently based on the split of contextual representation vectors learning by one layer recurrent neural networks. Considering the distribution of labels in the data and the difference of the difficulty for different granularity predictions, a weighted fusion loss function is introduced to the model. The experiments on two real clinical datasets produced competitive results and proved the effectiveness of our model.

2 Related Work

Automatic ICD coding is an attractive research point in biomedical information field. It has been studied for more than two decades [9]. Machine learning methods and deep learning networks have been widely used and have received expected performances. As the earliest conventional machine learning method, Larkey and Croft [9] presented an ensemble model by combining three statistical classifiers to assign ICD codes to discharge summaries. While Perotte et al. [3] proposed two models based on support vector machine (SVM), a "flat" SVM and "hierarchical" SVM, for ICD-9 coding on MIMIC-II dataset. Similarly, Koopman et al. [10] tried to assign cancer-related codes to death certificates automatically by using SVM.

Recently, deep learning methods have been widely used in ICD coding. Some of them are based on convolutional neural networks (CNN) [6,7,11–14]. Some others employ recurrent neural networks (RNN) [4,8,15,16] to learn the representation of clinical texts. With the flourishing of transformer in natural language processing [17], some research utilize transformer [18,19] as the encoder of input text. By introducing the attention mechanism, those deep learning models achieved state-of-the-art(SOTA) performance in ICD coding task. In addition to discharge summaries, some studies combined extra knowledge information (e.g. knowledge graph, medical concept, co-occurrence relation) [12,14,20,21] in order to improve the prediction results. Xie et al. [12] proposed a model based on Multi-Scale Feature Attention and Structured Knowledge Graph Propagation(MSATT-KG). Besides the n-gram feature vector captured by a multi-scale CNN, a graph convolutional neural network (GCN) [22] is introduced to capture the hierarchical relations among medical codes. Sonabend et al. [21] developed an unsupervised knowledge integration (UNITE) algorithm to automatically assign ICD codes for a specific disease by using Concept Unique Identifiers (CUIs) listed in the Unified Medical Language System (UMLS).

Although some studies utilized the hierarchical structure to produce the representation vector of an ICD code, most of them are interested in only one granularity of ICD codes. We tried to explore a multitask model which can capture contextual features related to multiple label granularity by training several classifiers at the same time. To assign ICD codes based on the injection of hierarchical relation feature.

3 Approach

ICD coding can be formally defined as a multi-label classification based on text. Suppose an input clinical note is X, it is labeled with a set of ICD codes $Y_t \subseteq Y$, Y is the whole label set. Y_t can be represented as a multi-hot vector $[y_1, y_2, \cdots, y_j, \cdots, y_C], y_j \in \{0,1\}$,$C$ denotes the size of Y. The task is to learn a function $f : X \rightarrow Y$ that maps input words sequence into labels. We proposed a multi-task model to learn K functions $\{f^1, f^2, \cdots, f^k, \cdots, f^K\}$ simultaneously based on input clinical notes aiming to capture the features related to labels of different granularities.

The overall framework of our proposed model is shown in Fig. 1, which mainly consists of four modules. Firstly, the representation of the word sequence of a clinical note is captured by a bidirectional LSTM network. Secondly, this contextual feature representation is split into several parts to learn label-specific weighted feature for each label level by using label-wise attention mechanism. Then, a reinforcement strategy is employed based on the label-specific weighted features. The finer-grained feature is reduced and concatenated to coarser-grained feature as the final input of classifier. We now describe each parts of the model in detail.

3.1 Representation of Clinical Note

Here, a recursive neural network is employed to capture longer dependency of clinical text. In particular, an one-layer bidirectional LSTM network is used to extract the contextual information and learn a latent semantic feature vectors for input words. Given a clinical note (i.e. discharge summary) X is a word sequence consisting of n words $(x_1, x_2, \cdots x_i, \cdots x_n)$. Each word x_i is vectorized to a word embedding vector e_{x_i} with the dimension of d_e by Word2vec. The input words sequence is transferred to an embedding matrix $\mathbf{E} = [\mathbf{e_{x_1}}, \mathbf{e_{x_2}}, \cdots, \mathbf{e_{x_i}}, \cdots, \mathbf{e_{x_n}}] \in \mathbb{R}^{d_e \times n}$. The output of bidirectional LSTM for each input word is a latent feature vector $\mathbf{h_i}$, which is the concatenation of two directional hidden states of LSTMs. Given the dimension of the LSTM hidden states is d_h, the feature representation of input clinical text is a matrix $\mathbf{H} = [\mathbf{h_1}, \mathbf{h_2}, \cdots, \mathbf{h_n}] \in \mathbb{R}^{2d_h \times n}$.Layer normalization [23] is used to make the training process more stable and prevent overfitting.

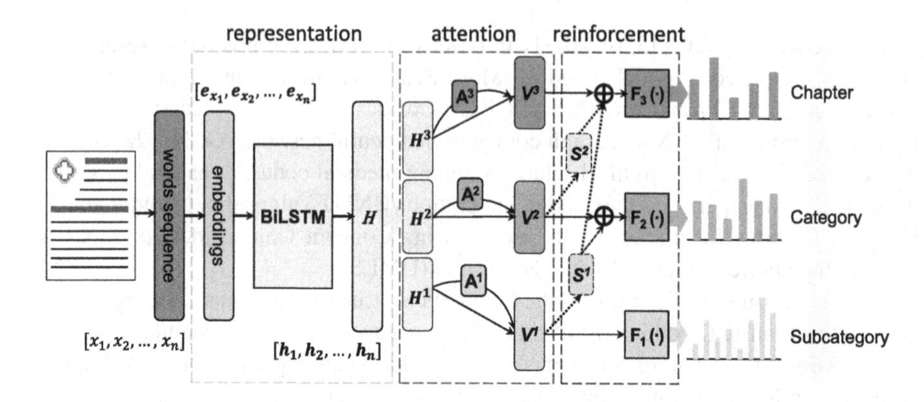

Fig. 1. The workflow of our multitask model for three different granularity ICD coding. Dotted line arrows indicate the reinforcement from fine-grained level to coarse-grained level without backpropagation.

3.2 Multi-granularity Attention Layer

Label-wise attention has been proved helpful in ICD coding and used in recent works [7,11,14,15]. At the multi-granularity attention layer, we split the output of neural networks $\mathbf{H}$ into K equal parts firstly. Each part $\mathbf{H^k} = [\mathbf{h_1^k}, \mathbf{h_2^k}, \cdots, \mathbf{h_n^k}] \in \mathbb{R}^{d_s \times n}$ represents a feature that is responsible for a classifier of one kind of granularity. Here d_s is the dimension of $\mathbf{h}_i^k$, which equals to $\frac{2d_h}{K}$. The label-specific attention weights of each label granularity is computed independently as follows,

$$\mathbf{Z}^k = tanh(\mathbf{W}^k \mathbf{H}^k) \tag{1}$$

$$\mathbf{A}^k = softmax((\mathbf{U}^k)^{\mathrm{T}} \mathbf{Z}^k) \tag{2}$$

$$\mathbf{V}^k = (\mathbf{A}^k)^{\mathrm{T}} \mathbf{H}^k \tag{3}$$

a hyperbolic tangent activation function is used to activate $\mathbf{H}^k$, where $\mathbf{W}^k \in \mathbb{R}^{d_s \times d_s}$, then attention score matrix $\mathbf{A^k} = [\alpha_\mathbf{1}^\mathbf{k}, \alpha_\mathbf{2}^\mathbf{k}, \cdots, \alpha_\mathbf{n}^\mathbf{k}] \in \mathbb{R}^{C^k \times n}$ is obtained by a *softmax* function based on the activated vector matrix $\mathbf{Z}^k \in \mathbb{R}^{d_s \times n}$. Here, C^k is the total number of labels at the kth level, the weight matrix $\mathbf{U}^k \in \mathbb{R}^{d_s \times C^k}$ is randomly initialized. The label-specific feature at the $k-th$ label granularity level $\mathbf{V}^k \in \mathbb{R}^{d_s \times C^k}$ is prepared for classifier by equation (3). $\mathbf{V^1}, \mathbf{V^2}, \cdots, \mathbf{V^k}, \cdots, \mathbf{V^K}$ is the label-specific representation which capture the important information relating to certain labels within each label granularity, respectively.

3.3 Reinforcement from Fine-Grained to Coarse-Grained

To utilize the hierarchical relations between different label granularity, we employ a reinforcement strategy by injecting finer-grained feature into coarser-grained classification, considering the nature that subcategories contain some common features belonging to their ancestors. For each weighted feature representation $\mathbf{V}^k$, we reduce its dimension by mean pooling and produce a single vector $\mathbf{s}^k \in \mathbb{R}^{d_u}$. The input of the kth classifier is the concatenation of $\mathbf{V}^k$ and projections of previous k-1 levels $\mathbf{s}^{k-1}, \mathbf{s}^{k-2}, \cdots, \mathbf{s}^1$. The probability of labels produced by classifier can be denoted as follows,

$$\hat{Y}^k = F^k(\mathbf{V}^k \oplus (\mathbf{s}^{k-1}, \mathbf{s}^{k-2}, \cdots, \mathbf{s}^1)) \tag{4}$$

here, the symbol '$\oplus$' denotes the operation of concatenation, F^k is a single-layer feed-forward network with bias, following by a *sigmoid* activation. The gradients flow of one classifier is propagated only along its own feature dimensions during backward passing stage, in case the features of finer-grained classifier will be passed and biased towards coarser-grained prediction.

3.4 Training and Optimization

The proposed multitask model is to train K multi-label classifiers synchronously. Candidates labels of each granularity are produced with the optimization of model aiming to minimize the total loss of all classifiers. Therefore, we use an adaptive multitask total loss calculation method. For each classifier, the training objective is to minimize the binary cross-entropy loss,

$$L^k_{BCE}(X, Y^k) = -\sum_{l=1}^{C^k} y^k_l \log(\hat{y}^k_l) + (1 - y^k_l) \log(1 - \hat{y}^k_l) \tag{5}$$

Considering the distribution of labels and the difficulties at different label granularities, we calculate the total loss value of K classification tasks by a joint weighted method,

$$L_{total} = \sum_{k=1}^{K} (\frac{C^k}{\sum C^k} + w_k) L^k_{BCE}(X, Y^k) \tag{6}$$

here, $w_k \in w_1, w_2, \cdots, w_K$ is a weight for each loss which can be learned by a *softmax* function during training.

4 Experiments and Results

4.1 Datasets

In order to verify the effectiveness of the proposed model, we conducted experiments on two real-world EMR datasets. MIMIC-III is a public dataset in English [24], which is the benchmark dataset of recent work. Each admission was tagged with a set of ICD-9 codes, including diagnosis codes and procedure codes. We produced two experimental datasets, MIMIC-III-full and MIMIC-III-50 (with 50 highest frequency codes), with the split same as previous works [7,12,13,15]. Xiangya is a private dataset in Chinese, which has been tagged with ICD-10 codes. The statistical details of the two source datasets are shown in Table 1. Two experimental datasets are prepared same as MIMIC-III: Xiangya-full and Xiangya-50. The splitting sets for training, validation ant testing are same as [25]. Discharge summaries on both datasets are the input notes, which record every detail of patients in hospital course. Pre-processing has been carried out on Chinese text, such as word segmentation, removing stop words.

Table 1. Basic Information of Datasets

Statistic items	MIMIC-III	Xiangya
# of records	52,722	7,732
# of codes	8,929	1,177
Avg # of codes per record	15.9	3.6
Max # of codes per record	71	14
Avg length of input notes(# of words)	1511.6	444.7
Max length of input notes(# of words)	8,772	1,629

4.2 Baselines

The proposed model is compared against the following recent baseline models, which performed experiments on same datasets and without extra knowledge information.

DeepLabeler. DeepLabeler [6] predicts ICD codes by combining local features and global features of a clinic note. The local features and global features are captured by a multi-scale CNN and a doc2vec model separately.

CAML&DR-CAML. A single layer CNN with an attention layer constructs the Convolutional Attention network for Multi-Label(CAML) classification model [7]. Description Regularized CAML(DR-CAML) is an extension of CAML proposed by Mullenbach et al. at the same time, which introduce the description of an ICD code to the loss function as a regularization term. Both models have achieved high performance on MIMIC datasets.

MultiResCNN. Multi-Filter Residual CNN [13] contains a multi-filter convolutional layer to capture various text patterns with different lengths. A residual convolutional layer is used to enlarge the receptive field. The model achieved SOTA results on the MIMIC-II full dataset and in-line SOTA results on the MIMIC-III full dataset.

MVC-(R)LDA. MVC-LDA [11] uses multi-view CNN and label-wise attention mechanism to extract the fearure of input notes. Based on the occurrence of different ICD codes, a shift bias strategy is used to learn different thresholds in the prediction function. In MVC-RLDA, the same regularization term based on ICD title description as DR-CAML is introduced to the loss function.

LAAT&JointLAAT. Label Attention model (LAAT) [15] is a label attention model based a bidirectional LSTM encoder. Each label-specific vector is used to build a binary classifier for a given label. JointLAAT is an extension of LAAT with hierarchical joint learning. Both of them achieved SOTA results on MIMIC-II dataset and MIMIC-III dataset.

4.3 Metrics and Parameters

Following previous works, F1 and AUC metrics of both micro and macro are used to evaluate the performance. P@n indicates the proportion of the correctly-predicted codes in the top-n predicted codes. The experiments are conducted on a server with NVIDIA GeForce Titan X Pascal CUDA GPU 12GB. To fine tune the hyper-parameters, a large number of experiments are performed, and the final parameter settings are listed in Table 2. Due to the large dimension of label space and the limitation of our device, the batch size for MIMIC-III-full is 8, while other parameters are shared on MIMIC-III-full and MIMIC-III-50. The training data is shuffled for each epoch. We train our models by using AdamW optimizer. Early stopping mechanism is used by monitoring the micro-F1 value. The training process is stopped when the metric of validation set has no improvement in 10 epoches. We ran our model 10 times with the same hyper-parameters. Different random seeds is used for initialization at each time. The performance is reported with the average value over the 10 runs (Table 3 and Table 4).

Table 2. Parameter settings

Parameter	MIMIC-III	Xiangya
Embedding size	100	100
Size of LSTM hidden layer	384	384
Size of feature projection	128	128
Max length	4000	1500
Dropout rate	0.3	0.3
Learning rate	0.001	0.001
Batch size	16/8	16

4.4 Comparison with Baselines

Since the split of data is aligned with previous work. The results are directly comparable to those in related literature. As for private Xiangya datasets, we ran LAAT and JointLAAT on two Chinese experimental datasets 10 times and reported the average value too, by using the same pre-training embeddings and hyper-parameters. The performances of other baseline models on Xiangya datasets is from previous work [25]. Table 3 and Table 4 show the results of subcategory codes prediction (i.e. the finest-grain labels) on four experimental datasets of MIMIC-III and Xiangya, respectively, comparing with all the aforementioned baseline models. It is obvious that the feature fusion strategy of different label granularity yields improvements over other models in F1 scores. The results of micro-F1 are outperform the best results of baseline models by a slight margin on all experimental datasets. Additionally, macro-F1 scores is improved by 0.2%, 0.3%, 0.4% and 1.8% than the best results among baselines on MIMIC-III-50, Xiangya-50 and Xiangya-full dataset, respectively. Since macro average metrics is effected mainly by rare-label samples, the model seems better for dealing with long-tail distribution dataset. Due to long-tail problem is more serious in Xiangya-full dataset, nearly 40% subcategory codes have only one sample, the improvement of macro-F1 on Xiangya-full dataset is more apparent than in others.

4.5 Analysis of Reinforcement

In order to evaluate the contribution of the reinforcement form fine-grained to coarse-grained, we tried to carry out the model without reinforcement or with a reverse reinforcement method. Comparison experiments are performed on MIMIC-III-50 dataset and Xiangya-50 dataset instead of the full label space datasets because of efficiency issue. The results on main metric of F1 score is shown in Table 5. The reinforcement from fine-grained level to coarse-grained level produced the best F1 scores in all granularity predictions. It is obvious that the model without reinforcement between different granularity gets lower F1 scores. Furthermore, it is noteworthy that the reverse reinforcement distinctively trailed in subcategory code prediction, although it shows

Table 3. Comparison with baseline models on MIMIC-III datasets for subcategory code prediction.

Model	MIMIC-III-full					MIMIC-III-50				
	AUC		F1		P@8	AUC		F1		P@5
	Mic	Mac	Mic	Mac		Mic	Mac	Mic	Mac	
DeepLabeler	97.6	–	40.1	–	56.7	93.8	–	63.4	–	62.3
CAML	98.6	89.5	53.9	8.8	70.9	90.9	87.5	61.4	53.2	60.9
DR-CAML	98.5	89.7	52.9	8.6	69.0	91.6	88.4	63.3	57.6	61.8
MultiResCNN	98.6	91.0	55.2	8.5	73.4	92.8	89.9	67.0	60.6	64.1
MVC-LDA	–	–	54.3	–	70.5	–	–	66.8	–	64.4
MVC-RLDA	–	–	55.0	–	71.2	–	–	67.4	–	64.1
LAAT	98.8	91.9	57.5	9.9	73.8	**94.6**	**92.5**	71.5	66.6	**67.5**
JointLAAT	98.8	92.1	57.5	10.7	73.5	**94.6**	**92.5**	71.6	66.1	67.1
Ours	98.7	90.5	**57.6**	**10.9**	**74.0**	94.5	**92.5**	**71.7**	**66.9**	67.3

Table 4. Comparison with baseline models on Xiangya datasets for subcategory code prediction.

Model	Xiangya-full					Xiangya-50				
	AUC		F1		P@8	AUC		F1		P@5
	Mic	Mac	Mic	Mac		Mic	Mac	Mic	Mac	
DeepLabeler	96.4	–	60.7	–	33.3	98.4	–	75.1	–	45.7
CAML	97.2	–	73.0	–	37.2	98.5	–	80.3	-	46.3
DR-CAML	95.4	–	70.9	–	35.7	98.2	–	80.4	–	46.0
MultiResCNN	96.8	–	74.3	–	37.8	98.7	–	81.7	–	46.9
MVC-LDA	96.8	–	71.1	–	37.1	98.4	–	79.8	–	46.6
MVC-RLDA	96.8	–	69.7	–	36.8	98.4	–	79.7	–	46.7
LAAT	98.1	**90.5**	74.8	9.9	38.6	99.1	98.0	83.5	82.0	52.9
JointLAAT	98.0	90.4	74.3	10.2	38.5	99.1	98.1	83.0	81.1	52.8
Ours	**98.3**	**90.5**	**75.0**	**12.0**	**38.9**	99.0	98.0	**83.8**	**82.4**	**53.0**

better results than the model without reinforcement on coarse granularity predictions. The gaps between two direction strategies of micro-F1 and macro-F1 scores are readily apparent (up to 1.4% and 2.1% differences in MIMIC-III-50 dataset, 0.9% and 1.3% differences in Xiangya-50 dataset). The label-specific feature of upper level may introduce bias to lower level prediction. Quite the contrary, the performance of coarse-grained level classifiers are enhanced by the injection of fine-grained level feature. That confirmed the fine-grained feature could help coarser-grained prediction to a certain extend.

Table 5. Comparison of reinforcement methods.

Dataset	Reinforcement	Subcategory code		Category code		Chapter	
		Mic-F1	Mac-F1	Mic-F1	Mac-F1	Mic-F1	Mac-F1
MIMIC-III-50	Fine to coarse	**71.7**	**66.9**	**74.5**	**69.4**	**79.3**	**73.6**
	w/o	71.4	66.4	74.1	68.3	78.9	72.9
	Coarse to fine	70.3	64.8	74.4	69.3	79.0	73.3
Xiangya-50	Fine to coarse	**83.8**	**82.4**	**91.3**	**87.3**	**94.6**	**90.7**
	w/o	83.2	81.9	90.8	86.0	94.2	89.2
	Coarse to fine	82.9	81.1	91.0	86.3	94.1	90.2

5 Conclusion

Automatic ICD coding has been expected to support professionals assigning codes accurately and efficiently. In this paper, we presented a multitask ICD coding model according to the hierarchical structure of ICD taxonomy. The model learns label-specific feature of different granularity independently, based on the contextual feature representation captured by a bidirectional LSTM network. A reinforcement strategy from fine-grained to coarse-grained is utilized to train multiple classifiers simultaneously for different label granularity. The results on four experimental datasets show the advantages of our model, which demonstrate the superiority of multi-grained features collaboration. However, label embedding is not involved in our model, which may achieve better label attentive weight based on the semantic information of the label. Additionally, pre-training on a large medical corpus may improve the generalization performance too.

Acknowledgements. This work is partly supported by the Science and Technology Major Project of Changsha (No. kh2202004) and Natural Science Foundation of Hunan Province via Grant 2021JJ40783.

References

1. ICD Homepage. https://www.who.int/classifications/classification-of-diseases
2. O'malley, K.J., Cook, K.F., Price, M.D., et al.: Measuring diagnoses: ICD code accuracy. Health Serv. Res. **40**(5p2), 1620–1639 (2005)

3. Perotte, A., Pivovarov, R., Natarajan, K., et al.: Diagnosis code assignment: models and evaluation metrics. JAMIA **21**(2), 231–237 (2013)
4. Shi, H., Xie, P., Hu, Z., et al.: Towards automated ICD coding using deep learning. arXiv preprint, arXiv:1711.04075 (2017)
5. Baumel, T., Nassour-Kassis, J., Cohen, R., et al.: Multi-label classification of patient notes: case study on ICD code assignment. In: Proceedings of the AAAI Workshop on Health Intelligence, pp. 409–416 (2018)
6. Li, M., Fei, Z., Zeng, M., et al.: Automated icd-9 coding via a deep learning approach. IEEE/ACM Trans. Comput. Bio. Bioinf. **16**(4), 1193–1202 (2019). https://doi.org/10.1109/TCBB.2018.2817488
7. Mullenbach, J., Wiegreffe, S., Duke, J., et al.: Explainable prediction of medical codes from clinical text. In: Proceedings of the Conference of the North American Chapter of the Association for Computational Linguistics: Human Language Technologies (NAACL-HLT), pp. 1101–1111 (2018)
8. Yu, Y., Li, M., Liu, L., et al.: Automatic ICD code assignment of Chinese clinical notes based on multilayer attention BiRNN. J. Bio. Inf. 103114 (2019)
9. Larkey, L.S., Croft, W.B.: Combining classifiers in text categorization. In: Proceedings of SIGIR, vol. 96, pp. 289–297 (1996)
10. Koopman, B., Zuccon, G., Nguyen, A., et al.: Automatic ICD-10 classification of cancers from free-text death certificates. Int. J. Med. Informatics **84**(11), 956–965 (2015)
11. Sadoughi, N., Finley, G.P., Fone, J., et al.: Medical code prediction with multi-view convolution and description-regularized label-dependent attention. ArXiv preprint arXiv:1811.01468 (2018)
12. Xie, X., Xiong, Y., Yu, P.S., et al.: EHR Coding with Multi-scale Feature Attention and Structured Knowledge Graph Propagation. In: Proceedings of the 28th ACM international conference on information and knowledge management (CIKM), pp. 649–658 (2019)
13. Li, F., Yu, H.: ICD Coding from Clinical Text Using Multi-Filter Residual Convolutional Neural Network. In Proceedings of the AAAI Conference on Artificial Intelligence **34**(05), 8180–8187 (2020)
14. Cao, P., Chen, Y., Liu, K., et al.: Hypercore: Hyperbolic and co-graph representation for automatic icd coding. In: Proceedings of the 58th Annual Meeting of the Association for Computational Linguistics(ACL), pp. 3105–3114 (2020)
15. Vu, T., Nguyen, D.Q., Nguyen, A.: A label attention model for icd coding from clinical text. arXiv preprint arXiv:2007.06351 (2020)
16. Dong, H., Suárez-Paniagua, V., Whiteley, W., et al.: Explainable automated coding of clinical notes using hierarchical label-wise attention networks and label embedding initialisation. J. Biomed. Inform. **116**, 103728 (2021)
17. Vaswani, A., Shazeer, N., Parmar, N., et al.: Attention is all you need. In: Proceedings of the 31st Conference on Neural Information Processing Systems (NeurIPS 2017), Long Beach, CA, USA, pp. 5998–6008 (2017)
18. Zhou, T., Cao, P., Chen, Y., et al.: Automatic ICD coding via interactive shared representation networks with self-distillation mechanism. In: Proceedings of the 59th Annual Meeting of the Association for Computational Linguistics and the 11th International Joint Conference on Natural Language Processing, vol. 1: Long Papers, pp. 5948–5957 (2021)
19. Biswas, B., Pham, T-H., Zhang, P.: TransICD: Transformer Based Code-wise Attention Model forExplainable ICD Coding. Preprint at, (2021) https://arxivorg/abs/210410652
20. Wu, Y., Zeng, M., Fei, Z., et al.: Kaicd: A knowledge attention-based deep learning framework for automatic icd coding. Neurocomputing **469**, 376–383 (2022)
21. Sonabend, A., Cai, W., Ahuja, Y., et al.: Automated ICD coding via unsupervised knowledge integration (UNITE). Int. J. Med. Informatics **139**, 104135 (2020)

22. Kipf, T.N., Welling, M.: Semi-Supervised Classification with Graph Convolutional Networks. In: Proceedings of the International Conference on Learning Representations (ICLR) (2017)
23. Ba, J., Kiros, J., Hinton, G.E.: Layer normalization. ArXiv, vol.abs/ arXiv: 1607.06450 (2016)
24. EW Johnson, A., Pollard, T.J., Shen, L., et al.: MIMIC-III, a freely accessible critical care database. Sci. Data, **3**(1), 1–9 (2016)
25. Wu, Y., Zeng, M., Yu, Y., et al.: A Pseudo Label-wise Attention Network for Automatic ICD Coding. arXiv preprint arXiv:2106.06822 (2021)

Effectively Training MRI Reconstruction Network via Sequentially Using Undersampled k-Space Data with Very Low Frequency Gaps

Tian-Yi Xing[1,2], Xiao-Xin Li[1,2(✉)], Zhi-Jie Chen[1,2], Xi-Yu Zheng[1,2], and Fan Zhang[1,2(✉)]

[1] College of Computer Science and Technology, Zhejiang University of Technology, Hangzhou 310023, China

[2] Key Laboratory of Visual Media Intelligent Processing Technology of Zhejiang Province, Hangzhou, China

mordecai@163.com, zf@zjut.edu.cn

Abstract. Convolutional Neural Networks (CNNs) have achieved great advances on Magnetic Resonance Imaging (MRI) reconstruction. However, CNNs are still suffering from significant aliasing artifacts for undersampled data with high acceleration rates. This is mainly due to the *huge gap* between the highly undersampled k-space data and its fully-sampled counterpart. To mitigate this problem, we constructed a series of well-organized undersampled k-space data, each of which has very small frequency gap with its neighbors. By sequentially using these undersampled data and their fully-sampled ones to train a given CNN model $\mathcal{N}$, the model $\mathcal{N}$ can gradually *know* how to fill the progressively increased frequency gaps and thus reduce the aliasing artifacts. Experiments on the MSSEG dataset demonstrated the effectiveness of the proposed training method.

Keywords: MRI reconstruction · Frequency gap · Effective training · Sequentially training

1 Introduction

Magnetic Resonance Imaging (MRI) has been widely used for a variety of clinical applications such as angiography, neuroimaging, cardiac examinations, and etc. A major limitation of MRI is the relatively slow imaging speed, and thus the patient must lie in the MRI scanner for long time [15]. This not only causes

X-X. Li and F. Zhang—This work was supported in part by the Zhejiang Provincial Natural Science Foundation of China under Grants LGF22F020027, GF22F037921 and LGF20H180002, in part by the program of the Education Department of Zhejiang Province under No. Y202147723 and Y202147457, in part by the National Natural Science Foundation of China uncer Grant 62271448.

M. S. Bansal et al. (Eds.): ISBRA 2022, LNBI 13760, pp. 30–40, 2022.
https://doi.org/10.1007/978-3-031-23198-8_4

discomfort to the patient, but also leads to artifacts due to the movement of the patient during the long acquisition time.

Various efforts have been devoted to accelerate MRI. Compressed sensing yields up to a 12.5x acceleration rate by undersampling and inferring missing k-space information [5,10]. Convolutional Neural Networks (CNNs) [4,11,14,20, 21,25] are now widely used in CS-based MRI reconstruction. However, images generated at high acceleration rates (e.g., 8x or 12x) are still difficult to use in practice [17]. This is because the highly undersampled MRI images lose too much information and the reconstructed MRI images are still suffering from significant aliasing artifacts. Moreover, collecting a large-scale dataset is very difficult due to a number of reasons [7], such as the cost of data acquisition, the patient privacy. This further increases the difficulty of training CNNs.

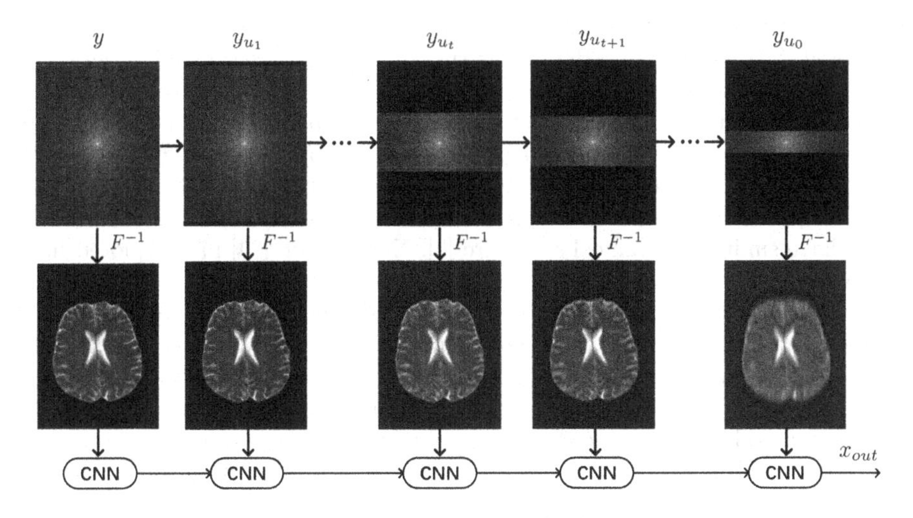

Fig. 1. A schematic illustration of the proposed subdivision strategy. The whole k-space data is divided along the phase-encoding direction into a lot of k-space units and the undersampled k-space data fed to CNN can be sequentially created by masking-out parts of these k-space units in an accumulative manner.

Based on the idea of progressive network [12,13,16], we introduce an effective training method via sequentially using a series of undersampled k-space data, each of which has very low frequency gaps with its neighbors. To this end, we first finely divide the whole k-space data along the phase-encoding direction into a lot of k-space units and the undersampled k-space data can be sequentially created by masking-out parts of these k-space units, as shown in Fig. 1. We refer to such a training data generation procedure as a k-space *subdivision strategy*. By using the subdivision strategy, the acceleration rate of the network is gradually increased during training until the target acceleration rate is reached. This forces the network to learn more information at high acceleration rates. We explore three k-space subdivision strategies: the uniform subdivision

(US) strategy, the reverse uniform subdivision (rUS) strategy and the gradually narrowing subdivision (GNS) strategy. We experimentally demonstrate that the GNS strategy could achieve the best performance.

2 Related Works

2.1 MRI Reconstruction

In the field of MRI reconstruction, deep learning methods can lead to higher reconstruction quality and higher acceleration rates than traditional reconstruction methods. For example, Yang *et al.* [25] applied Generative Adversarial Networks (GANs) to MRI reconstruction and obtained better reconstruction results. However, GANs suffer from vanishing gradient [23] and poor training stability. The Wasserstein GAN (WGAN) [1] could alleviate these problems. Eo *et al.* [6] pioneered the dual-domain MRI reconstruction and proposed the KIKI network, which alternately uses the k-space reconstruction subnet and the image-domain reconstruction subnet. The dual-domain network outperforms the networks only operating in the image domain. Due to the different clinical requirements of MRI, it is often necessary to generate MRI images with multiple contrasts, such as T1-weighted, T2-weighted and proton density images. Images in different contrasts have similar anatomical structure [9]. Xiang *et al.* [24] first explored how to use fully sampled T1-weighted image as an assisted contrast to guide the reconstruction of the undersampled T2-weighted images. Other studies [18, 22] using undersampled images as assisted contrasts to guide reconstruction. While using assisted contrasts can supplement some of the missing information, it may also allow the information in one mode to leak into another [18]. Even though all of these methods can improve the quality of reconstructed images, images reconstructed at high acceleration rates are currently still difficult to use in clinical practice.

2.2 Progressive Network

Progressive networks are widely used in the field of image reconstruction. Lyu *et al.* [16] proposed two-level progressive neural network, which cascades two 2x subnets in order to achieve better reconstruction results for 4x acceleration rates. The method progressively improves image quality and reduces the difficulty of training each network. Saharia *et al.* [19] used an iterative refinement method that gradually adds noises into the input images before performing super-resolution (SR) through a stochastic iterative denoising process, and achieved better SR results. The active acquisition method proposed by Zhang *et al.* [26] can actively sample the poorly reconstructed k-space lines in each iteration of the inference stage until the stopping criterion is reached. All of these methods decompose the difficulty of network training into multiple simple steps.

3 Proposed Method

In this section, we will deep into the proposed three k-space subdivision strategies and illustrate their underlying mechanisms and detailed implementations.

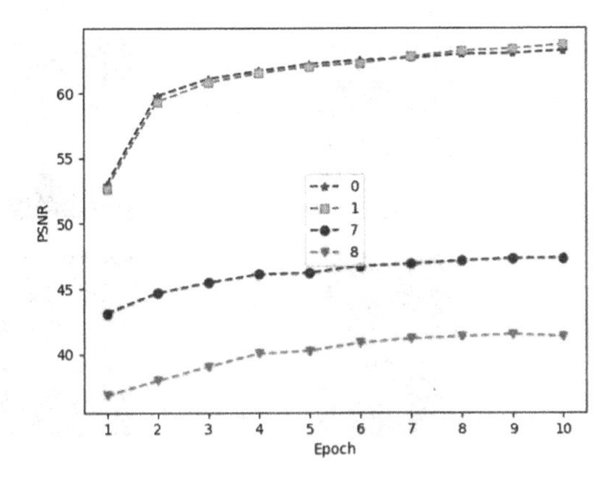

Fig. 2. Comparison of the learning difficulties of low- and high- frequency parts of the k-space data. The numbers, 0, 1, 7 and 8, indicate the location of the undersampled region of the k-space. The larger the number, the closer to the low frequency region.

3.1 Learning Difficulties of Different Parts of k-Space

The contributions of the information at different regions of the k-space are very different to the reconstruction performance. This can be demonstrated by the experiments in Fig. 2. We train the reconstruction network D5C5 [21] with the same undersampling rate of 0.9. We uniformly divided the fully-sampled k-space into ten parts. When the 10% undersampling parts are respectively in the high frequency regions of #0 and #1, the reconstruction performance of the network can achieve over 60 dB after 10 training epochs and, specially, the performance climbs up very quickly in the first two epochs. In contrast, when the 10% undersampling parts are located in the low frequency regions of #7 and #8, the reconstruction performance can only achieve 47 dB and 41 dB, respectively. Moreover, the reconstruction performance increases very slowly during the whole training epochs. It means that, to achieve better reconstruction results, the network should consume more training epochs for reconstructing the undersampled information in the low frequency region.

3.2 Subdivision Strategies

According to the observations in Subsect. 3.1, we propose three subdivision strategies, namely, the uniform subdivision (US) strategy, the reverse uniform subdivision (rUS) strategy and the gradually narrowing subdivision (GNS) strategy. Figure 3 gives an overview.

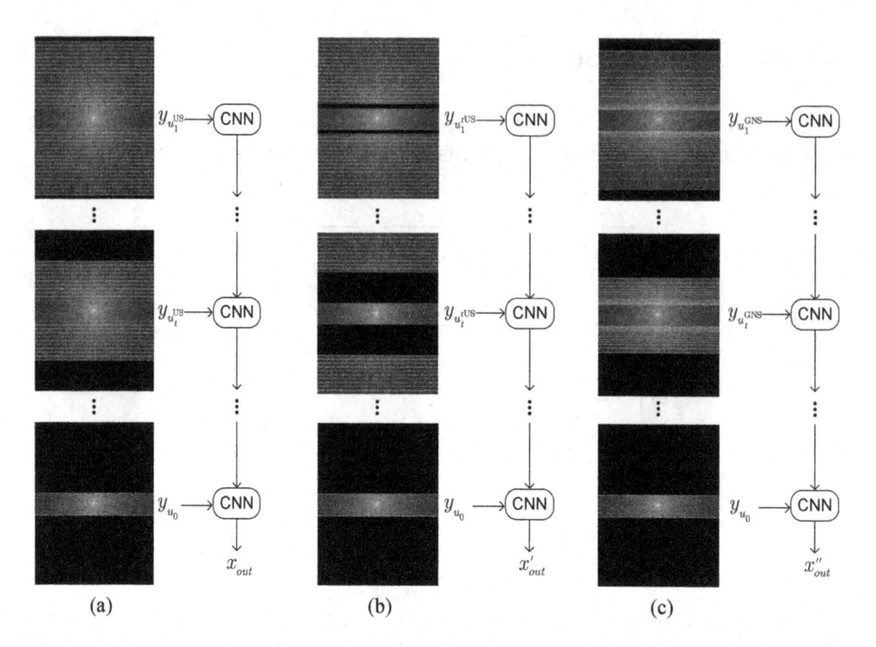

Fig. 3. The overview of the three subdivision strategies of the k-space: (a) the uniform subdivision (US) strategy, (b) the reverse uniform subdivision (rUS) strategy and (c) the gradually narrowing subdivision (GNS) strategy.

We first focus on the US strategy, which is illustrated in Fig. 3(a). We first uniformly subdivide the fully-sampled k-space data y along the phase-encoding direction into $T \geq 100$ parts, each of which we call a *data cell.* The US strategy undersamples the whole k-space data T times from two sides to the central region and the whole undersampled region accumulatively increases one data cell each time. For the t-th time, the acceleration rate can be calculated as follows

$$u_t^{\mathrm{US}} = \frac{T}{T + t \times \left(\frac{1}{u_0} - 1\right)}, 1 \leq t \leq T. \tag{1}$$

Figure 3(b) shows the rUS strategy. rUS differs from the US strategy only on the masking-out order of the data cells. rUS, contrary to US, undersamples the whole k-space data T times from the central region to its two sides. That is, the masked-out parts of k-space data gradually expands from low-frequency region to the high frequency region. The acceleration rate can be calculated in a similar manner of Eq. (1).

Figure 3(c) introduces the GNS strategy. As can be seen, the whole k-space data y is also subdivided along the phase-encoding direction into $T \geq 100$ data cells. However, unlike the previous two strategies, the width of the data cells are not equal to each other but gradually decreases from two sides to the central region. Specifically, the t-th subdivision acceleration rate u_t can be calculated as follows

$$u_t^{\mathrm{GNS}} = 1 + \frac{t}{T} \times (u_0 - 1), 1 \leq t \leq T. \tag{2}$$

Note that the procedure of GNS can also be reversed. However, with the reverse GNS strategy, the difference between the undersampled input and the fully-sampled target image is too small at the beginning of the training, which makes the network very difficult to be well trained.

3.3 Training Strategies Adapted to Subdivision Strategies

Suppose we have a set of training samples $\mathcal{D} = \{x^1, x^2, \cdots, x^n\}$. By using each subdivision strategy $\mathrm{s} \in \{\mathrm{US}, \mathrm{rUS}, \mathrm{GNS}\}$ and each acceleration rate u_t ($t \in \{1, 2, \cdots, T\}$), we can obtain a series of training set

$$\mathcal{D}_{u_t^{\mathrm{s}}} = \left\{ \left(x_{u_t^{\mathrm{s}}}^1, x^1\right), \left(x_{u_t^{\mathrm{s}}}^2, x^2\right), \cdots, \left(x_{u_t^{\mathrm{s}}}^n, x^n\right) \right\},$$

where $x_{u_t^{\mathrm{s}}} = F^{-1}\left(y_{u_t^{\mathrm{s}}}\right) = F^{-1}\left(M_{u_t^{\mathrm{s}}} y\right)$. Here, $M_{u_t^{\mathrm{s}}}$ is a 0–1 mask matrix corresponding to the acceleration rate u_t^{s}. That is, we have T training sets for each subdivision strategy. To make our reconstruction network gradually adapt to highly undersampled data, we usually adopt a very large T, e.g., $T \geq 100$, to ensure that the neighboring training data sets have very low frequency gap. Thus, we have a very large training data set $\mathcal{D}^{\mathrm{s}} = \left\{\mathcal{D}_{u_1^{\mathrm{s}}}, \mathcal{D}_{u_2^{\mathrm{s}}}, \cdots, \mathcal{D}_{u_T^{\mathrm{s}}}\right\}$.

How to effectively use these training samples and how to train the reconstruction network effectively are very important problems for applying our subdivision strategies into real-world optimization. For effective training, we suggest also using the *coarse-to-fine* training strategy. Specially, for each training set $\mathcal{D}_{u_t^{\mathrm{s}}}$ with $1 \leq t < T$, we train the reconstruction network within only a few, e.g., ≤ 3, epochs. For each training set $\mathcal{D}_{u_T^{\mathrm{s}}}$ with the target acceleration rate u_0, we train the reconstruction network by using a large number of, e.g., ≥ 50, epochs.

4 Experimental Results

4.1 Experimental Settings

All experiments were conducted on the same GPU server which has a Ubuntu 16.04 system with a 3.3 GHz Intel Xeon E5-2680 CPU (32G RAM) and an NVIDIA RTX 2080Ti GPU (11G RAM). The Adam solver with momentum of 0.9 was used to train the model. The batch size is fixed to 5. The learning rate and training epochs were set according to the specific model we used in our experiments. The convolution kernel size was fixed to 3×3, the pooling window size was set to 2×2, and the activation function was selected to be the Leaky ReLU. We use Peak Signal to Noise Ratio (PSNR) and Structural Similarity (SSIM) [8] as evaluation metrics.

As the deep cascade network [21] has faster convergence and fewer number of parameters, we chose the D5C5 network proposed in [21] as our baseline network, which contains $n_c = 5$ subnets and each subnet consists of $n_d = 5$ convolutional layers.

Our experiments were performed on the MSSEG-2016 dataset [2,3]. MSSEG-2016 was created for the MICCAI lesion segmentation challenge in 2016. The dataset consisted of 53 patients: 15 for training and 38 for testing. All data acquired on 1.5T or 3T MRI scanners at four medical centers, respectively. In this work, the pre-processed MRI data acquired at medical center 08 was selected as the training and test datasets, and only the T2 modality was used. The training and test sets were divided by 4 to 1, and five cross-validations were performed. The MRI images were normalized to the range of [0,1] and the slices with a lot of zero pixel values are deleted to ensure that the network can be effectively trained.

4.2 Effects of the Subdivision Number and Subdivision Strategies Against Reconstruction Performance

In this section, we explore the effects of the subdivision number T and the subdivision strategy against the reconstruction performance. The subdivision number (i.e., the number of the data cells) T was set to be 10, 50 and 100. To ensure the total number of the training epochs under different values of T are identical, the training epochs e for each acceleration rate of each subdivision number should be set to be different. Specifically, we set $e = 10$ for $T = 10$, $e = 4$ for $T = 50$ and $e = 2$ for $T = 100$, respectively. Also, we set $e = 50$ for the target acceleration rate. The learning rate was initially set to be 0.001, remained unchanged for the first 200 epochs, and divided by 10 at the 40th training epochs of the target acceleration rate.

Table 1. Quantitative evaluation of reconstruction performance caused by three different subdivision strategies (US, rUS, GNS) against three different acceleration rates (4, 8, 12), and three different subdivision numbers (10, 50, and 100).

	Acc = 4		Acc = 8		Acc =1 2	
	PSNR	SSIM	PSNR	SSIM	PSNR	SSIM
US-10	39.26	0.9942	33.37	0.9769	30.60	0.9578
US-50	39.26	0.9942	33.57	0.9778	30.92	0.9606
US-100	39.26	0.9942	33.61	0.9778	30.98	0.9609
rUS-10	39.35	0.9943	33.54	0.9780	30.72	0.9581
rUS-50	39.37	0.9943	33.58	**0.9782**	31.02	0.9610
rUS-100	39.35	0.9942	33.53	0.9781	30.96	0.9609
GNS-10	39.37	0.9942	33.58	0.9778	31.05	0.9610
GNS-50	39.40	**0.9944**	**33.63**	**0.9782**	**31.11**	**0.9613**
GNS-100	**39.39**	**0.9944**	33.61	0.9781	31.07	**0.9613**

Comparison experiments of the three subdivision strategies are reported in Table 1. As can be seen, for the US strategy, finer granularity leads to better performance. However, for the rUS and GNS strategies, too fine granularity might decrease the reconstruction performance. To sum up, the GNS-50 achieves the best reconstruction results.

4.3 Comparison with the Single-Scale Training Strategy

In this section, we compare the effects of the single-scale training strategy with our subdivision-based training strategy. The zero-filling method is also chosen as the baseline method in this experiment. The single-scale training means that we only uses the dataset with the target acceleration rate, i.e., the dataset $\mathcal{D}_{u_T^s}$ to train the target network, i.e., D5C5. The training epochs are set to the same ones with our training strategies.

Table 2. Quantitative evaluation of 1/4, 1/8, and 1/12 T2WI reconstruction performance by single scale and multi-scale training methods.

	Acc = 4		Acc = 8		Acc = 12	
	PSNR	SSIM	PSNR	SSIM	PSNR	SSIM
Zero-filling	33.03	0.9730	28.97	0.9278	27.17	0.8952
Single-scale	39.19	0.9941	33.30	0.9762	30.69	0.9584
Ours	39.40	0.9944	33.63	0.9782	31.11	0.9613

The PSNR and SSIM for 4x, 8x and 12x acceleration rates are reported in Table 2. The results of our method are set to be the best results in Table 1. The proposed subdivision sequential training method achieves the best reconstruction performance, with PSNR/SSIM improvements of 0.21/0.0003, 0.33/0.0020, and 0.42/0.0029 for the 4x, 8x, and 12x acceleration rates, respectively. Compared to the results of the single-scale training strategy, our training strategy can achieve better improvements at higher acceleration rates.

Figure 4 visually compares the reconstruction quality of different methods. We can see that our subdivision-based training method can significantly improve the quality of the reconstruction image and recover more details from k-space data at 8x acceleration rate. Currently, our method can only achieve satisfying reconstruction results for the 8x acceleration rate.

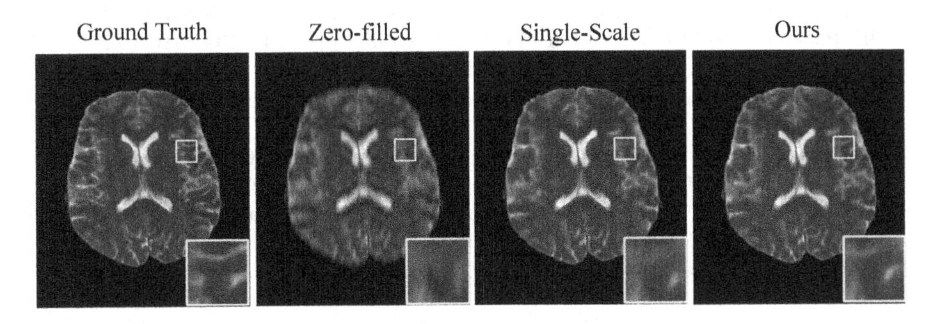

Fig. 4. Visualization of the reconstruction results of different training strategies against the 1/8 undersampled T2 images.

4.4 Comparison with Other Progressive Models

Table 3. Comparisons of our methods with other progressive methods.

	Acc = 4		Acc = 8		Acc = 12	
	PSNR	SSIM	PSNR	SSIM	PSNR	SSIM
Lyu's	39.38	0.9944	33.20	0.9754	30.41	0.9573
Ours	39.40	0.9944	33.63	0.9782	31.11	0.9613

We compare our results with the progressive model of Lyu *et al.* [16]. For fair comparison with the D5C5 network we used to apply our subdivision strategy, we used the same ℓ_2 loss function and added the data consistency layer between the two sub-UNets in the model of Lyu *et al.* [16]. The results are shown in Table 3. As can be seen, our method achieves higher PSNR and SSIM. At the 4x acceleration rate, the reconstruction performance of the two compared methods are very close to each other. However, with the increase of the acceleration rates, our method greatly outperforms the progressive model of Lyu *et al.* [16].

5 Conclusion

In this work, we propose a novel training strategy via subdividing the k-space into different subregions to obtain as many undersampled samples as possible. At first the network performs a simple task with a low acceleration rate, gradually increasing the difficulty to allow the network to adapt to the higher target acceleration rate. The comparison study with the single-scale method and the existing progressive model indicate that our method can obtain greater performance gains for higher acceleration rates.

References

1. Arjovsky, M., Chintala, S., Bottou, L.: Wasserstein generative adversarial networks. In: International Conference On Machine Learning, pp. 214–223. PMLR (2017)
2. Commowick, O., Cervenansky, F., Ameli, R.: Msseg challenge proceedings: multiple sclerosis lesions segmentation challenge using a data management and processing infrastructure. In: MICCAI (2016)
3. Commowick, O., et al.: Multiple sclerosis lesions segmentation from multiple experts: The MICCAI 2016 challenge dataset. Neuroimage **244**, 118589 (2021)
4. Defazio, A., Murrell, T., Recht, M.: MRI banding removal via adversarial training. Adv. Neural. Inf. Process. Syst. **33**, 7660–7670 (2020)
5. Donoho, D.L.: Compressed sensing. IEEE Trans. Inf. Theory **52**(4), 1289–1306 (2006)
6. Eo, T., Jun, Y., Kim, T., Jang, J., Lee, H.J., Hwang, D.: Kiki-net: cross-domain convolutional neural networks for reconstructing undersampled magnetic resonance images. Magn. Reson. Med. **80**(5), 2188–2201 (2018)

7. Fabian, Z., Heckel, R., Soltanolkotabi, M.: Data augmentation for deep learning based accelerated MRI reconstruction with limited data. In: International Conference on Machine Learning, pp. 3057–3067. PMLR (2021)
8. Hore, A., Ziou, D.: Image quality metrics: Psnr vs. ssim. In: 2010 20th International Conference On Pattern Recognition, pp. 2366–2369. IEEE (2010)
9. Huang, J., Chen, C., Axel, L.: Fast multi-contrast MRI reconstruction. Magn. Reson. Imaging **32**(10), 1344–1352 (2014)
10. Jaspan, O.N., Fleysher, R., Lipton, M.L.: Compressed sensing MRI: a review of the clinical literature. Br. J. Radiol. **88**(1056), 20150487 (2015)
11. Jun, Y., Shin, H., Eo, T., Hwang, D.: Joint deep model-based mr image and coil sensitivity reconstruction network (joint-ICNet) for fast MRI. In: Proceedings of the IEEE/CVF Conference on Computer Vision and Pattern Recognition, pp. 5270–5279 (2021)
12. Lai, W.S., Huang, J.B., Ahuja, N., Yang, M.H.: Deep laplacian pyramid networks for fast and accurate super-resolution. In: IEEE Conference on Computer Vision and Pattern Recognition, pp. 624–632 (2017)
13. Lai, W.S., Huang, J.B., Ahuja, N., Yang, M.H.: Fast and accurate image super-resolution with deep laplacian pyramid networks. IEEE Trans. Pattern Anal. Mach. Intell. **41**(11), 2599–2613 (2019)
14. Li, X.-X., Chen, Z., Lou, X.-J., Yang, J., Chen, Y., Shen, D.: Multimodal MRI Acceleration via deep cascading networks with peer-layer-wise dense connections. In: de Bruijne, M., Cattin, P.C., Cotin, S., Padoy, N., Speidel, S., Zheng, Y., Essert, C. (eds.) MICCAI 2021. LNCS, vol. 12906, pp. 329–339. Springer, Cham (2021). https://doi.org/10.1007/978-3-030-87231-1_32
15. Lustig, M., Donoho, D., Pauly, J.M.: Sparse MRI: The application of compressed sensing for rapid mr imaging. Mag. Res. Med. Off. J. In. Soci. Mag. Res. Med. **58**(6), 1182–1195 (2007)
16. Lyu, Q., et al.: Multi-contrast super-resolution MRI through a progressive network. IEEE Trans. Med. Imaging **39**(9), 2738–2749 (2020)
17. Muckley, M.J., et al.: Results of the 2020 fastMRI challenge for machine learning MR image reconstruction. IEEE Trans. Med. Imaging **40**(9), 2306–2317 (2021)
18. Polak, D., et al.: Joint multi-contrast variational network reconstruction (jvn) with application to rapid 2d and 3d imaging. Magn. Reson. Med. **84**(3), 1456–1469 (2020)
19. Saharia, C., Ho, J., Chan, W., Salimans, T., Fleet, D.J., Norouzi, M.: Image super-resolution via iterative refinement. arXiv preprint arXiv:2104.07636 (2021)
20. Schlemper, J., Caballero, J., Hajnal, J.V., Price, A., Rueckert, D.: A deep cascade of convolutional neural networks for mr image reconstruction. In: Niethammer, M., et al. (eds.) IPMI 2017. LNCS, vol. 10265, pp. 647–658. Springer, Cham (2017). https://doi.org/10.1007/978-3-319-59050-9_51
21. Schlemper, J., Caballero, J., Hajnal, J.V., Price, A.N., Rueckert, D.: A deep cascade of convolutional neural networks for dynamic mr image reconstruction. IEEE Trans. Med. Imaging **37**(2), 491–503 (2018)
22. Sun, L., Fan, Z., Fu, X., Huang, Y., Ding, X., Paisley, J.: A deep information sharing network for multi-contrast compressed sensing MRI reconstruction. IEEE Trans. Image Process. **28**(12), 6141–6153 (2019)
23. Wiatrak, M., Albrecht, S.V., Nystrom, A.: Stabilizing generative adversarial networks: A survey. arXiv preprint arXiv:1910.00927 (2019)
24. Xiang, L., et al.: Deep-learning-based multi-modal fusion for fast mr reconstruction. IEEE Trans. Biomed. Eng. **66**(7), 2105–2114 (2019)

25. Yang, G., et al.: DAGAN: Deep de-aliasing generative adversarial networks for fast compressed sensing MRI reconstruction. IEEE Trans. Med. Imaging **37**(6), 1310–1321 (2018)
26. Zhang, Z., Romero, A., Muckley, M.J., Vincent, P., Yang, L., Drozdzal, M.: Reducing uncertainty in undersampled MRI reconstruction with active acquisition. In: Proceedings of the IEEE/CVF Conference on Computer Vision and Pattern Recognition, pp. 2049–2058 (2019)

Fusing Label Relations for Chinese EMR Named Entity Recognition with Machine Reading Comprehension

Shuyue Liu[1,2], Junwen Duan[1,2](✉), Feng Gong[1,2], Hailin Yue[1,2], and Jianxin Wang[1,2]

[1] School of Computer Science and Engineering, Central South University, Changsha, China
{liu_shuyue,jwduan,gongfeng,yuehailin,jxwang}@csu.edu.cn

[2] Hunan Provincial Key Lab on Bioinformatics, Central South University, Changsha, China

Abstract. Chinese electronic medical records named entity recognition (NER) is a core task in medical knowledge mining, which is usually viewed as a sequence labeling problem. Recent works introduce the machine reading comprehension (MRC) framework into this task and extract named entities in a question-answering manner, resulting in state-of-the-art performance. However, they extract entities of different types independently, ignoring the fact that entities presented in the context might highly correlate with each other. To address this issue, we extend the MRC-based model and propose Fusion Label Relations with MRC (FLR-MRC). The method implicitly models the relations between labels through graph attention networks and fuse label information with text for named entity recognition. Experimental results on the benchmark datasets CMeEE and CCKS2017-CNER demonstrate FLR-MRC outperform existing clinical NER methods, with F1-score reaching 0.6652 and 0.9101, respectively.

Keywords: Medical named entity recognition · Label relations · Machine reading comprehension

1 Introduction

Electronic medical record named entity recognition is to recognize disease, symptom, drug, treatment and other entities from the unstructured medical records. Named entity recognition in electronic medical records is an important foundation work for disease prevention, medication analysis and other application fields, and the effectiveness of its recognition can directly affect the subsequent research. Efficient medical entity recognition can not only serve medical professionals for clinical decision support, but also be applied to public health data analysis and provide personalized medical health management services for patients. In order to fully exploit the potential value contained in the large and growing volume of electronic medical record data and to lay the foundation for the next applications, it is the significance of this study to explore effective methods for medical entities in Chinese electronic medical records.

Previous NER approaches can be broadly classified into four main categories: 1) sequence-labeling approaches, 2) sequence-to-sequence approaches, 3) span-based

M. S. Bansal et al. (Eds.): ISBRA 2022, LNBI 13760, pp. 41–51, 2022.
https://doi.org/10.1007/978-3-031-23198-8_5

approaches, and 4) MRC-based approaches. Most existing works formalized the task of NER as a sequence-labeling task [2,4,14], which assigns a single label (tag) to each token in the sequence. However, as different NER tasks have different entity types, it is difficult to design a labeling scheme that fits all NER subtasks. Recently, Yan et al. [23] proposed a sequence-to-sequence (Seq2Seq) model, but this model may suffer from decoding efficiency problems and certain common drawbacks of Seq2Seq architectures, such as exposure bias. The span-based approach [11,13] enumerates all possible spans and performs span-level classification. However, span-based models may be limited by the maximum span length and can lead to considerable model complexity due to the nature of enumeration.

The machine reading comprehension (MRC) framework has recently been applied to task, which address the problem in a question-answer manner [12,24]. However, the training and prediction phases are very inefficient, because they have to create a question for each entity type independently, overlooking the relations among different entity types. Empirically, an entity is more likely to present if another entity type exists in the sentence. As shown in Fig. 1, in sentence 1, the body part type entity "延髓" and "脊髓" and the clinical symptom type entity "延髓和脊髓受损" "瘫痪" appear simultaneously.

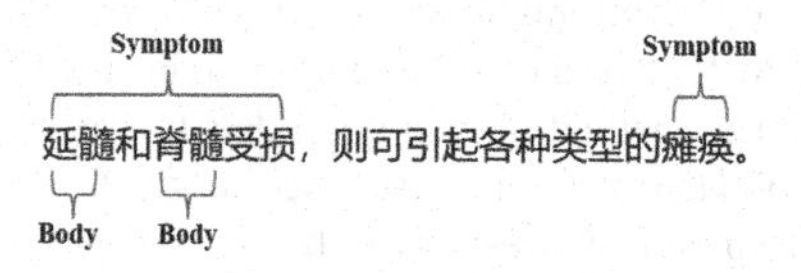

Fig. 1. Example of medical records

In response to the above observations, we propose an MRC-based NER model that can fuse label relations and capture the interactions between labels. In particular, the text and the questions with prior knowledge are first encoded independently with the pre-training model BERT. We consider the potential connections between labels and labels and model the relations between them through a Graph Attention Network (GAT) [19]computation. We compute the attention from both text-to-question direction and question-to-text directions. Finally, we extract the entities regarding all entity types simultaneously, improving entity extraction's efficiency.

We evaluate our model on the Chinese electronic medical record datasets CMeEE and CCKS2017-CNER. The experimental results show that our approach achieves 0.6652 on dataset CMeEE, which outperforms the Chinese natural language processing baseline CBLUE results. In addition, our model reaches 0.9101 on dataset CCKS2017-CNER, which is higher than the results of the SOTA model compared. We summarize the contributions of this paper as follows:

- We propose a NER framework FLR-MRC for Chinese electronic medical records, which implictly model the dependency among different entity types.
- Compared to the baselines, we achieved the best results on the benchmark datasets CMeEE and CCKS2017-CNER.

2 Related Work

NER can be roughly divided into four main categories, i.e., Sequence labeling based, Sequence-to-Sequence based, Span-based and MRC-based [10].

Sequence Labeling Based. Traditional sequence labeling models use CRF [9,18] as the backbone of NER. Hammerton [6] tried to solve the problem using LSTMs. Later work such as CNN [3,17], bidirectional LSTM [7,10] and Transformer [22] have also emerged. Recent pre-training methods for large-scale language models, such as BERT [4]and ELMo [15] further enhance the performance of NER, yielding state-of-the-art performance. Nevertheless, the sequence labeling approach does not fully leverage the knowledge of labels, because it treats entity types as independent, meaningless one-hot vectors. Recently, self-attention-based pre-trained language models such as BERT are widely used to boost the NER task.

Sequence-to-Sequence Based. Gillick et al. [5] used the Seq2Seq model to solve the NER problem, taking sentences as input and outputting the start position, span length, and labels of all entities. Later researchers used the Seq2Seq architecture for overlapping NER. Yan et al. [23] handled unified NER by a Seq2Seq model with a pointer network to generate a sequence of all possible entity start-end indexes and types. However, the Seq2Seq architecture suffers from decoding efficiency and exposure bias problems.

Span-Based. Several studies have categorized NER into span-level categories, and these methods enumerate all possible ranges and determine whether they are valid words and types [13,20,21]. Yu et al. [25] used biaffine attention to measure the likelihood of mentioning text span. However, due to their exhaustive nature, these methods suffer from maximum span and considerable model complexity, especially for large span entities.

MRC-Based. Li et al. [12] propose a new framework that, instead of treating the NER task as a sequence labeling problem, view it as a MRC task. Yang et al. [24] improve the operational efficiency of the reading comprehension model by incorporating the label information into the text information through an attention-based mechanism approach. Although the method extracts all entity words at once and simultaneously, it essentially views each extraction independently. They ignore there might be dependency between entity types.

Based on the above problems, we propose a new model FLR-MRC, which models the relations between labels through graph attention networks and fuses labels information with text as a way to train and predict.

3 Problem Definition

We formulated the following task: given an input text $X = (x_1, x_2, \ldots, x_n)$ consisting of n tokens, find all candidate entity spans in X and assign a label $c \in C$ to each span, where C is a predefined label type. Each entity type is characterized by a natural language query, and entities are extracted by answering these queries given the contexts.

We use the QA formalism for NER because it exploits labeling knowledge. Typically, questions are generated by a manually designed preprocessing step, and previous work [12] used annotations of each category (called label annotations) as questions. We follow this setting in our work.

4 Methods

In this section, we first give a general description of our model architecture. Our model consists of four key modules: an embedding layer, a layer for modeling label relations based on graph attention networks, a label fusion layer, and an output layer, as shown in Fig. 2. Our architecture takes as input the text X containing the entity type and the entity label annotation Y. GAT is then used to model the generated label embeddings, capturing the implicit relations between the labels. Then the semantic fusion module fuses the text embeddings generated by the shared encoder with the label relations output using GAT to obtain the label knowledge-enhanced text embeddings of the text. Finally, a linear layer is used to predict whether each token is the beginning or the end of a certain entity class.

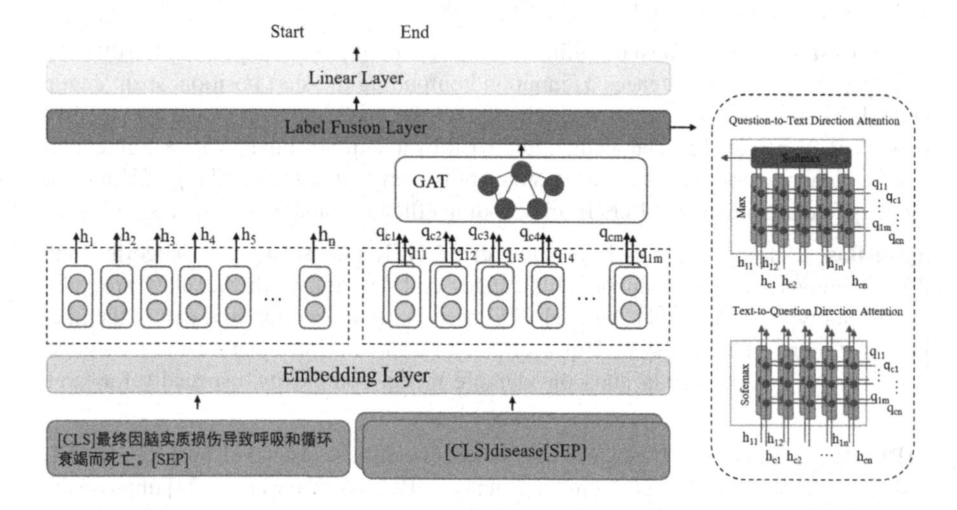

Fig. 2. Illustration of our model

4.1 BERT-Based Semantic Embedding Layer

The role of the semantic embedding layer is mainly to encode text and label information, and the pre-training model BERT can dynamically encode the embedding vectors according to different given contexts, which has certain advantages over static word vectors such as Word2vec. Referring to the previous work of Li et al. [12], we adopt the annotation information of each entity type label as the prior problem for the reading comprehension task. We then encode the input text X and label annotations Y into embeddings h_X and h_Y by BERT.

4.2 Label Relation Layer

There are certain relations between entity types, for example, body part entities and clinical symptom entities in the electronic medical record dataset CMeEE often appear together. We want to perform entity recognition with full consideration of the relations between entity labels, and detect those entities that are more difficult to identify by such implicit features. We can use GAT to allow the model to automatically capture the connections between entity types. We construct a graph attention network for labels, and the nodes of the graph represent different label types.

GAT implements feature transformation of nodes by stacking multiple graph attention layers.The input to the graph attention layer is defined as $\boldsymbol{h} = \{\boldsymbol{h}_1, \boldsymbol{h}_2, \boldsymbol{h}_3, \ldots, \boldsymbol{h}_N\}, \boldsymbol{h}_i \in R^F$, where N is the number of nodes and F is the dimensionality of node features. After a Graph Attention Layer, a new feature vector is output, and the dimension of the node feature of this feature vector is F', which can be expressed as $\boldsymbol{h}' = \{\boldsymbol{h'}_1, \boldsymbol{h'}_2, \boldsymbol{h'}_3, \ldots, \boldsymbol{h'}_N\}$. In the graph attention layer, the features of the node input first go through a linear transformation: $W\boldsymbol{h}_i$, where $W \in R^{F' \times F}$. That is, after a linear transformation, the feature dimension of all nodes changes from F to F'.

We perform a self-attention mechanism on the nodes to calculate the attention factor,

$$e_{ij} = a\left(W\boldsymbol{h}_i, W\boldsymbol{h}_j\right) \tag{1}$$

here e_{ij} represents the importance of features of node j to node i. With the above formula, we can calculate the importance of all node features to node i in the graph. To calculate the attention coefficients between two node feature vectors, they are first linearly transformed W to change their features from F at input to F', and then the two vectors are joined together into a vector of dimension $2F$, followed by a feedforward neural network α and a LeakyRELU activation function to finally obtain e_{ij}, a number used to measure the importance of the node j features to the node i,

$$e_{ij} = LeakyRELU\left(\boldsymbol{a}^T \left[W\boldsymbol{h}_i \| W\boldsymbol{h}_j\right]\right) \tag{2}$$

where $\|$ indicates the connection operation. Finally, all attention coefficients e_{ij} are normalized using the softmax function and the normalized score α_{ij}:

$$\alpha_{ij} = softmax_j\left(e_{ij}\right) = \frac{\exp\left(e_{ij}\right)}{\sum\limits_{k \in N_i} \exp\left(e_{ik}\right)} \tag{3}$$

Then the features of all neighboring nodes j of node i are linearly weighted using α_{ij} to obtain the features $\boldsymbol{h}'_i$ of the updated node i:

$$\boldsymbol{h}'_i = \sigma\left(\sum_{j \in N_i} \alpha_{ij} W \boldsymbol{h}_j\right) \tag{4}$$

4.3 Label Fusion Layer

The label fusion layer aims to provide a priori knowledge using semantic information about labels. Inspired by Seo et al. [16], to more fully fuse label relations and input sentence embeddings, we compute attention weights in two directions: the text-to-question direction and the question-to-text direction, respectively. The weighted sentence embeddings and label relations are then stitched together as a new textual representation.

Text-to-Question Attention. In this step, we need each token in the sentence to be able to compute attention weights with the label text. Let x_i represent the i-th character of

X, $1 \leq i \leq n$, and let y_j^c represent the j-th character of the c-th entity type of Y, $1 \leq j \leq m, 1 \leq c \leq k$. By dot product calculate the information of the joint sentence text and tag text, and then apply the softmax function to calculate the attention weights α_{x_i,y_j^c} for the i-th token in the sentence and the j-th token in the c-th tag annotation:

$$\alpha_{x_i,y_j^c} = \frac{\exp(h_{x_i} \cdot h_{y_j^c})}{\sum_{c=1}^{k}\sum_{j=1}^{m} \exp\left(h_{x_i} \cdot h_{y_j^c}\right)} \tag{5}$$

where h_{x_i} is the embedding vector of the i-th sentence token and $h_{y_c^j}$ represents the embedding vector of the j-th character in the c-th label relation. Then we obtain the selected tag information by multiplying the weights and label embeddings, fusing this information into the sentence embedding as follows:

$$H_{x_i}^c = h_{x_i} + \sum_{j=1}^{m} \alpha_{x_i,y_j^c} \cdot h_{y_j^c} \tag{6}$$

The final obtained sentence embedding containing label relation information is $H_{x_i}^c$, and the final sentence fusing all label relation information is represented as $\hat{H}_{x_i} = (H_{x_i}^1, \cdots, H_{x_i}^{|C|})$

Question-to-Text Attention. In this step, we calculate the attention weights in the question-to-text direction in conjunction with the label relations, and then multiply the weights with the sentence embedding to get the weighted sentence embedding:

$$\alpha_{y_j^c,x_i} = \frac{\exp(h_{x_i} \cdot \max_{1<j<m} h_{y_j^c})}{\sum_{i=1}^{n} \exp\left(h_{x_i} \cdot \max_{1<j<m} h_{y_j^c}\right)} \tag{7}$$

Similarly, we fuse this part of the sentence information into the label embedding to obtain the weighted label embedding H_y^c:

$$H_y^c = h_{y^c} + \sum_{i=1}^{n} \alpha_{y^c,x_i} \cdot h_{x_i} \tag{8}$$

The final label that incorporates all the sentence information is represented as $\hat{H}_y = (H_{y^1}, \cdots, H_{y^{|C|}})$. We stitch the final label-relational representation and the sentence representation together as the input to the next layer.

We extract the target by outputting the starting and ending positions of the answer fragments, specifically, for each token on the input, we compute its probability value as the starting and ending position of the answer:

$$start_{x_i} = sigmoid\left(M_s \circ h_{x_i} + b_s\right) \tag{9}$$

$$end_{x_i} = sigmoid\left(M_e \circ h_{x_i} + b_e\right) \tag{10}$$

where M_s, M_e, b_s, and b_e are all learnable parameters, h_{x_i} represents the i-th token in the final text representation of the final text representation, and$\circ$ represents the element multiplication.

4.4 Loss Function

Given a text $X = \{x_1, x_2, ..., x_n\}$, the text contains n tokens and $|C|$ label categories. For each category $c \epsilon C$ define a vector $S^c \epsilon \{0,1\}^n$,where $S^c_{x_i} = 1$ only if the i-th token is the starting position of the real label. Similarly define $E^c \epsilon \{0,1\}^n$to represent the endpoint vector of real labels. $start_{x_i}$ and end_{x_i} refer to the definition in 4.3. Finally, the loss functions L_s and L_e are defined for the prediction of the answer starting position and the prediction of the answer ending position as follows:

$$L_s = \frac{1}{n} \sum_{c \in C} \sum_{1 \leq i \leq n} CE\left(start^c_{x_i}, S^c_{x_i}\right) \tag{11}$$

$$L_e = \frac{1}{n} \sum_{c \in C} \sum_{1 \leq i \leq n} CE\left(end^c_{x_i}, E^c_{x_i}\right) \tag{12}$$

where CE represents the cross-entropy function. The final loss $L = L_s + L_e$.

5 Experiments

5.1 Datasets

We evaluate our model on the Chinese Medical Entity Extraction dataset [26] (CMeEE) and the CCKS2017-CNER dataset. CMeEE is a public dataset of Chinese medical texts. In CMeEE, medical text named entities are classified into nine categories. The details are shown in Table 1. CCKS2017-CNER is a dataset proposed in the Chinese electronic medical record named entity recognition evaluation task published by China Conference on Knowledge Graph and Semantic Computing(CCKS).The dataset was manually labeled with five categories of entities (including signs and symptoms, examinations and tests, diseases and diagnoses, treatments, and body parts). The details are shown in Table 2.

Table 1. Statistics of different types of entities in the CMeEE dataset

CMeEE	dis	sym	dru	equ	pro	bod	mic	dep	ite
Training set	16002	12271	3937	902	6447	17708	1942	356	2608
Dev set	4965	4132	1442	245	2087	5892	607	115	927

5.2 Baselines

To validate the effectiveness of joint label and text modeling method proposed in this paper, we compare with the following baseline method. The baselines include classical sequence-labeling based (BiLSTM-CRF, Lattice-LSTM, MUSA-BiLSTM-CRF), MRC-based (MRC NER, LEAR) and Bert-based (MacBERT-large) approaches.

BiLSTM-CRF [7]: The BiLSTM layer learns the contextual information of the sequence and the CRF layer learns the dependency information between the tags.

Lattice-LSTM [27]: A character information-based entity recognition method with the addition of subword information constructs a word-character lattice for Chinese NER.

MRC NER [12]: Consider NER as an MRC/QA task, which is the SOTA approach on flat and nested NER.

MacBERT-Large [26]: Mac-BERT is an improved BERT with novel MLM as a correction pre-training task.

MUSA-BiLSTM-CRF [1]: A method using a multi-headed attention mechanism based on BiLSTM-CRF model and an external dictionary. We compare the results of the single model without the external dictionary in the paper.

LEAR [24]: This is an MRC method that introduces a priori knowledge of the label.

Table 2. Statistics of different types of entities in the CCKS2017-CNER dataset

CCKS2017-CNER	Signs	Check	Disease	Treatment	Body
Training set	7831	9546	722	1048	10719
Test set	2311	3143	553	465	3021

5.3 Parameter Settings

The hyperparameter setting of the paper is shown in Table 3. In the training, a moderate batch size yields better performance, and we set it to 2 empirically. We use Adam [8]for optimization. We use a strict evaluation metrics that an entity is confirmed correct when the entity boundary and the entity categorical label are correct simultaneously. We employ precision (P), recall (R) and F1-score (F1) to evaluate the performance.

Table 3. Experimental hyperparameter setting

Parameter	CMeEE	CCKS2017-CNER
Maximum input length	512	512
Batch size	2	2
Epoches	20	30
Learning rate of BERT layer	2e−5	3e−5
Learning rate of other layers	4e−5	6e−5

5.4 Results and Discussion

The main experimental results on the CMeEE dataset and the CCKS 2017 dataset are shown in Table 4. We can find that the F1 scores of our method on the two datasets are higher than that of the baseline methods, and the possible reasons are: 1)Our approach is more likely to identify some longer entities that are more difficult to identify for baseline methods. 2) We introduced GAT to implicitly model the dependency among different entity types.

We introduce GAT modeling labeling relations on the dataset CMeEE with more entity types to enable our model to identify entity types more comprehensively through relations. Compared with the traditional MRC method, the precision of our method becomes a little bit lower, but the recall is much higher because we identify the entity types that were not identified before by considering the relations between labels. Finally, our method of introducing GAT modeling labeling relations obtains the highest F1-score, proving our method's effectiveness.

Table 4. Performance of the models on the CMeEE and CCKS2017.

Datasets	Model	P	R	F1
CMeEE	Bilstm-CRF	0.6039	0.5135	0.5550
	Lattice-LSTM	0.6302	0.5843	0.6064
	MRC NER	0.6906	0.5492	0.6119
	MacBERT-large	–	–	0.6240
	LEAR	0.6578	0.6581	0.6579
	FLR-MRC (ours)	**0.6679**	**0.6625**	**0.6652**
CCKS2017	Bilstm-CRF	0.8116	0.8237	0.8177
	Lattice-LSTM	0.8031	0.7629	0.7825
	MRC NER	0.8522	0.8856	0.8686
	LEAR	0.9131	0.8987	0.9058
	MUSA-BiLSTM-CRF	0.9056	0.9091	0.9073
	FLR-MRC (ours)	**0.9098**	**0.9104**	**0.9101**

6 Conclusion

In this paper, we proposed a MRC-based NER framework FLR-MRC, to address the problem that existing MRC methods extract entity types independently and ignore the correlations among different entity types in Chinese electronic medical records. Our model first encoded text and label annotations independently and used GAT to model the relations of entity labels. Then the semantic fusion module was used to fuse the label relations and text representations. We achieved the best results on the CMeEE and the CCKS2017-CNER dataset, and also validated the effectiveness of the method through comparison experiments and ablation experiments.

Acknowledgements. This work is partly supported by the Science and Technology Major Project of Changsha (No. kh2202004) and Natural Science Foundation of Hunan Province via Grant 2021JJ40783.

References

1. An, Y., Xia, X., Chen, X., Wu, F.X., Wang, J.: Chinese clinical named entity recognition via multi-head self-attention based BiLSTM-CRF. Artif. Intell. Med. **127**, 102282 (2022)
2. Chiu, J.P., Nichols, E.: Named entity recognition with bidirectional LSTM-CNNs. Trans. Assoc. Comput. Linguis. **4**, 357–370 (2016)
3. Collobert, R., Weston, J., Bottou, L., Karlen, M., Kavukcuoglu, K., Kuksa, P.: Natural language processing (almost) from scratch. J. Mach. Learn. Res. **12**(ARTICLE), 2493–2537 (2011)
4. Devlin, J., Chang, M.W., Lee, K., Toutanova, K.: Bert: pre-training of deep bidirectional transformers for language understanding. arXiv preprint arXiv:1810.04805 (2018)
5. Gillick, D., Brunk, C., Vinyals, O., Subramanya, A.: Multilingual language processing from bytes. arXiv preprint arXiv:1512.00103 (2015)
6. Hammerton, J.: Named entity recognition with long short-term memory. In: Proceedings of the Seventh Conference on Natural Language Learning at HLT-NAACL 2003, pp. 172–175 (2003)
7. Huang, Z., Xu, W., Yu, K.: Bidirectional LSTM-CRF models for sequence tagging. arXiv preprint arXiv:1508.01991 (2015)
8. Kingma, D.P., Ba, J.: Adam: a method for stochastic optimization. arXiv preprint arXiv:1412.6980 (2014)
9. Lafferty, J., McCallum, A., Pereira, F.C.: Conditional random fields: probabilistic models for segmenting and labeling sequence data (2001)
10. Lample, G., Ballesteros, M., Subramanian, S., Kawakami, K., Dyer, C.: Neural architectures for named entity recognition. arXiv preprint arXiv:1603.01360 (2016)
11. Li, F., Lin, Z., Zhang, M., Ji, D.: A span-based model for joint overlapped and discontinuous named entity recognition. arXiv preprint arXiv:2106.14373 (2021)
12. Li, X., Feng, J., Meng, Y., Han, Q., Wu, F., Li, J.: A unified MRC framework for named entity recognition. arXiv preprint arXiv:1910.11476 (2019)
13. Luan, Y., Wadden, D., He, L., Shah, A., Ostendorf, M., Hajishirzi, H.: A general framework for information extraction using dynamic span graphs. arXiv preprint arXiv:1904.03296 (2019)
14. Ma, X., Hovy, E.: End-to-end sequence labeling via bi-directional LSTM-CNNs-CRF. arXiv preprint arXiv:1603.01354 (2016)
15. Peters, M.E., et al.: Deep contextualized word representations. In: Proceedings of the 2018 Conference of the North American Chapter of the Association for Computational Linguistics: Human Language Technologies, Volume 1 (Long Papers), pp. 2227–2237. Association for Computational Linguistics, New Orleans, Louisiana (2018). https://doi.org/10.18653/v1/N18-1202, https://aclanthology.org/N18-1202
16. Seo, M., Kembhavi, A., Farhadi, A., Hajishirzi, H.: Bidirectional attention flow for machine comprehension. arXiv preprint arXiv:1611.01603 (2016)
17. Strubell, E., Verga, P., Belanger, D., McCallum, A.: Fast and accurate entity recognition with iterated dilated convolutions. arXiv preprint arXiv:1702.02098 (2017)
18. Sutton, C., McCallum, A., Rohanimanesh, K.: Dynamic conditional random fields: Factorized probabilistic models for labeling and segmenting sequence data. J. Mach. Learn. Res. 8(3) (2007)
19. Veličković, P., Cucurull, G., Casanova, A., Romero, A., Lio, P., Bengio, Y.: Graph attention networks. arXiv preprint arXiv:1710.10903 (2017)
20. Xu, M., Jiang, H., Watcharawittayakul, S.: A local detection approach for named entity recognition and mention detection. In: Proceedings of the 55th Annual Meeting of the Association for Computational Linguistics (Volume 1: Long Papers), pp. 1237–1247 (2017)

21. Yamada, I., Asai, A., Shindo, H., Takeda, H., Matsumoto, Y.: Luke: deep contextualized entity representations with entity-aware self-attention. arXiv preprint arXiv:2010.01057 (2020)
22. Yan, H., Deng, B., Li, X., Qiu, X.: Tener: adapting transformer encoder for named entity recognition. arXiv preprint arXiv:1911.04474 (2019)
23. Yan, H., Gui, T., Dai, J., Guo, Q., Zhang, Z., Qiu, X.: A unified generative framework for various ner subtasks. arXiv preprint arXiv:2106.01223 (2021)
24. Yang, P., Cong, X., Sun, Z., Liu, X.: Enhanced language representation with label knowledge for span extraction. arXiv preprint arXiv:2111.00884 (2021)
25. Yu, J., Bohnet, B., Poesio, M.: Named entity recognition as dependency parsing. arXiv preprint arXiv:2005.07150 (2020)
26. Zhang, N., et al.: CBLUE: a chinese biomedical language understanding evaluation benchmark. arXiv preprint arXiv:2106.08087 (2021)
27. Zhang, Y., Yang, J.: Chinese NER using lattice LSTM. arXiv preprint arXiv:1805.02023 (2018)

Private Epigenetic PaceMaker Detector Using Homomorphic Encryption - Extended Abstract

Meir Goldenberg(✉), Sagi Snir, and Adi Akavia

University of Haifa, Haifa, Israel
meirgold@hotmail.com, ssagi@research.haifa.ac.il, akavia@cs.haifa.ac.il

Abstract. The Epigenetic Pacemaker (EPM) model uses DNA methylation data to predict human epigenetic age. The methylation values are collected from different individuals and are considered to be of medical importance. Sharing this data publicly among labs and other third parties for model calculation purposes may violate the privacy of personal medical records. The use of standard encryption approaches can prevent the exposure of these personal records to third parties, when at rest, but running computations on the data requires decrypting it first, and thus exposing the data to the computing party. This work proposes computing EPM while limiting data exposure by employing cryptographic secure computing techniques including homomorphic encryption. Our protocol has rigorous privacy guarantees against computationally bounded adversaries in the two-server model. We implemented a relaxed version of the protocol showing good correlation with low accuracy error between the model computed with and without encryption.

1 Introduction

Background on the Epigenetic Pacemaker Model (EPM) [20–22]. DNA methylation is a well-studied epigenetic mark that functions to define the states of cells as they undergo developmental changes [19]. The Epigenetic Pacemaker Model is generated based on the input of methylation values from CpG sites. The model is based on the molecular clock (MC) [10] which estimates the chronological age based on the methylation values and also on the Universal Pacemaker (UPM) [15,23,24,26] which defines the rate change in epigenetic aging. The epigenetic pacemaker model (EPM) [20–22] uses a conditional expectation maximization (CEM) algorithm [14] to calculate the combined model in two steps: the *site step* uses the same parameters as the linear MC model for calculating the initial methylation values and change rate per site, while the *time step* calculates the epigenetic age as defined by the UPM.

The Need for Privacy. The input to EPM consists of private medical data on individuals: methylation values from different CpG sites, and an initial epigenetic

This work was supported in part by the Israel Science Foundation grant 3380/19 and 3291/21, and by the Israel National Cyber Directorate via the Haifa, BIU and Tel-Aviv Cyber Centers.

M. S. Bansal et al. (Eds.): ISBRA 2022, LNBI 13760, pp. 52–61, 2022.
https://doi.org/10.1007/978-3-031-23198-8_6

age. Exposing this personal medical information to the entity calculating the model is often undesired. To avoid such exposure, a straw-man solution would be for each patient (or clinic) to run the EPM [20] in isolation, while relying solely on the methylation data in its possession. The problem however is that the EPM algorithm requires methylation data from *many individuals* to produce accurate predictions, where each patient has the methylation data from only a single individual (similarly, each clinic typically has data only from few individuals). A better solution, from a statistical perspective, is for all individuals or clinics to join their data, and execute the EPM algorithm on the union of all their data. However, privacy, business, and even legal concerns generally forbid this kind of transparent data-sharing arrangement.

Our Contribution. In this work we propose the first privacy-preserving solution for EPM. Our solution supports computing the result for the union of the data, but while protecting the secrecy of the raw data from all parties (including those contributing data or participating in the computation). Introducing a privacy-preserving solution on top of the EPM algorithm may have impact on the model accuracy; our preliminary results (from a relaxed version of the new solution) show a high accuracy level of less than 3% change in epigenetic age value for the majority of individuals.

Related Work. Prior work on privacy-preserving genome analysis focused on Genome Wide Association (GWAS) [3,4,6,13,18], i.e., on statistically associating innate genomic variability in single nucleotide polymorphism (SNPs) with a risk for a disease or a particular trait. In contrast, we focus on genomic alterations –within a particular genome– that are occurring throughout the lifetime of the individual (methylation), and propose a privacy preserving solution for inferring epigenetic age from such alternations.

2 Methods

2.1 Epigenetic Pacemaker (EPM) Algorithm

Epigenetic Pacemaker (EPM) [22]. Let $s_1,...,s_n$ be genome *sites* that undergo methylation. Each site s_i starts at birth with an *initial level* of methylation, denoted s_i^0 and undergoes methylation over life. According to the EPM model, each site s_i is associated with a *rate* r_i in which methylation events occur, but this rate may vary over time (arbitrarily, and independently of other individuals). The *EPM property* mandates that the site methylation rates change proportionally throughout lifetime over all sites of the same individual. That is, at any point in time, if a change in rate occurred in site i of individual j, then the rates in *all* sites i' in j, are simultaneously changed and by the same factor. This methylation rate is correlated with the *aging rate*, providing a good estimate on the *epigenetic age (e-age)* of an individual (as opposed its chronological age) [17]. We denote by t_j the weighted average e-age of individual j, accounting for the rate changes an individual has undergone through life. The algorithmic task under the EPM model is to: *Find the maximum likelihood values of s_i^0, r_i, and t_j, when given the observed methylation levels in n genome sites as measured in m individuals.* This input is denoted by $(\hat{s}_{i,j})_{m \cdot n}$ where $\hat{s}_{i,j}$ denotes the methylation measured in individual j at site i.

The EPM Algorithm. Snir [20] presented an algorithm to this EPM problem, which provably converges to a local optima of the maximum likelihood function. Furthermore, [20] presented an experimental evaluation validating the concrete good efficiency of this algorithm. This algorithm is the starting point of our solution.

To describe the algorithm we first organize the observed variables $\hat{s}_{ij}$ as well as the unknown variables t_j, r_i and s_i^0 as follows. Let X be a $mn \times 2n$ matrix whose kth row is all zero except for the value t_j in the ith entry of its first half and 1 in the ith entry of its second half. Let β be a column vector whose first n entries are $r_1, \ldots, r_n$ and the last n entries are $s_1^0, \ldots, s_n^0$. Let y be the column vector whose $im + j$ entry contains $s_{i,j}$. The algorithm of [20] consists of several iterations, each composed of two main components: a site step and a time step. In the *site step* values for t_j's are fixed to be the values obtained from the previous iteration (on the first iteration, they are initialized to random or provided values), and the algorithm solves the linear regression problem system specified by X and y (where X is with the said values for t_j's) to obtain values for r_i and s_i^0 (i.e., for β). In the *time step*, r_i and s_i^0 are fixed to the values obtained in the site step (of the current iteration), and the individual's e-age estimates t_j's are set to their maximum likelihood values, which as proved by [20], is given by the following closed form rational function:

$$t_j = \frac{\sum_{i=1}^{n} r_i(\hat{s}_{i,j} - s_i^0)}{\sum_{i=1}^{n} r_i^2} \qquad (1)$$

Furthermore, [20] proves that at every such step an increase in the likelihood is guaranteed, and so, a local optimum is eventually reached. The iterations can proceeds until the improvement in the Residual Sum of Square (RSS) falls below a threshold δ given as a parameter to the algorithm.

Our goal in this work is to compute the epigenetic age in a *privacy preserving* fashion. Therefore, We must not reveal even the number of iterations required for convergence, because this could potentially reveal significant information on the input (see [1] for discussion of such attacks). We therefore slightly modify the algorithm of [20], in specifying the number of iterations in advance, by a user-defined parameter denoted iter.

2.2 Private EPM Computing Over Federated Data: The Problem

As stated above, this work proposes a privacy-preserving solution for the EPM problem, that supports computing the result for the union of the data, while protecting the secrecy of the raw data from all parties (including those contributing data or participating in the computation).

Formally, we consider a setting in which there are m individuals, called *Data Owners* and denoted by $\mathsf{DO}_1, \ldots, \mathsf{DO}_m$. Each data owner DO_j holds observed methylation levels $\hat{s}_{1,j}, \ldots \hat{s}_{n,j}$ in n sites $s_1, \ldots, s_n$.[1] The data owners wish to compute the epigenetic age estimator specified by the EPM [20]) on their joint data, but without revealing information on their individual data. Following [12], we focus on the *two-server model* in which the data-owners are aided by two *non-colluding* servers Srv_1 and Srv_2, so that the servers execute the bulk of the computation, while complexity of the data owners is proportional only to the size of their individual input (in encrypted form). The goal is to compute the same epigenetic age estimation as outputted by the EPM algorithm when executed on the union of the individual data, but without exposing any information on the raw data beyond what can be inferred from the designated output and leakage profile. The security requirement is to guarantee correctness and privacy against all passive computationally-bounded adversaries in the two-server model. That is, the adversaries we consider may corrupt any subset of the data owners and at most one server (two-server model); parties controlled by the adversary follow the protocol specification, albeit they may collude to infer as much information as possible from their view of the interaction (passive adversary); and all parties are restricted to performing probabilistic polynomial time computations (computationally bounded).

2.3 Homomorphic Encryption

Homomorphic Encryption (HE) is a central tool that we use in order to avoid the exposure of sensitive medical data while maintaining the ability to compute on the data. HE is a public-key encryption scheme that allows one to perform arithmetic computations on encrypted data, without exposing these values and without knowledge of the secret decryption key. See a tutorial in [9].

3 Results

Our main result in this work is introducing *the first protocol for securely computing the EPM model.* We give an overview of our protocol in Sect. 3.1; the formal specification and analysis of our protocol is deferred to the full version of this paper. We implemented a relaxed version of our protocol, and conducted an empirical evaluation showing it produces predictions attaining a high level of accuracy comparable to the predictions of the original EPM Algorithm without privacy preservation. These results are discussed in Sect. 3.2.

3.1 The Theoretical Privacy Preservation Protocol for EPM

We present an intuitive description of our privacy preserving protocol for the EPM model (formal details to appear in the full version).

[1] All measurements are for a known and identical set of genome sites.

Designing a secure protocol for the EPM algorithm [20] requires two main components: a privacy preserving computation of linear regression for the *site step* of the EPM ([20], Site Step), and a privacy preserving computation of the rational function computing in the *time step* of the EPM ([20], Time Step). Our starting idea for constructing such components is the idea to use the privacy preserving linear regression protocol of Giacomelli et al [8] for the site step, and homomorphic computation to perform the time step. This naive approach however does not work due to two main problems, as detailed next.

The first problem with the naive approach emanates from the fact that the EPM algorithm consists of several iterations, each executing both the site and the time steps, where to preserve privacy it is imperative not to reveal any intermediate computation results (see discussion of concrete attacks in [1]). But plugging into the site step the protocol by Giacomelli et al [8] would completely expose all the intermediate models, because [8] outputs the regression model in the clear. Namely, the naive approach is insecure.

Our first attempt for resolving the problem of intermediate models exposure is to follow Akavia et al [1], who recently addressed a similar issue in the context of privacy preserving feature selection. The high level approach of [1] is to modify the protocol of [8] (as well as a subsequent protocol [2]) to output the model in encrypted form in order to avoid such exposure of intermediate models. But –as they show– this requires introducing several new tricks, in order to evade the need for overly expensive computations on the encrypted model. Particularly, the homomorphic computations in [2,8] are over the ring of integers modulo N (denoted, $\mathbb{Z}_N$), and to map the results of the computation in $\mathbb{Z}_N$ to the actual regression model –which is computed over the rational numbers–, a so called rational reconstruction [7,25] is required. Unfortunately, computing rational reconstruction over encrypted data is too expensive to be practical, and so outputting the result of the homomorphic computation in encrypted form will not be practically usable. To resolve this issue, [1] proposed another modification of the [8] where they produce a *scaled* regression model, for which they prove that it is readily also a scaled version of the regression model over the rationals; namely, it requires no rational reconstruction. In their context, producing a scaled model is acceptable (because they use the intermediate models only for ranking their weights, which is invariant under scaling). In contrast, in our context, we cannot use this scaled model, because we use the concrete model weights to update the values for the next iteration, and so using a scaled version will not produce the correct updates.

Our approach for resolving the problem with using a scaled model is to simultaneously consider both the site and the time steps (rather than analyzing each of them in isolation). Concretely, we prove that we may use the scaled model in the site step *as long as we properly modify the time step* as to correct the bias caused by this scaling. That is, we carefully introduce two modifications –a scaling of the site step and a modified algebraic computation in the time step– that cancel each other out. We prove that these modifications achieve the following desired invariant: at the termination of each iteration (i.e., of both the site and the time steps), our doubly modified computation produces the exact same updated values as produced in the EPM Algorithm (i.e., with no modifications at all). Moreover, the modifications we introduce eliminate the main bottlenecks for homomorphic computation in the EPM algorithm, leading to a considerable performance speed up compared to the naive approach. We note that our approach relies on diving into the precise linear algebra in the EPM algorithm together with the details of the scaling factor in [1].

The second problem we face is that the time step requires computing a rational function over encrypted data. However computing division over encrypted data is typically too expensive to be practical, thus making the naive approach not practically applicable.

Our solution to this problem again relies on a careful interplay with the algebraic formulas in the different steps of the EPM. Concretely, we first change the time step to compute the numerator and denominator of the ratio function *as a pair of values* rather than computing their ratio. We then show how to scale the linear regression instance that we solve in the site step –using a function of the denominator from the time step– to obtain a new linear regression instance. We show that solving the new instance produces the same solution as in the instance that is obtained when using the original time step from the EPM algorithm.

In summary, via a series of careful algebraic manipulations and algorithmic modification we are able to propose a protocol that computes the exact same output as the EPM algorithm, but does so while avoiding the complexity bottlenecks associated with homomorphic computation.

3.2 Empirical Results

The Relaxed Protocol. As a preliminary empirical evaluation we implemented a relaxed version of our protocol of Sect. 3.1. The relaxed version computes each site step by employing [8] as is, with none of the modifications we introduced in Sect. 3.1. As discussed above, this naive approach leaks all the intermediate models produced at each site step. The relaxed version then computes the time step, in the clear, while using the cleartext intermediate models. We stress that this relaxed version is not secure; rather, it serves as a first step toward implementing our protocol of Sect. 3.1. This relaxed protocol allows using a homomorphic encryption schemes that supports only computing addition over encrypted data (rather than supporting both addition and multiplication); such encryption schemes are called *linearly homomorphic encryption (LHE)*. This relaxation was taken –for a first prototype– because libraries supporting linearly homomorphic encryption are much simpler and faster to use from a programming point of view. Concretely, we use the Paillier cryptosystem [16] for the LHE, as implemented in the *phe* library version 1.5.0 [5]. We note that we intend to extend this empirical evaluation into a full implementation of the protocol in future work.

Our system implementing the above relaxed protocol consists of three components: The Data Owners (DO), one server named the Machine Learning Engine (MLE), and another server named the Crypto Service Provider (CSP). The DOs use the public encryption key provided by the CSP to encrypt the methylation and age values. The encrypted data is then sent to the MLE which calculates the model with assistance from the CSP.

Data Preparation. The methylation values [11] and ages are provided as floating point numbers with many digits in the fractional part. The LHE requires us to work with integers, therefore we need to convert the age and methylation values to integers. In order to minimize the loss of accuracy, we tested the loss on various rounding values. Our tests showed that rounding the numbers to 2 digits in the fraction part of the number will cause an accuracy loss of up to 3% for the majority of the results.

We used Pearson Correlation for a preliminary feature selection phase, as common in machine learning, where we remove sites with low correlation with the target ages in the training data. Several tests were conducted on the provided training data in order to optimize the correlation coefficient. Results show the following: when correlation was set to be greater than 90%, 91%, and 92% the algorithm returned 42, 24, and 12 sites (columns of X) respectively. To maintain a moderate number of sites we set the threshold at Pearson coefficient of 91%. This trims the data matrix to a matrix with $d = 24$ sites at maximum. The precision (number of digits) is set to be $\ell = 2$, with values scaled to range $[-\delta, \delta]$ for $\delta = 100$ at maximum. The number of individuals is $m = 472$. The above parameters settings yields: $N = 3.883e^{366}$ (for the common parameter N as described in [8]).

Libraries and Hardware. The protocol was implemented in Python 3.8 using the *phe* library version 1.5.0 for Paillier LHE and numpy library version 1.22.4 for matrix operations. Computation was executed on a cloud server with 4 Intel virtual CPUs running at 2 GHz, 8 GB of RAM and Ubuntu 20.0 OS.

Runtime. The runtime of the model calculation on 472 individuals with 24 CpG sites (after optimization) was: 2 h and 50 min. The runtime of the EPM algorithm without encryption was: 3 s.

Accuracy. In order to measure the accuracy of the relaxed protocol, the model was calculated based on the training data using two separate algorithms:

1. The original EPM algorithm, with the linear algebra operations, and without privacy preservation.
2. The EPM algorithm with the relaxed privacy preserving protocol.

Each of the algorithms was run on the training data extracted from [11], with number of individuals and sites equal 24 and 472 respectively. In both cases, the CEM algorithm converged after 4 iterations and the model values were recorded.

As described in [20], the EPM model consists of S^0 and $rate$ parameter values per methylation site. We compared the percentage of change in values for these parameters (defined as *Error Percentage*) between the two models. The change in values was measured per site where the model from the original algorithm was considered to contain the "golden" values. The results are depicted in Fig. 1.

For each site, we took the "true" value (rate, starting value) as inferred from the unsecured algorithm versus the"estimated" value as produced from the relaxed model, and calculated the error percentage. The plots in the figure describe the distribution of the error among the sites. We observe the following: For the S^0 values, the majority of sites (16 out of 24) have a maximum difference of 0.22% between the models. The largest difference observed for two of the sites is 0.62% For the *rate* values, the majority of sites (19 out of 24) have a maximum difference of 0.41% between the models. The largest difference observed for four of the sites is 1.21%.

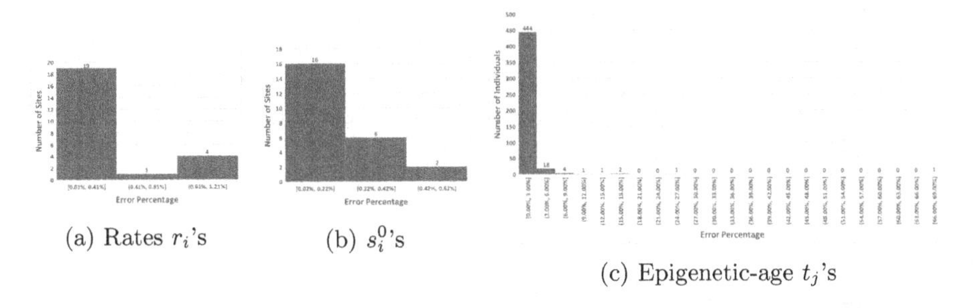

(a) Rates r_i's (b) s_i^0's (c) Epigenetic-age t_j's

Fig. 1. The histograms specify the number of sites (a-b) or individuals (c), in which the values predicted by our privacy preserving implementation differ from the output of the EPM [20] by the percentage indicated by the x-axes.

4 Conclusions and Future Work

In this work we have presented the first privacy preserving solution for computing the epigenetic pacemaker (EPM) model. Our solution comprises of two components: A theoretical rigorous protocol with privacy guarantees (against computationally-bounded adversaries in the two-server model), and a relaxed version that we implemented. We show that the implemented version attains very high accuracy compared to the true values (i.e., the output of the baseline algorithm that has no privacy guarantee). Specifically, for the site model parameters *rates* and *methylation starting value* error rate of under 1.21% is obtained. For the *epigenetic age* model parameter, a slightly higher error rate of 3% is attained but with a very small deviation, suggesting that with high probability we can expect at most this error. While our current implementation is for a less secure variant of our protocol, where intermediate models are leaked to the adversary, we regard the novelty of a first theoretical solution to the problem, and a practical implementation with high model parameters inference accuracy, as a significant first step in this topical direction. We leave for future work the task of implementing and evaluating our protocol with no such relaxations. This would entail employing a fully homomorphic encryption in the implementation, rather than the linearly homomorphic encryption that we have used in our relaxed implementation for providing a first demonstration of our approach.

References

1. Akavia, A., Galili, B., Shaul, H., Weiss, M., Yakhini, Z.: SIR: privacy preserving feature selection for sparse linear regression. Manuscript (2022)
2. Akavia, A., Shaul, H., Weiss, M., Yakhini, Z.: Linear-regression on packed encrypted data in the two-server model. In: Proceedings of the 7th ACM Workshop on Encrypted Computing & Applied Homomorphic Cryptography, pp. 21–32 (2019)
3. Blatt, M., Gusev, A., Polyakov, Y., Goldwasser, S.: Secure large-scale genome-wide association studies using homomorphic encryption. Proc. Natl. Acad. Sci. **117**(21), 11608–11613 (2020)
4. Bonte, C., Makri, E., Ardeshirdavani, A., Simm, J., Moreau, Y., Vercauteren, F.: Towards practical privacy-preserving genome-wide association study. BMC Bioinform. **19**(1), 1–12 (2018)
5. Data61, C.: Python paillier library (2013). https://github.com/data61/python-paillier
6. Dong, C., et al.: Maliciously secure and efficient large-scale genome-wide association study with multi-party computation. IEEE Trans. Dependable Secure Comput. (2022)
7. Fouque, P.-A., Stern, J., Wackers, G.-J.: CryptoComputing with rationals. In: Blaze, M. (ed.) FC 2002. LNCS, vol. 2357, pp. 136–146. Springer, Heidelberg (2003). https://doi.org/10.1007/3-540-36504-4_10
8. Giacomelli, I., Jha, S., Joye, M., Page, C.D., Yoon, K.: Privacy-preserving ridge regression with only linearly-homomorphic encryption. In: Preneel, B., Vercauteren, F. (eds.) ACNS 2018. LNCS, vol. 10892, pp. 243–261. Springer, Cham (2018). https://doi.org/10.1007/978-3-319-93387-0_13
9. Halevi, S.: Homomorphic encryption. In: Lindell, Y. (ed.) Tutorials on the Foundations of Cryptography, pp. 219–276. Springer, Cham (2017). https://doi.org/10.1007/978-3-319-57048-8_5
10. Horvath, S.: DNA methylation age of human tissues and cell types. Genome Biol. **14**(10), 1–20 (2013)
11. Jaffe, A.E., et al.: Mapping DNA methylation across development, genotype and schizophrenia in the human frontal cortex. Nat. Neurosci. **19**(1), 40–47 (2016)
12. Kamara, S., Mohassel, P., Raykova, M.: Outsourcing multi-party computation. Cryptology ePrint Archive, Report 2011/272 (2011)
13. Lu, W.J., Yamada, Y., Sakuma, J.: Privacy-preserving genome-wide association studies on cloud environment using fully homomorphic encryption. In: BMC Medical Informatics and Decision Making, vol. 15, pp. 1–8. Springer (2015). https://doi.org/10.1186/1472-6947-15-S5-S1
14. Meng, X.L., Rubin, D.B.: Maximum likelihood estimation via the ECM algorithm: a general framework. Biometrika **80**(2), 267–278 (1993)
15. Muers, M.: Genomic pacemakers or ticking clocks? Nat. Rev. Genet. **14**(2), 81–81 (2013)
16. Paillier, P.: Public-key cryptosystems based on composite degree residuosity classes. In: Stern, J. (ed.) EUROCRYPT 1999. LNCS, vol. 1592, pp. 223–238. Springer, Heidelberg (1999). https://doi.org/10.1007/3-540-48910-X_16
17. Pinho, G.M., et al.: Hibernation slows epigenetic ageing in yellow-bellied marmots. Nature Ecol. Evol. **6**(4), 418–426 (2022)
18. Simmons, S., Berger, B.: Realizing privacy preserving genome-wide association studies. Bioinformatics **32**(9), 1293–1300 (2016)

19. Smith, Z.D., Meissner, A.: DNA methylation: roles in mammalian development. Nat. Rev. Genet. **14**(3), 204–220 (2013)
20. Snir, S.: Epigenetic pacemaker: closed form algebraic solutions. BMC Genomics **21**(2), 1–11 (2020)
21. Snir, S., Pellegrini, M.: An epigenetic pacemaker is detected via a fast conditional expectation maximization algorithm. Epigenomics **10**(6), 695–706 (2018)
22. Snir, S., vonHoldt, B.M., Pellegrini, M.: A statistical framework to identify deviation from time linearity in epigenetic aging. PLoS Comput. Biol. **12**(11), e1005183 (2016)
23. Snir, S., Wolf, Y.I., Koonin, E.V.: Universal pacemaker of genome evolution. PLoS Comput. Biol. **8**(11), e1002785 (2012)
24. Snir, S., Wolf, Y.I., Koonin, E.V.: Universal pacemaker of genome evolution in animals and fungi and variation of evolutionary rates in diverse organisms. Genome Biol. Evol. **6**(6), 1268–1278 (2014)
25. Wang, P.S., Guy, M., Davenport, J.H.: P-adic reconstruction of rational numbers. ACM SIGSAM Bull. **16**(2), 2–3 (1982)
26. Wolf, Y.I., Snir, S., Koonin, E.V.: Stability along with extreme variability in core genome evolution. Genome Biol. Evol. **5**(7), 1393–1402 (2013)

NIDN: Medical Code Assignment via Note-Code Interaction Denoising Network

Xiaobo Li[1], Yijia Zhang[1(✉)], Xingwang Li[1], Jian Wang[2], and Mingyu Lu[1]

[1] School of Information Science and Technology, Dalian Maritime University, Liaoning 116024, China
zhangyijia@dlmu.edu.cn

[2] School of Computer Science and Technology, Dalian University of Technology, Liaoning 116026, China

Abstract. Clinical records are files that contain detailed information about a patient's health status. Clinical notes are typically complex, and the medical code space is large, so medical code assignment from clinical text is a long-standing challenge. The traditional manual coding method is inefficient and error-prone. Incorrect coding may lead to adverse consequences. With machine learning and computer hardware development, the deep neural network model has been widely applied in the medical care domain. However, noise in lengthy documents, complex code association, and the imbalanced class problem urgently need to be solved. Therefore, we propose a **N**ote-code **I**nteraction **D**enoising **N**etwork (NIDN). We exploit the self-attention mechanism to identify the most relevant context of the medical code in the clinical document. We leverage the label attention mechanism to learn code-specific text representation. We utilize the correlation between labels in multi-task learning to assist the model in prediction. To better learn from lengthy texts and improve the performance of long-tail distribution, we develop a denoising module to reduce the influence of noise in medical code prediction. Experimental results show that our proposed model outperforms competitive baselines on a real-world MIMIC-III dataset.

Keywords: Medical Code Assignment · Attention Mechanism · Denoising Mechanism

1 Introduction

Medical code assignment has become systematic research in medical language processing and clinical decision support tasks. The international classification of diseases (ICD) is a unified international classification of various diseases. Medical code assignment is also called ICD coding. ICD is the basis for determining global health trends and statistics. It contains about 55000 unique codes related to injuries, diseases and causes of death, enabling health practitioners to exchange health information worldwide in a common language. Trained professional coders write these ICD codes according to clinical notes of discharge summaries, which contain past medical history, medications

M. S. Bansal et al. (Eds.): ISBRA 2022, LNBI 13760, pp. 62–74, 2022.
https://doi.org/10.1007/978-3-031-23198-8_7

on admission, discharge status, history of present illness and laboratory data. However, clinical notes usually include complex professional terms, polysemy and spelling errors. Since traditional medical coding is costly and inaccurate, approaches based on the deep neural network are proposed for medical code assignment task.

J. Mullenbach et al. [1] designed a Convolutional Attention network for Multi-Label classification (CAML) to identify the most relevant parts of the code. Li and Yu [2] developed a Multi-Filter Residual Convolutional Neural Network (MultiResCNN) to capture various text patterns from multiple perspectives and expand the receptive field. Thanh et al. [3] proposed a new label attention model (LAAT), which uses the joint hierarchical mechanism to deal with the long-tail distribution problem. B. Biswas et al. [4] applied a transformer-based framework to obtain the interdependence among texts and used a code-wise attention mechanism to get code-specific representations.

However, medical code assignment still has many challenges, mainly in the following three aspects:

Noisy and Lengthy Clinical Text: Clinical notes include many professional medical vocabularies, complex semantic information and noise, such as polysemy and spelling errors. Furthermore, clinical documents are lengthy, usually accompanied by thousands of words. Thus, it is difficult to capture the long-term dependence between words.

Code Association: There are many internal connections between medical codes. For example, "365.12" and "365.89" represent "Low tension glaucoma" and "Other specified glaucoma", respectively, which can be classified as "Glaucoma". However, in the existing models, researchers pay more attention to the word association and note-code interaction and do not consider code association. Therefore, it is necessary to attempt to incorporate code association information into our model.

Long-tail Distribution Problem: The long-tail distribution phenomenon is caused by the class imbalance problem. In the MIMIC-III dataset, ICD code distribution has a severe imbalanced class problem. More than half of the medical codes never appear. A small part of the code appears more than 1000 times. About 4000 codes appear between 1 and 10 times. Therefore, we learn more about common diseases in the model training, which declines the prediction performance of rare diseases.

To solve the above three challenges, we propose a novel model called **N**ote-code **I**nteraction **D**enoising **N**etwork (NIDN). This paper introduces BiGRU and **R**ecalibrated **A**ttention **M**odule (RAM) to extract rich semantic information from clinical texts. And this paper uses a note-code interaction module to obtain the direct relationship between clinical text and medical code. At last, we leverage the denoising module to alleviate the class imbalance problem. Our contributions are as follows:

- This paper designs a joint learning mechanism to fuse the self-attention matrix and label attention matrix to deal with the long-term dependency problem in clinical documents and extract code-specific text representation.
- Multi-task learning is introduced to obtain code association to assist medical code prediction.
- The proposed model combines the focus loss function and the truncation loss function to design a denoising module. We leverage the focus loss function to give different weights to high-frequency labels and low-frequency labels which can alleviate the

class imbalance problem. We use the truncation loss function to discard noisy samples to obtain cleaner ones.

- We conduct comprehensive experiments on the MIMIC-III dataset to demonstrate the validity of the proposed model.

2 Method

We propose a Note-Code Interaction Denoising Network (NIDN) for medical code assignment, which consists of a text learning module, note-code interaction module and joint denoising of focus loss and truncation loss. We show the whole architecture of NIDN in Fig. 1.

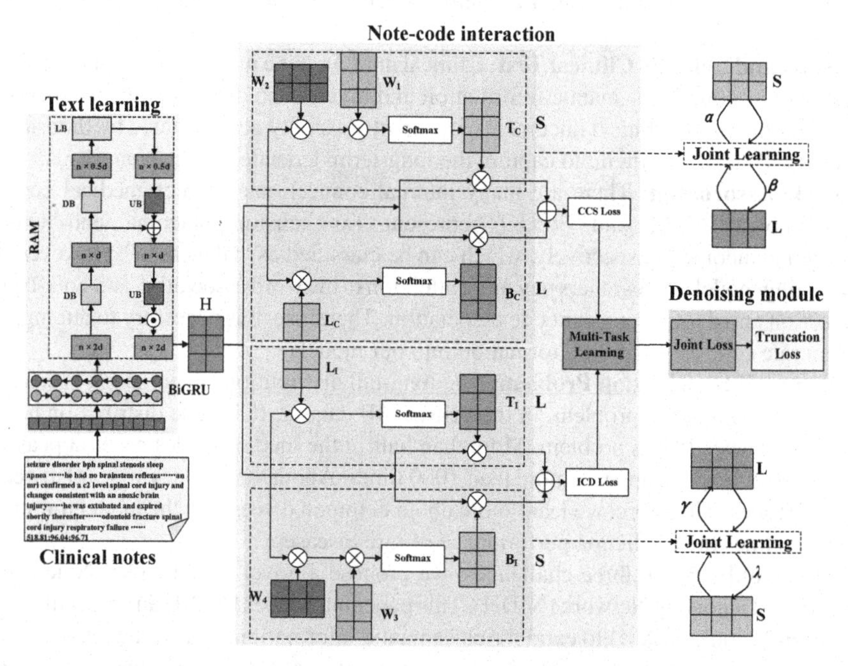

Fig. 1. Schematic overview of NIDN

2.1 Text Learning Module

We split the clinical text into n tokens. Our model uses a token list $c = \{c_1, c_2, \ldots, c_n\}$ as input. We use word2vec to train clinical documents into an embedding matrix E, denoted as $E = \{e_1, e_2, \ldots, e_n\}$, each word vector $e_n \in R^{d_e}$, where d_e is the embedding dimension. Then, we feed word embedding matrix E $\in R^{n \times d_e}$ into the Bi-GRU layer to capture the long-term dependency from medical document. Based on Bi-GRU, we utilize **R**ecalibrated **A**ggregation **M**odule (RAM) to learn better abstract feature representation, which can selectively stimulate and suppress features of different importance according to global information. We can get the final output $H \in R^{n \times 2d_r}$.

We stacked small convolution filters block (kernel size = 3) to obtain a good receptive field. In details, we regard the operations of "Down-node Block", "Lateral-node Block" and "Up-node Block" as $F_{DB}(\cdot)$, $F_{LB}(\cdot)$ and $F_{UB}(\cdot)$. We can represent the three blocks as follows:

$$F_{DB}(\cdot) = F^{l}_{3\times3}\left(\tanh\left(F^{d}_{3\times3}(\cdot)\right)\right) \tag{1}$$

$$F_{LB}(\cdot) = F^{l}_{3\times3}\left(\tanh\left(F^{l}_{3\times3}(\cdot)\right)\right) \tag{2}$$

$$F_{UB}(\cdot) = F^{l}_{3\times3}\left(\tanh\left(F^{u}_{3\times3}(\cdot)\right)\right) \tag{3}$$

The whole computation process is denoted as:

$$D_1 = F_{DB}(O) \tag{4}$$

$$D_2 = F_{DB}(D_1) \tag{5}$$

$$L_1 = F_{LB}(D_2) \tag{6}$$

$$U_1 = F_{UB}(L_1) \oplus D_1 \tag{7}$$

$$H = F_{UB}(U_1) \oplus O \tag{8}$$

where O is the output of Bi-GRU and $H \in R^{n \times 2d_r}$ denotes the final output of RAM.

2.2 Note-Code Interaction Module

The whole ICD coding process is to traverse the clinical notes, find the part of the text most related to the code, and match the corresponding medical code for this part of the text. We utilize the self-attention mechanism to identify the most relevant context of the medical code in the clinical document. We utilize the label attention mechanism to learn code-specific text representation. We leverage the correlation between labels in multi-task learning to assist the model prediction, which can improve the model's generalization ability.

2.2.1 Self-attention Mechanism

In the discharge summary, a text description may contain many labels, so it can be marked by multiple medical codes. Moreover, each code may correspond to various components, but each component should contribute to the code differently. Each clinical note should have most relevant to its clinical context.

We adopt a self-attention mechanism to capture the contribution of different components to each text. The attention score of clinical text ($T^S \in R^{l \times n}$) can be calculated as:

$$T^S = softmax(W_1 \tanh(W_2 H)) \tag{9}$$

where $W_2 \in R^{2d_r \times d}$ and $W_1 \in R^{l \times d}$ are the parameters that need training. The d denotes a hyperparameter that we can set. Finally, the clinical text matrix $M^{(S)} \in R^{l \times 2d_r}$ is calculated as follows.

$$M^{(S)} = T^S H^T \tag{10}$$

2.2.2 Label Attention Mechanism

The attention in the self-attention mechanism is based on the clinical text, while label attention pays more attention to the internal relationship between code and a specific text. As we all know, each medical code has a specific definition in the dataset. We obtain a label attention matrix (L) through the pretraining code and its corresponding text definition, which can be represented as $L \in R^{l \times 2d_r}$.

$$B^{(l)} = LH^T \tag{11}$$

where $B^{(l)} \in R^{l \times n}$ indicates the relation between code definition text and codes. We calculate the dot product between $B^{(l)}$ and H^T to obtain the text representation of the whole context for the code. The formula is as follows:

$$M^{(l)} = B^{(l)} H^T \tag{12}$$

Finally, the result can be represented by $M^{(l)} \in R^{l \times 2d_r}$.

2.2.3 Joint Learning Mechanism

We generate self-attention matrix S and label attention matrix L. In the code prediction, we need to balance the contribution of the two attention matrices to the model's performance. Thus, we design a joint learning mechanism to extract the most critical information from the two attention matrices.

We introduce two hyperparameters W_3 and W_4 to multiply S and L, respectively. Furthermore, input their results into the sigmoid activation function. Then we generate two weight vectors α and β.

$$Sigmoid(x) = \frac{1}{1 + e^{-x}} \tag{13}$$

$$\alpha = Sigmoid(SW_3) \tag{14}$$

$$\beta = Sigmoid(LW_4) \tag{15}$$

W_3 and W_4 are the parameters to be trained. α and β denote the contribution of the self-attention matrix and label attention matrix to the final feature representation, respectively. Then, we limit the two weight vectors to between 0 and 1.

$$\alpha = \frac{\alpha}{\alpha + \beta} \tag{16}$$

$$\beta = 1 - \alpha \tag{17}$$

$$0 < \alpha_i + \beta_i \leq 1 \tag{18}$$

Next, we multiply α and β by the self-attention matrix and label attention matrix respectively, to obtain the final label prediction matrix. Finally, we introduce the learnable parameters W_5 and bias to generate the final prediction value of the label. The formula is as follows:

$$lpm = \alpha * S + \beta * L \tag{19}$$

$$y_{hat} = W_5 * lpm + bias \tag{20}$$

where W_5 and bias are the learnable parameters of the fully connected layer.

2.2.4 Multi-task Learning

Multitask learning refers to the joint training of multiple related subtasks to improve each task's performance and generalization ability. We introduced Clinical Classifications Software (CCS) into our model, which is a popular International medical classification tool. CCS allows researchers to better interpret clinical conditions by classifying diseases in the ICD and analyzing various diagnostic and procedural data types. Furthermore, CCS and ICD codes have a one-to-many relationship. CCS converts the most common 50 labels and all labels in EMR into 38 and 295 labels for training, which significantly compresses the label space.

2.3 Denoising Module

WE introduce the focus loss function to solve the class imbalance problem. The main goal of the focus loss function is to reweight BCE loss to make rare labels get reasonable "attention". The focus loss of each medical code branch FL_b can be denoted as:

$$FL_b = \sum_{i=1}^{d_c} \left[-\alpha \left(1 - \hat{y}\right)^{\gamma} log\widehat{y_i} - (1 - \alpha)\widehat{y_i}^{\gamma} \log\left(1 - \widehat{y_i}\right) \right] \tag{21}$$

where I is the i-th label and d_i represents all labels in the c-th coding system. We introduce two hyperparameters α and γ to balance the model's learning ability for high-frequency labels and low-frequency labels.

We calculate the joint loss of the two subtasks as follows:

$$FL_M = \lambda_d FL_d + \lambda_s FL_s \tag{22}$$

where FL_d and FL_s are the focal loss for the ICD and CCS coding branches, and λ_d and λ_s are their corresponding weights, respectively.

We improve the focal loss function and design the truncation loss function. Specifically, we introduce a dynamic threshold to discard the samples with large loss in each iteration. The formula is as follows:

$$T_{loss}(y, \tilde{y}) = \begin{cases} 0, & F_{loss}(y, \tilde{y}) > \varepsilon \cup (\tilde{y} = 1) \\ F_{loss}, & Otherwise, \end{cases} \tag{23}$$

where ε is the pre-defined threshold and F_{loss} denotes the focal loss. If the loss through the focus loss function is greater than the threshold, it will be regarded as noise samples, and its loss will be reset to 0, so the model will not be disturbed by these noise samples. However, a fixed threshold cannot be suitable for all training processes. As the number of iterations increases, training loss will be reduced. Therefore, to adapt to the loss change in the whole process, we introduce a dynamic threshold function to adjust the threshold in the training process dynamically.

$$D_T = \min(\kappa T, D_{max}), \tag{24}$$

where T represents the number of training iterations, D_{max} denotes the upper bound, and κ is a hyperparameter to monitor the speed to arrive at the maximum drop rate.

In the denoising process, we get cleaner samples, which significantly improves the model's prediction performance.

3 Results and Analysis

In this section, we introduce the dataset required for experiment, parameters setting and indicators, experimental results, and the detailed analysis of our model.

3.1 Datasets

MIMIC-III: The third version of Medical Information Mart for Intensive Care (MIMIC-III) is a free, open-source dataset for intensive care. The database records the relevant data of patients in the intensive care unit of Beth Israel women's Dickens medical center from 2001–2012. It has the medical records and health data of more than 40000 patients.

We take the discharge summary part as the input of the model. There are 52722 discharge summaries in the MIMIC-III-full dataset, of which 8921 different ICD-9-CM codes appear. Then, we preprocess the data according to the frequency of code occurrence, separate the top 50 codes with the highest frequency from the whole code set, and generate the MIMIC-III-50 dataset.

3.2 Settings

3.2.1 Evaluation Metrics

To Comprehensively Evaluate the MODEL's Performance, We Introduced Many Indicators and Conducted Experiments on the Test Set. The Evaluation Indicators Include Macro-averaged and Micro-averaged F1, Macro-averaged and Micro-averaged AUC-ROC (the Area Under the ROC Curve), and Precision at k (Where k $\in$ {5,8,15}).

3.2.2 Hyper-parameter Tuning

For the NIDN, We Set the Learning Rate as 0.001, and the Drop Rate is 0.2. When Calculating the Multi-task Joint Loss, We Set the CCS and ICD Loss Weights to 0.3 and 0.7, Respectively. In the Focus Loss Function, the Weight Factor α and Γ Are 1 and 2, Respectively. In the Truncation Loss Function, the Truncation Loss Threshold is 0.15. We Introduce an Early Stop Mechanism to Terminate Model Training by Monitoring Macro-averaged F Scores. If the MODEL's Score Does not Rise Within 10 Rounds, the Training Will Be Stopped Automatically. This Mechanism Will Avoid Overfitting and Ineffective Training of the Model.

3.3 Performance Comparison with Other Models

In Table 1 and Table 2, we compare the experimental results of other model and our proposed model (NIDN) on the MIMIC-III-50 and MIMIC-III-full datasets, respectively. Experiments clearly show that the NIDN outperforms the other baseline (CNN, BiGRU, CAML, DR-CAML, MultiResCNN and JLAN) in major evaluation metrics. In particular, the performance of our proposed model improves significantly in F1 scores. F1 score is an important indicator to evaluate the performance of rare label allocation. The higher F1 score in our experiments shows that our model can better solve the class imbalance problem.

Table 1. MIMIC-III-50 dataset results (in %).

Models	AUC		F1		P@5	R@5
	Macro	Micro	Macro	Micro		
CNN	87.6	90.7	57.6	62.5	62.0	-
BiGRU	82.8	86.8	48.4	54.9	59.1	-
CAML	87.5	90.9	53.2	61.4	60.9	-
DR-CAML	88.4	91.6	57.6	63.3	61.8	-
MultiResCNN	89.9	92.8	60.6	67.0	64.1	62.1
JLAN	92.6	94.1	66.5	69.7	66.8	63.8
NIDN	**92.7**	**94.6**	**68.0**	**72.0**	**67.0**	**65.3**

Table 2. MIMIC-III-full dataset results (in %).

Models	AUC		F1		P@8	P@15
	Macro	Micro	Macro	Micro		
CNN	80.6	96.9	4.2	41.9	58.1	44.3
BiGRU	82.2	97.1	3.8	41.7	58.5	44.5
CAML	89.5	98.6	8.8	53.9	70.9	56.1
DR-CAML	89.7	98.5	8.6	52.9	69.0	54.8
MultiResCNN	91.0	98.6	8.5	55.2	73.4	58.4
JLAN	91.8	98.8	9.7	56.7	74.1	57.9
NIDN	91.6	98.6	**10.5**	**57.6**	**74.6**	**59.1**

3.4 Ablation Study

In this section, we evaluate each module in our model. We set up the following three groups of comparative experiments to verify the effectiveness of the **N**ote-code **I**nteraction **M**odule (NIM) and **D**enoising **M**odule (DM) on the MIMIC-III-50 dataset.

3.4.1 Compatibility of Main Modules and Different Basic Models

We selected three baseline models, and we added the main module (NIM, DM) of our model NIDN to these three base models. Figure 2 shows that the two modules can improve the performance of the basic model. In particular, the BiGRU model with a note-code interaction module and denoising module performs better than CAML and MultiResCNN.

3.4.2 Effect of the Note-Code Interaction Module

To verify the importance of the note-code interaction module in the training process, we removed this module and conducted experiments on the MIMIC-III-50 dataset. The note-code interaction module mainly uses the self-attention mechanism, label attention mechanism and multi-task learning. 'L' is label attention, 'S' is self-attention, and 'M' is multi-task learning. We use MacroAUC, MicroAUC, MacroF1, MicroF1 and P@5 to measure performance.

It can be seen from Table 3 that when we use the self-attention mechanism or label attention mechanism alone, the performance of the model is poor. On the contrary, the performance of the model using a joint learning mechanism is improved. Specifically, when we combine these three parts (L, S, M), we can get the best performance.

3.4.3 Effect of the Denoising Module

To verify the importance of the denoising module in the training process, we removed this module and conducted experiments on the MIMIC-III-50 dataset. We use MacroAUC, MicroAUC, MacroF1, MicroF1 and P@5 to measure performance.

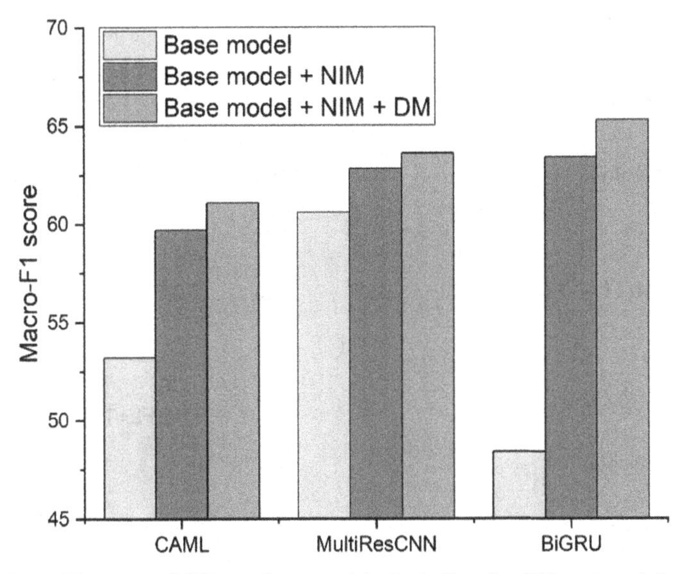

Fig. 2. Macro-F1 scores of different base models, including the different modules of NIDN

Table 3. Result of the ablation experiment.

Models	AUC		F1		P@5
	Macro	Micro	Macro	Micro	
L	92.2	94.3	63.5	69.1	66.3
S	92.3	94.1	66.7	70.8	65.7
L + S	92.6	94.5	66.8	71.4	66.4
L + S + M	92.7	94.6	**68.0**	**72.0**	**67.0**

The denoising module mainly introduces two sub-modules: focus loss function and truncation loss function. In Table 4, experimental results indicate that the model with focus loss function or truncation loss function has better prediction performance than the traditional binary cross entropy loss function (BCE-loss). The model combining focus loss and truncation loss function has the best prediction effect. Therefore, the two submodules are complementary to each other. In Fig. 3, we can also see that the denoising model can reduce the model loss and accelerate the convergence speed of the model.

Table 4. Results of the denoising module experiment.

Models	AUC		F1		P@5
	Macro	Micro	Macro	Micro	
BCE-loss	92.6	94.5	66.2	70.4	66.4
TL	92.8	94.6	66.9	71.0	67.0
FL	92.6	94.7	66.7	71.6	67.1
DM(TL + FL)	92.7	94.6	**68.0**	**72.0**	67.0

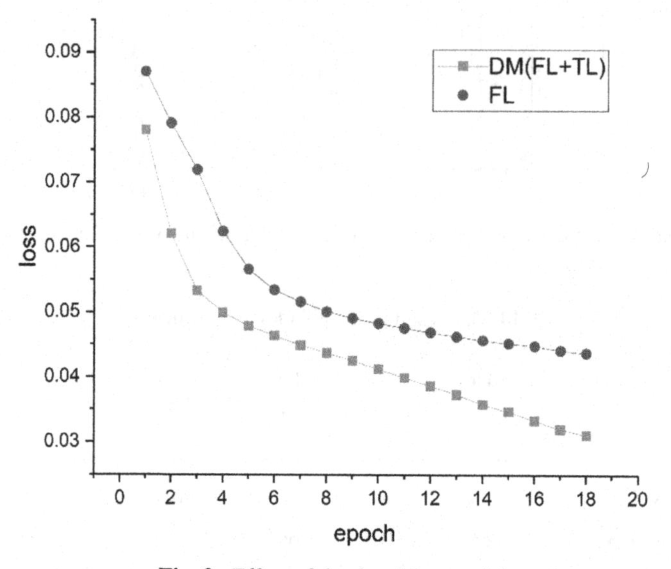

Fig. 3. Effect of the denoising model

4 Future Work

In recent years, the transformer-based pretraining model has been widely used in NLP tasks. The model based on transformer architecture can extract rich semantic information and solve the long-term dependency between texts. Among them, the Bert model based on the transformer framework has made significant progress in various fields of NLP. However, the improvement effect in text classification tasks is not obvious, and they also need strong computing power to train parameters. In the next stage of research, we can try to introduce domain prior knowledge and use the pretraining biomedical language representation model.

5 Conclusion

In this paper, we propose a novel model NIDN to solve three challenges in the field of medical code assignment: noisy and lengthy clinical text, code association, and long-tail distribution problem. We introduce BiGRU and Recalibrated Attention Module (RAM) to extract rich semantic information from clinical texts. We design the note-code interaction module to obtain the direct relationship between clinical text and medical code. The denoising module helps our model to alleviate the class imbalance problem and discard noisy samples to obtain cleaner ones. Experiments show that our proposed model exceeds competitive baseline models on the real-world MIMIC-III database.

Acknowledgment. This work is supported by grant from the Natural Science Foundation of China (No. 62072070).

References

1. Mullenbach, J., Wiegreffe, S., Duke, J., Sun, J., Japa, E.J.: Explainable prediction of medical codes from clinical text (2018)
2. Li, F., Yu, H.: ICD coding from clinical text using multi-filter residual convolutional neural network. In: Proceedings of the AAAI Conference on Artificial Intelligence, pp. 8180–8187 (2020)
3. Vu, T., Nguyen, D.Q., Japa, N.A.: A label attention model for ICD coding from clinical text (2020)
4. Biswas, B., Pham, T.-H., Zhang, P.: Transicd: Transformer based code-wise attention model for explainable ICD coding. In: Tucker, A., Henriques Abreu, P., Cardoso, J., Pereira Rodrigues, P., Riaño, D. (eds.) AIME 2021. LNCS (LNAI), vol. 12721, pp. 469–478. Springer, Cham (2021). https://doi.org/10.1007/978-3-030-77211-6_56
5. Sun, W., Ji, S., Cambria, E., Japa, M.P.: Multitask balanced and recalibrated network for medical code prediction (2021)
6. Li, X., Zhang, Y., Dong, D., Wei, H., Lu, M.J.B.: Medical code prediction via joint learning attention networks and denoising mechanism. JLAN 22(1), 1–21 (2021)
7. Hu, J., Shen, L., Sun, G.: Squeeze-and-excitation networks. In: Proceedings of the IEEE Conference on Computer Vision and Pattern Recognition, pp. 7132–7141 (2018)
8. Dong, H., Wang, W., Huang, K., Coenen, F.: Automated social text annotation with joint multilabel attention networks. IEEE Trans. Neural Netw. Learn. Syst. **32**(5), 2224–2238 (2020)
9. Luo, J., Xiao, C., Glass, L., Sun, J., Ma, F.: Fusion: towards automated ICD coding via feature compression. In: Findings of the Association for Computational Linguistics: ACL-IJCNLP 2021, pp. 2096–2101 (2021)
10. Zhou, T., Cao, P., Chen, Y., Liu, K., Zhao, J., Niu, K., et al.: Automatic ICD coding via interactive shared representation networks with self-distillation mechanism. In: Proceedings of the 59th Annual Meeting of the Association for Computational Linguistics and the 11th International Joint Conference on Natural Language Processing (Volume 1: Long Papers), pp. 5948–5957 (2021)
11. Wang, W., Feng, F., He, X., Nie, L., Chua, T.-S.: Denoising implicit feedback for recommendation. In: Proceedings of the 14th ACM International Conference on Web Search and Data Mining, pp. 373–381 (2021)
12. Xin, J., Tang, R., Yu, Y., Lin, J.: BERxiT: early exiting for BERT with better fine-tuning and extension to regression. In: Proceedings of the 16th Conference Of The European Chapter of the Association for Computational Linguistics: Main Volume, pp. 91–104 (2021)
13. Byrd, J., Lipton, Z.: What is the effect of importance weighting in deep learning? In: International Conference on Machine Learning: PMLR, pp. 872–881 (2019)

14. Ji, S., Pan, S., Japa, M.P.: Medical code assignment with gated convolution and note-code interaction (2020)
15. Bai, T., Vucetic, S.: Improving medical code prediction from clinical text via incorporating online knowledge sources. In: The World Wide Web Conference, pp. 72–82 (2019)
16. Zhao, S., Liu, T., Zhao, S., Wang, F.: A neural multi-task learning framework to jointly model medical named entity recognition and normalization. In: Proceedings of the AAAI Conference on Artificial Intelligence, pp. 817–824 (2019)
17. Xie, X., Xiong, Y., Yu, P.S., Zhu, Y.: Ehr coding with multi-scale feature attention and structured knowledge graph propagation. In: Proceedings of the 28th ACM International Conference on Information and Knowledge Management, pp. 649–658 (2019)
18. Wu, Y., Zeng, M., Fei, Z., Yu, Y., Wu, F.-X., Li, M.J.N.: Kaicd: a knowledge attention-based deep learning framework for automatic ICD coding 469, pp. 376–383 (2022)
19. Cao, P., Chen,Y., Liu, K., Zhao, J., Liu, S., Chong, W.: Hypercore: hyperbolic and co-graph representation for automatic icd coding. In: Proceedings of the 58th Annual Meeting of the Association for Computational Linguistics, pp. 3105–3114 (2020)
20. Alsentzer E, et al.: Publicly available clinical BERT embeddings (2019)
21. Bai, S., Kolter, J.Z., Japa, K.V.: An empirical evaluation of generic convolutional and recurrent networks for sequence modeling (2018)
22. Kavuluru, R., Rios, A., Lu, Y., Aiim, J.: An empirical evaluation of supervised learning approaches in assigning diagnosis codes to electronic medical records. Artif. Intell. Med. **65**(2), 155–166 (2015)

Research on the Prediction Method of Disease Classification Based on Imaging Features

Yu Sheng, Shengyi Yang, Huirong Hu, and Guihua Duan(✉)

School of Computer Science, Central South University, Changsha, China
duangh@csu.edu.cn

Abstract. The medical imaging diagnosis report records the imaging signs and diagnostic conclusions in the medical imaging service in detail. The diagnostic conclusion is the conclusive opinion made by the doctor based on the imaging signs and his own experience, and generally includes the corresponding disease category. Using imaging signs to intelligently predict the corresponding disease categories can provide effective support for reducing doctors' conclusion errors and improving the quality of medical students' diagnostic reports. Aiming at the text characteristics of medical imaging diagnosis reports, this paper proposes a multi-dimensional feature fusion capsule network model based on structural domain dictionary to predict disease categories for imaging signs. Our model first uses the medical imaging domain dictionary constructed based on our own dataset to improve the accuracy of text representation, and then constructs a low-dimensional feature fusion layer to mix and extract the local features and contextual semantic features of words to realize the multi-level semantic modeling of imaging signs, and finally uses the capsule network to obtain location information and perform feature clustering, which can capture high-level semantic features to predict diagnostic conclusions. Experiments show that the model proposed in this paper outperforms baselines on the imaging diagnosis report dataset, the precision of the model reaches 97.30%, and F1 is 2.49% higher than the best performing baseline.

Keyword: Medical imaging diagnosis report · Imaging signs · Disease prediction · CapsNet

1 Introduction

As an important part of medical data, medical imaging diagnosis reports record the imaging findings and imaging diagnosis opinions including the location, size and shape of the lesion in detail. It is not only important data for clinical disease prediction, decision support and medication pattern mining, but also an important evidence in medical disputes and judicial certification [1].

Writing medical imaging diagnosis report correctly and normatively is a basic skill that medical students and clinical imaging doctors must learn and master. According to the data of the annual meeting of the Chinese Medical Doctor Association, the current annual growth rate of imaging data in China is 30%, while the growth rate of imaging

M. S. Bansal et al. (Eds.): ISBRA 2022, LNBI 13760, pp. 75–87, 2022.
https://doi.org/10.1007/978-3-031-23198-8_8

physicians is only 4.1%. There is an obvious professional talent gap [2]. The medical imaging online experimental platform developed by our research team provides an online imaging reading training environment for the training of imaging doctors. Medical students can write diagnostic reports by reading the imaging resources provided by the platform, and the platform will intelligently score the diagnostic reports written by students. Among them, disease prediction is an important means to judge the correctness of imaging signs in the diagnostic report written by students. The diagnostic conclusions predicted by the imaging signs written by medical students are compared with those written by professional imaging doctors in the system to measure the logical correlation between the imaging performance written by students and the diagnostic conclusions, and the standardization and correctness of imaging performance. Therefore, the realization of diagnosis conclusion prediction based on medical imaging signs is of great significance to improve the writing ability of medical students' imaging diagnosis reports, and promote the standardization and high-quality quantitative development of medical imaging diagnosis reports.

In recent years, the methods used in applying medical text to achieve disease prediction and diagnosis mainly include machine learning algorithms and deep learning. Miguel [3] used RF algorithm to predict the existence of breast cancer based on blood sample data; [4] using liver text dataset, SVM, Bayesian etc. are used for liver disease diagnosis. Although machine learning algorithms have been used in many experimental studies, they have some problems, such as over fitting, low classification accuracy, high computational complexity, slow convergence speed and so on. As for deep learning, Han [5] applied the TextCNN model to establish a diagnostic model for SPECT nuclear medicine diagnosis text to assist doctors in diagnosis; [6] proposed an improved GRU framework LS-GRU to solve the problem of using chest CT imaging report text to diagnose emphysema and pneumonia. Yang [7] proposed a single label classification model based on Bert-TextCNN to identify disease subtypes in clinical trials, and proved the effect of this method. [8] used a model based on the fusion of Bert-BiLSTM-Attention to diagnose and classify the text of traditional Chinese medicine medical records; Blinov [9] et al. proposed a new full connection layer combination method and used Bert model only on domain data to improve the accuracy of disease prediction in solving the problem of predicting clinical diagnosis of EMRs data. The pooling operation in the traditional convolutional neural network will cause information loss and cannot retain the position semantic information of vocabulary, therefore, Fan [10] applied the capsule network to the EMR disease auxiliary diagnosis of small samples. It can be seen that most studies using medical texts for disease diagnosis will use text classification methods in general fields, which have a certain effect on disease category prediction. However, the medical imaging diagnosis reports record the doctor's diagnosis conclusions and imaging signs including clinical manifestations, lesion characteristics which contain a large number of professional terms and vocabulary combinations of imaging and medicine, and there may be a variety of lesions and signs in different parts, for example, "左侧颈部、双侧锁骨区见明显糖代谢异常增高灶及肿大淋巴结 (abnormal hyperglycemia and swollen lymph nodes can be seen in the left neck and bilateral clavicular areas)". Therefore, the existing research methods can not be fully applicable to the medical imaging diagnosis report text.

In view of the above text structure characteristics of medical imaging diagnosis report, this paper proposes a multi-dimensional feature fusion capsule network model based on domain dictionary to realize disease prediction for medical imaging diagnosis report. The main work is as follows: (1) Due to the lack of professional medical imaging report dictionaries in China, we structure the medical imaging report dataset provided by a medical big data platform from a university and build a domain dictionary so as to improve the accuracy of text representation in the word embedding layer; (2) The attention mechanism [11] is used to dynamically fuse the word level local information and the contextual semantic information between words in the text; (3) In order to ensure the invariance of the relationship between words and sentences, grasp the positional semantic information in the imaging text, we use the three-layer capsule network to cluster the low-dimensional fusion features, and further obtain the high-level semantic features to achieve disease prediction and classification.

2 Materials and Methods

2.1 Materials

The dataset consists of PET-CT diagnosis reports provided by a medical big data platform in a university from 2014 to 2017, with a total of 12000 copies. We select 4925 brain PET-CT diagnostic reports including 1556 brain atrophy, 908 leukoencephalopathy, 631 cerebral infarction, 530 brain tumor, 300 other brain diseases and 1000 normal and 3868 lung PET-CT diagnostic reports including 601 tuberculosis, 518 pneumonecta, 400 pneumonia, 1143 lung cancer, 206 other lung diseases and 1000 normal as experimental datasets for disease prediction models. After desensitization, each PET-CT imaging report text only retains two parts: imaging signs and diagnostic conclusions, as shown in Table 1.

Table 1. Schematic diagram of PET-CT imaging diagnosis report

Imaging signs	Diagnostic conclusions
脑部PET-CT图像示右侧小脑半球见囊实性占位病变, 实性部分可见糖代谢增高, SUVmax为26.57, 大小约为37 × 30 mm, 边缘尚清楚, 周围无明显水肿带… (The PET-CT imaging of the brain showed cystic and solid space-occupying lesions in the right cerebellar hemisphere, and increased glucose metabolism in the solid part. The SUVmax was 26.57, the size was about 37 × 30 mm, the edge was still clear, and there was no obvious edema around it …)	1. 脑肿瘤(Brain tumor); …

2.2 Methods

The overall architecture of the model proposed in this paper is shown in Fig. 1, including an embedding layer using Word2vec, a Multi-level feature fusion layer consisting of Multi-channel TextCNN and BiLSTM, a feature clustering layer composed of two layers of capsule network and a layer of fully connected capsule network.

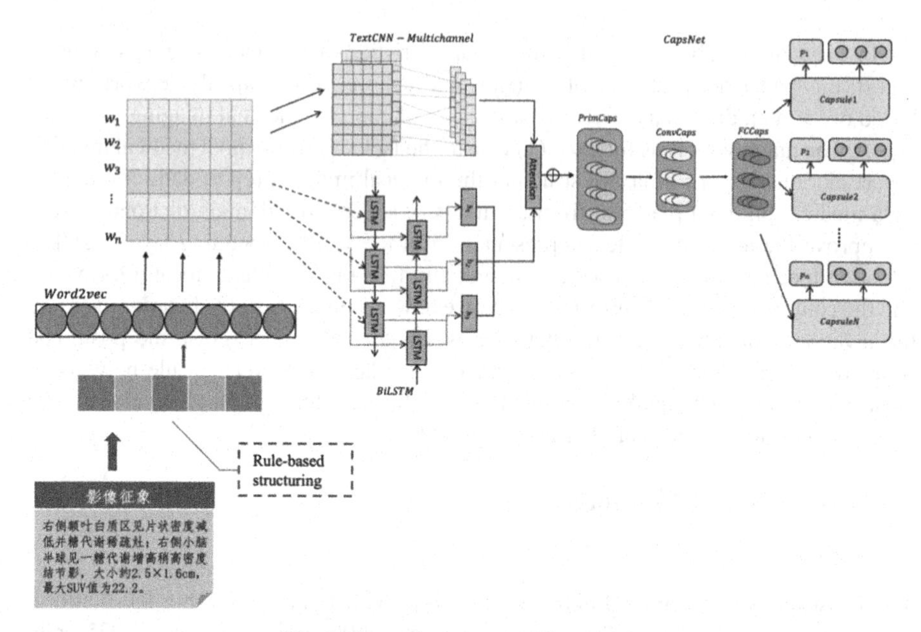

Fig. 1. Overall structure of the model.

2.3 Embedding

A medical imaging diagnosis report can be up to 1,000 words in length. In order to improve the accuracy of the semantic representation of the imaging description text, avoid the error propagation caused by too long text, semantic ambiguity and word segmentation errors in the disease prediction task, which will affect the overall performance of the model and increase the training time, this section first performs structured preprocessing on medical imaging report text and then Word2vec is integrated into the self-built dictionary for word embedding.

Self-built Medical Imaging Report Dictionary. In this paper, an iterative method is used to construct a medical imaging report dictionary. After discussions with radiologists, we identified 4 key entities in the medical imaging diagnostic report, namely examination site (BP), site properties (BPC), predicate (PW), and signs and symptoms (DI). Using the 1320 words in the Chinese-English noun comparison table in the appendix of the medical imaging textbook of the third edition of the People's Medical Publishing House [12] as the basic dictionary, the named entity recognition model proposed by our research team [13] was used to process 12,000 PET-CT diagnosis reports, and we obtained a medical imaging vocabulary containing 4 types of entities. Finally, with the help of radiologists, we manually revised the vocabulary, added the initial basic dictionary and got the medical imaging report dictionary. The contents of some dictionaries are shown in Fig. 2.

Medical Imaging Diagnosis Report Text Structure. In the medical imaging diagnosis report, the description of a complete inspection item generally starts with the inspection item (BP[+BPC]), and ends with the corresponding inspection result ([PW +]DI).

检查部位(Examination site)	属性(Attributes)	谓词(predicate)	症状征象(Symptomatic signs)
脑实质(Brain parenchyma)	密度(Density)	见(See)	钙化灶(Calcification)
脑皮质(Cerebral cortex)	大小(Size)	未见(Not seen)	糖代谢 (Glycometabolism)
肺上叶(Upper lobes of the lung)	体积(Volume)	无(Not)	局灶性浓聚(Focal concentration)
眼内肌(Intraocular muscles)	外形(Appearance)	弥漫(Diffuse)	水肿带(Edema)
枕叶(Occipital)	短径(Short diameter)	呈现(Present)	膨胀(Expand)
胆囊(Gallbladder)	轮廓(Outline)	存在(Exist)	斑点状(Spotted)
胸锁乳突肌 (Sternocleidomastoid muscle)	直径(Diameter)	显示(Show)	占位病变(Space occupying lesion)
小脑半球(Cerebellar hemispheres)	形态(Morphology)	伴有(Accompany)	减低(Reduce)
脑室三角区(Triangle of Ventricular)	结构(Structure)	不伴有(Without)	楔形(Wedge)
...	...	...	...

Fig. 2. Part of the medical imaging report dictionary

For example, in expression "胆囊密度均匀 (uniform gallbladder density)", it starts with "胆囊密度 (BP + BPC)" and ends with "均匀 (DI)". Based on the sequence results of entity recognition above, our research team designed a structured method of rule extraction to extract a whole imaging sign text into multiple two-tuples in the form of <检查项 (check item), 检查结果 (check result)> to facilitate subsequent text analysis. The structured method proposed by our research team has been compared with structured methods such as manual rules and dependency analysis, and we confirmed that our method has a higher recognition accuracy [13]. Combined with the rule extraction method, the imaging report can better integrate syntactic features.

The self-built medical imaging report dictionary is used as a custom dictionary for Jieba word segmentation, and the structured text is further segmented to divide the symptoms that are too long, as shown in Table 2. Red represents BP, green represents BPC, blue represents DI, and yellow represents PW.

Word Embedding. We use Word2vec [14] which is pretrained on a dataset of 12,000 medical imaging diagnostic reports to implement word embedding. As shown in formula (1), $E \in R^{|V| \times d}$, $|V|$ is the size of the vocabulary and d is the dimension of the word vector.

$$w_i = E(x_i) \tag{1}$$

2.4 Multi-level Feature Fusion

TextCNN. TextCNN has a strong ability to extract word-level features of text. Therefore, in order to fully extract word-level features in medical imaging diagnosis reports, use the prior knowledge in medical imaging reports as well as make it convenient to capture features associated with disease diagnosis results, we use a Multi-channel TextCNN

Table 2. Text structure and word segmentation of medical imaging diagnosis report

Medical Imaging Diagnosis Report	Structural and word segmentation
CT平扫示脑实质密度均匀，未见异常密度灶和糖代谢增高灶；脑室系统扩大，形态如常，脑沟、脑裂增宽，脑中线结构居中。(CT scan showed that the density of the brain parenchyma was uniform, and no abnormal density foci and foci with increased glucose metabolism were found; the ventricular system was enlarged and the shape was normal, The sulci and fissures are widened, and the midline structure of the brain is centered.)	***P*₁**:<脑实质(brain parenchyma), 密度(density), 均匀(uniform)>, <脑实质(brain parenchyma), 未见(not seen), 异常(abnormal), 密度灶(density foci), 糖代谢(glucose metabolism), 增高灶(height increase stove)> ***P*₂**:<脑室系统(the ventricular system), 扩大(expand)>, <脑室系统(the ventricular system), 形态(morphology), 如常(normal)> ***P*₃**:<脑沟(sulci), 脑裂(cerebral fissure), 增宽(widen)> ***P*₄**:<脑中线(midline of the brain), 结构(structure), 居中(Centered)>

to obtain word-level features of the diagnostic report. One channel is set to no longer change the word vector during the training process; the other channel uses the same initialization method to set the word vector and change with the network training. In order to avoid the loss of information caused by the pooling operation which may affect the subsequent model performance, this section only obtains the feature representation after convolution, and removes the pooling layer.

BiLSTM. Due to the gradient disappearance of RNN and other problems while LSTM cannot capture the bidirectional semantic dependence of the text, we use BiLSTM to extract the overall structural features of the text in order to better analyze the internal relationship between words and words in medical imaging diagnosis reports. For an input sentence of imaging diagnosis report text, we obtain its vector representation processed by forward LSTM and backward LSTM respectively, and the vector containing the global structure information of words is obtained by splicing, as shown in formula (2).

$$V_b = [F_{\overrightarrow{LSTM}}(w_1, w_2, w_3, \ldots w_i); F_{\overleftarrow{LSTM}}(w_1, w_2, w_3, \ldots w_i)] \tag{2}$$

Attention. In order to obtain the key pattern information, the attention mechanism is used to dynamically fuse the word level feature vector extracted by convolution operation with the output vector of BiLSTM. As shown in formula (3–4), σ^T represents the transpose of the parameters. The new vector V_T contains the phrase level feature of the text and the context structure information of the word.

$$\delta_i = \frac{\exp(\sigma^T v_i)}{\sum_i \exp(\sigma^T v_i)} \tag{3}$$

$$V_T = \delta_a V_a + \delta_b V_b \tag{4}$$

Feature Clustering Layer. In order to fully extract the positional semantic information of the description words in the whole sentence about the morphological position of the lesion and the symptoms in the medical imaging diagnosis report, Capsule network is used to perform feature clustering on low-dimensional features to further mine high-level semantic features of imaging diagnosis reports, as shown in the Fig. 3. PrimCaps receives the fused feature vectors from TextCNN and BiLSTM as input vectors, and converts scalars into vectors to retain the position and semantic information of words. ConvCaps uses a shared transformation matrix internally to extract local capsules. In order to ensure that the feature vector output from each layer of capsule can be correctly input to the specific capsule of the next layer, a dynamic routing algorithm [15] is used between PrimCaps and ConvCaps to cluster features.

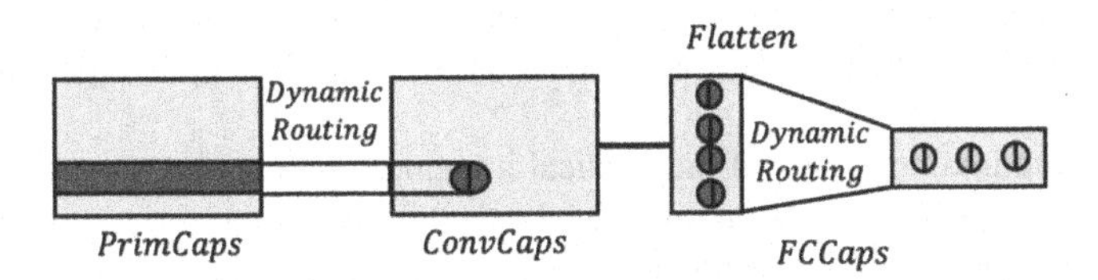

Fig. 3. Capsule network for feature aggregation

For the fused feature vector v_t, after affine transformation, the shallow capsule is obtained first, and the initial routing weight b_{ij} is constantly updated according to the dot product between the predicted output vector $\hat{u}_{j|i}$ and the actual product vector v_j, as shown in formula (5–6). All capsules in the current layer are weighted and summed, as in formula (7), to obtain the total input of the capsules s_j in the next layer. Finally, calculate the weight sum of the predicted output of each capsule in the next layer, as shown in formula (8). The result v_j is the output vector of the capsule, which forms a textual feature representation that can be used for classification.

$$b_{ij} = b_{ij} + \hat{u}_{j|i} v_j \tag{5}$$

$$c_{ij} = \frac{\exp(b_{ij})}{\sum_k \exp(b_{ik})} \tag{6}$$

$$s_j = \sum_i c_{ij} \hat{u}_{j|i} \tag{7}$$

$$v_j = \frac{||s_j||^2}{1+||s_j||^2} \frac{s_j}{||s_j||} \tag{8}$$

Full Connection Layer. v_j is the high-level feature vector obtained from the low-level fusion features after feature aggregation as the input of FCCaps, as shown in Fig. 4, which is transformed into one-dimensional output through flatten transformation. The loss function of the capsule network is calculated as formula (9). v_k is the discriminant vector representing the class, while $||v_k||$ is the length of the capsule representing the probability of the category. m^+ represents the probability corresponding to the correct category, and m^- represents the probability corresponding to the wrong category. It is the proportion of two parts of losses.

$$L_k = T_k \max\left(0, m^+ - ||v_k||\right)^2 + \beta(1 - T_k)\max\left(0, ||v_k|| - m^-\right)^2 \tag{9}$$

3 Experiments and Results

3.1 Experimental Setup

We use CUDA 10.0 as the deep learning computing platform. The experimental dataset is divided into training set, test set and validation set according to 8:1:1, and the longest sequence length of the text is set to 100, covering more than 95% of the input sentence length. The routing times of the capsule network is 3, and the hidden nodes are set to 32. During model training, Adam is selected as the optimizer update parameter, and Dropout is set to 0.5.

Because the imaging signs of the medical imaging report are organized and hierarchical, each diagnostic conclusion given by the doctor corresponds to a certain imaging segment. Therefore, we use the processed imaging text of a single disease to experiment, and apply the model hierarchically on the report containing multiple diseases to get complete diagnosis conclusions.

3.2 Experimental Results and Analysis

Experimental Evaluation Index. The standard metrics, including precision, recall, F1 measure are used to evaluate the performance of the proposed model. Since the disease classification in this experiment includes multiple diseases, the experimental evaluation index adopts the macro average of various indexes.

Comparative Experiment. In order to evaluate the effect of our model proposed in this paper, we selected LR [16], SVM [17], TextCNN [18], BERT [19], TextRNN [20], BiLSTM-Attention [21], FastText [22] as baseline models for comparative experiments on the same training and test sets with the same processed input text. We used grid search to determine the best combination of parameters for each model on the experimental set. The experimental results of each model are shown in Table 3. It can be seen that our method has the best experimental results and the F1 is 94.75%, which is 2.49% higher than the best performing baseline model TextCNN.

Table 3. The average prediction performance between our model and the baselines

Model	Precision (%)	Recall (%)	F1 (%)
LR	89.88	86.57	88.06
SVM	90.77	88.15	89.29
TextCNN	94.85	91.11	92.26
BERT	92.06	91.25	91.37
TextRNN	92.46	92.03	92.15
BiLSTM-Attention	92.19	90.33	91.11
FastText	92.50	89.37	91.65
Our method	**97.30**	**93.17**	**94.75**

Ablation Experiment. In order to further verify the effect of the hybrid model proposed in this paper, ablation experiments are carried out. We experimented on the TextCNN, BiLSTM, CapsNet separately and the hybrid models TextCNN + BiLSTM, TextCNN + CapsNet, BiLSTM + CapsNet. The results are as shown in Table 4. The hybrid model proposed in this paper has the best effect in presicion, recall and F1. It can be seen that multi-level feature extraction and feature clustering have a good effect on imaging diagnosis report text classification.

Table 4. Experimental results of seven groups of ablation experiments

Model	Precision (%)	Recall (%)	F1 (%)
TextCNN	94.85	91.11	92.26
BiLSTM	92.09	89.76	90.06
CapsNet	92.92	89.20	90.37
TextCNN + BiLSTM	95.35	91.52	92.82
TextCNN + CapsNet	96.45	92.37	93.85
BiLSTM + CapsNet	94.07	88.75	91.45
Our method	**97.30**	**93.17**	**94.75**

Effect of Changing the Number of Channels of TextCNN. Under the condition that other parameters are same, the pretrained word vectors are used in the Single-Channel experiment with the settings fine tuned during the training process. The Multi-Channel adds a static channel, which the pretrained word vectors are directly used, and not change during training. The experimental results are shown in Fig. 4. The F1 of Multi-Channel TextCNN is 1.23% higher than the Single-Channel TextCNN, which confirms the Multi-Channel alleviates the over fitting situation of only using the fine-tuning method to a certain extent, and makes the word vector continuously fine-tuning during the training

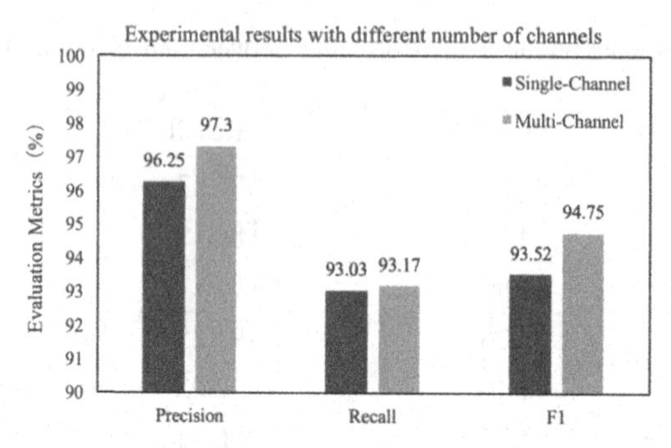

Fig. 4. Experimental results with different number of channels

process more suitable for the disease prediction task, so as to effectively improve the classification performance.

Influence of Word Embedding with Different Granularity. In order to further verify the validity of Word2vec based on domain dictionary, TextCNN, TextRNN, FastText and the model proposed in this paper are selected to experiment on different granularity of embedding layer. The experimental results are shown in the Table 5. Compared with dividing the text by character level granularity, using Word2vec to implement word embedding has significantly improved the precision, recall and F1 results. After adding the domain dictionary to Wordvec, the F1 has increased by 1.09%, which shows that the domain dictionary makes the professional vocabulary division and semantic modeling in the medical imaging reports more accurate which is helpful to improve the overall performance of the disease prediction classification model.

Table 5. The experimental results of different granularity embedding vectors

Model	Character vector			Word2vec			Word2vec with dictionary		
	P(%)	R(%)	F1(%)	P(%)	R(%)	F1(%)	P(%)	R(%)	F1(%)
TextCNN	90.72	88.58	89.33	93.16	90.62	91.62	94.85	91.11	92.26
TextRNN	90.07	87.21	88.46	92.11	90.08	91.96	92.46	92.03	92.15
FastText	89.75	86.93	87.05	91.85	88.78	90.81	92.50	89.37	91.65
Our method	**92.70**	**89.11**	**90.47**	**95.46**	**92.25**	**93.66**	**97.30**	**93.17**	**94.75**

Generalization Evaluation. In order to verify the validity of the model for predicting the diagnosis conclusions of different parts, we used pulmonary PET-CT diagnostic reports for generalization evaluation. The experimental results are shown in Table 6. The F1 of the model reaches 92.41%. For common lung diseases such as lung cancer and pneumonia, the F1 reaches more than 92%, indicating the applicability of the model in this paper for medical imaging diagnosis reports.

Table 6. The experimental results of different lung diseases

Diseases	Precision (%)	Recall (%)	F1 (%)
Lung cancer	96.81	95.65	96.22
Pneumonectasis	96.77	81.08	88.24
Tuberculosis	91.18	88.57	89.96
Pneumonia	93.56	91.03	92.06
Normal	99.37	99.37	99.37
Other diseases	84.09	93.62	88.60
Macro average	93.63	91.55	92.41

The confusion matrices of our method and the best performing baseline model TextCNN on brain diseases and lung diseases are shown in Fig. 5 respectively. According to the confusion matrix, it can be intuitively seen that our method has a certain improvement in the prediction and classification effect of the best performing baseline model.

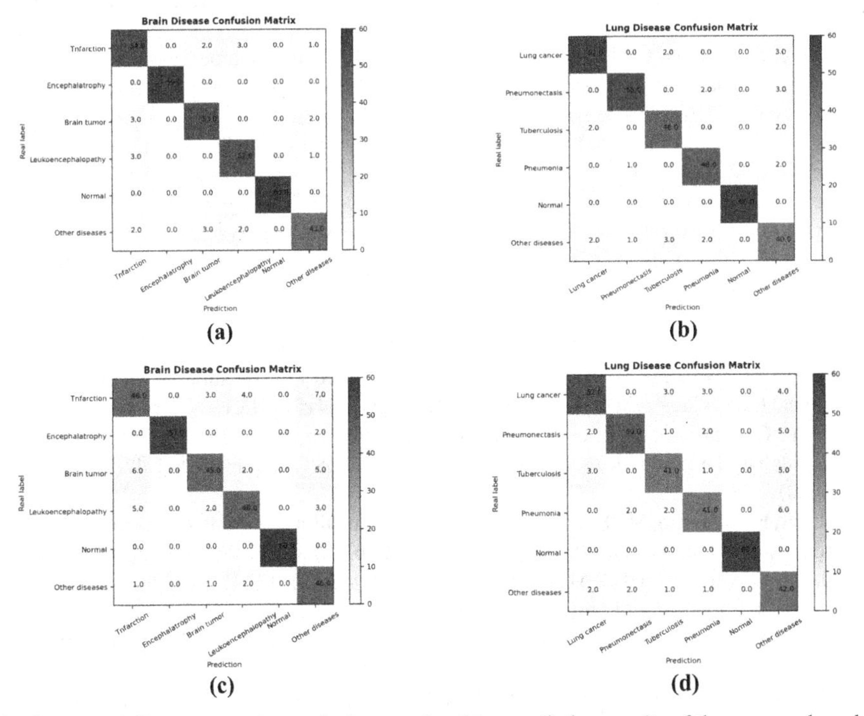

Fig. 5. (a) and (b) represent the confusion matrix of the prediction results of the proposed model in brain and lung diseases, (c) and (d) respectively represent the confusion matrices of the best performing baseline model TextCNN for brain and lung diseases.

4 Conclusion

In this paper, a multi-dimensional feature fusion capsule network model based on structured imaging signs and the domain dictionary is proposed to predict the diagnosis conclusion of the imaging signs in the medical imaging diagnosis report. We apply a structured method to construct a domain dictionary to solve the problem of professional terminology recognition in medical imaging report, which improves the accuracy of text embedding. And then we fully mine the semantic information of medical imaging diagnosis report through multi-level feature fusion and feature clustering. Finally, a series of experiments are carried out to confirm the effectiveness of the model proposed in this paper, and it has a good predictive performance for the diagnosis conclusion on the real medical imaging diagnosis report dataset.

Acknowledgement. This work is supported in part by the National Natural Science Foundation of China (No. 61877059).

References

1. Bai, B.H.: Research and Discussion on standardized writing of medical image diagnosis report. Contemporary Med. **22**(25), 8–10 (2016)
2. Zheng, L.F., Quan, Q.M., Xu, J.B., et al.: Standardization of radiology report writing can improve clinical skills of medical students and residents. Progress Modern Biomed. **15**(32), 6377–6380 (2015)
3. Patricio, M., et al.: Using Resistin, glucose, age and BMI to predict the presence of breast cancer. BMC Cancer **18**(1), 29–33 (2018)
4. Ramana, B.V., Babu, M.S.P., Venkateswarlu, N.B.: A critical study of selected classification algorithms for liver disease diagnosis. Int. J. Database Manage. Syst. **3**(2), 101–114 (2011)
5. Han, C.C., Lin, Q., Man, Z.X., et al.: Mining nuclear medicine diagnosis text for correlation extraction between lesions and their representations. Comput. Sci. **47**(S2), 524–530 (2020)
6. Li, Q.,Li, Y.K., Xia, S.Y., et al.: An improved medical text classification model: LS-GRU. J. Northeastern Univ. (Natural Sci.) **41**(07), 938–942+961 (2020)
7. Yang, F.H., Wang, X.W., Li, J.: BERT-TextCNN-based classification of short texts from clinical trials. Chinese J. Med. Library Inf. Sci. **30**(1), 54–59 (2021)
8. Du, l., Cao, D., Lin, S.Y., et al.: Extraction and automatic classification of TCM medical records based on attention mechanism of BERT and Bi-LSTM. Comput. Sci. **47**(S2), 416–420 (2020)
9. Blinov, P., Avetisian, M., Kokh, V., Umerenkov, D., Tuzhilin, A.: Predicting clinical diagnosis from patients electronic health records using BERT-based neural networks. In: Michalowski, M., Moskovitch, R. (eds.) AIME 2020. LNCS (LNAI), vol. 12299, pp. 111–121. Springer, Cham (2020). https://doi.org/10.1007/978-3-030-59137-3_11
10. Fan, H.J., Li, X.D., Ye, S.T.: Aided disease diagnosis method for EMR semantic analysis. Comput. Sci. **49**(1), 153–158 (2022)
11. Wu, H.Y., Yan, J., Huang, S.B., et al.: CNN_BiLSTM_Attention hybrid model for TextClassification. Comput. Sci. **47**(S02), 23–27+34 (2020)
12. Jin, Z.Y., Gong, Q.Y.: Medical Imaging. People's Medical Publishing House, Beijing (2015)

13. Sheng, Y., Hu, H.R., Wang, C.C., Yang, S.Y.: Research on the Structured Method of Medical Imaging Diagnosis Report. Data Analysis and Knowledge Discovery (2022). http://libdb.csu.edu.cn:80/rwt/CNKI/NNYHGLUDN3WXTLUPMW4A/kcms/detail/10.1478.G2.20220613.0831.006.html
14. Mikolov, T., Sutskever, I., Chen, K., Corrado, G.S., Dean, J.: Distributed Representations of Words and Phrases and their Compositionality. NIPS (2013)
15. Sabour, S., Frosst, N., Hinton, G.E.: Dynamic routing between capsules. arXiv preprint arXiv: 1710.09829 (2017)
16. Crowder, M., Hand, D.J.: Practical longitudinal data analysis. Technometrics **39**(1), 112 (1996)
17. Xiao, L., Yao, N.: Research on Chinese classification based on TF-IDF. International Conference on Neural Networks, Information and Communication Engineerin, Qingdao, China, pp. 59–64 (2021)
18. Kim, Y.: Convolutional neural networks for sentence classification. In: EMNLP, Doha, Qatar, pp. 1746–1751 (2014)
19. Devlin, J., Chang, M.-W., Lee, K., Toutanova, K.: BERT: pre-training of deep bidirectional transformers for language understanding. arXiv preprint arXiv:1810.04805 (2019)
20. Liu, P., Qiu, X., Huang, X.: Recurrent neural network for text classification with multi-task learning. arXiv preprint arXiv:1605.05101 (2016)
21. Wu, H., He, Z., Zhang, W., Hu, Y., Wu, Y., Yue, Y.: Multi-class text classification model based on weighted word vector and BiLSTM-attention optimization. In: Huang, D.-S., Jo, K.-H., Li, J., Gribova, V., Bevilacqua, V. (eds.) ICIC 2021. LNCS, vol. 12836, pp. 393–400. Springer, Cham (2021). https://doi.org/10.1007/978-3-030-84522-3_32
22. Joulin, A., Grave, E., Bojanowski, P., Mikolov, T.: Bag of tricks for efficient text classification. In: EACL, Valencia, Spain, pp. 427–431 (2017)

M-US-EMRs: A Multi-modal Data Fusion Method of Ultrasonic Images and Electronic Medical Records Used for Screening of Coronary Heart Disease

Bokai Yang[1,2], Yingnan Zuo[1,2], Shunxiang Yang[1,3], Genqiang Deng[4], Suisong Zhu[4], and Yunpeng Cai[1](✉)

[1] Shenzhen Institute of Advanced Technology, Chinese Academy of Sciences, Shenzhen 518055, China
yp.cai@siat.ac.cn
[2] University of Chinese Academy of Sciences, Beijing 100049, China
[3] Shandong University, Jinan 250100, China
[4] Huazhong University of Science and Technology Union Shenzhen Hospital, Shenzhen 518060, China

Abstract. Coronary heart disease (CHD) is the leading cause of death and disease burden in China and world-wide. The accurate screening for CHD can be of great value to guiding and facilitating the treatment. Traditional methods, such as computed tomography (CT) and coronary computed tomography angiography (CCTA), are costly and harmful to humans. To address these issues, we proposed a multi-modal data fusion method of ultrasonic (US) images and electronic medical records (EMRs) named M-US-EMRs model to automatically screen CHD. Comparing to traditional methods, this model features cheap, harmless to humans, easy to implement and independent of doctors. The experiment result shows that, M-US-EMRs model reached an overall classification AUC of 79.19%. Furthermore, the interpretable models that we proposed lead to an increased trust from the people who use them. Our project developed effective tools with good performance for CHD screening among a Chinese population that will help to improve the primary prevention and management of cardiovascular disease. Our method lays the groundwork for using automated interpretation to support serial patient tracking and scalable analysis of millions of ultrasonic images and EMRs archived within healthcare systems.

Keywords: Ultrasound images · Electronic medical records · Multi-modal · Convolutional neural network

1 Introduction

Cardiovascular diseases cause approximately one-third of deaths worldwide [1]. Among cardiovascular illnesses, coronary heart disease (CHD) ranks as the most

M. S. Bansal et al. (Eds.): ISBRA 2022, LNBI 13760, pp. 88–99, 2022.
https://doi.org/10.1007/978-3-031-23198-8_9

prevalent [2], which is the number one cause of death, disability, and human suffering globally. In medical practice, risk assessment based on electronic medical records (EMRs) is a fundamental component of prevention and screening of CHD. More aggressive risk factor modification (such as medical imaging examination) should be required among individuals with a predicted high risk. A common medical imaging tool for CHD screening is echocardiography, which is a noninvasive, relatively inexpensive, radiation-free imaging modality that is an indispensable part of modern cardiology.

The risk prediction equations based on EMRs will help clinicians guide preventive approaches, individualized counseling, and treatment decisions for CHD by more accurate estimation of 10-year risk of CHD. The Framingham Heart Study identified several risk factors and developed the first coronary heart disease (CHD) risk equations in 1976 [3]. Wu et al [4] developed 10-year risk prediction models for ischemic CHD, including ischemic stroke and coronary events, using derivation and validation cohorts with 9903 and 17,329 subjects, respectively. All standard CHD risk assessment models based on EMRs make an implicit assumption that each risk factor is related in a linear fashion to CHD outcomes [5]. Such models may thus oversimplify complex relationships which include large numbers of risk factors with non-linear interactions. Before the advent of deep learning, a wealth of techniques had been developed to extract clinically relevant information from cardiovascular images. Early algorithms typically required significant manual tuning to transform an input image into the desired output [6]. Deep learning, specifically using convolutional neural networks (CNNs), is a cutting-edge machine learning technique that has proven "unreasonably" successful at learning patterns in images and has shown great promise helping experts with image-based diagnosis in radiology, pathology, and dermatology. It is worth mentioning the work by Betancur et al., an end-to-end DL model, estimating per-vessel CAD probability without any assumed subdivision of the input coronary territories from imaging data [7]. The authors in Wolterink et al. built a coronary artery calcification (CAC) detector, also based on DL trained on raw CT images [8]. Echocardiography has a central role in the diagnosis and management of cardiovascular disease [9]. However, deep learning has not yet been widely applied to echocardiography because Ultrasonic diagnosis depends on doctors' flexible operation and switching methods combined with medical history, which leads to great differences in ultrasound images [10].

With the extensive application of computer technology and digital imaging technology in the medical domain, medical case has become digitalized and multi-modal. However, artificial intelligence based on deep learning has not been widely used in the analysis of echocardiography, and multimodal analysis combined with medical images and EMRs is even less. In this paper, we intend to conduct a multi-modal early screening of CHD by using ultrasound images and EMRs. These images and EMRs have their own advantages as well as disadvantages, and different modality medical images and EMRs can provide complementary information for medical diagnosis. First, we use pre training models to extract features from ultrasound images. Then, the features of EMR data are extracted manually. At last, we fuse the extracted features and propose a multi-modal classification method for CHD. To the best of our knowledge, we are the first to employ the combination of EMRs and ultrasound images to obtain an AI model for screening of CHD. Experimental results showed that our method can reach high classification AUC value.

In particular, we make the following contributions:

- We construct a large-scale electronic medical records and corresponding cardiac ultrasonic image dataset.
- We propose a multi-modal data fusion method of ultrasonic images and electronic medical records named M-US-EMRs model to automatically screen CHD. Our method lays the groundwork for using automated interpretation to support serial patient tracking and scalable analysis of millions of ultrasonic images and EMRs archived within healthcare systems.
- We demonstrate that M-US-EMRs outperforms single-modal model on screening CHD tasks.

2 Methods

2.1 Problem Formulation

The early screening of CHD based on ultrasound images and EMRs belongs to the problem of classification. Given ultrasound images, EMRs and their corresponding CHD labels, the target is that judging the ultrasound images and EMRs belong to CHD or non-CHD. In this paper, we aim to learn a deep representation for ultrasound images and EMRs, and use them for CHD classification. For simplicity, we define $D = \{x1_i, x2_i, y_i\}$ as ultrasound images and electronic medical records data set. Where $x1_i$ indicates the i-th person's ultrasonic images data, and $x2_i$ indicates the ith person's EMRs data. $y_i \in \{0, 1\}$ denotes the corresponding category of $x1_i$, $x2_i$, 0 represents non-CHD and 1 represents CHD. N is the total number of samples. The proposed model in this paper accepts $x1_i$, $x2_i$ as input and y_i as output. Mathematically, it can be described by minimizing the cross-entropy between the reference labels and outputs.

2.2 Model Architecture

The architecture of M-US-EMRs is designed and built drawing on the experiences of Inception-v3 [11], a mature neural network which has been proved to be effective in solving a variety of problems in computer vision field. M-US-EMRs is composed of 4 streams of CNNs and classifier after them, as shown in Fig. 1. For the image part of CNN network architecture, we have 3 traditional inception modules at the 35×35 with 288 filters each. This is reduced to a 17×17 grid with 768 filters using the grid reduction technique. The output size of each module is the input size of the next one. We have marked the convolution with 0-padding, which is used to maintain the grid size. 0-padding is also used inside those Inception modules that do not reduce the grid size. All other layers do not use padding.

For each type of image, we apply the pre-trained Inception-v3 network as a new starting point, where the Inception-v3 network with fine-tuning is faster and easier to train than training the network from scratch. The Inception-v3 model is pretrained on the ImageNet dataset, which can divide more than one million natural images into 1000 categories. We use simplified-Inception-v3 after transfer learning to extract features. The simplification strategy of the simplification model is to delete some inception blocks in the network structure. Its purpose is to overcome the problem of over fitting in the training process of complex networks. Before input to the final classifier module, the four type of images' output feature maps from the feature extraction module and the features from EMRs are concatenated. Compared with sum and prod operation, concatenation preserves individual module independent features and provides more information. The classifier module is XGBoost [12], which is an implementation of gradient boosted decision trees designed for speed and performance. XGBoost can perform the three major gradient boosting techniques, that is Gradient Boosting, Regularized Boosting, and Stochastic Boosting. It also allows for the addition and tuning of regularization parameters, making it stand out from other libraries.

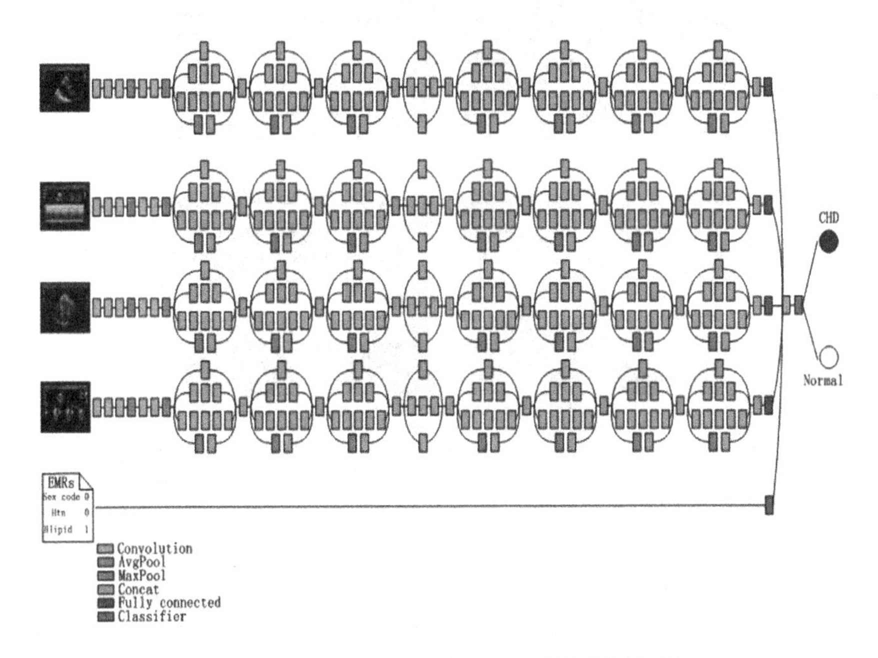

Fig. 1. The network architecture of M-US-EMRs.

3 Experiment

In this section, data details and implementation details are given.

3.1 Data Description

Data used in this paper were obtained from Huazhong University of Science and Technology Union Shenzhen Hospital. Ultrasonic images were obtained from the Picture Archiving and Communication System (PACS). EMRs (including diagnostic labels) were obtained from the PACS and Hospital Information System (HIS). Data fusion was performed across systems for patients with the same personal IDs. According to the statistics of the data set, we have got a total of 27,398 patients' information from Shenzhen Nanshan Hospital which include ultrasound images and EMRs. The ultrasonic images data can be classified into four main categories: 2D echocardiography, M-mode echocardiography, color Doppler and spectral Doppler. The four main categories of images are shown in Fig. 2.

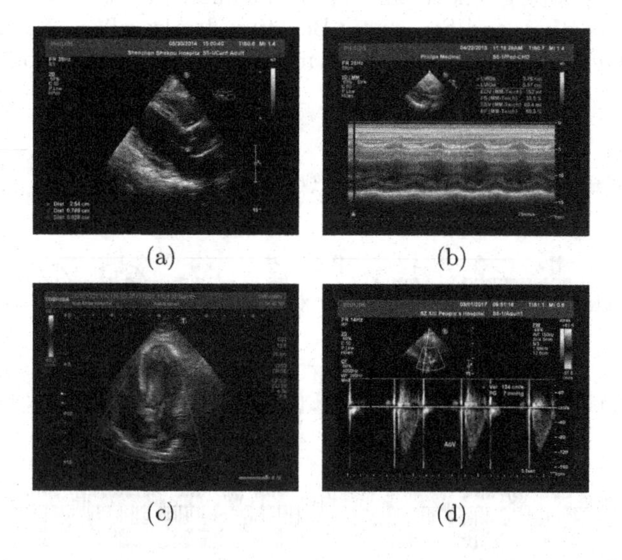

Fig. 2. The four main categories of images: (a) 2D echocardiography, (b) M-mode echocardiography, (c) Color Doppler, and (d) Spectral Doppler.

All the EMRs information is stored in two tables. First, we associate PACS and HIS tables with the same personal IDs. Then, we select the fields we need from the EMRs. 12 fields are selected in our experiment (excluding label), among which date_of_birth and reg_date are unstructured data and others are structured data. The explanation of the EMRs' fields is shown in Table 1. After excluding 462 patients without ultrasound images, the analysis cohort consisted of 26936 patients, which include more than 500,000 ultrasound images. Furthermore, the echocardiograms used in this study were drawn randomly from

Table 1. Explanation of the EMRs' fields.

List of fields	Examination	Remarks
Date_of_birth	Date of birth	
Sex_code	Gender code	0 male and 1 female
Reg_date	Date of registration	
Htn	Hypertension	0 no 1 yes
Dm	Diabetes mellitus	0 no 1 yes
Hlipid	Hyperlipidemia	0 no 1 yes
Huric	Hyperuricemia	0 no 1 yes
AR	Arrhythmia	0 no 1 yes
AF	Atrial fibrillation	0 no 1 yes
PB	Premature beat	0 no 1 yes
Cpain	Chest distress or chest pain	0 no 1 yes
Hache	Headache or dizziness	0 no 1 yes
Type1	Label of coronary heart disease	0 no 1 yes

real echocardiograms acquired for clinical purposes, from patients with a range of ages, sizes, and hemodynamics. There are a total of 4,530 patients suffering from CHD (CHD patients), accounting for 16.82% of the total data set. CHD includes the following diseases: coronary atherosclerotic heart disease, myocardial infarction, ischemic cardiomyopathy, ischemic heart disease, acute coronary syndrome, asymptomatic myocardial ischemia and angina pectoris. There are a total of 22406 patients don't suffering from CHD (non-CHD patients), accounting for 83.18% of the total data set. Due to the large deviation in age and gender between the patients with CHD and non-CHD, and considering that the main incidence population of CHD is the middle-aged and the elderly, this study only included patients aged 40–79, and 1:2 stratified sampling was conducted for the CHD population and non-CHD population according to age (40–59, 60–79) and gender. In the end, we got 11106 patients in the research cohort who met the above conditions, CHD patients 3702, non-CHD patients 7404. Furthermore, because all medical records were collected during routine clinical activities and the obtained data were anonymous, following the Guidelines of the WMA Declaration of Helsinki term 32, a waive-of-consent protocol was adopted and approved by the Institutional Review Board of Shenzhen Institute of Advanced Technology, Chinese Academy of Sciences (No. SIAT-IRB-151115-H0084).

3.2 Implementation Details

Figure 3 provides a detailed schematic of the experimental framework.

The first step was data preprocessing. On the one hand, we process the original images into a size acceptable to the model because the size of the original images are varied. In order to ensure the stability of the subsequent model, we randomly rotate the processed image by 0 to 10°C. On the other hand, we pre-

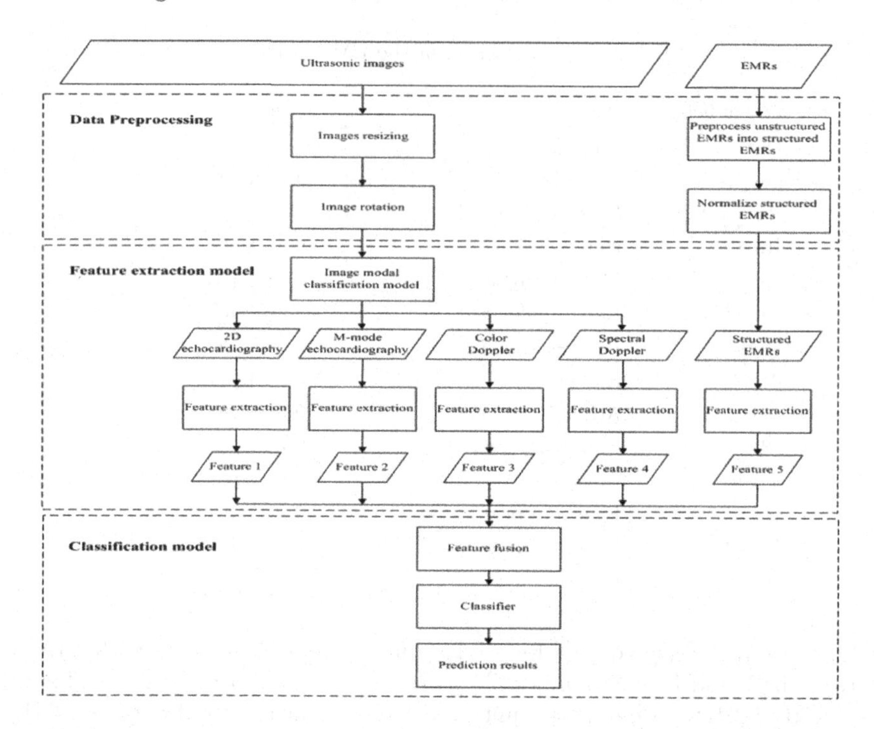

Fig. 3. Experimental framework of multi-modal classification of CHD.

process and normalize unstructured EMRs fields (date_of_birth and reg_date) and get three new structured fields. Age was defined as the age from birth to present. REG_DAYS was defined as the number of days from the first day of registration to the present. DURING_DAYS was defined as the number of days from the first registration to the last registration. At last, 10 structured fields and the three new structured fields were normalized.

The second step was to extract feature. First, we label 50 persons without CHD (1216 images) and 50 persons with CHD (967 images) manually and train a four classification CNN model of image category. Next, we used the selected images and deep learning model simplified-Inception-v3, Inception-v3 and ResNet-50 to extract the features of each image mode. we also used Inception-v3 and ResNet-50 [13] as baselines to extract the features of each image mode. At the same time, we extract the required structured data from EMRs.

The final step was classification model. We fuse the features learned by the model from ultrasound images and the features learned by the model from electronic medical records. At last, we send the fused features to a classifier and obtain the final results. As a control, we take the prediction result of ultrasound images for CHD as a baseline, the prediction result of EMRs for CHD as another baseline. Specifically, we directly use simplified-Inception-v3, Inception-v3 and ResNet-50 to classify CHD and non-CHD rather than feature extraction. At the same time, we used XGBoost, CNN and Fully-connected neural network to classify CHD and non-CHD on the processed EMRs data. For all models, the people are randomly split into a training, validation, and test set. 80% of the labeled data is used as the training set, 10% as validation set and 10% as test set. We select the same number of people as the CHD patients in our experiment to ensure that positive and negative samples are balanced. While splitting, we ensure that people' notes stay within the set, so that all people' notes in the test set are from patients not previously seen by the model. The reported numbers across different models are obtained from testing on the same test set. The validation set is used to choose the hyper parameters for the models.

3.3 Evaluation Criteria

We evaluate area under the ROC curve (AUC) of all models as a quantitative measure. The details are as follows:

$$AUC = \frac{\sum_{i \in positive} rank_i - \frac{M \times (M+1)}{2}}{M \times N} \tag{1}$$

In formula 1, Where M represents the number of positive samples, N is the number of negative samples, and rank i is the order of probability from high to low for positive examples. Therefore, a larger AUC value indicates a better classification result of the learner and a better prediction effect of the model.

4 Results

4.1 Experimental Environment

The experiments were performed using the prevailing deep learning framework TensorFlow version 2.3.1 with the Windows Server 2012 operating system. The TensorFlow platform was deployed on a high-performance computing server equipped with 8-core Intel Xeon E5-2690 V4 processors with 64 GB of memory. In addition, the computing server was equipped with NVIDIA Tesla P100 GPU with 16 GB of memory. The programing language used was Python 3.6.

4.2 Classification Performance

In this section, we provide details of the classification results of our experiments. We have done three groups of experiments. EMRs group contains 3 groups of binary experiments, Ultrasonic images group contains 3 groups of binary experiments and Multi-modal group contains 3 groups of binary experiments and the results are shown in Table 2. It is worth noting that this paper recorded the average performance of five repeated classification tasks on each data set.

Table 2. The classification results of CHD.

	Method	AUC
EMRs	XGBoost	0.7340
	CNN	0.7070
	Fully-connected neural network	0.6851
Image	simplified-Inception-v3	0.7571
	Inception-v3	0.7466
	ResNet-50	0.7310
EMRs + Image	simplified-Inception-v3 + XGBoost	**0.7919**
	Inception-v3 + XGBoost	0.7792
	ResNet-50 + XGBoost	0.7694

4.3 Interpretability

The feature importance of the XGBoost model measures the relative contribution of the feature variables in the process of building the decision trees. Figure 4 depicts the top-ranked features selected by the XGBoost model. The days since first registration is the most important risk factor, which is even more important than age. In addition to traditional risk factors, several other features representing the dynamic trends of medical activities also played important roles in risk prediction. For example, duration of illness would be linked to a higher risk of CHD onset. In addition, the core risk variables would provide additional information to the risk scores.

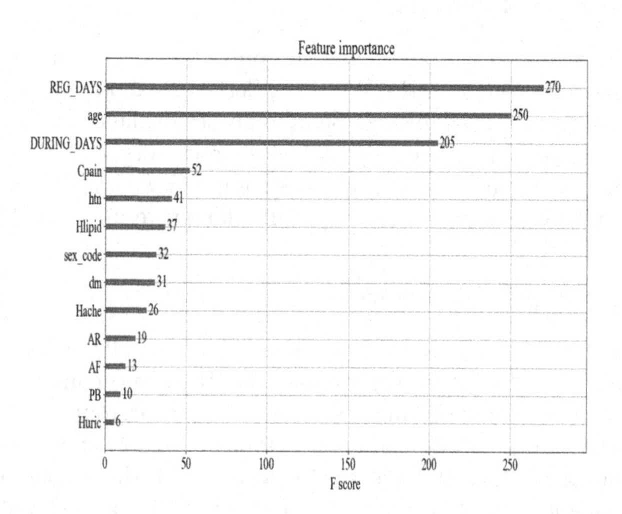

Fig. 4. The top-ranked features of EMRs selected by the XGBoost model.

It is acknowledged that the "black-box" nature of machine-learning algorithms, in particular neural networks, can be difficult to interpret. The inability of humans to understand predictions of complex machine learning models poses a difficult challenge when such approaches are used in healthcare. Therefore, we consider it crucial that well-performing approaches should be understood and trusted by those who use them. However, improvements in data visualization methods have improved understanding of these models, illustrating the importance areas that network connections focus on. We use OpenCV to generate several images that overlay the original image on the heatmap, and the images are shown in Fig. 5.

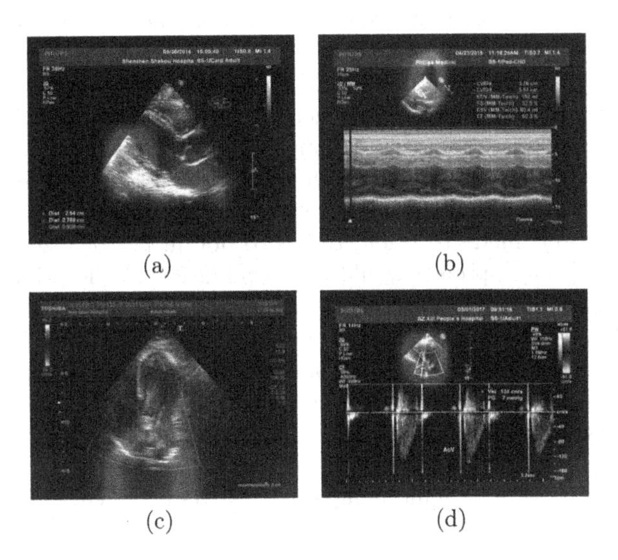

Fig. 5. Ultrasonic images' heatmap: (a) 2D echocardiography, (b) M-mode echocardiography, (c) Color Doppler and (d) Spectral Dopplere.

5 Conclusion and Discussion

This paper presents an early screening method of coronary heart disease based on multimodal fusion of ultrasonic images and EMRs. First, we use models simplified-Inception-v3 to extract features from ultrasound images. Then, the features of EMR data are extracted manually. At last, we fuse the extracted features and propose a multi-modal classification method named M-US-EMRs for CHD. The study was performed in a large cohort of patients in Shenzhen Nanshan Hospital evaluated by calculating ROC curves and AUC. In the detailed results, ultrasound images with simplified-Inception-v3 can get an AUC value of 0.7571, which is better than EMRs with XGBoost (0.7340). AUC value of multimodal classification of CHD with M-US-EMRs is 0.7919. Multi-modal classification performed the best, with AUC value improving by 3.48%. This is an encouraging step forward. In addition, the visualization results of the simplified-Inception-v3 model show that the model has learned to associate the important parts of ultrasound images with CHD, since those parts also were focused by doctors during screening CHD. The visualization of XGboost model shows the feature importance ranking of EMRs. To our knowledge, this is the first investigation applying machine-learning to routine data in patients' ultrasound image and EMRs, demonstrating improved screenings of CHD risk in a large general population. The deep learning approach presented in this paper could further be used to assist clinicians during diagnosing. Unlike established approaches to risk prediction, the machine-learning methods used are not limited to a small set of risk factors, and incorporated more pre-existing medical conditions, like ultrasound images and EMRs. This may assist the drive towards personalised medicine, by better tailoring risk management to individual patients. In fact, we see this work as taking a step toward augmenting clinical practice rather than replacing current approaches. Specifically, we would like to see more measurements taken when patients are asymptomatic but at risk of CHD, with quantitative comparisons made to prior studies to obtain personalized longitudinal trajectories. Such an approach would shift evaluation to the primary care setting, with data collected by nonexperts-and the resulting initiation and tailoring of care would hopefully reduce the alarming increase in CHD incidence that has taken place in recent decades. An automated method to interpret echocardiograms could help democratize echocardiography, shifting evaluation of the heart to the primary care setting and rural areas. Our method lays the groundwork for using automated interpretation to support serial patient tracking and scalable analysis of millions of ultrasonic images and EMRs archived within healthcare systems. We anticipate validation of our proposed approach in other multi-modal data such as physiological signals and gene data to identify patients at risk of CHD. In conclusion, we are optimistic that our approach will have a broad clinical impact by (1) introducing relatively low-cost screening method of CHD into clinical practice, (2) laying the groundwork for using automated interpretation to support serial patient tracking and scalable analysis of millions of ultrasonic images and EMRs archived within healthcare systems available in large medical information repository, and (3) enabling causal insights that require systematic longitudinal tracking of patients.

Acknowledgments. This work is supported by the Strategic Priority Research Program of Chinese Academy of Sciences (XDB38050100), Shenzhen Science and Technology Program (SGDX20201103095603009).

References

1. Ralapanawa, U., Sivakanesan, R.: Epidemiology and the magnitude of coronary artery disease and acute coronary syndrome: a narrative review. J. Epidemiol. Global Health **11**(2), 169 (2021)

2. Roth, G.A., Mensah, G.A., Johnson, C.O., et al.: Global burden of cardiovascular diseases and risk factors, 1990–2019: update from the GBD 2019 study. J. Am. Coll. Cardiol. **76**(25), 2982–3021 (2020)
3. Kannel, W.B., McGee, D., Gordon, T.: A general cardiovascular risk profile: the Framingham Study. Am. J. Cardiol. **38**(1), 46–51 (1976)
4. Wu, Y., Liu, X., Li, X., et al.: USA-PRC Collaborative study of cardiovascular and cardiopulmonary epidemiology research group. Estimation, 10
5. Obermeyer, Z., Emanuel, E.J.: Predicting the future-big data, machine learning, and clinical medicine. N. Engl. J. Med. **375**(13), 1216 (2016)
6. Polonsky, T.S., McClelland, R.L., Jorgensen, N.W., et al.: Coronary artery calcium score and risk classification for coronary heart disease prediction. JAMA **303**(16), 1610–1616 (2010)
7. Betancur, J., Hu, L.H., Commandeur, F., et al.: Deep learning analysis of upright-supine high-efficiency SPECT myocardial perfusion imaging for prediction of obstructive coronary artery disease: a multicenter study. J. Nucl. Med. **60**(5), 664–670 (2019)
8. Wolterink, J.M., Leiner, T., de Vos, B.D., et al.: Automatic coronary artery calcium scoring in cardiac CT angiography using paired convolutional neural networks. Med. Image Anal. **34**, 123–136 (2016)
9. Kusunose K, Haga A, Abe T, et al. Utilization of artificial intelligence in echocardiography. Circ. J. CJ-19-0420 (2019)
10. Zhang, J., Gajjala, S., Agrawal, P., et al.: Fully automated echocardiogram interpretation in clinical practice: feasibility and diagnostic accuracy. Circulation **138**(16), 1623–1635 (2018)
11. Szegedy, C.,Vanhoucke, V., Ioffe, S., et al.: Rethinking the inception architecture for computer vision In: Proceedings of the IEEE Conference on Computer Vision and Pattern Recognition, pp. 2818–2826 (2016)
12. Chen, T., He, T., Benesty, M., et al.: Xgboost: extreme gradient boosting. R package version 0.4-2, **1**(4), 1–4 (2015)
13. He, K., Zhang, X., Ren, S., et al.: Deep residual learning for image recognition. In: Proceedings of the IEEE Conference on Computer Vision and Pattern Recognition, pp. 770–778 (2016)

Transposition Distance Considering Intergenic Regions for Unbalanced Genomes

Alexsandro Oliveira Alexandrino[1](✉), Andre Rodrigues Oliveira[1], Géraldine Jean[2], Guillaume Fertin[2], Ulisses Dias[3], and Zanoni Dias[1]

[1] Institute of Computing, University of Campinas, Campinas, Brazil
{alexsandro,andrero,zanoni}@ic.unicamp.br
[2] Nantes Université, CNRS, LS2N, UMR 6004, 44000 Nantes, France
{geraldine.jean,guillaume.fertin}@univ-nantes.fr
[3] School of Technology, University of Campinas, Limeira, Brazil
ulisses@ft.unicamp.br

Abstract. In seminal works of genome rearrangements, the distance between two genomes is measured by the minimum number of rearrangements (e.g., reversals, transpositions, DCJs, or combination of them) needed to transform a given permutation into another, where permutations represent gene orders of genomes with the same gene content. For the past few years, researchers have been extending the traditional models of genome rearrangement distance by either considering unbalanced genomes or adding more features to the representation of the genomes to be compared. In this work, we make progress in this direction by studying the intergenic transposition distance on unbalanced genomes, which also considers insertions and deletions as non-conservative rearrangements in the set of possible rearrangements to compute the distance. The best previously known result for this problem is a 4.5-approximation using breakpoints. In this paper, we use an adaptation of the breakpoint graph, a structure used in the literature on genome rearrangements, to present a new lower bound for the distance and a 4-approximation algorithm.

Keywords: Transpositions · Indels · Intergenic regions

1 Introduction

A class of problems motivated by the Computational Biology field seeks the minimum sequence of operations that transforms one genome into another. These operations are called genome rearrangements, which are global events that alter

This work was supported by the National Council of Technological and Scientific Development, CNPq (grants 425340/2016-3 and 202292/2020-7), the Coordenação de Aperfeiçoamento de Pessoal de Nível Superior - Brasil (CAPES) - Finance Code 001, and the São Paulo Research Foundation, FAPESP (grants 2013/08293-7, 2015/11937-9, and 2019/27331-3).

M. S. Bansal et al. (Eds.): ISBRA 2022, LNBI 13760, pp. 100–113, 2022.
https://doi.org/10.1007/978-3-031-23198-8_10

the relative order of genes or change the gene content in a genome. In these problems, called genome rearrangement problems, one of the main difficulties is how to represent genomes. Initially, permutations were used for this representation, where each element in the permutation represents a gene or a specific segment of the genome [5,13]. However, this representation requires the same gene content in both genomes and does not allow gene repetition. Later, researchers started representing genomes through strings, so that the existence of repeated genes and genomes with different sets of genes could also be represented [14,18].

Several rearrangement events have been proposed and studied in the literature: reversals [13], which reverse a segment of the genome, transpositions [5], which exchange two adjacent segments of the genome, and the Double-Cut-and-Join (DCJ) [19], which generalize several rearrangement events. A *model* is the set of rearrangements allowed to transform one genome into another, and may contain a single kind of rearrangement or several of them together. The term *distance* denotes the minimum number of rearrangements from a model needed to transform one genome into another. All rearrangements mentioned so far are called *conservative*, as they do not change the number of genes in the genome, only their relative order. Two strings are called *balanced* if they have the same set of elements and if each element appears the same number of times in both strings; they are called *unbalanced* otherwise. In the above case where conservative rearrangements are considered, since we need to transform one permutation or string into another, the pair of permutations (resp. strings) must have same size (resp. must be balanced).

For a more realistic representation, new studies began to assume that genomes could also have distinct sets of genes. Thus, it became necessary to add *non-conservative* rearrangements to the models, such as insertions and deletions (*indels*), to be able to transform one genome into another. In this case, strings could be unbalanced since elements missing in one of the strings can be added by insertions (or removed from the other string by deletions) [2,4,8,12].

More recently, researchers also argued that using only the relative order neglects some information also present in the genome, and that using the size of the intergenic regions, i.e., regions between each pair of genes, results in more realistic distance estimations [6,7]. In particular, in [6,7], experiments were performed using algorithms that considered intergenic regions and algorithms that neglected them. The authors concluded that the algorithms considering intergenic regions had better behavior in most situations.

In the original formulation of the DCJ distance that does not consider intergenic regions, it was proved that the solution space is very large and difficult to analyze. Bulteau *et al.* [11] proved that the DCJ and Indels distance, in which indels act only on intergenic regions, can restrict the solution space.

After these studies, several rearrangement problems were investigated considering the representation that uses intergenic regions together with the relative order of genes: DCJs, reversals, transpositions, reversals and transpositions, as well as the combination of these rearrangements with indels [1,3,9–11,16]. In particular, an NP-hardness proof and a 4.5-approximation algorithm, based on breakpoints, were presented for the model that takes into account the transformation of unbalanced strings by transpositions and indels [3].

In this work, we present a new lower bound for the Transposition and Indel Distance considering intergenic regions and we improve the approximation factor to 4, using an adaptation of the breakpoint graph structure.

2 Definitions

We represent the information of a genome using two structures: a string and a list of numbers. The sequence of genes is mapped into a string preserving the relative order of genes in the genome, where each element of the string represents a gene, and the size distribution of intergenic regions is mapped into a list of non-negative integers, where the size of each region (number of nucleotides) is mapped into an element of the list.

Given a source genome $\mathcal{G}_1$ and a target genome $\mathcal{G}_2$, the sequence of genes in $\mathcal{G}_2$ is mapped using a string called *identity string* $\iota^n = (1\ 2\ \dots\ n)$ and the sequence of genes in $\mathcal{G}_1$ is mapped using the string $A = (A_1\ A_2\ \dots\ A_m)$, with labels according to the mapping of the sequence of genes from $\mathcal{G}_2$ into ι^n. Since we are studying a model with only transpositions and indels, the orientation of genes is disregarded because these operations cannot change orientation of genes.

A continuous segment of genes present only in the target genome is represented by a single character in ι^n, because an insertion can add this whole segment at once. Furthermore, a continuous segment of genes present only in the source genome is represented by the character α, since this segment is removed regardless of which genes it contains. Note that in this way there is no continuous sequence of two or more elements in A that are not in ι^n and vice versa. The *alphabets* Σ_A and Σ_{ι^n} are the sets of characters of A and ι^n, respectively.

The sizes of the intergenic regions from $\mathcal{G}_1$ are represented by the list $\breve{A} = (\breve{A}_1, \breve{A}_2, \dots, \breve{A}_{m+1})$, where $\breve{A}_i$ is the size (number of nucleotides) of the intergenic region between genes A_{i-1} and A_i, for $2 \leq i \leq m$, and $\breve{A}_1$ and $\breve{A}_{m+1}$ are the sizes of the intergenic regions at the extremities of the genome $\mathcal{G}_1$. Similarly, we use the list $\breve{\iota}^n = (\breve{\iota}^n_1, \breve{\iota}^n_2, \dots, \breve{\iota}^n_{n+1})$ for the sizes of intergenic regions from $\mathcal{G}_2$.

For the following definitions, we consider a genome $\mathcal{G} = (A, \breve{A})$ such that A has m elements and $\breve{A}$ has $m+1$ elements. We use the notation $\mathcal{G} \cdot \beta$ to represent the resulting genome after applying β to $\mathcal{G}$, where β can be a single rearrangement or a sequence of rearrangements. Analogously, we use $A \cdot \beta$ and $\breve{A} \cdot \beta$ to denote the resulting string and list of intergenic regions, respectively.

A *transposition* $\tau^{(i,j,k)}_{(x,y,z)}$, with $1 \leq i < j < k \leq m+1$, $0 \leq x \leq \breve{A}_i$, $0 \leq y \leq \breve{A}_j$, and $0 \leq z \leq \breve{A}_k$, affects a genome $\mathcal{G} = (A, \breve{A})$ as follows: $\mathcal{G} \cdot \tau^{(i,j,k)}_{(x,y,z)} = \mathcal{G}' = (A', \breve{A}')$, where $A' = (A_1\ \dots\ A_{i-1}\ \underline{A_j\ \dots\ A_{k-1}}\ \underline{A_i\ \dots\ A_{j-1}}\ A_k\ \dots\ A_m)$ and $\breve{A}' = (\breve{A}_1, \dots, \breve{A}_{i-1}, x + y', \underline{\breve{A}_{j+1}, \dots, \breve{A}_{k-1}}, z + x', \underline{\breve{A}_{i+1}, \dots, \breve{A}_{j-1}}, y + z', \breve{A}_{k+1}, \dots, \breve{A}_{m+1})$, with $x' = \breve{A}_i - x$, $y' = \breve{A}_j - y$, and $z' = \breve{A}_k - z$.

An *insertion* $\phi^{(i,S,\breve{S})}_{(x)}$, where $0 \leq i \leq m$, $0 \leq x \leq \breve{A}_{i+1}$, S is a sequence of characters, and $\breve{S}$ is a list of non-negative integers with $|\breve{S}| = |S| + 1$, affects a genome $\mathcal{G} = (A, \breve{A})$ as follows: $\mathcal{G} \cdot \phi^{(i,S,\breve{S})}_{(x)} = \mathcal{G}' = (A', \breve{A}')$, where $A' = (A_1 \ \ldots \ A_i \ \underline{S_1 \ \ldots \ S_{|S|}} \ A_{i+1} \ \ldots \ A_m)$ and $\breve{A}' = (\breve{A}_1, \ldots, \breve{A}_i, \underline{x + \breve{S}_1, \breve{S}_2, \ldots, \breve{S}_{|S|}, \breve{S}_{|\breve{S}|} + x'}, \breve{A}_{i+2}, \ldots, \breve{A}_{m+1})$, with $x' = \breve{A}_{i+1} - x$.

A deletion $\psi^{(i,j)}_{(x,y)}$, with $1 \leq i \leq j \leq m+1$, $0 \leq x \leq \breve{A}_i$, and $0 \leq y \leq \breve{A}_j$, affects a genome $\mathcal{G} = (A, \breve{A})$ as follows: $\mathcal{G} \cdot \psi^{(i,j)}_{(x,y)} = \mathcal{G}' = (A', \breve{A}')$, where $A' = (A_1 \ \ldots \ A_{i-1} \ A_j \ \ldots \ A_m)$ and $\breve{A}' = (\breve{A}_1, \ldots, \breve{A}_{i-1}, \underline{x + y'}, \breve{A}_{j+1}, \breve{A}_{j+2}, \ldots, \breve{A}_{m+1})$, with $y' = \breve{A}_j - y$.

An insertion $\phi^{(i,S,\breve{S})}_{(x)}$ can only affect a single intergenic region if the sequence of characters S is empty. A deletion $\psi^{(i,j)}_{(x,y)}$ can only affect a single intergenic region if $i = j$. In this case, we have the constraint that $0 \leq x \leq y \leq \breve{A}_i$.

Let $\mathcal{I} = (\mathcal{G}_1, \mathcal{G}_2)$ be an instance for the genome rearrangement problem, where $\mathcal{G}_1$ is the source genome and $\mathcal{G}_2$ is the target genome. The goal of the *Transposition and Indel Distance* problem is to find a minimum length sequence of transpositions and indels that transforms $\mathcal{G}_1$ into $\mathcal{G}_2$. The size of such a sequence is called the *distance* between $\mathcal{G}_1$ into $\mathcal{G}_2$ and it is denoted by $d(\mathcal{I})$.

3 Labeled Intergenic Breakpoint Graph

In the following definitions, we use an instance $\mathcal{I} = (\mathcal{G}_1, \mathcal{G}_2)$, where $\mathcal{G}_1 = (A, \breve{A})$ is the source genome and $\mathcal{G}_2 = (\iota^n, \breve{\iota}^n)$ is the target genome.

Let π^A be a string containing the elements of A different from α preserving the relative order they appear on A, in other words, π^A is equal to the string A excluding any α element. The string π^A contains the extra elements $\pi^A_0 = 0$ and $\pi^A_{m+1} = n + 1$, where $|\pi^A| = m$ is the number of elements in π^A without considering the extra elements 0 and $n + 1$.

In a similar way, let π^ι be a string containing the elements of ι^n that belong to $\Sigma_{\iota^n} \cap \Sigma_A$ preserving the relative order they appear on ι^n. Note that π^ι also has size $|\pi^\iota| = m$ and both π^ι and π^A only have elements that belong to $\Sigma_{\iota^n} \cap \Sigma_A$. The string π^ι also contains the extra elements 0 and $n + 1$.

The Labeled Intergenic Breakpoint Graph $G(\mathcal{I}) = (V, E, w, \ell)$ is a graph where $V = \{+\pi^A_0, -\pi^A_1, +\pi^A_1, \ldots, -\pi^A_m, +\pi^A_m, -\pi^A_{m+1}\}$ is the set of vertices, $E = E_s \cup E_t$ is the set of edges that is divided into the set of *source edges* E_s and the set of *target edges* E_t, $w : E \rightarrow \mathbb{N}$ is a weight function, and $\ell : E \rightarrow (\Sigma_{\iota^n} \setminus \Sigma_A) \cup \{\alpha, \emptyset\}$ is an edge labeling function.

The set of source edges is defined as $E_s = \{e_i = (+\pi^A_{i-1}, -\pi^A_i) : 1 \leq i \leq m + 1\}$, that is, source edges connect vertices that correspond to adjacent elements in π^A. We say that the source edge $e_i = (+\pi^A_{i-1}, -\pi^A_i)$ has index i. If π^A_{i-1} and π^A_i are adjacent in A, then e_i has empty label (i.e., $\ell(e_i) = \emptyset$) and the weight $w(e_i)$ is equal to the size of the intergenic region between π^A_{i-1} and π^A_i in $\mathcal{G}_1 = (A, \breve{A})$.

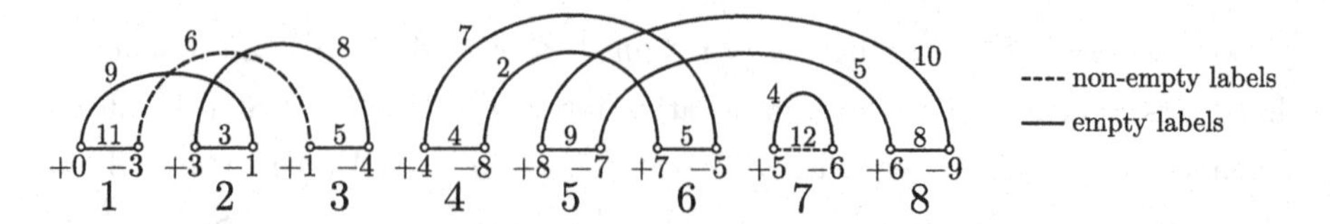

Fig. 1. An example of the Labeled Intergenic Breakpoint Graph $G(\mathcal{I})$ for $A = (3\ 1\ 4\ 8\ 7\ 5\ \alpha\ 6)$, $\breve{A} = (11, 3, 5, 4, 9, 5, 6, 6, 8)$, $\iota = (1\ 2\ 3\ 4\ 5\ 6\ 7\ 8)$, and $\breve{\iota} = (9, 3, 3, 8, 7, 4, 5, 2, 10)$. Source edges are drawn as horizontal lines and target edges as arcs. There are only two edges with non-empty labels, represented by dashed lines. The label of the target edge $(+1, -3)$ is 2 and the label of the source edge $(+5, -6)$ is α. A number above each edge represents its weight; numbers below vertices represent the indexes of source edges above them. $G(\mathcal{I})$ has four cycles: $C_1 = (3, 1, 2)$ is a labeled positive oriented cycle; $C_2 = (6, 4)$ is a clean balanced 2-cycle; $C_3 = (7)$ is a labeled negative trivial cycle; and $C_4 = (8, 5)$ is a clean negative 2-cycle.

If π^A_{i-1} and π^A_i are not adjacent in A, then e_i has label $\ell(e_i) = \alpha$ and the weight $w(e_i)$ is equal to the sum of the intergenic region between the pair (π^A_{i-1}, α) and the intergenic region between the pair (α, π^A_i) in $\mathcal{G}_1 = (A, \breve{A})$.

The set of target edges is defined as $E_t = \{e'_i = (+\pi^\iota_{i-1}, -\pi^\iota_i) : 1 \leq i \leq m{+}1\}$, that is, target edges connect vertices that correspond to adjacent elements in π^ι. Similarly, we say that the target edge $e'_i = (+\pi^\iota_{i-1}, -\pi^\iota_i)$ has index i. The weight $w(e'_i)$ is equal to the sum of the intergenic regions between the elements π^ι_{i-1} and π^ι_i in $\mathcal{G}_2 = (\iota^n, \breve{\iota}^n)$. If $\pi^\iota_i = \pi^\iota_{i-1} + 1$, then the label of e'_i is empty (i.e., $\ell(e'_i) = \emptyset$), otherwise, the target edge e'_i has label $\ell(e'_i) = \pi^\iota_{i-1} + 1$.

Figure 1 shows an example of a Labeled Intergenic Breakpoint Graph. The graph $G(\mathcal{I})$ has $2(m+1)$ vertices and $2(m+1)$ edges, which are divided into $m+1$ source edges and $m+1$ target edges. Each vertex is incident exactly to one source edge and one target edge and, therefore, there exists an unique decomposition of G into cycles. Furthermore, any cycle C has alternating edge types, that is, any pair of consecutive edges of a cycle C have different types (source or target).

We follow a pattern used in similar structures, such as the breakpoint graph and the weighted cycle graph [5,15], when drawing the Labeled Intergenic Breakpoint Graph, which is shown in Fig. 1. The vertices are drawn in a horizontal line from left to right following the sequence $+\pi^A_0, -\pi^A_1, +\pi^A_1, \ldots, -\pi^A_m, +\pi^A_m, -\pi^A_{m+1}$. Source edges are drawn as horizontal lines and target edges as arcs.

We represent edges in a cycle C using their indexes. We use o_i to represent the index of the i-th source edge of C and t_i to represent the index of the i-th target edge of C. We list the sequence of edge indexes $(o_1, t_1, o_2, t_2, \ldots, o_k, t_k)$ assuming that we traverse the cycle beginning with the rightmost vertex in C and that the first edge is the source edge e_{o_1} (i.e., $o_1 > o_i$, for $1 < i \leq k$, and the source edge e_{o_1} is traversed from right to left). In the examples, we represent a cycle C using just the list of source edge indexes $(o_1, o_2, \ldots, o_k)$, since the target edges can be inferred by each pair of consecutive source edges. For definitions, statements, and proofs, we use both the indexes of source and target edges to make notations more clear.

A cycle $C = (o_1, t_1, o_2, t_2, \ldots, o_k, t_k)$ is called a k-cycle. A 1-cycle is a *trivial* cycle, a 2-cycle is a *short* cycle, and a k-cycle with $k \geq 3$ is a *long* cycle.

We classify cycles according to three characteristics: position of its vertices in the strings; labels of its edges; and weight of its source and target edges.

A p-cycle $C = (o_1, t_1, \ldots, o_p, t_p)$ is *non-oriented* if $o_i > o_{i+1}$, for $1 \leq i < p$; otherwise, C is *oriented*. Note that all oriented cycles are long cycles. For any oriented cycle $C = (o_1, t_1, o_2, t_2, \ldots, o_p, t_p)$, there exists a triple (o_i, o_j, o_k) such that $i < j < k$, $o_i > o_k > o_j$, and $k = j + 1$ [5, Lemma 3.3]. Such triple is called an *oriented triple*.

Given two tuples of elements $(x_1, x_2, \ldots, x_k)$ and $(y_1, y_2, \ldots, y_k)$, we say they are *interleaving* if either $x_1 > y_1 > x_2 > y_2 > \ldots > x_k > y_k$ or $y_1 > x_1 > y_2 > x_2 > \ldots > y_k > x_k$. We will use the concept of interleaving triples in some proofs.

We say that an (source or target) edge e is *labeled* if it has a non-empty label (i.e., $\ell(e) \neq \emptyset$); otherwise, e is *clean*. A cycle C is *labeled* if any of its edges is labeled. If all edges of C are clean, we say that C is *clean*.

Furthermore, a cycle $C = (o_1, t_1, \ldots, o_k, t_k)$ is *balanced* if the sum of the weights from its target edges is equal to the sum of the weights from its source edges (i.e., $\sum_{i=1}^{k} w(e'_{t_i}) = \sum_{i=1}^{k} w(e_{o_i})$); otherwise, C is *unbalanced*. We further classify unbalanced cycles into *positive* or *negative*. A positive cycle $C = (o_1, t_1, \ldots, o_k, t_k)$ is such that $\sum_{i=1}^{k} w(e'_{t_i}) > \sum_{i=1}^{k} w(e_{o_i})$. A negative cycle $C = (o_1, t_1, \ldots, o_k, t_k)$ is such that $\sum_{i=1}^{k} w(e'_{t_i}) < \sum_{i=1}^{k} w(e_{o_i})$.

We say that a cycle C is *good* if it is balanced and clean, otherwise we say that C is a *bad* cycle. Note that the graph $G(\mathcal{I})$ has only trivial good cycles if, and only if, $\mathcal{G}_1 = \mathcal{G}_2$. In this way, we can interpret transforming $\mathcal{G}_1$ into $\mathcal{G}_2$ as transforming the graph $G(\mathcal{I})$ into a graph containing only trivial good cycles.

Let $c(\mathcal{I})$ be the number of cycles in $G(\mathcal{I})$ and let $c_g(\mathcal{I})$ be the number of good cycles in $G(\mathcal{I})$. Given a rearrangement β, let $\Delta c(\beta, \mathcal{I}) = (|\pi^A| + 1 - c(\mathcal{I})) - (|\pi^A \cdot \beta| + 1 - c(\mathcal{I}'))$, where $\mathcal{I}' = (\mathcal{G}_1 \cdot \beta, \mathcal{G}_2)$, in other words, $\Delta c(\beta, \mathcal{I})$ denotes the variation in the number of cycles in the graph relative to the number of source edges in the graph. Given a rearrangement β, let $\Delta c_g(\beta, \mathcal{I}) = (|\pi^A| + 1 - c_g(\mathcal{I})) - (|\pi^A \cdot \beta| + 1 - c_g(\mathcal{I}'))$, where $\mathcal{I}' = (\mathcal{G}_1 \cdot \beta, \mathcal{G}_2)$. If the instance $\mathcal{I}$ can be inferred from the context, we just use $\Delta c(\beta)$ and $\Delta c_g(\beta)$ as a notation simplification.

We note that the Labeled Intergenic Breakpoint Graph is analogous to the breakpoint graph structure [5] when we disregard the weight and label functions. Therefore, in our proofs, we can use known results on how transpositions affect cycles in a breakpoint graph.

A transposition τ, which exchanges the adjacent segments $A_i, \ldots, A_{j-1}$ and $A_j, \ldots, A_{k-1}$ in A, acts on three source edges with indexes i', j', k' such that: either $e_{i'}$ is incident to $-A_i$ or $A_i = \alpha$ and the label $\ell(e_{i'})$ comes from the element A_i; either $e_{j'}$ is incident to $-A_j$ or $A_j = \alpha$ and the label $\ell(e_{j'})$ comes from the element A_j; either $e_{k'}$ is incident to $-A_k$ or $A_k = \alpha$ and the label $\ell(e_{k'})$ comes from the element A_k.

Lemma 1. *For any transposition τ and genomes $\mathcal{G}_1$ and $\mathcal{G}_2$, we have that $\Delta c_g(\tau) \leq 2$.*

Proof. Note that a transposition cannot remove labels of target edges and it can only remove a label of a source edge by transferring α to another source edge. We divide the proof considering the number of cycles affected by τ [5].

If τ affects three cycles C_1, C_2, and C_3, then the vertices of these cycles are joined into a single cycle C'. In the best scenario, C_1, C_2, and C_3 are unbalanced and clean, and C' is balanced and clean ($\Delta c_g(\tau) = 1$).

If τ affects two cycles C_1 and C_2, then the vertices of these cycles are rearranged into two new cycles C'_1 and C'_2. In the best scenario, C_1 and C_2 are unbalanced and clean, and C'_1 and C'_2 are balanced and clean ($\Delta c_g(\tau) = 2$).

If τ affects one cycle C, in the best case, this transposition splits C into three new cycles. If C is good, in the best scenario, the new cycles are also good ($\Delta c_g(\tau) = 2$). Note that the total weight of source and target edges from C is distributed into the new cycles. If C is unbalanced or labeled, then at least one of the new cycles is unbalanced or labeled and, consequently, at most two of the new cycles are good cycles ($\Delta c_g(\tau) \leq 2$). □

Lemma 2. *For any genomes* $\mathcal{G}_1 = (A, \breve{A})$ *and* $\mathcal{G}_2 = (\iota^n, \breve{\iota}^n)$, *we have that* $d(\mathcal{I}) \geq \frac{|\pi^A|+1-c_g(\mathcal{I})}{2}$.

Proof. For any indel β, we have that $\Delta c_g(\beta) \leq 1$ [1], and for any transposition τ, we have that $\Delta c_g(\tau) \leq 2$ (Lemma 1). Since $\mathcal{G}_1 = \mathcal{G}_2$ if, and only if, $|\pi^A| + 1 - c_g(\mathcal{I}) = 0$, any sequence that transforms the source genome into the target genome has size of at least $(|\pi^A| + 1 - c_g(\mathcal{I}))/2$. Therefore, $d(\mathcal{I}) \geq (|\pi^A| + 1 - c_g(\mathcal{I}))/2$. □

4 4-Approximation Algorithm

If we consider only operations acting on good cycles, the results from intergenic problems that do not consider indels can be directly applied.

Lemma 3. *If* $G(\mathcal{I})$ *has a good oriented cycle* C, *then there exists a sequence* S *of at most three transpositions such that* $\Delta c_g(S) = 2$ *(Oliveira et al. [17, Lemma 4.6]).*

Lemma 4. *If* $G(\mathcal{I})$ *has a good non-oriented long cycle* C, *no oriented cycles, and all cycles are good, then there exists a sequence* S *of at most seven transpositions such that* $\Delta c_g(S) = 4$ *(Oliveira et al. [17, Lemma 4.7]).*

Lemma 5. *If* $G(\mathcal{I})$ *has only trivial or short cycles such that all cycles are good, then there exists a sequence* S *of two transpositions such that* $\Delta c_g(S) = 2$ *(Oliveira et al. [17, Lemma 4.8]).*

In the following, we show how indels can turn trivial bad cycles into good cycles.

Lemma 6. *If* $G(\mathcal{I})$ *has a trivial bad cycle* C *such that either* C *is clean or* C *is non-negative with a clean source edge, then there is an indel* β *that turns* C *into a balanced cycle, i.e.,* $\Delta c_g(\beta) = 1$.

Proof. Suppose that $C = (o_1, t_1)$ is clean. Then a deletion of $w(e_{o_1}) - w(e'_{t_1})$ nucleotides, if $w(e_{o_1}) > w(e'_{t_1})$, or an insertion of $w(e'_{t_1}) - w(e_{o_1})$ nucleotides, if $w(e_{o_1}) < w(e'_{t_1})$, applied only to the intergenic region corresponding to o_1 turns C into a balanced cycle. Therefore, $\Delta c_g(\beta) = 1$, where β represents the operation applied.

Now, suppose that $C = (o_1, t_1)$ is non-negative with a clean source edge but $\ell(e'_{t_1})$ is non-empty. Note that if $\ell(e'_{t_1})$ is empty, then C is clean and we can use an indel as discussed in the above paragraph. Let $-A_i$ and $+A_{i-1}$ be the two elements connected by edges e_{o_1} and e'_{t_1}. By definition of the graph, we know that $\ell(e'_{t_1}) = |A_{i-1} + 1| = |A_i - 1|$. Let x and y be the sizes of the intergenic regions between the pairs of elements $(A_{i-1}, \ell(e'_{t_1}))$ and $(\ell(e'_{t_1}), A_i)$ in the target genome $\mathcal{G}_2$, respectively. Therefore, an insertion $\phi^{(i-1,S,S')}_{\min(w(e_{o_1}),x)}$, where $S = (\ell(e'_{t_1}))$, $S' = (x', y')$, $x' = x - \min(w(e_{o_1}), x)$ and $y' = y - (w(e_{o_1}) - \min(w(e_{o_1}), x))$, turns C into two good trivial cycles C_1 and C_2, such that C_1 has vertices $(+A_{i-1}, -\ell(e'_{t_1}))$ and C_2 has vertices $(+\ell(e'_{t_1}), -A_i)$. Note that the source edges of C_1 and C_2 have weight x and y, which are the same weights as their target edges. Since the size of π^A increases by one and the number of balanced cycles increases by two, we have that $\Delta c_g(\beta) = 1$, where β represents the operation applied. □

Lemma 7. *If $G(\mathcal{I})$ has a trivial bad cycle $C = (o_1, t_1)$ such that C is negative or the source edge e_{o_1} is labeled, then there is a sequence S of at most two indels such that $\Delta c_g(S) = 1$.*

Proof. Let $-A_i$ and $+A_j$ be the vertices connected by the source edge e_{o_i}. Note that if e_{o_i} is labeled then $j \neq i + 1$. Consider the deletion $\psi^{(i+1,j)}_{(x,y)}$, where x and y are chosen so that $w(e_{o_1}) \leq w(e'_{t_1})$ in the new graph (note that this is always possible, for example we can choose $x = 0$ and $y = \breve{A}_j$ to turn $w(e_{o_1})$ to 0). This deletion turns the cycle C into a good cycle or into a non-negative cycle with a clean source edge. In the first case we already have a sequence $S = (\psi^{(i+1,j)}_{(x,y)})$ such that $\Delta c_g(S) = 1$ and $|S| = 1$. In the second case, we can apply another indel according to Lemma 6 to turn C into good cycle(s) and, then, we have a sequence of indels S such that $\Delta c_g(S) = 1$ and $|S| = 2$. □

Lemma 8. *If $G(\mathcal{I})$ has a positive cycle $C = (o_1, t_1, \ldots, o_k, t_k)$ that is clean, then there exists an insertion that transforms C into a good cycle.*

Proof. An insertion of $\sum_{i=1}^{k} w(e'_{t_i}) - \sum_{i=1}^{k} w(e_{o_i})$ nucleotides in the intergenic region corresponding to the source edge e_{o_1} turns C into a good cycle. □

If there is an oriented labeled cycle or at least two labeled cycles, then the next lemmas can be used with Lemma 6 to create new trivial good cycles in the graph.

Lemma 9. *If $G(\mathcal{I})$ has an oriented bad cycle C, then there is a transposition that turns C into three cycles such that one of them is a trivial non-negative cycle with a clean source edge.*

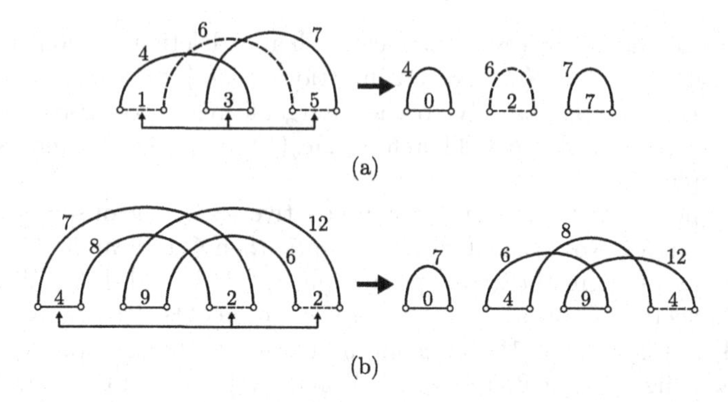

Fig. 2. Examples of transpositions applied to non-trivial cycles. **(a)** an oriented bad cycle (labeled and positive): a transposition transforms this cycle into three cycles such that one is a trivial cycle with a clean source edge. **(b)** two non-trivial bad cycles (both are labeled, one is positive and the other is negative): a transposition applied to three of its source edges transforms these cycles into two cycles such that one of them is a trivial cycle with a clean source edge.

Proof. Let (o_i, o_j, o_k) be an oriented triple of C, such that $k = j + 1$. A transposition applied on these three source edges o_i, o_j, o_k turns C into three cycles such that at least one of them is trivial, since $k = j + 1$. We can always choose the transposition in a way that the trivial cycle is non-negative and has a clean source edge; for this we just need to move any α element and the exceeding weight from the source edge where the trivial cycle is formed to one of the other two cycles. □

Lemma 10. *If $G(\mathcal{I})$ has at least two non-trivial bad cycles C and D that are non-oriented, then a transposition τ can turn these cycles into C' and D' such that one of these two cycles is a trivial non-negative cycle with a clean source edge.*

Proof. Let $C = (o_1^c, t_1^c, o_2^c, t_2^c \dots, o_x^c, t_x^c)$ and $D = (o_1^d, t_1^d, o_2^d, t_2^d, \dots, o_y^d, t_y^d)$. Suppose, without loss of generality, that $o_1^c < o_1^d$. A transposition applied on the source edges o_1^c, o_x^c, and o_1^d transforms these cycles into a trivial cycle C', which contains the target edge t_x^c, and a non-trivial cycle D' of size $y + x - 1$. We can move any α element and the exceeding weight from the source edge where the trivial cycle is formed to the cycle D'. In this way, cycle C' is a trivial non-negative cycle with a clean source edge. □

Figure 2 shows two examples of transpositions applied to non-trivial cycles according to Lemmas 9 and 10.

Lemma 11. *If $G(\mathcal{I})$ has only one bad short cycle $C = (o_1^c, t_1^c, o_2^c, t_2^c)$, all other cycles are good, and all cycles of the graph are non-oriented, then it is possible to find a sequence S such that $|S|/\Delta c_g(S) \leq 2$.*

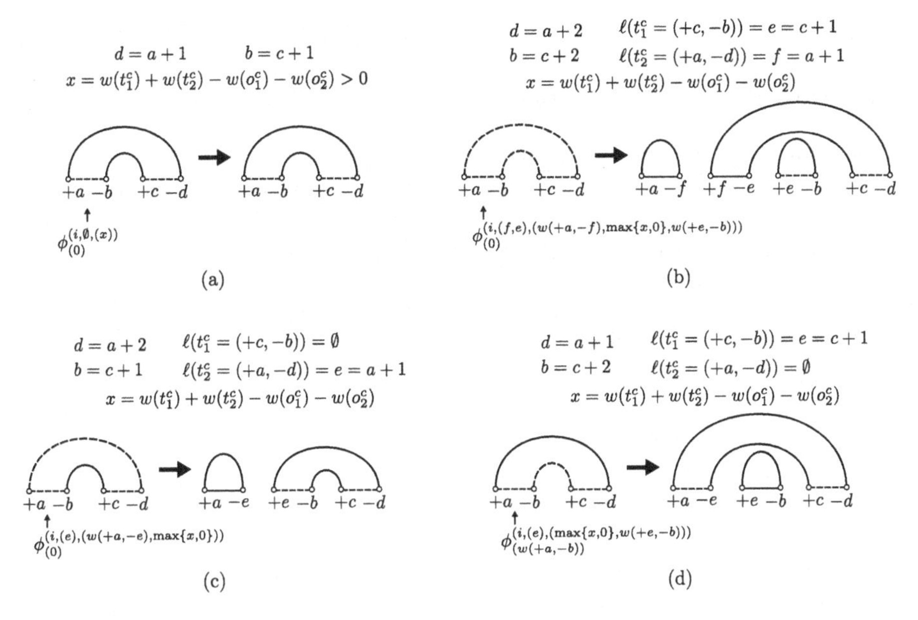

Fig. 3. Four examples of an insertion applied to a short cycle C. **(a)** if C is clean and positive, we just add the amount of nucleotides needed to balance it. **(b)** if both target edges of C are labeled, we add two elements that generate one good trivial cycle plus one non-positive trivial cycle, and a sequence of nucleotides to the resulting 2-cycle to transform it into a non-positive cycle. **(c-d)** if only one of the target edges of C is labeled, we add one element that generates one good trivial cycle and a sequence of nucleotides to the resulting 2-cycle to transform it into a non-positive cycle with no labeled target edges.

Proof. First, we need to apply indels according to Fig. 3 to transform C into good cycles. In the worst case, all edges of C are labeled. We apply an insertion according to Fig. 3b to clean both target edges of C. Note that the leftmost trivial cycle can always be a good cycle, since we can move any α element from the leftmost source edge to the rightmost cycle. In this case, the insertion created cycles C' (trivial), C'' (short) and C''' (trivial), such that only C' is good, but all of them are non-positive, since we can add weight to the leftmost source edge of C'' and the only source edge of C'''. Note that the leftmost source edge of C'' was created by the insertion and it has any weight we want and it is clean. Now, we just need to apply a deletion in each of the two labeled source edges to turn C'' and C''' into good cycles.

So far, in the worst case, we use a sequence S' with three indels such that $\Delta c_g(S') = 1$. After applying this sequence, all cycles of the graph are good. So, we can either apply Lemmas 4 or 5 to the graph. Using Lemma 4, we end up with a sequence S with 10 operations such that $\Delta c_g(S) = 5$. And, using Lemma 5, we end up with a sequence S with 5 operations such that $\Delta c_g(S) = 3$. □

Lemma 12. *If $G(\mathcal{I})$ has only one bad cycle $C = (o_1^c, t_1^c, \dots, o_k^c, t_k^c)$, with $k > 2$, all other cycles are good, and all cycles of the graph are non-oriented, then it is possible to find a sequence S such that $|S|/\Delta c_g(S) \leq 2$.*

Proof. Suppose that there is another non-oriented cycle $D = (o_1^d, t_1^d, \dots, o_k^d, t_k^d)$, which is a good cycle according to this lemma statement, such that there are interleaving triples (o_i^c, o_j^c, o_k^c) and (o_r^d, o_s^d, o_t^d). Bafna and Pevzner [5] showed that a transposition applied to (o_r^d, o_s^d, o_t^d) creates new cycles C' and D' with the same set of vertices of C and D, respectively, such that C' is an oriented cycle and its oriented triple interleaves with a triple of D'. We can use a sequence S_C of size at most two, using Lemmas 9 and 6, applied to the oriented triple of C' to increase the number of good cycles in the graph (i.e., $\Delta c_g(S_C) \geq 1$). Since the oriented triple interleaves with a triple of D', after the transposition from Lemma 9, the cycle D' is transformed into an oriented cycle D'' [5]. We can then apply Lemma 3 to D''. Then, we have a sequence S with six operations and $\Delta c_g(S) \geq 3$.

If the above condition is not true, Bafna and Pevzner [5] and Oliveira *et al.* [17] showed that there must exist two non-trivial cycles D and E (which must be good cycles according to this lemma statement) such that there is a transposition applied to D and E creating two other cycles D' and E' and turning C into an oriented cycle C', such that the number of balanced cycles remains the same. The cycle C' contains the same set of vertices and edges as C. Bafna and Pevzner [5] also showed that one of the following two cases must occur: (i) one of the cycles D' or E' is also oriented; or (ii) there is an oriented triple in C' such that a transposition applied to this triple will turn D' or E' into an oriented cycle.

In the first case, we can apply Lemma 9 together with Lemma 6 on C', and we can apply Lemma 3 on the other oriented cycle to form a sequence S with $|S|/\Delta c_g(S) \leq 2$.

The second case is similar, since we can use the transposition from Lemma 9 to apply to the oriented triple C' that also turns D' or E' into an oriented cycle. □

Using these results, we present Algorithm 1. At each step, this algorithm increases the number of good cycles in the Labeled Intergenic Breakpoint Graph of the instance, which eventually will be formed only by trivial good cycles and $|\pi^A| + 1 - c_g(\mathcal{I}) = 0$. This means that the source genome was transformed into the target genome. Every iteration of the algorithm uses a sequence of operations S such that $|S|/\Delta c_g(S) \leq 2$. Therefore, in order to obtain $|\pi^A|+1-c_g(\mathcal{I}) = 0$, it takes at most $2(|\pi^A|+1-c_g(\mathcal{I}))$ operations. Using the lower bound of Lemma 2, we conclude that Algorithm 1 is a 4-approximation algorithm.

The time complexity of Algorithm 1 is $O(n^4)$: constructing the Labeled Intergenic Breakpoint Graph $G(\mathcal{I})$ and determining its cycles can be done in $O(n)$ time, since the graph has $O(n)$ vertices and edges; applying indels from Lemmas 6 to 8 or Lemma 11 takes $O(n)$ time; finding all oriented triples for a transposition from Lemma 9 or three source edges for a transposition from

Algorithm 1: 4-approximation algorithm for computing a minimum length sequence of transpositions and indels that transforms $\mathcal{G}_1$ into $\mathcal{G}_2$.

Input: Instance $\mathcal{I} = (\mathcal{G}_1, \mathcal{G}_2)$
Output: A sequence of transpositions and indels that transforms $\mathcal{G}_1$ into $\mathcal{G}_2$

```
1  Let S_β ← ∅
2  while G1 ≠ G2 do
3      if G(I) has an unbalanced or labeled trivial cycle C = (o1, t1) then
4          Use Lemmas 6 or 7 to find a sequence S with |S|/Δc_g(S) ≤ 2.
5      else if G(I) has an oriented good cycle then
6          Find a sequence S, using Lemma 3, such that |S|/Δc_g(S) ≤ 3/2.
7      else if G(I) has a positive clean cycle then
8          Find an insertion φ, using Lemma 8, such that Δc_g(φ) = 1, and let
           S = {φ}.
9      else if G(I) has an oriented bad cycle then
10         Find a sequence S, such that |S|/Δc_g(S) ≤ 2, using Lemmas 9 and 6.
11     else if G(I) has at least two non-trivial bad cycles then
12         Find a sequence S, such that |S|/Δc_g(S) ≤ 2, using Lemmas 10 and 6.
13     else if G(I) has only one non-oriented bad cycle then
14         Find a sequence S, such that |S|/Δc_g(S) ≤ 2, using Lemmas 11 or 12.
15     else
16         In this case all cycles are good, so we can use Lemmas 3 to 5 to find a
           sequence S such that |S|/Δc_g(S) ≤ 7/4.
17     Apply S to G1 and append it to S_β.
18 return the sequence S_β
```

Lemmas 10 and 12 takes up to $O(n^3)$ time; and the loop of line 2 iterates up to $O(n)$ times, resulting in an overall time complexity of $O(n^4)$. The Labeled Intergenic Breakpoint Graph $G(\mathcal{I})$ takes $O(n)$ space in memory, since its number vertices and edges are $n + 1$ and $2(n + 1)$, respectively. Since Algorithm 1 only requires to store the sequence of operations applied, which is also $O(n)$, and the graph $G(\mathcal{I})$, altogether Algorithm 1 has space complexity of $O(n)$.

5 Conclusion

In this work, we presented a 4-approximation for the Transposition and Indel Distance problem considering gene order and intergenic region sizes. Our results rely on the properties of a structure called Labeled Intergenic Breakpoint Graph, which is capable of representing both gene order and the distribution of intergenic region sizes of the two input genomes. This is an extension of previous models in the literature that also used some variation of the breakpoint graph, but considered only genomes with the same set of genes. Our algorithm has an improvement on the approximation factor compared to the 4.5-approximation algorithm that used the concept of breakpoints [3].

For future works, we intend to decrease the approximation factor of the algorithm by trying to find better sequences of operations that increase the number of good cycles in the Labeled Intergenic Breakpoint Graph. We intend to expand the model to allow reversals and, consequently, to consider gene orientation. We also plan to implement several algorithms dealing with genome rearrangements considering intergenic regions from the literature (including Algorithm 1), and run diverse experiments in order to evaluate their respective performances.

References

1. Alexandrino, A.O., Brito, K.L., Oliveira, A.R., Dias, U., Dias, Z.: Reversal distance on genomes with different gene content and intergenic regions information. In: Martín-Vide, C., Vega-Rodríguez, M.A., Wheeler, T. (eds.) AlCoB 2021. LNCS, vol. 12715, pp. 121–133. Springer, Cham (2021). https://doi.org/10.1007/978-3-030-74432-8_9
2. Alexandrino, A.O., Oliveira, A.R., Dias, U., Dias, Z.: Genome rearrangement distance with reversals, transpositions, and indels. J. Comput. Biol. **28**(3), 235–247 (2021)
3. Alexandrino, A.O., Oliveira, A.R., Dias, U., Dias, Z.: Incorporating intergenic regions into reversal and transposition distances with indels. J. Bioinform. Comput. Biol. **19**(06), 2140011 (2021)
4. Alexandrino, A.O., Oliveira, A.R., Dias, U., Dias, Z.: Labeled cycle graph for transposition and indel distance. J. Comput. Biol. **29**(03), 243–256 (2022)
5. Bafna, V., Pevzner, P.A.: Sorting by transpositions. SIAM J. Discret. Math. **11**(2), 224–240 (1998)
6. Biller, P., Guéguen, L., Knibbe, C., Tannier, E.: breaking good: accounting for fragility of genomic regions in rearrangement distance estimation. Genome Biol. Evol. **8**(5), 1427–1439 (2016)
7. Biller, P., Knibbe, C., Beslon, G., Tannier, E.: Comparative genomics on artificial life. In: Beckmann, A., Bienvenu, L., Jonoska, N. (eds.) CiE 2016. LNCS, vol. 9709, pp. 35–44. Springer, Cham (2016). https://doi.org/10.1007/978-3-319-40189-8_4
8. Braga, M.D., Willing, E., Stoye, J.: Double cut and join with insertions and deletions. J. Comput. Biol. **18**(9), 1167–1184 (2011)
9. Brito, K.L., Jean, G., Fertin, G., Oliveira, A.R., Dias, U., Dias, Z.: Sorting by genome rearrangements on both gene order and intergenic sizes. J. Comput. Biol. **27**(2), 156–174 (2020)
10. Brito, K.L., Oliveira, A.R., Alexandrino, A.O., Dias, U., Dias, Z.: A new approach for the reversal distance with indels and moves in intergenic regions. In: Jin, L., Durand, D. (eds.) Comparative Genomics (RECOMB CG 2022). LNCS, vol. 13234, pp. 205–220. Springer, Cham (2022)
11. Bulteau, L., Fertin, G., Tannier, E.: Genome rearrangements with indels in intergenes restrict the scenario space. BMC Bioinform. **17**(14), 426 (2016)
12. El-Mabrouk, N.: Genome rearrangement by reversals and insertions/deletions of contiguous segments. In: Giancarlo, R., Sankoff, D. (eds.) CPM 2000. LNCS, vol. 1848, pp. 222–234. Springer, Heidelberg (2000). https://doi.org/10.1007/3-540-45123-4_20
13. Hannenhalli, S., Pevzner, P.A.: Transforming cabbage into turnip: polynomial algorithm for sorting signed permutations by reversals. J. ACM **46**(1), 1–27 (1999)
14. Kolman, P., Waleń, T.: Reversal distance for strings with duplicates: linear time approximation using hitting set. In: Erlebach, T., Kaklamanis, C. (eds.) WAOA 2006. LNCS, vol. 4368, pp. 279–289. Springer, Heidelberg (2007). https://doi.org/10.1007/11970125_22

15. Oliveira, A.R., Brito, K.L., Dias, Z., Dias, U.: Sorting by weighted reversals and transpositions. In: Alves, R. (ed.) BSB 2018. LNCS, vol. 11228, pp. 38–49. Springer, Cham (2018). https://doi.org/10.1007/978-3-030-01722-4_4
16. Oliveira, A.R., et al.: Sorting signed permutations by intergenic reversals. IEEE/ACM Trans. Comput. Biol. Bioinf. **18**(6), 2870–2876 (2021)
17. Oliveira, A.R., Jean, G., Fertin, G., Brito, K.L., Dias, U., Dias, Z.: Sorting permutations by intergenic operations. IEEE/ACM Trans. Comput. Biol. Bioinf. **18**(6), 2080–2093 (2021)
18. Willing, E., Zaccaria, S., Braga, M.D., Stoye, J.: On the inversion-indel distance. BMC Bioinform. **14**, S3 (2013)
19. Yancopoulos, S., Attie, O., Friedberg, R.: Efficient sorting of genomic permutations by translocation, inversion and block interchange. Bioinform. **21**(16), 3340–3346 (2005)

An SMT-Based Framework for Reasoning About Discrete Biological Models

Boyan Yordanov[1(✉)], Sara-Jane Dunn[2], Colin Gravill[1], Hillel Kugler[3(✉)], and Christoph M. Wintersteiger[2]

[1] Scientific Technologies, London, UK
byyordanov@gmail.com, colin@scientifictechnologies.com
[2] Microsoft Research, Cambridge, UK
sjdunn@deepmind.com, cwinter@microsoft.com
[3] Bar-Ilan University, Ramat Gan, Israel
hillelk@biu.ac.il

Abstract. We present a framework called the **Reasoning Engine**, which implements Satisfiability Modulo Theories (SMT) based methods within a unified computational environment to address diverse biological analysis problems. The reasoning engine was used to reproduce results from key scientific studies, as well as supporting new research in stem cell biology. The framework utilizes an intermediate language for encoding partially specified discrete dynamical systems, which bridges the gap between high-level domain specific languages (DSLs) and low-level SMT solvers. We provide this framework as open source together with various biological case studies, illustrating the synthesis, enumeration, optimization and reasoning over models consistent with experimental observations to reveal novel biological insights.

1 Introduction

Formulating hypotheses (e.g. as mathematical models) based on experimental observations to study a given system (e.g. biological, physical, chemical, etc.) is at the core of the scientific method. However, deriving mechanistic models that are provably consistent with experimental observations and known system properties remains challenging. To address this challenge while studying cellular decision-making, in [14] a domain-specific language (DSL) was developed to capture partial mechanistic knowledge (e.g. genetic interactions between various transcription factors) that had been uncovered through years of painstaking experimentation, together with various constraints encoding experimental observations about the activity of genes and transcription factors in different chemical contexts. By encoding computational analysis queries as Satisfiability Modulo Theories (SMT) problems, solvable using off-the-shelf tools [4,33], the resulting

S.-J. Dunn is now at DeepMind, but completed this work while at MSR Cambridge.
H. Kugler: Supported by the Horizon 2020 research and innovation programme for the Bio4Comp project under grant agreement number 732482 and by the ISRAEL SCIENCE FOUNDATION (Grant No. 190/19).

M. S. Bansal et al. (Eds.): ISBRA 2022, LNBI 13760, pp. 114–125, 2022.
https://doi.org/10.1007/978-3-031-23198-8_11

methodology [41] allowed consistent models to be synthesized, and the behavior of a system under untested experimental conditions to be predicted [14].

This approach [12,14,15,41] offered several advantages over how biological models had been constructed and analyzed previously. Together with its various extensions [17,38] and related SMT-based methodologies, these techniques have so far been applied to study stem cell decision-making [13,14], sea urchin development [34], neuron maturation [38], epidermal commitment [32], genetic motifs and function [11,26], synthetic biology [42] and DNA computing [43]. In each of these diverse areas, discrete dynamical systems provided a suitable and powerful modeling framework, which together with SMT-based analysis led to novel insights. These similarities suggested that a unified computational reasoning framework would help expand the existing studies, while facilitating future development of the methodology and its application to novel domains. Here, we introduce such a framework called the **Reasoning Engine**, which is made available with many of the case studies listed above as an open source project [44].

The Reasoning Engine was implemented using F# [20,40] with Z3 [33] as a built-in SMT-solver and includes the DSL and reasoning methodologies supporting `RE:IN` [14,41] (available so far only as a stand-alone tool), `RE:SIN` [38] and `RE:MOTE` [11,26], which were previously unreleased. The resulting library [44] can be used to develop novel stand-alone tools and libraries using .NET or can be accessed from within Jupyter [22] or .NET Interactive [1] notebooks. This enables a powerful computational science workflow, where complete case studies including code, text, references and figures are developed together to improve accessibility and reproducibility as demonstrated in the provided case studies. Recently the approach has been used effectively in new stem cell studies modeling the germline stem cells in C. *elegans* [2] and in expanding models of mouse embryonic stem cells initially presented in [13,14].

2 Reasoning Engine

Effectively utilizing SMT-based reasoning requires significant expertise, making this powerful technology inaccessible to researchers working in other domains (e.g. biology, chemistry). DSLs, on the other hand, facilitate the formalization and codification of domain expertise but are limited without built-in analysis algorithms. The Reasoning Engine bridges the gap between DSLs developed specifically for codifying biological knowledge and existing analysis technologies such as SMT solvers (Fig. 1).

At the core of the Reasoning Engine is a data structure and a corresponding DSL (**Reasoning Engine Intermediate Language** or **REIL**) for representing discrete dynamical systems that are entirely or only partially known (Fig. 1). `REIL` provides a convenient compilation target for higher-level DSLs that are more suitable for expressing the systems and constraints emerging from a specific application area. Roughly speaking, a `REIL` program specifies the *problem*

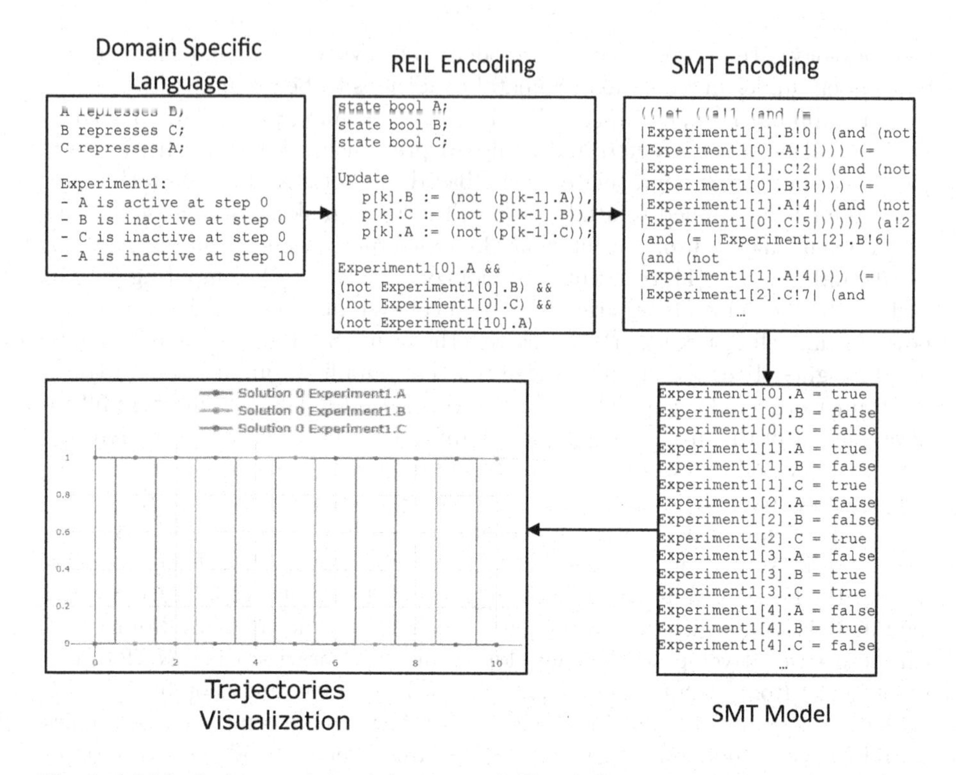

Fig. 1. A biological network involving genes A, B and C is expressed as a DSL program together with some experimental observations. The program is compiled to REIL and is then encoded as an SMT problem (shown partially). An SMT solver decides if this problem is satisfiable, returning a model (shown partially) with concrete variable values, which are used to reconstruct and visualize satisfiable trajectories of the system.

variables of a discrete dynamical system, which can be of integer (`int`), non-negative integer (`nat`) or Boolean (`bool`) type possibly further bounded to a specific range. Each variable is either a *system*, *path* or *state* variable. Path variables are replicated for each trajectory of the system that is considered as part of the analysis and can be used, for example, to represent biological perturbations in a particular experiment, e.g. knockout or over expression of a gene. State variables are replicated for each experiment and at every discrete time step, while system variables are not replicated.

Update rules are defined to represent how state variables change their values over executions of the system and can encode either synchronous or asynchronous, as well as deterministic or non-deterministic updates. Constraints over the variables of the system (e.g. a given state variable at a given step of a named trajectory was "observed" as a specific value) are also part of the REIL program. In the current implementation only finite trajectories are considered, however supporting infinite trajectories with properties specified as temporal logic is a natural extension [18]. Additional details about the REIL syntax and semantics are illustrated in the examples made available together with its implementation [44].

The Reasoning Engine encodes a `REIL` program as an SMT problem using an approach inspired by Bounded Model Checking (BMC) [5,41]. The problem variables from a `REIL` program are encoded as SMT variables of an appropriate type, together with additional constraints to ensure that all variables are indeed in the specified ranges. Here, REIL provides some abstraction over the low-level SMT and allows several encoding strategies to be compared, for example by using either bit-vectors or integers with suitable constraints to encode the same variables. Following the BMC approach, trajectories of the system are "unrolled" for a given finite number of steps, which can be specified as a parameter. The trajectories to be represented explicitly are determined by the constraints specified as part of the `REIL` program (i.e. one trajectory is generated for each *experiment* named in the constraints). A "fresh" copy is created for each state variable for each experiment (trajectory) and a single copy of each path variable is created for each experiment. System variables are unique and are encoded directly without creating any copies. Once trajectories of the system are "unrolled", the constraints can be encoded in a straightforward fashion.

The implementation provided as part of this work [44] uses the Z3 solver [33] through its .NET wrapper as an SMT solver, but other solvers can be targeted using the standardized SMT-LIB representation. Applying the SMT solver produces an SMT model when the problem is satisfiable or a certificate that no solutions exist otherwise. Due to the inherent complexity of the problem, UNKNOWNs may occur for discrete, finite-state systems due to computation limits. Note that the models encoded in REIL are decidable, thus UNKNOWNs are due to timeouts and not due to the problem being undecidable. The SMT model returned by the solver specifies values for all problem variables that make the problem satisfiable, which are then structured to represent a concrete instance of the dynamical system (e.g. biological model) and its trajectories. This concrete model is guaranteed to satisfy all experimental constraints and if the initial model was not completely specified (e.g. some genetic interactions were only hypothesized), then the unknown system parameters and states have now been instantiated. The framework also supports enumeration or reasoning over the complete set of satisfying models. Each solution is represented simply as a map of variables to satisfying values and returned to the higher-level domain-specific tool for further analysis. The domain-specific tool is then responsible for structuring this solution (e.g. to reconstruct the problem structure from the flat variables-to-values map), visualizing the results and presenting them to the researcher to reveal insights about the structure and function of the system under investigation (see Sect. 3).

A schematic figure illustrating how the Reasoning Engine framework bridges the gap between high-level DSLs and low-level SMT-based solvers to support the synthesis and analysis of biological models is shown in (Fig. 2). We also provide a simple, illustrative example of a `REIL` program and representative solutions generated by the Reasoning Engine (Fig. 3). This example is also available as

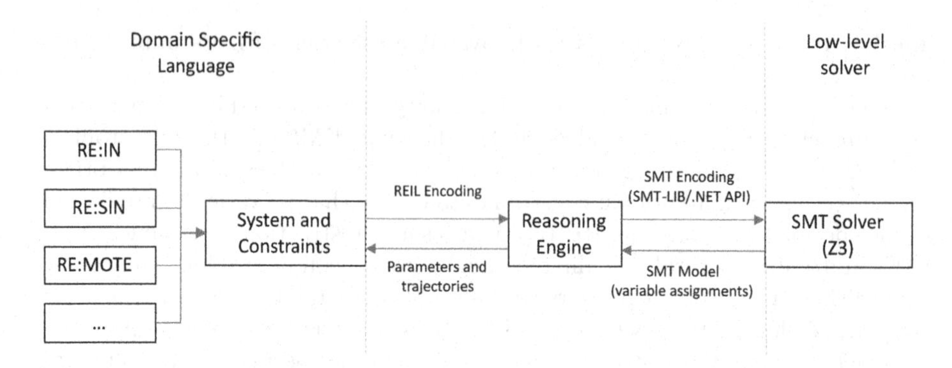

Fig. 2. Reasoning Engine workflow. A DSL is used to formalize and encode domain knowledge. For example, the program is a model of a given biological system together with various constraints based on experimental observations. This description is then transformed into a representation in a common, lower-level modeling language (Reasoning Engine Intermediate Language or REIL), which is further encoded as an SMT-problem and solved using an existing SMT solver. The resulting SMT model provides values for all variables of the problem, which are structured to represent parameters and trajectories of the system being studied.

part of the interactive notebooks and case studies provided as part of the open source project [44].

3 Applications

In this section, we illustrate the compilation of high-level programs into `REIL` and the transformation of solutions back into the domain level. We highlight the biological DSLs and case studies included as part of the Reasoning Engine, where such a reasoning approach has helped uncover novel biological insights.

We first consider the simple example from Fig. 1. The starting point is a high-level program in a "biological" DSL capturing a subset of Boolean genetic networks. The translation to `REIL` defines the semantics of the DSL (i.e. genes are represented as Boolean variables and genetic interactions as update rules) and specifications are translated directly between the Boolean network DSL and `REIL`. The resulting `REIL` program is encoded and solved as an SMT problem as discussed in Sect. 2 and the solution is represented as a map of variables to satisfying values. In this case, the solution specifies the states of the three genes at every step of a satisfying trajectory corresponding to "Experiment1" and can be used to reconstruct a concrete, high-level model or to visualize the findings.

Figure 1 provides a simple example of synthesis (i.e. find concrete parameter values or system trajectories that satisfy a given specification) but other analysis problems including enumeration of consistent models, optimization and the rigorous testing of predictions against the entire set of consistent models are also supported by the framework [12,14,15,41]. In the following sub-sections we briefly highlight the extended features of the Reasoning Engine framework

a

```
unique state int temperature;
unique state bool heatIsOn;
update
    p[k].temperature :=
        if (p[k-1].heatIsOn) then
               (p[k-1].temperature + 1)
        else
               (p[k-1].temperature - 1),
    p[k].heatIsOn := p[k-1].temperature < 18;

#test[0].temperature < 20 & #test[0].temperature >15;
#test[10].temperature < 18;
```

b

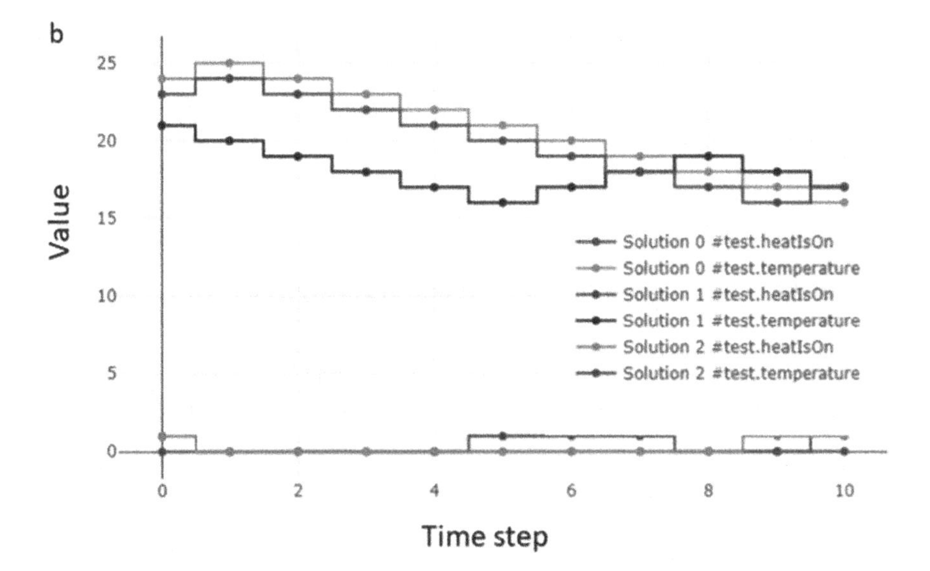

Fig. 3. A basic temperature control system model in REIL. (a) We assume that the temperature increases (decreases) by one degree per time step whenever the heating is on (off). To capture this, we use two state variables: an integer `temperature` (since no continuous quantities are currently supported) and a Boolean `heatIsOn`. We define the system dynamics such that the heating is switched on whenever the temperature drops below 18°. We are interested in checking whether temperatures below 18° can be reached at time step 10 if the initial room temperature is somewhere between 20 and 25 degrees. (b) Using the Reasoning Engine, we find out that the constraints are satisfiable and generate all three different solutions, which correspond to different executions of the system. Additional details and other examples are available in [44].

through biological examples from case studies published previously. Detailed versions of these case studies were prepared as computational notebooks and are made available as part of the Reasoning Engine project [44].

Uncovering Biological Programs. In [14], an SMT-based analysis approach called `RE:IN` was developed and used to uncover the minimal *biological program*

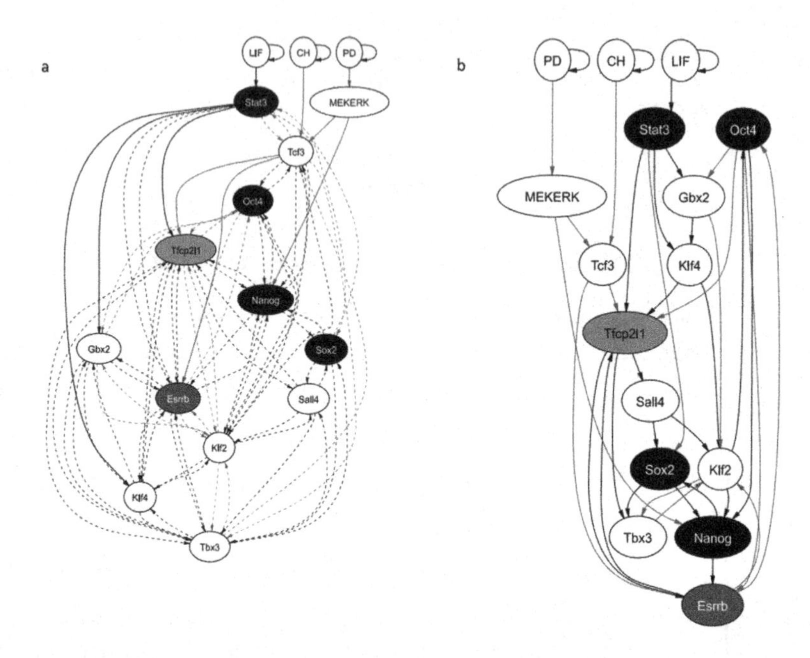

Fig. 4. Synthesizing models of pluripotency in mouse embryonic stem cells. (a) an abstract model of the pluripotency network in mouse embyonic stem cells as developed in [14], dashed interactions represent uncertainty about the existence of certain interactions (possible interactions), which the SMT solver can resolve based on experimental constraints specifying system dynamics. (b) Example of a concrete network synthesized, all interactions are now definite, represented as solid lines, here one solution out of potentially a large set of different networks is displayed.

governing stem cell pluripotency and predict the effects of genetic perturbations, leading to novel biological insights that were then validated experimentally [13,14]. The `RE:IN` DSL is similar to the one from Fig. 1 but can represent partially known genetic networks, where interactions are either *definite* or *possible*. Definite interactions are well characterized and have strong experimental support, while possible interactions are hypothesized and may have only partial experimental support. To replicate the main results from [14] within the Reasoning Engine framework presented here, see Fig. 4, we define a `RE:IN` DSL that compiles to `REIL` (see Sect. 2). Roughly speaking, we introduce additional choice system variables for possible interactions and *regulation conditions* (how different regulators combine to affect a target) as part of the `REIL` translation, which also leads to more biologically plausible update rules. The resulting network with definite and possible interactions, and set of regulation conditions for each component is called an Abstract Boolean Network (ABN).

We make available the case study replicating the main results from [14] as the `RE:IN` notebook example in [44]. This notebook illustrates how an initial ABN and experimental observations is loaded as a RE:IN program and how it is analyzed to identify required and disallowed interactions, find minimal satisfying models or formulate and test predictions using the Reasoning Engine. The notebook infrastructure is particularly suitable for such analysis and allows for different genetic networks and results to be visualized and extended with text, explanations and references besides archiving the code used as part of the study.

We also provide a notebook that replicates the main results from [41] as part of [44]. These examples illustrate how the RE:IN methodology within the Reasoning Engine can help synthesize and analyze both synchronous and asynchronous Boolean network models of biological systems and identify previously unknown signals, transcription factors or interactions. These case studies are based on existing biological models of myeloid progenitor differentiation [25], the cell cycle [30] and the murine cardiac gene regulatory network [21].

Biological Switching Networks. When studying how known genetic interactions [39] affect neuron maturation and fate specification in the mammalian cortex, it was observed that the fixed network interactions assumption from [14,41] was too restrictive for this biological system. An extension called `RE:SIN` (Reasoning Engine for Switching Interaction Networks) was proposed [38] to allow the encoding of biological networks with different topology for different cell types observed during fate specification. The `RE:SIN` approach was later used to study how stem cells differentiate in the epidermis (epidermal commitment) [32], revealing that a switching network might be required in this system.

As part of the Reasoning Engine project [44], we provide a `RE:SIN` DSL, which compiles to `REIL`, enabling the synthesis and analysis of switching biological networks. The compilation process is similar to the one used for `RE:IN` but fresh copies of all choice variables are created for each individual cell type and additional state variables are encoded to represent the cell type along system executions. We used the Reasoning Engine to reproduce the main results from [32] and make this case study available as the `RE:SIN` notebook example in [44].

Biological Network Motifs. The constraints supported by `RE:IN` or `RE:SIN` were limited in terms of the ability to reason jointly about the frequently appearing structural patterns and their influence on the function of biological networks. Yet, it is known that a recurring set of *network motifs* serve as building blocks of diverse biological networks, give rise to certain dynamical behaviors and are important for understanding complex biological programs [31]. An extension of the previous methodologies called `RE:MOTE` (Reasoning Engine for Interaction Networks with Motif Constraints) was proposed in [11,26] to enable reasoning over the requirement for specific network motifs as a way of explaining how certain dynamical behaviors arise. The `RE:MOTE` DSL which compiles to `REIL` is made available as part of the Reasoning Engine project [44]. The compilation of the genetic networks encoded as part of a `RE:MOTE` program is the same as for `RE:IN` programs but additional constrains specifying the presence or absence of certain network sub-structures (i.e. motifs) are supported as part of the DSL and are compiled to suitable SMT constraints (see [11,26] for details). We reproduce the main results from [11,26] as part of the Reasoning Engine and make this case study available as an example in [44]. As part of this notebook, we analyze the motifs involved in typical biological behaviors such as sign-sensitive delay and pulse generation and demonstrate the scalability and biological relevance of the approach by studying the motifs required within other gene interaction networks.

4 Related Work

Synthesis approaches for biological modeling are becoming an important area of research and applications, see for example [8,9,19,23,24,35–37] and references within. A number of verification tools including [6,7,10,27,28,33] have been used to analyze biological models, many of which support translating high-level biological models to the input format of the verification solver. Compared to many of these approaches, here we use a simplified intermediate language as a common compilation target for several DSLs expressing different biological problems. This allows us to reproduce the results from a large number of published biological studies, which we make available as interactive computational notebooks. Verification tools for non biological applications also often use a common intermediate language, e.g. [3,16,29], future work can explore translations to such formats opening the way to the application and evaluation of powerful verification tools.

5 Conclusion

Several DSLs useful for reasoning about interaction networks and cellular decision-making are made available as part of the current framework [44]. Developing novel languages for additional domains amounts to defining the syntax, implementing a compiler targeting the intermediate `REIL` language (thereby defining the semantics) and providing a method for reconstructing the domain structure from the flat variables-to-values maps returned by the solver.

The Reasoning Engine framework introduced here could facilitate the application of formal verification techniques in biology and other domains of the computational sciences where discrete dynamical systems provide a suitable modeling framework. It allows researchers to encode their domain knowledge and problems using DSLs without having to develop specialized analysis algorithms. The approach is also beneficial for researchers developing novel tactics, simplifications or verification techniques, which can be also implemented directly as part of the Reasoning Engine, making them available for all application domains. Once integrated as part of the Reasoning Engine, novel analysis techniques could be tested on the range of biological examples available as part of the framework.

Acknowledgments. Hillel Kugler's research was supported by the Horizon 2020 research and innovation programme for the Bio4Comp project under grant agreement number 732482 and by the ISRAEL SCIENCE FOUNDATION (Grant No. 190/19).

References

1. .NET interactive (2022). https://github.com/dotnet/interactive
2. Amar, A., Hubbard, E.J.A., Kugler, H.: Modeling the C. elegans germline stem cell genetic network using automated reasoning. Biosystems **217**, 104672 (2022). https://doi.org/10.1016/j.biosystems.2022.104672, https://www.sciencedirect.com/science/article/pii/S0303264722000624
3. Barnett, M., Chang, B.-Y.E., DeLine, R., Jacobs, B., Leino, K.R.M.: Boogie: a modular reusable verifier for object-oriented programs. In: de Boer, F.S., Bonsangue, M.M., Graf, S., de Roever, W.-P. (eds.) FMCO 2005. LNCS, vol. 4111, pp. 364–387. Springer, Heidelberg (2006). https://doi.org/10.1007/11804192_17
4. Barrett, C., Stump, A., Tinelli, C.: The SMT-LIB standard - version 2.0. In: Proceedings of the 8th International Workshop on Satisfiability Modulo Theories (SMT'10), Edinburgh, Scotland (2010)
5. Biere, A., Cimatti, A., Clarke, E., Zhu, Y.: Symbolic model checking without BDDs. In: Cleaveland, W.R. (ed.) TACAS 1999. LNCS, vol. 1579, pp. 193–207. Springer, Heidelberg (1999). https://doi.org/10.1007/3-540-49059-0_14
6. Caspi, P., Halbwachs, N., Pilaud, D., Plaice, J.: LUSTRE: a declarative language for programming synchronous systems, pp. 178–188 (1987)
7. Cavada, R., et al.: The nuXmv symbolic model checker. In: Biere, A., Bloem, R. (eds.) CAV 2014. LNCS, vol. 8559, pp. 334–342. Springer, Cham (2014). https://doi.org/10.1007/978-3-319-08867-9_22
8. Chevalier, S., Froidevaux, C., Paulevé, L., Zinovyev, A.: Synthesis of Boolean networks from biological dynamical constraints using answer-set programming. In: 2019 IEEE 31st International Conference on Tools with Artificial Intelligence (ICTAI), pp. 34–41 (2019). https://doi.org/10.1109/ICTAI.2019.00014
9. Chevalier, S., Noël, V., Calzone, L., Zinovyev, A., Paulevé, L.: Synthesis and simulation of ensembles of Boolean networks for cell fate decision. In: Abate, A., Petrov, T., Wolf, V. (eds.) CMSB 2020. LNCS, vol. 12314, pp. 193–209. Springer, Cham (2020). https://doi.org/10.1007/978-3-030-60327-4_11
10. Cimatti, A., Clarke, E., Giunchiglia, F., Roveri, M.: NUSMV: a new symbolic model checker. Int. J. Softw. Tools Technol. Transfer **2**(4), 410–425 (2000). https://doi.org/10.1007/s100090050046
11. Dunn, S., Kugler, H., Yordanov, B.: Formal analysis of network motifs links structure to function in biological programs. IEEE/ACM Trans. Comput. Bio. Bioinform. **18**, 261–271 (2019)
12. Dunn, S.J.: Automated formal reasoning to uncover molecular programs of self-renewal. In: Cahan, P. (ed.) Computational Stem Cell Biology. MMB, vol. 1975, pp. 79–105. Springer, New York (2019). https://doi.org/10.1007/978-1-4939-9224-9_4
13. Dunn, S.J., Li, M.A., Carbognin, E., Smith, A.G., Martello, G.: A common molecular logic determines embryonic stem cell self-renewal and reprogramming. EMBO J. **38**, e100003 (2018)
14. Dunn, S.J., Martello, G., Yordanov, B., Emmott, S., Smith, A.: Defining an essential transcription factor program for naïve pluripotency. Science **344**(6188), 1156–1160 (2014)
15. Dunn, S.-J., Yordanov, B.: Automated reasoning for the synthesis and analysis of biological programs. In: Liò, P., Zuliani, P. (eds.) Automated Reasoning for Systems Biology and Medicine. CB, vol. 30, pp. 37–62. Springer, Cham (2019). https://doi.org/10.1007/978-3-030-17297-8_2

16. Filliâtre, J.C., Paskevich, A.: Why3 — where programs meet provers. In: Felleisen, M., Gardner, P. (eds.) ESOP 2013. LNCS, vol. 7792, pp. 125–128. Springer, Heidelberg (2013). https://doi.org/10.1007/978-3-642-37036-6_8
17. Goldfeder, J., Kugler, H.: BRE:IN - a backend for reasoning about interaction networks with temporal logic. In: Bortolussi, L., Sanguinetti, G. (eds.) CMSB 2019. LNCS, vol. 11773, pp. 289–295. Springer, Cham (2019). https://doi.org/10.1007/978-3-030-31304-3_15
18. Goldfeder, J., Kugler, H.: Temporal logic based synthesis of experimentally constrained interaction networks. In: Chaves, M., Martins, M.A. (eds.) MLCSB 2018. LNCS, vol. 11415, pp. 89–104. Springer, Cham (2019). https://doi.org/10.1007/978-3-030-19432-1_6
19. Guziolowski, C., et al.: Exhaustively characterizing feasible logic models of a signaling network using answer set programming. Bioinformatics **29**(18), 2320–2326 (2013)
20. Harrop, J.: F# for Scientists. Wiley, Hoboken (2011)
21. Herrmann, F., Groß, A., Zhou, D., Kestler, H.A., Kühl, M.: A Boolean model of the cardiac gene regulatory network determining first and second heart field identity. PLOS ONE **7**(10), 1–10 (2012). https://doi.org/10.1371/journal.pone.0046798
22. Kluyver, T., et al.: Jupyter notebooks - a publishing format for reproducible computational workflows. In: Loizides, F., Scmidt, B. (eds.) Positioning and Power in Academic Publishing: Players, Agents and Agendas, pp. 87–90. IOS Press (2016). https://eprints.soton.ac.uk/403913/
23. Koksal, A.: Program Synthesis for Systems Biology. Ph.D. thesis, University of California at Berkeley (2018), Technical report No. UCB/EECS-2018-49
24. Koksal, A., Pu, Y., Srivastava, S., Bodik, R., Fisher, J., Piterman, N.: Synthesis of biological models from mutation experiments. In: SIGPLAN-SIGACT Symposium on Principles of Programming Languages. ACM (2013)
25. Krumsiek, J., Marr, C., Schroeder, T., Theis, F.J.: Hierarchical differentiation of myeloid progenitors is encoded in the transcription factor network. PLOS ONE **6**(8), 1–10 (2011). https://doi.org/10.1371/journal.pone.0022649
26. Kugler, H., Dunn, S.-J., Yordanov, B.: Formal analysis of network motifs. In: Češka, M., Šafránek, D. (eds.) CMSB 2018. LNCS, vol. 11095, pp. 111–128. Springer, Cham (2018). https://doi.org/10.1007/978-3-319-99429-1_7
27. Kwiatkowska, M., Norman, G., Parker, D.: PRISM 4.0: verification of probabilistic real-time systems. In: Gopalakrishnan, G., Qadeer, S. (eds.) CAV 2011. LNCS, vol. 6806, pp. 585–591. Springer, Heidelberg (2011). https://doi.org/10.1007/978-3-642-22110-1_47
28. Larsen, K.G., Pettersson, P., Yi, W.: UPPAAL in a nutshell. Int. J. Softw. Tools Technol. Transfer **1**(1–2), 134–152 (1997)
29. Le, H.M., Große, D., Herdt, V., Drechsler, R.: Verifying SystemC using an intermediate verification language and symbolic simulation. In: Proceedings of the 50th Annual Design Automation Conference, pp. 1–6 (2013)
30. Li, F., Long, T., Lu, Y., Ouyang, Q., Tang, C.: The yeast cell-cycle network is robustly designed. Proc. Nat. Acad. Sci. **101**(14), 4781–4786 (2004). https://doi.org/10.1073/pnas.0305937101https://www.pnas.org/content/101/14/4781
31. Milo, R., Shen-Orr, S., Itzkovitz, S., Kashtan, N., Chklovskii, D., Alon, U.: Network motifs: simple building blocks of complex networks. Science **298**(5594), 824–827 (2002)
32. Mishra, A., et al.: A protein phosphatase network controls the temporal and spatial dynamics of differentiation commitment in human epidermis. Elife **6**, e27356 (2017). https://doi.org/10.7554/eLife.27356

33. de Moura, L., Bjørner, N.: Z3: an efficient SMT solver. In: Ramakrishnan, C.R., Rehof, J. (eds.) TACAS 2008. LNCS, vol. 4963, pp. 337–340. Springer, Heidelberg (2008). https://doi.org/10.1007/978-3-540-78800-3_24
34. Paoletti, N., Yordanov, B., Hamadi, Y., Wintersteiger, C.M., Kugler, H.: Analyzing and synthesizing genomic logic functions. In: Biere, A., Bloem, R. (eds.) CAV 2014. LNCS, vol. 8559, pp. 343–357. Springer, Cham (2014). https://doi.org/10.1007/978-3-319-08867-9_23
35. Razzaq, M., Kaminski, R., Romero, J., Schaub, T., Bourdon, J., Guziolowski, C.: Computing diverse Boolean networks from phosphoproteomic time series data. In: Češka, M., Šafránek, D. (eds.) CMSB 2018. LNCS, vol. 11095, pp. 59–74. Springer, Cham (2018). https://doi.org/10.1007/978-3-319-99429-1_4
36. Rosenblueth, D.A., Muñoz, S., Carrillo, M., Azpeitia, E.: Inference of Boolean networks from gene interaction graphs using a SAT solver. In: Dediu, A.-H., Martín-Vide, C., Truthe, B. (eds.) AlCoB 2014. LNCS, vol. 8542, pp. 235–246. Springer, Cham (2014). https://doi.org/10.1007/978-3-319-07953-0_19
37. Sharan, R., Karp, R.M.: Reconstructing Boolean models of signaling. J. Comput. Biol. **20**(3), 249–257 (2013)
38. Shavit, Y., et al.: Automated synthesis and analysis of switching gene regulatory networks. Biosystems **146**, 26–34 (2016)
39. Srinivasan, K., et al.: A network of genetic repression and derepression specifies projection fates in the developing neocortex. Proc. Nat. Acad. Sci. **109**(47), 19071–19078 (2012) https://doi.org/10.1073/pnas.1216793109, https://www.pnas.org/content/109/47/19071
40. Syme, D.: The early history of F#. Proc. ACM Program. Lang. **4**(75), 1–58 (2020)
41. Yordanov, B., Dunn, S.J., Kugler, H., Smith, A., Martello, G., Emmott, S.: A method to identify and analyze biological programs through automated reasoning. NPJ Syst. Bio. Appl. **2**(16010), 1–16 (2016)
42. Yordanov, B., Wintersteiger, C.M., Hamadi, Y., Kugler, H.: SMT-based analysis of biological computation. In: Brat, G., Rungta, N., Venet, A. (eds.) NFM 2013. LNCS, vol. 7871, pp. 78–92. Springer, Heidelberg (2013). https://doi.org/10.1007/978-3-642-38088-4_6
43. Yordanov, B., Wintersteiger, C.M., Hamadi, Y., Phillips, A., Kugler, H.: Functional analysis of large-scale DNA strand displacement circuits. In: Soloveichik, D., Yurke, B. (eds.) DNA 2013. LNCS, vol. 8141, pp. 189–203. Springer, Cham (2013). https://doi.org/10.1007/978-3-319-01928-4_14
44. Yordanov, B., Gravill, C., Dunn, S.J., Kugler, H., Wintersteiger, C.M.: Reasoning engine (2022). https://github.com/fsprojects/ReasoningEngine

ARGLRR: An Adjusted Random Walk Graph Regularization Sparse Low-Rank Representation Method for Single-Cell RNA-Sequencing Data Clustering

Zhen-Chang Wang, Jin-Xing Liu, Jun-Liang Shang, Ling-Yun Dai, Chun-Hou Zheng, and Juan Wang(✉)

School of Computer Science, Qufu Normal University, Rizhao 276826, China
wangjuansdu@163.com

Abstract. Researchers may now explore biological concerns at the cell level because of the advancement of single-cell transcriptome sequencing technologies. One of the primary applications of single-cell RNA-seq (scRNA-seq) data is to identify cell types by clustering to reveal cell heterogeneity. However, due to characteristics such as higher noise and lesser coverage of scRNA-seq, the accuracy of existing clustering methods is compromised. Here, we propose a method called Adjusted Random walk Graph regularization Sparse Low-Rank Representation (ARGLRR), a practical sparse subspace clustering method, to identify cell types. The basic Low-Rank Representation (LRR) model focuses primarily on the global structure of data. We add adjusted random walk graph regularization to the framework of LRR, which makes up for the lack of local structure capture capability of LRR. With this method, the local and global structure of the scRNA-seq data will be captured. By imposing the similarity constraint on the LRR model, the cell-to-cell similarity estimation process further enhances the capacity of the proposed model to capture the global structural relationships between cells. The results on nine published scRNA-seq datasets demonstrate that ARGLRR outperforms other advanced comparison methods. Our method improves 6.99% and 5.85% over the best-performing comparison method in NMI and ARI metrics on the scRNA-seq datasets clustering experiments, respectively. We also use UMAP to visualize the learned similarity matrix and find that the similarity matrix obtained by ARGLRR improves the separation of cell types.

Keywords: Manifold graph regularization · Low-rank representation · Random walk · Cell type identification · Spectral clustering

1 Introduction

The rapid and consistent advancement of next-generation sequencing technologies promotes the generation of numerous single-cell related data. Unsupervised learning of Biologically relevant information and identification of cell types from these single-cell RNA sequencing (scRNA-seq) datasets have become a popular research topic [1]. The

M. S. Bansal et al. (Eds.): ISBRA 2022, LNBI 13760, pp. 126–137, 2022.
https://doi.org/10.1007/978-3-031-23198-8_12

clustering of scRNA-seq datasets provides an efficient tool for identifying cell types from a significant number of cell sequencing data. However, cell heterogeneity characteristics and contamination caused by transcriptional processes expose clustering experiments to many disturbances. Unanticipated non-linear distortions occur during the transcription process, and some noise enters and contaminates the data during sequencing [2]. This leads to the usual 'dropout' occurrence, where a gene is identified in one cell but not in another with low expression levels. The interference caused by the aforementioned characteristics, as well as the existence of noise, make clustering of scRNA-seq data extremely difficult.

Due to the above factors, the selection of an appropriate clustering approach for scRNA-seq data clustering is quite challenging. With this in mind, numerous academics have introduced their strategies for coping with the problem. One of the most representative methods is the improved method based on low-rank representation (LRR) [3]. By imposing low-rank structures on the similarity matrix, Zheng et al. developed a similarity learning-based scRNA-seq cell type determination approach SinNLRR [4]. Zhang et al. proposed the SCCLRR [5] approach by combining normalized Euclidean distance and cosine similarity and using them to balance the intrinsic manifold in the local regularization term.

The above LRR and LRR-based approaches mainly focus on the data's global structure. Thus, they overlook the local geometrical information of the data which is crucial for subspace clustering [6]. To solve this issue, Lu et al. [7] combine the graph regularization with the LRR approach to encode the local structure information of data for hyperspectral image destriping. Du et al. proposed a facial recognition method called GLRSRR [8] by merging LRR and graphical regularization terms. The successful applications in these fields have inspired us to combine LRR with graph regularization on single-cell type detection.

In this paper, we impose the similarity constraint with adjusted random walk graph regularization constraints on the LRR model and propose a single-cell clustering framework called the Adjusted Random Walk Graph Regularization Sparse Low-Rank Representation (ARGLRR) model. This enables the model to capture local and global structure information of scRNA-seq data. ARGLRR introduces the adjusted random walk graph regularization, enabling graph regularization to be more robust to outliers, and improving the effect of graph regularization on smoothing manifold data. The adjusted random walk algorithm is used to update the weights between two vertices in the graph, which can obtain the adjacency matrix that fits the original structure. This enables the proposed algorithm to better reflect the potential relationships of local data points.

Previous studies [9] have demonstrated that calculating the distance or similarity matrix between cells is considered a key step in cell type identification. Hence, we also introduce the similarity constraint. Similarity constraints provide the method for better learning of similarity information between cells and hence better learning of the global structure of the data. We apply ARGLRR to the clustering experiments for performance validation. The experimental results on nine popular scRNA-seq datasets and seven related comparison methods comprehensively demonstrate the efficacy and robustness of ARGLRR in subspace clustering. Moreover, we utilize UMAP to generate 2-D visualization graphs to visualize what subpopulations the cells are differentiated into.

The visualization of scRNA-seq data further demonstrates the superior cell clustering performance of our method.

2 Related Work

In this section, we will go through Low-rank representation, Graph Regularization Based on Manifold, and Random Walk in more detail.

2.1 Low-Rank Representation (LRR)

LRR attempts to represent all vectors as the lowest rank representation of a linear combination of the data. Given a single-cell data matrix $\mathbf{X} = [\mathbf{X}_1, \mathbf{X}_2, ..., \mathbf{X}_n] \in \mathbf{R}^{m \times n}$, m and n are defined as the number of data samples and features, respectively. According to the low-rank model theory [10], data $\mathbf{X}$ originates from multiple single-subspace. Therefore, the matrix $\mathbf{X}$ is a set of observations from multiple low-dimensional subspaces. The object of LRR can be represented as:

$$\min_{\mathbf{Z}} \|\mathbf{Z}\|_* \qquad s.t.\ \mathbf{X} = \mathbf{AZ} \tag{1}$$

Specifically, $\mathbf{A} = [\mathbf{A}_1, \mathbf{A}_2, ..., \mathbf{A}_n]$ is often referred to as the dictionary matrix for representing the projection basis of the high-dimensional space to the low-dimensional space. $\mathbf{Z} = [\mathbf{Z}_1, \mathbf{Z}_2, ..., \mathbf{Z}_n]$ is the coefficient matrix. $\|\cdot\|_*$ represents the nuclear norm.

Choosing the matrix $\mathbf{X}$ as the dictionary can segment the data accurately. In real applications, the observation data are inevitably noisy. Considering possible noise and improving the robustness of the LRR, the improved formula can be defined as follows:

$$\min_{\mathbf{Z},\mathbf{E}} \|\mathbf{Z}\|_* + \gamma \|\mathbf{E}\|_l \qquad s.t.\ \mathbf{X} = \mathbf{XZ} + \mathbf{E}, \mathrm{Z} \geq 0 \tag{2}$$

Here, $\mathbf{E}$ is the error term that usually modeling different noises or outliers, $\lambda > 0$ is a parameter that balances the noise term $\mathbf{E}$ and low-rank representation $\mathbf{Z}$ according to the adjustment of the value.

2.2 Graph Regularization Based on Manifold

High-dimensional data usually has the property of residing on low-dimensional manifolds. It is quite challenging to obtain the complete local structure of the data manifolds and requires a vehicle such as a graph. Under the assumption of local invariance [11], two data points that are near in the original geometry structure in place will remain close in the reconstructed graph [12]. Research in manifold learning theory [13] has further shown that building a nearest neighbor graph on data points is an effective way to model local geometric structures. Thus, we choose the K-nearest neighbor (KNN) graph to model the manifold structure. In the K-NN graph, edges correspond to links to the neighbor vertex.

The data matrix $\mathbf{X}$ can be modeled by a graph $G = (V, E)$, which V is the set of the vertices and E is the set of the edges connecting two vertices in V. To this end, the weight of the edge connecting vertices, in general, can be defined as the weight matrix $\mathbf{W} \in \mathbf{R}^{n \times n}$. 0–1 weighting and heat kernel weighting, etc. are frequently chosen to construct the weight matrix. More accurately, the formula of the graph regularization is written as $tr(\mathbf{ZLZ}^T)$. $tr(\cdot)$ is the trace function of the data matrix. The final degree matrix $\mathbf{D}$ is a diagonal matrix, $\mathbf{L}$ denotes the graph Laplacian matrix based on the manifold and $\mathbf{L} = \mathbf{D} - \mathbf{W}$.

2.3 Random Walk

Random walk is a mathematical concept that essentially depicts a path that is formed by a sequence of random steps. The random walk is used to learn the structure of the data by randomly walking over the network until it reaches a labeled node and to compute similarities between nodes. During the walk on the graph, the random walker considers the many possible ways to reach a point in the graph and the statistical probability that it will eventually reach that point. Therefore, one of the most challenging tasks in random walking is to construct an accurate transfer probability matrix, which affects the likelihood that the random walker will transfer across nodes.

Relying on comprehensive consideration of global geometrical structure information, the random walk has become an effective measure of the nonlinear manifold [14]. The classical random walk transition matrix on a weighted undirected graph is calculated as: $\mathbf{P}_{RW} = \mu \mathbf{P}^t \mathbf{W} + (1 - \mu)\mathbf{P}$, where $\mathbf{P}$ is the initial transition probability and $\mathbf{W}$ is the graph adjacency matrix.

3 Adjusted Random Walk Graph Regularization Sparse Low-Rank Representation (ARGLRR)

In this section, we will provide a detailed description of the ARGLRR. Our goal is to efficiently exploit the global and local structure information of scRNA-seq data. We introduce the adjusted random walk graph regularization and similarity constraint to enhance the performance of LRR.

3.1 Algorithm Framework

As mentioned in the previous section, LRR is often used to obtain the global information of data but ignores the local structure of data. We attempt to apply the adjusted random walk to graph regularization to accurately obtain a feasible local manifold structure for the scRNA-seq data. By constructing an adjacency matrix as an undirected graph, we can better explain the initial relationship of data points. In the previous manifold learning theory [13], the nearest neighbor graph has been proved to be an efficient way to model local geometric structures. A data sample can be represented by its nearest neighbor samples using the attributes of the manifold structure. Therefore, we construct K-Nearest Neighbor (KNN) graphs to encode geometric structures. Using the grid search approach, we get the optimal number of neighbors and set the value of K to 5. We present adjusted random walk graph regularization and apply the adjusted random walk technique in the construction of KNN graphs to update the weights. We treat the data's basic clusters as graph nodes and define the weight of the connected edges of vertices i and j in the graph as follows.

$$\mathbf{w}_{ij} = \begin{cases} 1, & \textit{if } \mathbf{x}_i \in \mathbf{N}_K(\mathbf{x}_j) \textit{ or } \mathbf{x}_i \in \mathbf{N}_K(\mathbf{x}_j) \\ 0, & \textit{otherwises} \end{cases} \tag{3}$$

$\mathbf{N}_K(\mathbf{x}_i)$ denotes the set of k nearest neighbors of $\mathbf{x}_i$.

The adjusted random walk enables a more practical similarity measure to be calculated between two vertices. On manifold distributed data, this makes the random walk more robust and accurate. When the adjusted weight matrix $\mathbf{W}$ as the initial value of calculating $\mathbf{P}_{RW}$, the weight matrix between $\mathbf{X}_i$ and $\mathbf{X}_j$ can be derived as:

$$\hat{\mathbf{W}}_{ij} = \frac{(\mathbf{P}_{RW})_{ij} + (\mathbf{P}_{RW})_{ji}}{2} \tag{4}$$

The transition matrix is $\mathbf{P}_{RW} = \mu\mathbf{P}^t\mathbf{W} + (1 - \mu)\mathbf{P}$. Following the completion of the initial adjacency matrix, we integrate the scale information to improve the modeling effect. Specifically, the adjustment measure is to restrict the weight matrix $\hat{\mathbf{W}}_{ij}$ obtained by random wandering updates by using the original weight matrix $\mathbf{W}$. We define the adjusted weight matrix as $\mathbf{B}$.

$$\mathbf{B} = \mathbf{W} \odot \hat{\mathbf{W}}_{ij} \tag{5}$$

Here, $\odot$ is the Hadamard Product operator. The restricted weight matrix becomes sparser, preserving the local structure better. Furthermore, the derivation of the mathematical formula of the random walk graph regularization is as follows:

$$\begin{aligned} &\min_{\mathbf{Z}} \Sigma_{ij} \left\| \mathbf{z}_i - \mathbf{z}_j \right\|^2 \mathbf{B}_{ij} \\ &= \min_{\mathbf{Z}} tr\left(\mathbf{Z}(\mathbf{D} - \mathbf{B})\mathbf{Z}^T\right) \\ &= \min_{\mathbf{Z}} tr\left(\mathbf{Z}\mathbf{L}^{ARW}\mathbf{Z}^T\right) \end{aligned} \tag{6}$$

Here, $\mathbf{L}^{ARW}$ is defined as the graph Laplacian matrix in the adjusted random walk condition and $\mathbf{B}_{ij}$ is the weights of nodes $\mathbf{x}_i$ and $\mathbf{x}_j$ in the graph.

Suppose $\mathbf{Z}$ denotes the low-rank representation matrix, λ_1, λ_2, β, γ are adjustable balance parameters, $\mathbf{S}$ is the symmetric similarity matrix, $\|\cdot\|_F^2$ is matrix Frobenius norm, $\mathbf{E} \in \mathbf{R}^{m \times n}$ denotes the error term, $\|\cdot\|_1$ is the l_1-norm. The objective function of ARGLRR is defined as:

$$\begin{aligned} &\min_{Z,E} \|\mathbf{Z}\|_* + \lambda_1 \|\mathbf{Z}\|_1 + \beta tr\left(\mathbf{Z}\mathbf{L}^{ARW}\mathbf{Z}^T\right) + \gamma \|\mathbf{Z} - \mathbf{S}\|_F^2 + \lambda_2 \|\mathbf{E}\|_1 \\ &\qquad s.t. \quad \mathbf{X} = \mathbf{X}\mathbf{Z} + \mathbf{E} \end{aligned} \tag{7}$$

The similarity matrix $\mathbf{S}$ between pairs of data points is obtained by the Gaussian kernel function.

The framework of the method is shown in Fig. 1. As can be seen from Fig. 1, the main steps of our method are data preprocessing, calculation of the affinity matrix, and using the affinity matrix for spectral clustering and visualization. The final clustering results are then obtained using spectral clustering and visualized.

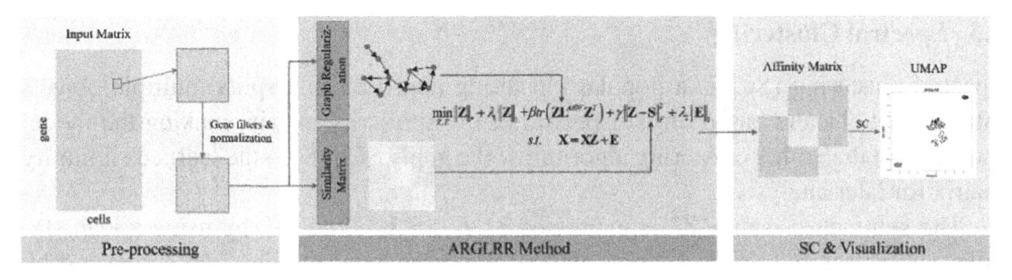

Fig. 1. Framework of ARGLRR.

3.2 Optimization

Because of the highly interdependent between **Z**, the problem (7) not solved directly but iteratively by Linearized Alternating Direction Method with Adaptive Penalty (LADMAP) [15] algorithm.

Given the auxiliary variables **J** to make the objective function separable, meets the condition of $\mathbf{Z} = \mathbf{J}$. The updated constraints are introduced and the corresponding Lagrange multipliers Y_1, Y_2 are added to the constraint formulas, respectively. Similarly, the corresponding constraint parameters μ and the corresponding F2 parametric constraint formulas are added. The final Lagrangian formula is as follows

$$L(\mathbf{Z}, \mathbf{E}, \mathbf{J}, Y_1, Y_2, \mu) = \|\mathbf{Z}\|_* + \lambda_1\|\mathbf{J}\|_1 + \beta tr\left(\mathbf{Z}\mathbf{L}^{ARW}\mathbf{Z}^T\right) + \gamma\|\mathbf{Z} - \mathbf{S}\|_F^2 + \lambda_2\|\mathbf{E}\|_1 + \langle Y_1, \mathbf{X} - \mathbf{X}\mathbf{Z} - \mathbf{E}\rangle + \langle Y_2, \mathbf{Z} - \mathbf{J}\rangle + \frac{\mu}{2}(\|\mathbf{X} - \mathbf{X}\mathbf{Z} - \mathbf{E}\|_F^2 + \|\mathbf{Z} - \mathbf{J}\|_F^2) \tag{8}$$

We summarize the process of optimization iterations in Algorithm 1.

Algorithm 1. LADMAP for solving (8)

Input: Data matrix X, Balance parameters λ_1, λ_2, β, γ, similarity matrix $\mathbf{S}$, Laplacian matrix $\mathbf{L}^{ARW}$

Output: matrix $\mathbf{Z}^*$

Repeat

1: Fix the others and update $\mathbf{Z}_{k+1}$ using $\mathbf{Z}_{k+1} = \theta\frac{1}{\eta_1}\left(\mathbf{Z}_k - \frac{\nabla_Z q(\mathbf{Z}_k)}{\eta}\right)$

2: Fix the others and update $\mathbf{E}_{k+1}$ using $\mathbf{E}_{k+1} = S_{\frac{\gamma}{\mu}}\left(\mathbf{X} - \mathbf{X}\mathbf{Z}_k + \frac{Y_k}{\mu_k}\right)$

3: Fix the others and update $\mathbf{J}_{k+1}$ using $\mathbf{J}_{k+1} = \max\left\{\theta_{\frac{\lambda}{\mu_k}}\left(\mathbf{Z}_{k+1} + \frac{Y_2^k}{\mu_k}\right), 0\right\}$

4: Fix the others and update Y_1, Y_2 using $Y_1^{k+1} = Y_1^k + \mu_1^k\left(\mathbf{X} - \mathbf{X}\mathbf{Z}_{k+1} - \mathbf{E}_{k+1}\right)$ and $Y_2^{k+1} = Y_2^k + \mu_2^k\left(\mathbf{Z}_{k+1} - \mathbf{J}_{k+1}\right)$

until convergence

3.3 Spectral Clustering

Spectral clustering (SC) is a popular clustering approach that splits multiple objects into distinct clusters based on the spectral of the similarity matrix, making the model more robust than other clustering algorithms. We apply SC [16] to the learned similarity matrix for later analysis.

The symmetric matrix $\mathbf{Z}^*$ is expressed as $\mathbf{Z}^* = \mathbf{U}^*\Sigma^*(\mathbf{V}^*)^T$ by using a lean singular value decomposition. $\mathbf{U}^*$, Σ^* and $\mathbf{V}^*$ are orthogonal matrices. $\mathbf{M}$ is denoted by $\mathbf{M} = \mathbf{U}^*(\Sigma^*)^{1/2}$ to get the i-th row of $\mathbf{m}_i \in$ M, and affinity matrix $\mathbf{H}$ is defined as:

$$\mathbf{H}_{ij} = \left(\frac{\mathbf{m}_i^T \mathbf{m}_j}{\|\mathbf{m}_i\|_2 \|\mathbf{m}_j\|_2}\right)^2 \tag{9}$$

SC will partition the dataset into clusters using the features of the corresponding Laplacian matrix based on the obtained affinity matrix $\mathbf{H}$. As a consequence, we will get the predicted clustering results.

4 Experiment

4.1 Experimental Data

We employ differently sized scRNA-seq datasets to verify the clustering performance of methods. We conduct comprehensive experiments on nine common single-cell datasets. These datasets consist of clusters from cell types including embryonic stem cells, blood cells, neural cells, skin cells, etc., which ensures as wide a range of cell types is used to verify the applicability of our method to different types and sizes of scRNA-seq datasets. We report the exhaustive statistics of these datasets in Table 1.

Table 1. Details of the datasets selected for the experiment.

Serial number	Datasets	Cell (n)	Genes (m)	Cell type
1	Darmanis	466	22085	9
2	Deng	259	22147	10
3	Pollen	249	6982	11
4	Treutlein	80	23129	5
5	Ting	114	14450	5
6	Kolod	704	13473	3
7	Leng	460	19084	4
8	Bre	957	20689	4
9	mECS	182	8989	3

We excluded cells with ambiguous labels, and utilize the pre-processing datasets to limit the influence of error terms. Gene filters and median normalization [17] are employed to process our datasets to eliminate incompletely expressed genes. Pseudo counts of 1 and log (base 2) transformations are adopted on all the datasets, as it normalizes the raw read counts of genes.

4.2 Evaluation Metrics

We use Normalized mutual information (NMI) and Adjusted Rand Index (ARI) as the evaluation measures for the clustering performance:

Normalized mutual information (NMI) [18]: The value of NMI can indicate the degree of similarity between two different data objects, and is also one of the variables used to evaluate the effect of clustering. When the value of NMI is equal to 0, it implies that there is no mutual information, and when it is equal to 1, it means that the clustering information and the real information are perfectly correlated.

Adjusted Rand Index (ARI) [19]: We also use the adjusted Rand index as an evaluation metric, which is considered a popular metric of similarity between true and expected clusters. The range of possible ARI values is $[-1, 1]$. Its result of 1 indicates that the tested cluster is perfect, while 0 implies that the labels are uncorrelated.

4.3 Parameters Selection

As stated in the formula (7), it needs to adjust four parameters to obtain the optimal value of ARGLRR. Parameter λ_1 and λ_2 is the balance parameter in the low-rank representation sparse framework. The parameter β is related to the local manifold structure information. The parameter γ is used to control the similarity constraint term. These four parameters have different values for data sets with different numbers and cell types. We use the grid search approach in this work to modify separately the four parameters on different datasets and choose the best parameter.

Our model performs consistently across all datasets when adjusting λ_1, regardless of which value in the interval it takes in Fig. 2 (A). In Fig. 2 (B), it performs similarly on all datasets except Bre, Darmanis, and Kolod during the λ_2 test. Specifically, they all show stable performance in the parameter in the rest of the datasets, except for two quite big datasets, Leng and Bre. As shown in Fig. 2 (C), the parameter β, while initially tiny for each dataset, rapidly attains ideal values and stabilizes as the selected values rise. Figure 2 (D) shows that the parameter γ obtains an ideal value at the start, then varies as the value taken grows, but also stabilizes at the end. Satisfactorily, none of the four parameters had any significant outliers, indicating excellent stability.

4.4 Result

We utilize seven popular clustering algorithms as reference methods: SinNLRR [4], SCCLRR [5], ECC [20], Corr [21], MPSSC [22], PCA + k-means [23], t-SNE + k-means [24] to assess the effectiveness of our methodology in actual applications. We measure the performance of clustering by NMI and ARI. The heatmaps of experimental results of NMI and ARI are reported in Fig. 3. Grids with darker colors signify higher values.

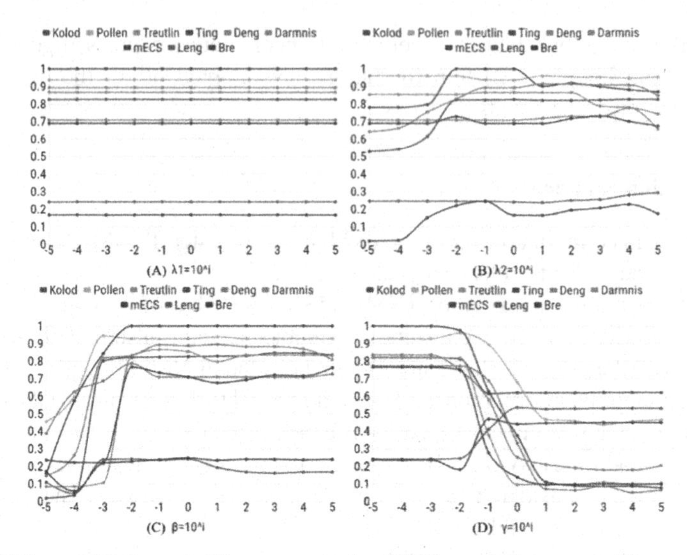

Fig. 2. (A) The variation curve of the parameter λ_1. (B) The variation curve of the parameter λ_2. (C) The variation curve of the parameter β. (D) The variation curve of the parameter γ.

Figure 3 shows that ARGLRR achieves the best results in different datasets. The clustering performance of ARGLRR is further proved by average NMI and ARI values of 0.8057 and 0.7577 in the nine datasets, respectively. Particularly, it improved the optimal NMI values of the comparison method by an average of 6.99% on the nine datasets, with significant gains in the Bre and Leng datasets. It improves 5.85% above the optimal result among the optimal comparison method in the ARI indicator. The impressive results indicate that ARGLRR has promising applications in cell clustering.

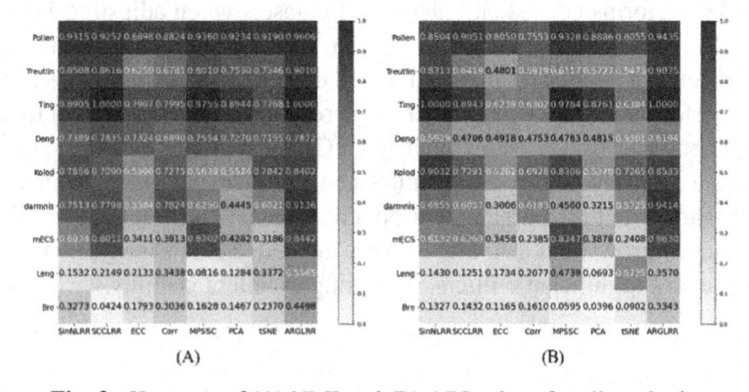

Fig. 3. Heatmap of (A) NMI and (B) ARI values for all methods.

4.5 Visualize Clustering Results Using UMAP

In comparison to t-SNE, Uniform Manifold Approximation and Projection (UMAP), a nonlinear dimensionality reduction method, keep a more consistent collection of cell subsets, making plots easier to understand [25]. Therefore, we visualize the cell clustering by UMAP to facilitate better obtaining an intuitive performance of ARGLRR in this section. For comparison, we select four common methods: t-SNE, ECC, Corr, and SinNLRR. Figure 4 illustrates the visualization results of the ARGLRR for the clustering task over the other methods on three datasets, Pollen, Ting, and Treutlin.

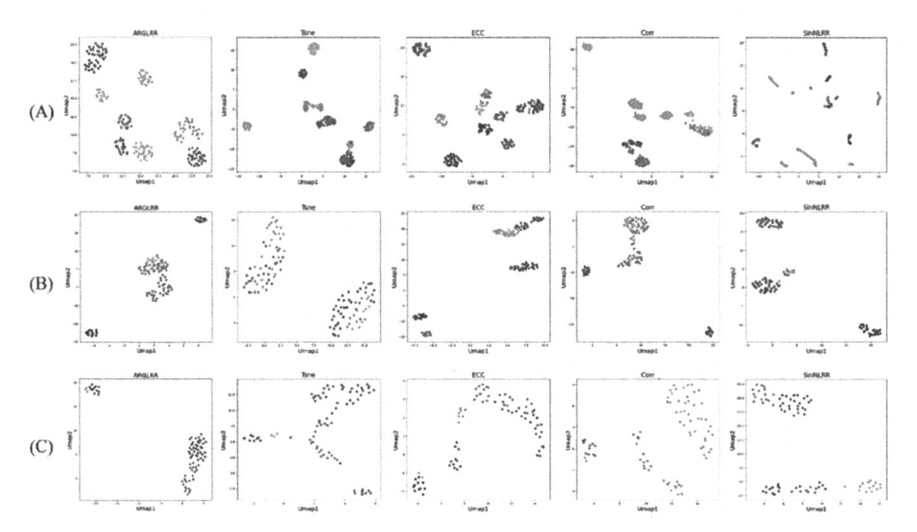

Fig. 4. UMAP effect on three datasets (A) Pollen, (B) Ting, and (C) Treutlin.

As can be seen from Fig. 4, ARGLRR may discriminate between various types of cells and preserve them in distinct clusters. Specifically, the same clusters are more compact in the cell, and clusters are separated from each other as much as possible. ARGLRR further enhances the intra cluster consistency and inter cluster heterogeneity. Even when the distance between clusters is tiny or multiple clusters are jumbled together in certain data sets, it performs much better than t-SNE, ECC, Corr, and SinNLRR. The superiority of visualization results illustrates the validity of the ARGLRR clustering results.

5 Conclusions and Future Work

In this paper, we propose a developed method for clustering single-cell sequencing data, ARGLRR, to overcome the neglect of local structure information by applying adjusted random walk manifold regularity constraints and to add more similarity information by introducing similarity constraints. The results of clustering experiments based on real datasets and subsequent visualization reveal that ARGLRR is more accurate than the commonly utilized scRNA-seq clustering methods in detecting cell types. Despite the good experimental results, ARGLRR still has some shortcomings that need to be improved. In the future, we plan to attempt additional downstream analyses to demonstrate the advantages of the approach. In addition, since each omics may include diverse information in single-cell multi-omics data, we also plan to expand the newly proposed method to multi-omics data clustering may be investigated.

Acknowledgment. This work is supported by the National Natural Science Foundation of China (Grant Nos. 62172253, 62172254, 61972226, and 61902215).

References

1. Wang, H.-Y., Zhao, J.-P., Zheng, C.-H., Su, Y.-S.: scCNC: a method based on capsule network for clustering scRNA-seq data. Bioinformatics, btac393 (2022). https://doi.org/10.1093/bioinformatics/btac393
2. Wang, C., Mu, Z., Mou, C., Zheng, H., Liu, J.: Consensus-based clustering of single cells by reconstructing cell-to-cell dissimilarity. Brief. Bioinform. **23**, bbab379 (2022). https://doi.org/10.1093/bib/bbab379
3. Liu, G., Lin, Z., Yan, S., Sun, J., Yu, Y., Ma, Y.: Robust recovery of subspace structures by low-rank representation. IEEE Trans. Pattern Anal. Mach. Intell. **35**, 171–184 (2013). https://doi.org/10.1109/TPAMI.2012.88
4. Zheng, R., Li, M., Liang, Z., Wu, F.-X., Pan, Y., Wang, J.: SinNLRR: a robust subspace clustering method for cell type detection by non-negative and low-rank representation. Bioinformatics **35**, 3642–3650 (2019). https://doi.org/10.1093/bioinformatics/btz139
5. Zhang, W., Li, Y., Zou, X.: SCCLRR: A robust computational method for accurate clustering single cell RNA-Seq data. IEEE J. Biomed. Health Inform. **25**, 247–256 (2021). https://doi.org/10.1109/JBHI.2020.2991172
6. Cheng, T., Wang, B.: Graph and total variation regularized low-rank representation for hyperspectral anomaly detection. IEEE Trans. Geosci. Remote Sens. **58**, 391–406 (2020). https://doi.org/10.1109/TGRS.2019.2936609
7. Lu, X., Wang, Y., Yuan, Y.: Graph-regularized low-rank representation for destriping of hyperspectral images. IEEE Trans. Geosci. Remote Sens. **51**, 4009–4018 (2013). https://doi.org/10.1109/TGRS.2012.2226730
8. Du, H., Zhang, X., Hu, Q., Hou, Y.: Sparse representation-based robust face recognition by graph regularized low-rank sparse representation recovery. Neurocomputing **164**, 220–229 (2015). https://doi.org/10.1016/j.neucom.2015.02.067
9. Zheng, R., Liang, Z., Chen, X., Tian, Y., Cao, C., Li, M.: An adaptive sparse subspace clustering for cell type identification. Front. Genet. **11**, 407 (2020). https://doi.org/10.3389/fgene.2020.00407
10. Candès, E.J., Li, X., Ma, Y., Wright, J.: Robust principal component analysis? J. ACM **58**, 1–37 (2011). https://doi.org/10.1145/1970392.1970395
11. Cai, D., Wang, X., He, X.: Probabilistic dyadic data analysis with local and global consistency. In: Proceedings of the 26th Annual International Conference on Machine Learning - ICML 2009, Montreal, Quebec, Canada, pp. 1–8. ACM Press (2009)
12. Dai, L.-Y., Feng, C.-M., Liu, J.-X., Zheng, C.-H., Yu, J., Hou, M.-X.: Robust nonnegative matrix factorization via joint graph Laplacian and discriminative information for identifying differentially expressed genes. Complexity **2017**, 1–11 (2017)
13. Belkin, M., Niyogi, P.: Laplacian eigenmaps and spectral techniques for embedding and clustering. In: Advances in Neural Information Processing Systems. MIT Press (2001)
14. Yin, H., Zaki, S.M.: A self-organising multi-manifold learning algorithm. In: Ferrández Vicente, J.M., Álvarez-Sánchez, J.R., de la Paz López, F., Toledo-Moreo, FcoJavier, Adeli, H. (eds.) IWINAC 2015. LNCS, vol. 9108, pp. 389–398. Springer, Cham (2015). https://doi.org/10.1007/978-3-319-18833-1_41

15. Lin, Z., Liu, R., Su, Z.: Linearized alternating direction method with adaptive penalty for low-rank representation. In: Advances in Neural Information Processing Systems 24 (2011)
16. von Luxburg, U.: A tutorial on spectral clustering (2007). http://arxiv.org/abs/0711.0189
17. Kiselev, V., et al.: SC3 - consensus clustering of single-cell RNA-Seq data. Bioinformatics (2016). https://doi.org/10.1101/036558
18. Strehl, A., Ghosh, J.: Cluster ensembles – a knowledge reuse framework for combining multiple partitions, 35. https://doi.org/10.5555/777092.777110
19. Hubert, L., Arabie, P.: Comparing partitions. J. Classif. **2**, 193–218 (1985)
20. Liu, H., Zhao, R., Fang, H., Cheng, F., Fu, Y., Liu, Y.-Y.: Entropy-based consensus clustering for patient stratification supplementary information, 32 (2017)
21. Jiang, H., Sohn, L.L., Huang, H., Chen, L.: Single cell clustering based on cell-pair differentiability correlation and variance analysis. Bioinformatics (2018). https://doi.org/10.1093/bioinformatics/bty390
22. Park, S., Zhao, H.: Spectral clustering based on learning similarity matrix. Bioinformatics **34**, 2069–2076 (2018). https://doi.org/10.1093/bioinformatics/bty050
23. Van der Maaten, L., Hinton, G.: Visualizing data using t-SNE. J. Mach. Learn. Res. **9**, 2579–2605 (2008)
24. Bro, R., Smilde, A.K.: Principal component analysis. Anal. Methods **6**, 2812–2831 (2014). https://doi.org/10.1039/C3AY41907J
25. Becht, E., et al.: Dimensionality reduction for visualizing single-cell data using UMAP. Nat. Biotechnol. **37**, 38–44 (2019). https://doi.org/10.1038/nbt.4314

An Efficient and User-Friendly Software for PCR Primer Design for Detection of Highly Variable Bacteria

Dongzheng Hu[1], Wubin Qu[2], Fan Tong[1], Xiangwen Zheng[1], Jiangyu Li[1], and Dongsheng Zhao[1](✉)

[1] Information Center, Academy of Military Medical Sciences, Beijing 100850, China
dszhao@bmi.ac.cn

[2] iGeneTech Bioscience Co., Ltd., Beijing 102206, China

Abstract. Objective: To design and implement a simple and highly integrated primer design software for detection of highly variable bacteria. **Methods**: Firstly, gene named entity recognition technology was applied to annotate the genes in the literature, thus the conservative gene knowledge base of highly variable bacteria could be established. Then, new primer design workflow was created by integrating Clustal Omega, Gblocks, Primer3, MFEprimer3.0, which achieves high-performance specificity detection with improved index algorithms and parallel computing. Finally, multiple primers could be obtained by combining primers designed by different conservative genes based on the greedy algorithm. **Results**: A web-based primer design software was implemented. Primers were designed for five highly variable bacteria, including Escherichia coli, Shigella, Listeria monocytogenes, Staphylococcus aureus, and Salmonella. By comparison, the single primers designed by our software were more specific and sensitive than primers in the literature for detecting the same conservative gene. The multiple primers designed for each bacterium have strong specificity, and the coverage rate of Escherichia coli primers is 93. 14% in terms of sensitivity, and the coverage rate of other bacterial primers is more than 99%. **Conclusion**: The software can be applied to the design of PCR primers for the detection of highly variable bacteria.

Keywords: Highly variable bacteria · Multiplex primer design · Specificity · Sensitivity · High-performance computing

1 Introduction

Mutation is one of the properties of biological organisms, and highly variable bacteria refer to bacteria that evolve rapidly or are highly diverse in biological populations, such as Escherichia coli, Shigella, etc. There are many subspecies and subtypes of highly variable bacteria, and new subtypes are constantly emerging as resistance genes and virulence factors spread through horizontal gene migration in bacteria. Therefore, infectious diseases caused by highly variable bacteria are difficult to detect and have many complications, posing a huge threat to human health [1].

M. S. Bansal et al. (Eds.): ISBRA 2022, LNBI 13760, pp. 138–147, 2022.
https://doi.org/10.1007/978-3-031-23198-8_13

Polymerase chain reaction (PCR) is one of the most commonly used methods for detecting pathogenic bacteria. In PCR, primer design is a critical step. Primers have two important characteristics: specificity and sensitivity, specificity requires primer to bind to target DNA, not to other DNA, and sensitivity requires primer to bind to all target DNA.

Currently, there are no primer design software and methods specifically for detection of highly variable bacteria. The routine primer design methods include the steps of literature query to obtain conservative genes, gene sequence acquisition, primer design, PCR amplification test to screen primers [2]. In order to detect highly variable bacteria accurately and sensitively, using multiple sequence alignment software to obtain conservative regions of conservative genes as templates for primer design can improve the specificity of primers to a certain extent [3, 4]. In order to reduce the number of tests reasonably and save the cost, researchers developed specific evaluation software to screen out good quality primers [5]. However, the existing methods and software have the following problems: manually searching for conservative genes through literature requires consulting a large number of documents, which is time-consuming and laborious; When evaluating and screening specificity of primers, only a few similar bacterial sequences are usually selected for detection of the specificity of primers, which is easy to cause bias. Primer design is a complex process and software functions are single. Researchers must master the skills of using multiple software to design the ideal primer, which is inefficient.

In this paper, a PCR primer design software was designed and implemented for detection of highly variable bacteria. Natural language processing technology was used to automatically extract conservative bacterial genes from a large number of literatures to guide the design of primers. A target bacterial database containing the gene sequences of all subtypes of bacteria was established to detect the sensitivity of primers, and an interference sequence database containing the whole species genome data was established to detect the specificity of primers, and index optimization algorithm and multi-thread parallel computing were used to accelerate the detection. Multiple software was integrated to realize the automated primer design process. Finally, greedy algorithm was used to combine primers reasonably to increase the sensitivity to target bacteria. Users only needs to enter gene sequences and primer parameters on the main page, wait for a while and get the desired primer. The operation is simple and requires no human intervention.

2 Materials and Methods

2.1 Software and Environment

The software designed in this study integrates 4 softwares:The multi-sequence comparison softwares Clustal Omega [6] (www.clustal.org). Aligned conserved sequence extraction software Gblocks [7] (molevol.cmima.csic.es/castresana/Gblocks.html).Primer design software Primer3 [8] (https://github.com/primer3-org/primer3). Primer specificity assessment software introduced in 《PCR Primer Design》, one of Springer's Protocol series, MFEprimer3.0 [5, 9] (github.com/quwubin/MFEprimer-3.0).

Software test environment: 48-core Intel(R) Xeon(R) CPU e5–2670 v3 server, 120GB memory, 1TB hard disk, CentOS Linux release 7. 9. 2009 OS.

2.2 Target Bacteria Database and Interference Sequence Database

The method used to assess primer sensitivity and specificity in this paper was to use MFEprimer3.0 to detect primer amplification of the target bacterial database and interference sequence database. The greater the number of target bacterial gene sequences that can be amplified, the higher the coverage of the target bacterial database indicates the better the sensitivity of the primers, and the non-amplification interference sequence indicates that the primers have strong specificity. There are many subtypes of highly variable bacteria. In order to verify the sensitivity of primers, this paper used all complete nucleotide sequences of target bacteria in GenBank containing all subtypes of target bacteria as the target bacteria database. For example, 3354 Escherichia coli sequences, 334 Shigella sequences, 820 Listeria monocytogenes sequences, 2936 Staphylococcus aureus sequences, and 2054 Salmonella sequences were downloaded as target bacterial databases for primer evaluation of five highly variable bacteria.

In addition to similar bacteria, it may also be disturbed by plankton, viruses, host cells and other species in the environment during PCR experiments. In order to fully verify the specificity of primers, the Nucleotide library of NCBI containing 384G gene data of the whole species was used as the interference sequence database, which could achieve higher specificity detection.

Over time, new subtypes of highly variable bacteria will be discovered, the Nucleotide library is constantly being updated, and the database used in this article will be updated periodically to re-evaluate the specificity and sensitivity of primers, and update and iterate on primers with decreased specificity and sensitivity.

2.3 Construction of a Knowledge Base of Bacterial Conservative Genes for Primer Design

In order to make full use of the existing prior knowledge in this field, this paper designs a software that uses web crawlers and gene naming entity recognition algorithms to assist in constructing a bacterial Conservative gene knowledge base. The basic idea is that the gene name that appears in a paper related to the design of a certain bacterial primer is likely to be a conservative gene suitable for primer design.

Using "target bacteria" and "primer design" as keywords, the web crawler software retrieves and downloads related literature from PubMed, including the title, author, abstract and full text of the document. The downloaded literature is pre-processed with word segmentation and sentence segmentation, and the named entity recognition algorithm that integrates deep neural network and traditional machine learning is used to automatically annotate the gene names that appear in literature [10], and perform manual confirmation. Merge, deduplication and sequence the conservative genes extracted from different literature to obtain the names of bacterial conservative genes sorted by frequency of use. Finally, the DNA sequence information of the gene will be downloaded from the NCBI nucleic acid database according to the gene name, and a bacterial conservative gene knowledge base including bacterial name, source literature, conservative gene name, DNA sequence and other information is formed to assist in subsequent primer design.

2.4 Primer Design Process

Primer design

Input: Bacterial name of primers to be designed, conservative gene knowledge base, conservative gene sequence files.

Output: Primer and amplification of the bacterial of interest.

Step1: Extract the DNA sequence of the conservative gene of the bacterium from the conservative gene repository according to the name of the bacterium;

Step2: Clustal Omega software was used to perform multiple sequence alignment of conservative gene DNA sequences to obtain aligned sequences;

Step3: Extract conservative sequences with Gblocks;

Step4: Set the parameters such as the position of primers before and after Primer3 and the number of primers is set to 5;

Step5: Primers were designed with Primer3, and 10 primers were obtained;

Step6: MFEprime3.0 was used for primer evaluation. The amplification of the primers to the target database and interference database was detected. If the interference sequence was not amplified and the coverage of the target database reached 90%, then Step7 was performed; otherwise, Step4 was performed. The parameters were adjusted to obtain a new set of candidate primers, and the coverage rate was re-detected. After three iterations, Step7 was entered;

Step7: Sort the primer pairs by coverage, with the primer pair with the highest coverage as the primer for testing the gene.

Fig. 1. Primer design process pseudocode

When using the software, users can not only use the conservative gene sequences in the knowledge base, but also upload the conservative gene sequence files obtained by themselves for primer design. The primer design process has the following three new features (see Fig. 1):

(1) The mechanism of primer evaluation, iteration and screening was established. MFEprimer3.0 was used to detect the coverage of the matched primers to the target bacteria database and the interference sequence database, and then the unqualified primers were discarded. The parameters of Primer3 were adjusted to re-design and evaluate primers, and the most sensitive pair of primers were selected as the primers for detecting this conservative gene;
(2) A collision-free hash function was designed by optimizing the index of MFEprimer3.0 software. The offset of primer position information in the index file is calculated according to the hash function, and the required data is read at one time, and the multi-thread parallel calculation is used to speed up the operation. The specificity and sensitivity of 10 primer pairs to nearly 400G of large-scale accounting sequence data can be tested in 45 min;
(3) The software automates the process without human intervention. The operation is very simple and convenient. Several processes are processed at one time to achieve efficient operation efficiency.

2.5 Primer Combination Algorithm

Users can upload several pairs of primers and select the target bacterial database. The software will combine primer pairs by greedy algorithm, check whether dimer and hairpin structure will be formed, evaluate the specificity of the combined multiple primers, and select the optimal combination of candidate primers (see Fig. 2).

Primer combination algorithm
Input: Primer pairs
Output: Multiple primers
Step1: The amplification of the input primer pair to the target database and interference database was detected, and the primer pair that amplified the interference sequence was deleted. Primer pairs that did not amplify interference sequences were ranked according to the coverage of target bacteria;
Step2: The primer pair was pairwise combined according to the coverage from high to low;
Step3: The combined primers were tested for dimer and hairpin structures, as well as the coverage of the target bacterial database. When 99% was reached, it was stopped and used as the final multiple primers, and when less than 99%, the remaining combinations were detected. If all combinations are less than 99%, enter Step4;
Step4: Add one primer, and combine it from high to low according to the multiple primer coverage obtained in Step3. Stop when the detection coverage reaches 99% or the addition of a primer increases the coverage by less than 1%. The combination with the highest coverage was used as the final multiplex primer. Otherwise, repeat Step4.

Fig. 2. Primer combination algorithm pseudocode

3 Results

Primers for the detection of five bacterial species(Escherichia coli, Shigella, Listeria monocytogenes, Staphylococcus aureus and Salmonella) were designed by using the conservative gene knowledge base and developed software. The specificity and sensitivity of these primers were tested and compared to published primers.

Table 1. Construction of conservative gene knowledge base.

Bacterial name	Number of crawled literature	Number of conservative genes
Escherichia coli	384	72
Shigella	73	32
Listeria monocytogenes	144	24
Staphylococcus aureus	348	59
Salmonella	173	48

3.1 Knowledge Base Build

A total of 1122 literature was crawled for gene naming entity labeling for 5 species of bacteria, and conservative genes were screened in combination with artificial confirmation methods, and the knowledge base construction is shown in Table 1.

3.2 Primer Evaluation and Comparison

Figure 3 shows the use page of the software. According to the constructed conservative gene knowledge base, 20 pairs of primers were designed for Escherichia coli and 10 pairs were designed for each of four bacteria, Shigella monocytogenes, Listeria monocytogenes, Staphylococcus aureus and Salmonella. In contrast to primers in literature, there were no duplicate primer sites. Due to space limitation, only the two pairs of primers with the best specificity for each bacterium are presented (see Fig. 3).

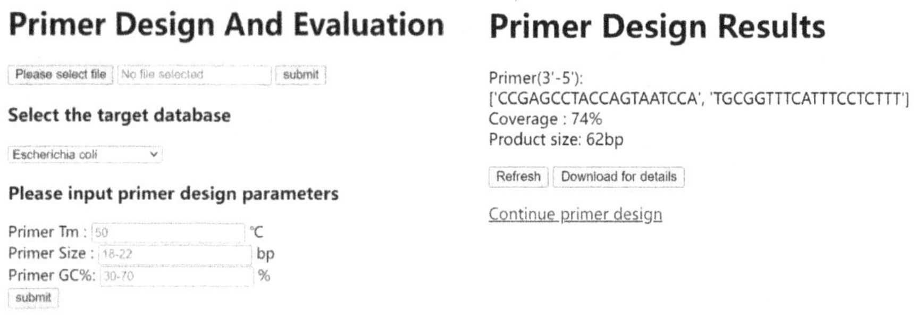

a. The parameter selection and file upload page b. The result page

Fig. 3. This is the usage page and the result page of the software.

When comparing specificity and sensitivity of primers with MFEprimer3.0, first of all, the binding sites in database that can be completely paired with 9 bases of the 3 ends of primers were found. Then, nearest-neighbour model was used to calculate the unchain temperature (Tm) of to calculate melting temperature of binding sites and primer sequences. If the Tm was higher than set Tm, it was considered as a stable binding site that could specifically bind to primers. In PCR experiments, commonly used annealing temperature is generally between 50 and 65. In this paper, Tm is set to 50 to detect specificity and sensitivity of primers designed in this paper and those in literature. None of the primers used in this paper amplified interference sequences, and the coverage of target bacterial database is shown in Table 2.

A total of 60 pairs of primers were designed in this paper, none of which amplified interfering sequence and had strong specificity. In terms of sensitivity comparison, 47 pairs of primers covered more than those in the literature, and 32 pairs of primer coverage differed by more than 10%;7 pairs of primers coverage is the same as that of primers in the literature, and only primers detecting TraT, F17, STA, and STB genes for Escherichia coli have slightly lower coverage of target database than primers in the literature.

Table 2. Amplification of the database by the two primers with the highest coverage of five species of bacteria and the primers in the literature.

Bacteria	Genes	Primers designed by this software		The most well-covered primers in the literature		
		Primers (5'-3')	Coverage of target database[%]	Primers (5'-3') and source literature		Coverage of target database[%]
Escherichia coli	ompT	GGAGGCCGAAAAGTCAGTCA ACCACCTCTGGCTGTAAAGC	57. 72	ATCTAGCCGAAGAAGGAGGC CCCGGGTCATAGTGTTCATC	[11]	14. 55
	chuA	CACATCCCAGCCCCAGATTT TCTGAGCGGTTTAGTGCGTT	48. 81	GACGAACCAACGGTCAGGAT TGCCGCCAGTACCAAAGACA	[12]	44. 39
Shigella	parC	TGAAGCGATGGTCCTGATGG CTGACCATCAACCAGCGGAT	98. 20	GTCTGAACTGGGCCTGAATGC AGCAGCTCGGAATATTTCGACAA	[13]	91. 62
	gadC	GCGGCAATCTGCTTAAGCTC CCCTGCGGAGAGGTTGATTT	96. 71	TCTGGGTCCGAGATGGGGATTTG CAGCGAT TGGTGAACGTCGACGCGGGTGCA GGAAGAA	[14]	96. 41
Listeria monocytogenes	inlAB	TAGCACCACTGTCGGGTCTA ACCTGCTAGGGGACTGATGT	97. 07	CTTCAGGCGGATAGATTAGG TTCGCAAGTGAGCTTACGTC	[15]	76. 83
	hly	CACTGATTGCGCCGAAGTTT AATGTGCGGCCAAGAAAAGG	85. 24	GTAAGCGGAAAATCTGTCTC ATTTCGTTACCTTCAGGATC	[16]	46. 95
Staphylococcus aureus	femA	GCTTTTGCTGATCGTGATGACA TTGCACTGCATAACTTCCGG	98. 64	AAAAAAGCACATAACAAGCG GATAAAGAAGAAACCAGCAG	[17]	45. 40
	lukS	ACATGGATGTCACTCATGCCA TGAGTCTTCCAATTGACCTCGT	76. 26	CAGGAGGTAATGGTTCATTT ATGTCCAGACATTTTACCTAA	[18]	21. 19
Salmonella	purE	CTGTTCAGCTTCGCCGAAAC GCGCCGGCAATAATCACTTG	99. 17	ATGTCTTCCCGCAATAATCC TCATAGCGTCCCCCGCGGATC	[19]	98. 59
	thrA	TCCTACCAGGAAGCGATGGA ACTTTGGCGCCGAAGTAAGA	98. 93	GTCACGGTGATCGATCCGGT CACGATATTGATATTAGCCCG	[19]	71. 62

Through combination of primers, the coverage of Escherichia coli database by five primers was finally achieved to reach 93.14%; The coverage of double primers to Shigella database reached 99.40%; The coverage of triple primers to Listeria monocytogenes database reached 99.15%; The coverage of double primers to Staphylococcus aureus database reached 100%; The coverage of double primers to Salmonella database reached 99.76%(see Table 3).

Table 3. Multiple primers lists of 5 species of bacteria.

Bacteria	Gene	Primers (5′-3′)	Amplicon size[bp]
Escherichia coli	ompT	GGAGGCCGAAAAGTCAGTCA ACCACCTCTGGCTGTAAAGC	336
	chuA	TCTGAGCGGTTTAGTGCGTT CACATCCCAGCCCCAGATTT	886
	TSPE4. C2	AGAGTTTATCGCTGCGGGTC TACCCGCGTTTCTGTCTCAC	109
	yjaA	TTGTTCTGCAACTCCACCCA GCGCGCTCACAACAATATCC	78
	fyuA	CTCTCCAGTCATCGGTCAGC GTTTGGCGACCAGGGTAAGA	83
Shigella	parC	TGAAGCGATGGTCCTGATGG CTGACCATCAACCAGCGGAT	61
	gyrA	CGTACTATACGCCATGAACGT GTCACCAACGACACGGGC	76
Listeria monocytogenes	hly	AATGTGCGGCCAAGAAAAGG CACTGATTGCGCCGAAGTTT	469
	iap	CGCAGGTGTAGTTGCTTGTG GCGCTGGTGTTGATAACAG	270
	inlF	GCGCCACCAGCATCAATTAA TTCCGCTCGTATTCCCTCC	67
Staphylococcus aureus	femA	GCTTTTGCTGATCGTGATGACA TTGCACTGCATAACTTCCGG	393
	Spa	TGTTGCCATCTTCTTTACCAGG CTGGCAAAGAAGACGGCAAC	69
Salmonella	purE	CTGTTCAGCTTCGCCGAAAC GCGCCGGCAATAATCACTTG	59
	agfB	TCGCATATGGCTATTCGCGT CTGCGAATGCTGCCACTTTT	105

4 Discussion

In this paper, a Web-based primer design software for the detection of highly variable bacteria is implemented by integrating natural language processing technology, primer specificity evaluation software, greedy algorithm and existing primer design technology. It can be seen from the experimental results that the primers designed by this software can amplify as many target bacterial sequences as possible without amplifying interference sequences, which greatly improves the sensitivity of primers.

However, the combined multiplex primers still failed to fully cover the target bacterial database, for example, 230 strain sequences were undetectable in Escherichia coli database and 5 strain sequences were undetectable in Salmonella database. On the one hand, this indicates the complexity of highly variable bacteria and the difficulty of PCR detection, and it is difficult to detect all subtypes of highly variable bacteria by multiplex PCR experiment with a set of universal primers. When necessary, primers specifically designed to detect a specific subtype should be designed for that subtype, that is, a combination of universal detection and specific detection. On the other hand, it is explained that this software still has certain limitations, and there is still room for further improvement in primer design.

The knowledge base constructed in this paper currently contains the conservative gene information of five species of bacteria, and there are other highly variable bacteria in nature. In the next work, we will continue to collect the conservative gene data of highly variable bacteria to expand the knowledge base and provide better services for related workers.

Acknowledgement. The authors would like to thank other people working in the same group for their help and guidance in this study.

References

1. Ling, T.K., Xiong, J., Yu, Y., Lee, C.C., Ye, H., Hawkey, P.M.: Multicenter antimicrobial susceptibility survey of gram-negative bacteria isolated from patients with community-acquired infections in the People's Republic of China. Antimicrob Agents Chemother **50**(1), 374–378 (2006)
2. Conter, C.C., et al.: PCR primers designed for new world Leishmania: a systematic review. Exp. Parasitol. **207**, 107773 (2019)
3. Bui, T.H., et al.: Multiplex PCR method for differentiating highly pathogenic Yersinia enterocolitica and low pathogenic Yersinia enterocolitica, and Yersinia pseudotuberculosis. J. Vet. Med. Sci. **83**(12), 1982–1987 (2021)
4. Bouju-Albert, A., Saltaji, S., Dousset, X., Prévost, H., Jaffrès, E.: Quantification of Viable Brochothrix thermosphacta in Cold-Smoked Salmon Using PMA/PMAxx-qPCR. Front Microbiol **12**, 654178 (2021)
5. Wang, K., et al.: MFEprimer-3.0: quality control for PCR primers. Nucleic Acids Res. **47**(1),W610–W613 (2019)
6. Sievers, F., et al.: Fast, scalable generation of high-quality protein multiple sequence alignments using Clustal Omega. Mol. Syst. Biol. **7**, 539 (2011)

7. Talavera, G., Castresana, J., Kjer, K., Page, R., Sullivan, J.: Improvement of phylogenies after removing divergent and ambiguously aligned blocks from protein sequence alignments. Syst. Biol. **56**(4), 564–577 (2007)
8. Untergasser, A., et al.: Primer3--new capabilities and interfaces. Nucleic Acids Res. **40**(15), e115 (2012)
9. Qu, W., Li, J., Cai, H., Zhao, D.: PCR primer design for the rapidly evolving SARS-CoV-2 genome. Methods Mol. Biol. **2392**, 185–197 (2021)
10. Tong, F., Luo, Z., Zhao, D.S.: A deep network based integrated model for dis-ease named entity recognition. IEEE **2017**, 618 (2017)
11. Velhner, M., Suvajdžić, L., Todorović, D., Milanov, D., Kozoderović, G.: Avian pathogenic Escherichia coli: diagnosis, virulence and prevention. Arch. Vet. Med. **11**(2), 21–31 (2019)
12. Hernández-Chiñas, U., et al.: Characterization of auto-agglutinating and non-typeable uropathogenic Escherichia coli strains. J. Infect. Dev. Ctries. **13**(6), 465–472 (2019)
13. Soltan Dallal, M.M., Yaghoubi, S., Dezhkam, A., Yavari, S., Jamee, A., Yaghoubi, S.: Rapid identification of mutations in quinolone-resistant Shigella isolates by scanning of gyrA and parC genes using high-resolution melting curve analysis. Online J. Health Allied Sci. **19**(1), 1–3 (2020)
14. Waterman, S.R., Small, P.L.C.: Identification of the promoter regions and σs-dependent regulation of the gadA and gadBC genes associated with glutamate-dependent acid resistance in Shigella flexneri. FEMS Microbiol Lett. **225**(1), 155–160 (2003)
15. Jung, Y.S., Frank, J.F., Brackett, R.E., Chen, J.: Polymerase chain reaction detection of Listeria monocytogenes on frankfurters using oligonucleotide primers targeting the genes encoding internalin AB. J. Food Prot. **66**(2), 237–241 (2003)
16. Du, J., Wu, S., Niu, L., Li, J., Zhao, D., Bai, Y.: A gold nanoparticles-assisted multiplex PCR assay for simultane-ous detection of Salmonella typhimurium, Listeria monocytogenes and Escherichia coli O157:H7. Anal. Methods **12**(2), 212–217 (2020)
17. Amin, B., Nasir, N.: Evaluation of femA gene and different primers for mecA gene for detection of methicillin-resistant Staphylococcus aureus (MRSA). Zanco J. Pure Appl. Sci. **31**(3), 16–22 (2019)
18. Al-Talib, H., et al.: A pentaplex PCR assay for the rapid detection of methicil-lin-resistant Staphylococcus aureus and Panton-Valentine Leucocidin. BMC Micro-biol **9**, 113 (2009)
19. Ben-Darif, E., et al.: Development of a multiplex primer extension assay for rapid detection of Salmonella isolates of diverse serotypes. J. Clin. Microbiol **48**(4), 1055–1060 (2010)

A Network-Based Voting Method for Identification and Prioritization of Personalized Cancer Driver Genes

Han Li[1], Feng Li[1](✉), Junliang Shang[1], Xikui Liu[2], and Yan Li[2]

[1] School of Computer Science, Qufu Normal University, Rizhao 276826, China
lifeng1028@qfnu.edu.cn

[2] Department of Electrical Engineering and Information Technology, Shandong University of Science and Technology, Jinan 250031, Shandong, China

Abstract. Identifying cancer driver genes is an important task in cancer research. Numerous methods that identify these driver genes have been proposed, most of which identify driver genes in entire patient cohorts. However, research has shown that cancers in different patients may be caused by different driver genes. Therefore, if we pay more attention to the individual level to identify personalized driver genes, patients can receive precise treatment and achieve better treatment results. Among the methods for identifying personalized cancer drivers, most of these methods only identify the coding driver genes, however, non-coding cancer drivers are also crucial for the initialization and development of cancer. Therefore, we develop an approach to identify coding drivers and non-coding drivers for individual patients named PerVote. In PerVote, the personalized network is firstly constructed based on expression data of mRNAs and miRNAs, then identifies cancer drivers by a voting approach in the network. To verify the performance of our method, we use five cancer datasets of TCGA and compare it with the state-of-the-art methods. The results show that PerVote outperforms other methods. Our method also predicts and prioritizes miRNA drivers, most of which are confirmed by OncomiR to be related to tumorigenesis. Therefore, PerVote can bring further help to the personalized treatment of cancer patients and is an effective method for the identification of cancer drivers.

Keyword: Cancer · Personalized driver genes · Interaction network · Voting mechanism

1 Introduction

Cancer is a heterogeneous disease that has a serious impact on human health and is one of the main causes of human mortality worldwide [1]. Cancer drivers are closely associated with initialisation and development of cancer [2]. Identifying cancer driver genes from a wide range of genomic data could help explain the mechanisms behind cancer and provide better treatment options [3]. With the rapid development of next-generation sequencing technology, such as The Cancer Genome Atlas (TCGA) [4], and

M. S. Bansal et al. (Eds.): ISBRA 2022, LNBI 13760, pp. 148–158, 2022.
https://doi.org/10.1007/978-3-031-23198-8_14

The International Cancer Genome Consortium (ICGC) [5] project has generated a large amount of genomic data. These data can provide a basis for cancer research [6, 7], thus many cancer driver identification methods have been developed.

These cancer driver gene identification methods can be divided into two categories according to the different features. One is to identify the driver gene at the population level, while the other is to identify the driver gene at the individual level. Among the methods for predicting driver genes at the population level, most of them are mutation-based methods, which mainly use the mutations data of genes to identify genes with significant mutations or mutated genes with significantly functional effects as driver genes. Particularly, MutSigCV [8] uses the significance of gene mutations to predict driver genes. OncodriveFM [9] uses the functional effects of gene mutations to identify driver genes. DriverML [10] uses the functional impact of gene mutations and a machine learning approach to discover driver genes. DriverSub [11] applies subspace learning framework and mutation data to predict driver genes. The other part is network-based methods, and these methods are to assess the role of genes or the effects of their on other genes to discover cancer drivers in the network. For example, DriverNet [12] discovers driver genes by integrating multiple omics data to assess the influence of genes on the network. CBNA [13] takes advantage of the controllability of complex network to find crucial nodes in the network and considers these crucial nodes as cancer drivers. These methods of identifying driver genes in entire patient cohorts ignore the characteristics of individual patients. Hence, based on the above questions, several methods to identify personalized driver genes at the individual level have been proposed. In this class of methods, DawnRank [14] predicts personalized driver genes by using gene expression data, PPI network, and PageRank algorithm according to the influence of genes in an interaction network. pDriver [15] uses gene expression data from each patient to build individual network for them, identifying crucial nodes as personalized cancer drivers in the network. Although these methods of identifying driver genes at the individual level can identify driver genes for individual patients, they all have some limitations. Thus, novel and effective methods for personalized driver gene identification need to be developed.

In this paper, we propose a method named PerVote to identify and prioritize personalized coding and non-coding cancer drivers (i.e., miRNA cancer drivers). Our method firstly constructs an individual personalized network for each patient based on gene expression data, then uses a voting approach to calculate the voting scores of each gene in the network and rank them according to the scores obtained from the voting process.

To verify the performance of our method, we apply our method to the five cancer datasets of TCGA to identify personalized cancer drivers and compare with state-of-the-art methods, including ActiveDriver [16], DawnRank, DriverML, MutSigCV, OncodriveFM, pDriver, SCS [17], and PNC [18]. The result shows that PerVote outperforms the other methods. We also use OncomiR [19], a database exploring miRNA dysregulation, to study miRNA drivers with high voting scores. In each cancer type, over 50% of miRNA drivers with high voting scores are validated to be involved in tumorigenesis of cancer. We also explore the number of druggable genes and actionable genes in the top 10 ranked driver genes predicted by PerVote for each patient, and the results show that our method can provide useful information for medicine.

2 Materials and Methods

2.1 Datasets

In our research, we use breast invasive carcinoma (BRCA), kidney renal clear cell carcinoma (KIRC), lung adenocarcinoma (LUAD), head and neck squamous cell carcinoma (HNSC), and lung squamous cell carcinoma (LUSC) datasets from TCGA. These datasets contain expression data of mRNAs, transcription factors (TFs), and miRNAs for 747, 445, 336, 240, and 475 samples, respectively. The TF list [20] is used to distinguish TFs in coding genes. We use several gene interaction databases, including PPIs [21], TarBase 7.0 [22], TargetScan 7.0 [23], miRWalk 2.0 [24], TransmiR 2.0 [25], and miRTarBase 6.1 [26] to refine the constructed personalized network and use CGC from the COSMIC [27] database as the gold standard for identified coding cancer drivers. We also obtain actionable genes and druggable genes from TARGET [28] and DGIdb 3.0 [29] to explore coding cancer drivers with high voting scores. The workflow of PerVote is shown in Fig. 1.

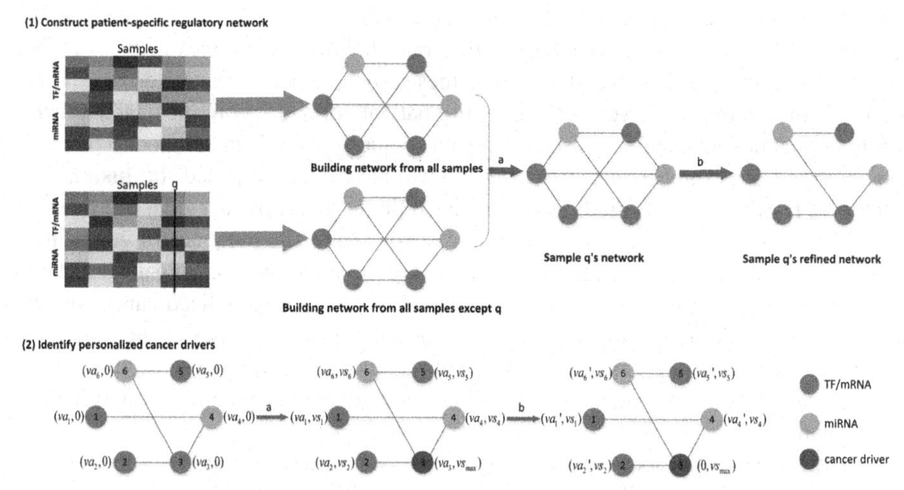

Fig. 1. Flowchart of PerVote. (1) Construct the patient-specific gene regulatory network: (a) Use the LIONESS method and gene expression data to construct the personalized network for each patient, (b) Use several gene interaction datasets to refine constructed network by removing unreal interactions, and removing the edges with lower weights. (2) Identify personalized cancer drivers: (a) Vote process according to the voting mechanism, (b) Update the voting ability of nodes in the network.

2.2 Evaluation Metrics

We use the F_1Score to evaluate the performance of our method. F_1Score is the harmonic mean of *Precision* and *Recall*, which can comprehensively show the performance of a method. The mathematical formula for F_1Score is as follows:

$$F_1Score = 2 * \frac{Precision * Recall}{Precision + Recall} \tag{1}$$

where *Precision* represents the fraction of validated personalized cancer drivers among all of the identified cancer drivers, *Recall* represents the fraction of validated personalized cancer drivers in the gold standard.

2.3 Construction of miRNA-TF-mRNA Network of Patient-Specific

In this study, we use LIONESS [30] to construct the personalized miRNA-TF-mRNA regulatory network. LIONESS considers that the network constructed by all samples is the average of the networks constructed by single sample. Therefore, LIONESS constructs a network for a single sample by using a linear framework. In the LIONESS model, $e_{ij}^{(\alpha)}$ is the weights of the edge between two nodes (i and j) in the network (α) constructed from all samples. LIONESS considers $e_{ij}^{(\alpha)}$ can be modeled as the linear combination of the weights of the edge between two nodes (i and j) in each individual network:

$$e_{ij}^{(\alpha)} = \sum_{s=1}^{N} w_s^{(\alpha)} e_{ij}^{(s)} \tag{2}$$

where $\sum_{s=1}^{N} w_s^{(\alpha)} = 1$.

In Eq. (2), $e_{ij}^{(s)}$ represents the edge weights between node i and node j in a sample (s). $w_s^{(\alpha)}$ represents the relative contribution of that sample (s) to the aggregated network.

Next, the edge weights in a network ($\alpha - q$) estimated using all but sample (q) of the samples can be modeled as:

$$e_{ij}^{(\alpha-q)} = \sum_{s \neq q}^{N} w_s^{(\alpha-q)} e_{ij}^{(s)} \tag{3}$$

where $\sum_{s \neq q}^{N} w_s^{(\alpha-q)} = 1$.

Considering that the ratio of the contribution for each sample to the two aggregated networks is the same, we have:

$$w_q^{(\alpha)} = 1 - w_s^{(\alpha)} / w_s^{(\alpha-q)} \tag{4}$$

In addition, by subtracting Eq. (3) from Eq. (2), we obtain:

$$e_{ij}^{(\alpha)} - e_{ij}^{(\alpha-q)} = w_q^{(\alpha)} e_{ij}^{(q)} - w_q^{(\alpha)} \sum_{s \neq q}^{N} w_s^{(\alpha-q)} e_{ij}^{(s)} \tag{5}$$

Then, the edge weights between node i and node j of the network of sample (q) can be written as:

$$e_{ij}^{(q)} = \frac{1}{w_q^{(\alpha)}} (e_{ij}^{(\alpha)} - e_{ij}^{(\alpha-q)}) + e_{ij}^{(\alpha-q)} \tag{6}$$

In our method, we assume that each sample makes the same contribution to the aggregated network obtained from all samples, we have:

$$w_q^{(\alpha)} = 1/N \tag{7}$$

Then, replacing $w_q^{(\alpha)}$ in Eq. (7) with $w_q^{(\alpha)}$ in Eq. (6), we get:

$$e_{ij}^{(q)} = N(e_{ij}^{(\alpha)} - e_{ij}^{(\alpha-q)}) + e_{ij}^{(\alpha-q)} \tag{8}$$

Through Eq. (8), we find that the sample-specific network can be obtained through the network constructed by all samples and the network constructed by all samples except one. We use Pearson, a common measure that represents correlations between variables, to calculate the edge weights in the aggregated network constructed with N samples and the aggregated network constructed with $N - 1$ samples. The nodes in the network represent mRNAs/TFs/miRNAs. As the constructed personalized network is a fully connected network of contains some unreal edges. We refine the network by removing edges that are not in PPIs [21], TarBase 7.0 [22], TargetScan 7.0 [23], miRWalk 2.0 [24], TransmiR 2.0 [25], and miRTarBase 6.1 [26], and removing edges that have a lower weight to make the constructed network more reliable and specific for each patient.

2.4 Voting Algorithm

The network we constructed for each patient can be represented as $G(V, E)$, where V represents the set of nodes in the network and E denotes the set of edges in the network. As cancer drivers have an important impact on other genes in the network, we then consider identifying key nodes as cancer drivers in patient-specific networks by using a voting mechanism. In patient-specific networks, a state tuple (vs_v, va_v) is attached to each node, which vs_v represents the voting score of node v and va_v represents the voting ability of node v. Based on the idea of HWVoteRank [31] considers the gene regulatory network as a heterogeneous network, we regard the personalized network constructed for each patient as a heterogeneous network. The details of the voting algorithm to identify key nodes are described as follows:

Initialize. The voting ability (va_v) of each node is set to 1 and the voting score (vs_v) is set to 0 in the network.

Vote Process. In each personalized network, each node can vote for its neighbors and can obtain a voting score from its neighbors. VoteRank++ [32] proposes that the node gets voting scores from its neighbors are related to its degree. WVoteRank [33] suggests that the voting scores a node gets from its neighbors are related to the weight between them. In the voting mechanism of our method, we consider both the degree of the node and the edge weights between nodes. Give nodes u and v, the voting proportion of u against v is defined as:

$$VP_{uv} = \frac{k_v}{\sum_{i \in \phi_u} k_i} + |\phi_v| * \frac{w(u, v)}{\sum_{u \in \phi_v} w(u, v)} \tag{9}$$

where k_v is the degree of node v, $w(u, v)$ represents the edge weights between node u and node v, $|\phi_u|$ is the number of neighbors of node u, and $|\phi_v|$ is the number of neighbors of node v.

Based on the voting proportion of node u to node v, we define the voting score that node v obtains from its neighbors as follows:

$$vs_v = \sqrt{|\phi_v| \sum_{u \in \phi_v} VP_{uv} \cdot va_u} \tag{10}$$

When a round of voting process has been completed, the node with the highest voting scores is selected as a key node in this round.

Update Voting Ability. When a node is selected as a key node, its voting ability is set to 0, and cannot participate in the subsequent voting process. In order to make the identified driver genes reasonably distributed in the personalized network, we suppress the voting ability of the 1-hop neighbors and 2-hop neighbors of the selected key node. As the gene regulatory network we constructed is a heterogeneous network, suppression exists only between nodes of the same type, the voting ability is suppressed by the following:

$$va_u = \lambda^{\frac{1}{p}} * va_u \tag{11}$$

where $\lambda \in [0, 1]$ is a suppressing factor, $p = 1$ for the first-order neighbor of the key node, and $p = 2$ for the second-order neighbor of the key node. In this study, we set $\lambda = 0.5$ in our experiment

Repeat. Start the next round of the voting process after updating the voting ability of nodes until all key nodes are selected.

3 Results

3.1 Performance Comparisons of PerVote with Other Methods for Identifying Coding Cancer Driver Genes

We identify personalized cancer drivers based on five TCGA cancer datasets, including gene expression data from 747 BRCA samples, 475 HNSC samples, 445 LUAD samples, 336 LUSC samples, and 240 KIRC samples. We evaluate the performance of our method by comparing it with eight other cancer driver identification methods. Since most of these methods only identify coding driver genes, we only compare coding drivers with these methods. We take the measures used by other methods to aggregate the identified personalized driver genes in each sample and compare the aggregated results. Our method is a voting process between nodes and their neighbors in the network, the higher the voting score a node gets from its neighbors, the higher it rank. We select the top 100 coding driver genes of each patient and aggregate their scores to get the results of the predicted driver genes. We obtain results from other methods under these five cancer datasets from the PNC paper. In the performance comparison of the methods, CGC is used as the gold

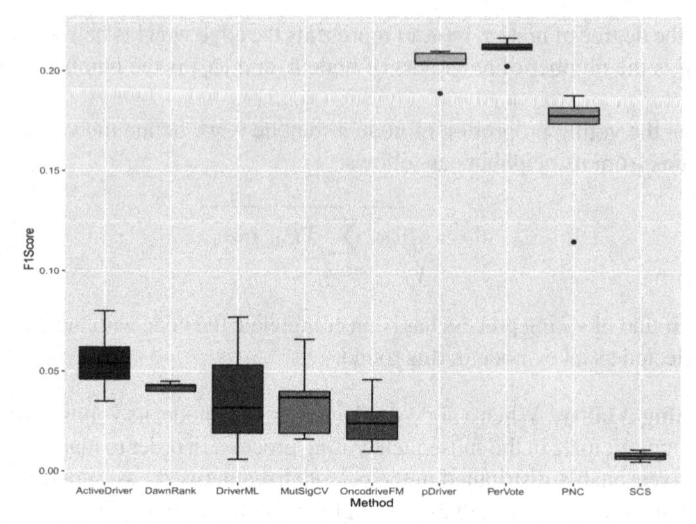

Fig. 2. Comparison of F_1Score of Pervote and other methods in five cancer datasets for identifying coding cancer drivers. The x-axis denotes the nine different methods and the y-axis denotes the F_1Score.

standard. Then, we calculate the F_1Score of nine methods for identifying coding cancer drivers.

The result is shown in Fig. 2. In this figure, PNC, pDriver, and PerVote have higher F_1Score than other methods, but PerVote has the highest F_1Score. According to the results in Fig. 2, it can be shown that our method is an effective driver gene identification method.

3.2 Discovery of miRNA Cancer Drivers

Our approach not only identifies coding drivers but also identifies and prioritizes miRNA cancer drivers. We use the OncomiR [19] database, a database of dysregulated miRNAs across different cancer types, to validate the miRNA drivers predicted by our method. We validate the top 20 miRNAs drivers with the OncomiR database.

The fractions of the top 20 miRNA drivers in OncomiR in these five cancers are shown in Fig. 3. We can see that over 50% of the miRNA drivers predicted by PerVote are related to tumorigenesis of cancer in each cancer type. This result suggests that PerVote is effective in identifying personalized miRNA drivers.

3.3 Exploration of Actionable Drivers and Druggable Drivers

PerVote calculates a score for personalized driver genes of each patient, and the higher the score of the gene, the higher the rank. In this section, we explore whether the gene with the higher rank in each patient could provide useful information for medicine, then help patients with treatment. We use two data sources to explore this idea, (i) DGIdb 3.0 [29], which contains genes for directed pharmacotherapy (druggable genes), we use only

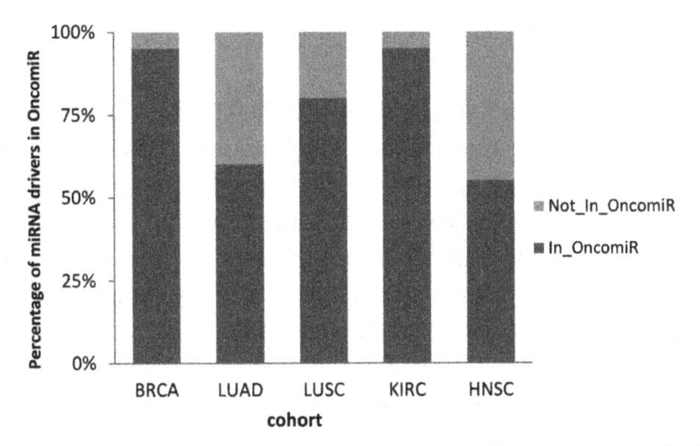

Fig. 3. The fractions of the top 20 miRNA drivers in OncomiR for five cancers. Each bar represents the fractions of predicted miRNA drivers in OncomiR and not in OncomiR.

200 genes of cancer-specific sources from DGIdb. (ii) TARGET [28], which contains 151 genes with genomic-driven therapy exists (actionable genes). We calculate the number of actionable genes and druggable genes in the top 10 ranked personalized cancer drivers for each patient. The results are shown in Fig. 4, we can find all patients harbor at least two druggable drivers, and most patients harbor actionable drivers. Therefore, PerVote can help personalize treatment for cancer patients and provide useful information for biomedicine.

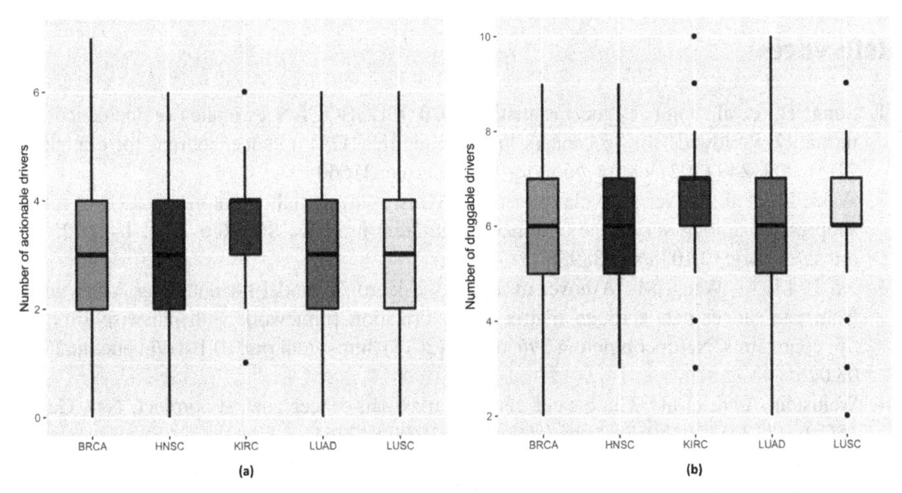

Fig. 4. The number of actionable drivers and druggable drivers. (a): The distribution of the number of druggable drivers in the top 10 cancer drivers predicted by PerVote for each patient. (b): The distribution of the number of actionable drivers in the top 10 cancer drivers predicted by PerVote for each patient.

4 Conclusions

In this work, we develop a new method named PerVote for identifying personalized driver genes through a voting mechanism, which mainly consists of two steps. In the first step, the personalized gene network is constructed based on gene expression data of cancer patients. In the second step, we calculate the voting scores of the nodes in the network based on a voting mechanism. The higher the voting score a node gets in the network, the higher its importance, and we consider the node with a higher score as a personalized driver. We evaluate the performance of PerVote by comparing it with other driver gene identification methods in five cancer types, the results show that our method has better performance. In the miRNA drivers that our method identified, most of them have been confirmed associated with cancer tumorigenesis by OncomiR. Furthermore, all patients have druggable genes in highly ranked driver genes, and most patients have actionable genes. These results suggest that PerVote can contribute to cancer driver gene research and the design of effective personalized cancer treatments.

In the future, we will apply PerVote to more cancer types to identify personalized driver genes for individual patients, and we also consider using more data types to improve Pervote.

Acknowledgment. This work has been supported by the National Natural Science Foundation of China (61902216, 61972236 and 61972226), and Natural Science Foundation of Shan-dong Province (No. ZR2018MF013).

References

1. Sung, H., et al.: Global cancer statistics 2020: GLOBOCAN estimates of incidence and mortality worldwide for 36 cancers in 185 countries. CA: a cancer journal for clinicians. **71**(3), 209–249 (2021). https://doi.org/10.3322/caac.21660
2. Waks, Z., et al.: Driver gene classification reveals a substantial overrepresentation of tumor suppressors among very large chromatin-regulating proteins. Sci. Rep. **6**(1), 1–12 (2016). https://doi.org/10.1038/srep38988
3. Xi, J., Li, A., Wang, M.: A novel unsupervised learning model for detecting driver genes from pan-cancer data through matrix tri-factorization framework with pairwise similarities constraints. Neurocomputing **296**, 64–73 (2018). https://doi.org/10.1016/j.neucom.2018.03.026
4. Weinstein, J.N., et al.: The cancer genome atlas pan-cancer analysis project. Nat. Genet. **45**(10), 1113–1120 (2013). https://doi.org/10.1038/ng.2764
5. Consortium, ICG: International network of cancer genome projects. Nature **464**(7291), 993 (2010). https://doi.org/10.1038/nature08987
6. Xi, J., Li, A., Wang, M.: HetRCNA: a novel method to identify recurrent copy number alternations from heterogeneous tumor samples based on matrix decomposition framework. IEEE/ACM Trans. Comput. Biol. Bioinf. **17**(2), 422–434 (2018). https://doi.org/10.1109/TCBB.2018.2846599
7. Xi, J., Li, A.: Discovering recurrent copy number aberrations in complex patterns via non-negative sparse singular value decomposition. IEEE/ACM Trans. Comput. Biol. Bioinf. **13**(4), 656–668 (2015). https://doi.org/10.1109/TCBB.2015.2474404

8. Lawrence, M.S., et al.: Mutational heterogeneity in cancer and the search for new cancer-associated genes. Nature **499**(7457), 214–218 (2013). https://doi.org/10.1038/nature12213
9. Gonzalez-Perez, A., Lopez-Bigas, N.: Functional impact bias reveals cancer drivers. Nucleic Acids Res. **40**(21), e169–e169 (2012). https://doi.org/10.1093/nar/gks743
10. Han, Y., et al.: DriverML: a machine learning algorithm for identifying driver genes in cancer sequencing studies. Nucleic Acids Res. **47**(8), e45–e45 (2019). https://doi.org/10.1093/nar/gkz096
11. Xi, J., et al.: Inferring subgroup-specific driver genes from heterogeneous cancer samples via subspace learning with subgroup indication. Bioinformatics **36**(6), 1855–1863 (2020). https://doi.org/10.1093/bioinformatics/btz793
12. Bashashati, A., et al.: DriverNet: uncovering the impact of somatic driver mutations on transcriptional networks in cancer. Genome Biol. **13**(12), 1–14 (2012). https://doi.org/10.1186/gb-2012-13-12-r124
13. Pham, V.V., et al.: CBNA: a control theory based method for identifying coding and non-coding cancer drivers. PLoS Comput Biol. **15**(12), e1007538 (2019). https://doi.org/10.1371/journal.pcbi.1007538
14. Hou, J.P., Ma, J.: DawnRank: discovering personalized driver genes in cancer. Genome Med. **6**(7), 1–16 (2014). https://doi.org/10.1186/s13073-014-0056-8
15. Pham, V.V., et al.: pDriver: a novel method for unravelling personalized coding and miRNA cancer drivers. Bioinformatics **37**(19), 3285–3292 (2021). https://doi.org/10.1093/bioinformatics/btab262
16. Reimand, J., Bader, G.D.: Systematic analysis of somatic mutations in phosphorylation signaling predicts novel cancer drivers. Mol Syst Biol. **9**(1), 637 (2013). https://doi.org/10.1038/msb.2012.68
17. Guo, W.-F., et al.: Discovering personalized driver mutation profiles of single samples in cancer by network control strategy. Bioinformatics **34**(11), 1893–1903 (2018). https://doi.org/10.1093/bioinformatics/bty006
18. Guo, W.-F., et al.: A novel network control model for identifying personalized driver genes in cancer. PLoS Comput. Biol. **15**(11), e1007520 (2019). https://doi.org/10.1371/journal.pcbi.1007520
19. Wong, N.W., Chen, Y., Chen, S., Wang, X.: OncomiR: an online resource for exploring pan-cancer microRNA dysregulation. Bioinformatics **34**(4), 713–715 (2018). https://doi.org/10.1093/bioinformatics/btx627
20. Lizio, M., et al.: Update of the FANTOM web resource: high resolution transcriptome of diverse cell types in mammals. Nucleic Acids Res. **45**, D737–D743 (2017). https://doi.org/10.1093/nar/gkw995
21. Vinayagam, A., et al.: A directed protein interaction network for investigating intracellular signal transduction. Sci. Signal. **4**(189), rs8–rs8 (2011). https://doi.org/10.1126/scisignal.2001699
22. Vlachos, I.S., et al.: DIANA-TarBase v7. 0: indexing more than half a million experimentally supported miRNA: mRNA interactions. Nucleic Acids Res. **43**(D1), D153–D159 (2015). https://doi.org/10.1093/nar/gku1215
23. Agarwal, V., Bell, G.W., Nam, J.-W., Bartel, D.P.: Predicting effective microRNA target sites in mammalian mRNAs. elife **4**, e05005 (2015). https://doi.org/10.7554/eLife.05005
24. Dweep, H., Gretz, N.: miRWalk2. 0: a comprehensive atlas of microRNA-target interactions. Nat. Methods. **12**(8), 697–697 (2015). https://doi.org/10.1038/nmeth.3485
25. Wang, J., Lu, M., Qiu, C., Cui, Q.: TransmiR: a transcription factor–microRNA regulation database. Nucleic Acids Res. **38**(suppl_1), D119–D122 (2010). https://doi.org/10.1093/nar/gkp803

26. Chou, C.-H., et al.: miRTarBase 2016: updates to the experimentally validated miRNA-target interactions database. Nucleic Acids Res. **44**(D1), D239–D247 (2016). https://doi.org/10.1093/nar/gkv1258
27. Forbes, S.A., et al.: COSMIC: exploring the world's knowledge of somatic mutations in human cancer. Nucleic Acids Res. **43**(D1), D805–D811 (2015). https://doi.org/10.1093/nar/gku1075
28. Van Allen, E.M., et al.: Whole-exome sequencing and clinical interpretation of formalin-fixed, paraffin-embedded tumor samples to guide precision cancer medicine. Nat. Med. **20**(6), 682–688 (2014). https://doi.org/10.1038/nm.3559
29. Cotto, KC., et al.: DGIdb 3.0: a redesign and expansion of the drug–gene interaction database. Nucleic Acids Res. **46**(D1), D1068–D1073 (2018). https://doi.org/10.1093/nar/gkx1143
30. Kuijjer, M.L., Tung, M.G., Yuan, G., Quackenbush, J., Glass, K.: Estimating sample-specific regulatory networks. Iscience. **14**, 226–240 (2019). https://doi.org/10.1016/j.isci.2019.03.021
31. Yu, D., Yu, Z.: HWVoteRank: a network-based voting approach for identifying coding and non-coding cancer drivers. Mathematics **10**(5), 801 (2022). https://doi.org/10.3390/math10050801
32. Liu, P., Li, L., Fang, S., Yao, Y.: Identifying influential nodes in social networks: a voting approach. Chaos, Solitons Fractals **152**, 111309 (2021). https://doi.org/10.1016/j.chaos.2021.111309
33. Sun, H.-L., Chen, D.-B., He, J.-L., Ch'ng, E.: A voting approach to uncover multiple influential spreaders on weighted networks. Phys. A **519**, 303–312 (2019). https://doi.org/10.1016/j.physa.2018.12.001

TDCOSR: A Multimodality Fusion Framework for Association Analysis Between Genes and ROIs of Alzheimer's Disease

Qi Zou, Yan Sun, Feng Li, Juan Wang, Jin-Xing Liu, and Junliang Shang(✉)

School of Computer Science, Qufu Normal University, Rizhao 276826, China
{sunyan225,lifeng1028,jlshang}@qfnu.edu.cn

Abstract. The complementary multimodality data fusion analysis provides a new perspective for revealing associations between genes and brain regions of interests (ROIs) of Alzheimer's disease (AD). In this paper, we proposed a multimodality fusion framework of AD gene-ROI association analysis, named the two-stage decision cluster optimal structure reduction (TDCOSR). Specifically, gene-ROI association pairs were constructed as fusion features, where the set of candidate genes was identified from samples with only genetic modality. Second, a series of back propagation neural networks (BPNNs) were trained in parallel and then divided into multiple decision clusters. Finally, the weights of decision clusters were balanced and predict disease states by a linear stacking approach. The weights of decision clusters provide a built-in interpretation mechanism to discover fusion features with strong discriminative power. We tracked the best discriminatory features to identify disease-specific gene-ROI associations. Experiments using the TDCOSR were performed on data from the Alzheimer's Disease Neuroimaging Initiative database, the results of which showed that the TDCOSR could identify more potential pathogenic genes and ROIs related to AD.

Keywords: Alzheimer's disease · Ensemble learning · Optimal feature extraction · Brain imaging genetics

1 Introduction

Alzheimer's disease (AD) is a highly heritable degenerative brain disorder that is considered the most common cause of cognitive decline [1]. Therefore, there is an urgent need for the early diagnosis and identification of effective treatment options for AD. Recently, brain imaging genomics has received unprecedented attention as an emerging research approach that can be used to discover potential pathogenic factors in AD [2].

Despite the many advantages of brain imaging genomics [3], there is a conflict between the data dependence of BIG and the cost of acquiring multimodality data. Current studies tend to directly discard samples with incomplete modalities, which further exacerbates the already serious problem of small samples. From another perspective, the dimensionality of the feature space involved in BIG is high (e.g., tens of thousands). With the feature space of such size, it is difficult for traditional models to converge to a stable

M. S. Bansal et al. (Eds.): ISBRA 2022, LNBI 13760, pp. 159–168, 2022.
https://doi.org/10.1007/978-3-031-23198-8_15

state. Hence, BIG requires more effective feature selection or dimensionality reduction methods to address this problem. Ensemble learning uses a set of decision-making systems that apply various strategies to combine classifiers to improve the predictive performance, which has excellent performance on data of all scales. Bi *et al.* [4] used a genetic evolution-based neural network cluster model to build a multitask framework to classify patients with brain diseases and extract important features.

Inspired by these promising studies of ensemble learning in brain imaging genetics, in this study, we presented a multimodality fusion framework for abnormal gene-region-of-interest (ROI) association analysis in AD, named two-stage decision cluster optimal structure reduction (TDCOSR). First, gene-ROI association pairs were constructed as fusion features, where candidate genes come from incomplete samples with only gene modalities data. Second, a series of back propagation neural networks (BPNNs) were trained in parallel, which was then divided into multiple decision clusters by using self-organizing mapping (SOM) and partitioned to the minimal optimal decision structure by using locality-sensitive hashing (LSH). Third, by assigning weights to decision clusters in a linear stacking manner, the ensemble achieved optimal classification performance. The weights of decision clusters provide a built-in interpretation mechanism to discover fusion features with strong discriminative power. We tracked the best discriminatory features to identify disease-specific gene-ROI associations. We used multimodality data from the Alzheimer's Disease Neuroimaging Initiative (ADNI) database to verify the reliability of the TDCOSR.

2 Methods

The detailed process of TDCOSR is shown in Fig. 1. We will describe TDCOSR in detail in the subsequent sections.

2.1 Multimodality Data Acquisition and Preprocessing

In this study, a total of 201 AD patients and 243 healthy controls (HCs) were collected from the ADNI database. Among them, 171 AD patients (male/female: 91/80; age: 75.42 $\pm$ 7.44) and 211 HCs (male/female: 113/98; age: 75.73 $\pm$ 4.89) were incomplete samples with only gene modalities, which were used for screening candidate genes in the study. Other samples, including 30 AD patients (male/female: 13/17; age: 73.43 $\pm$ 7.18) and 32 HCs (male/female: 14/18; age: 74.47 $\pm$ 6.25), had both genetic data and fMRI data, which were used for association analysis between genes and ROIs in the study.

The standard preprocessing steps were performed on functional magnetic resonance imaging (fMRI) data using the DPARSF tool. Finally, the AAL3 template was used to segment the functional image into 90 ROIs. Similarly, we used PLINK tool for gene quality control. Then, the MAGMA tool, along with the gene location obtained from the National Coalition Building Institute, was used to determine all gene coordinates. Finally, 486,199 SNPs were retained and annotated into 18,820 protein-coding genes.

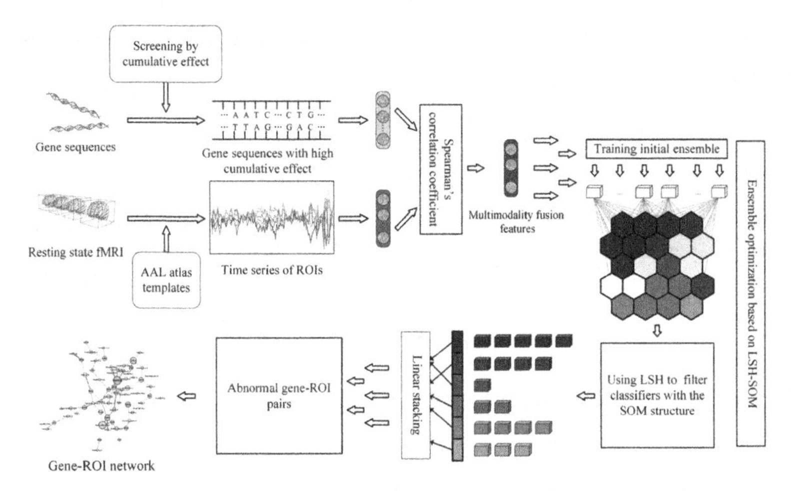

Fig. 1. The detailed process of TDCOSR.

2.2 The Construction of Multimodality Fusion Features

A candidate gene screening strategy was proposed to select risk genes. Specifically, the risk value of each gene is represented as the cumulative effect of SNPs within its physical location, and those genes with sufficiently high cumulative effects are screened out as candidate genes. Mathematically, the cumulative effect CE_g of a gene g can be defined as:

$$CE_g = \sum_{s \in S_g} MIC(s; l) \tag{1}$$

where S_g represents the set of SNPs located in g and $MIC(s; l)$ represents the pathogenic effect of SNP s on phenotype l, which can be measured by the maximal information coefficient (MIC).

These candidate genes are then coded according to its SNPs, with each SNP is coded as $\{0, 1, 2\}$ corresponding to a homozygous common genotype, heterozygous genotype, and homozygous minor genotype. The Spearman's correlation coefficient was employed to measure the associations between genes and ROIs.

2.3 Optimization of the Ensemble

Ensemble Optimization Based on the LSH-SOM Strategy

Dual perturbation of samples and features was utilized to train a series of BPNNs to constructed the initial ensemble. Further, for the BPNN BP_a, the decision bias $SV_a = [f_1^a, ..., f_i^a, ..., f_M^a]$, where $f_i^a \in \{0, 1\}$ if the i - th fusion feature is selected by BP_a, $f_i^a = 1$; otherwise, $f_i^a = 0$. Then, BPNNs can be clustered into different decision clusters using the SOM clustering method. For each decision bias SV_a at the t - th iteration, the neuron in the competitive layer closest to SV_a (winner neuron WN_a^t) is activated,

$$WN_a^t = \min_{i=1}^{\sqrt{B}} dist(SV_a, W_i(t)), \tag{2}$$

where $dist(\cdot)$ denotes the Manhattan distance and $W_i(t)$ is the weight vector of the i-th competitive layer neuron at the t - th iteration. Then, the neighboring neurons NN_a^t of WN_a^t are collected by,

$$NN_a^t = \{\psi | dist(SV_a, W_\psi(t)) < r(t)\}, \tag{3}$$

where $r(t)$ is the neighborhood radius. Finally, the weights of both NN_a^t and WN_a^t are updated

$$W_i(t+1) = W_i(t) + \theta(WN_a^t, i, t) \cdot \beta(t) \cdot dist(SV_a, W_i(t)), i \in WN_a^t \cup NN_a^t, \tag{4}$$

where β is a learning coefficient, $\theta(WN_a^t, i, t)$ is the gaussian nearest neighbor function. While the iteration is over, coarse-grained decision clusters of BPNNs are obtained. Suppose there are a total of D decision clusters, the center BPNN CEN_d and the optimal BPNN OPT_d in the d - th decision cluster are defined as,

$$CEN_d = \min_{c \in C_d} dist(SV_c, W_d),\ OPT_d = \max_{c \in C_d} acc_c, \tag{5}$$

where C_d represents BPNNs in the d - th decision cluster, W_d is the weight vector of the neuron that corresponds to the d - th decision cluster, and acc_c is the classification accuracy of BPNN c in the validation set.

After that, the LSH was introduced to clarify the decision boundaries. In brief, similar samples in their original expression space can be transformed by the LSH with a mapping to the hash index space, where these samples are adjacent with a high probability. Specifically, in the LSH, we use the classification confidence in the validation set rather than the decision bias to characterize the BPNN, i.e., $PC_a = [p_1^a, ..., p_j^a, ..., p_N^a]$, and $p_j^a \in [0, 1]$ [0, 1], meaning the probability of the BP_a predicts the j - th sample to an AD patient. Then, the LSH function can be expressed as,

$$HI(PC_a) = \begin{cases} 1 & PC_a \circ R < 0 \\ 0 & PC_a \circ R \geq 0 \end{cases}, \tag{6}$$

where $HI(PC_a)$ denotes the hash index of PC_a, and R is an N dimensional column vector with its elements being random values in $[-1, 1]$. For two BPNNs a and b, if $HI(PC_a) = HI(PC_b)$, then they are similar BPNNs. Since the LSH is a probability method, to reduce false positives and false negatives, the hash index table is constructed. Specifically, $\eta \times \tau$ LSH functions simultaneously act on PC_a and PC_b to form hash index tables HIT_a and HIT_b, where $HIT_a = [H_1^a, ..., H_i^a, ..., H_\eta^a]$ and $H_i^a = [HI_{i,1}(PC_a), ..., HI_{i,\tau}(PC_a)]$; then, the AND operator is used on H_i^a and H_i^b to generate η binary digits, which are computed by the OR operator to obtain 1 or 0, representing a is (or is not) similar to b.

Based on this LSH strategy, coarse-grained decision clusters can be optimized. In the d - th decision cluster, BPNNs satisfy,

$$C'_d = \{c|c \in C_d \cap (HIT_c = HIT_{OPT_d}) \cap (|PC_c - PC_{OPT_d}|^2 \le |PC_{CEN_d} - PC_{OPT_d}|^2)\}. \tag{7}$$

Weight of Decision Clusters

For a sample x, the probability of the d - th decision cluster classifying it to an AD patient can be written as,

$$P_d(x) = \sum_{c \in C'_d} P_c^{BPNN}(x)/|C'_d|, \tag{8}$$

where $P_c^{BPNN}(x)$ is the probability of BPNN c classifying x to an AD patient, and $|C'_d|$ is the scale of the d - th decision cluster. The probability of the ensemble classifying x to an AD patient is written as,

$$P_{ens}(x) = \sum_{d \in D} w_d \cdot P_d(x)/D \tag{9}$$

where w_d is the weight of the d - th decision cluster.

Eventually, decision clusters are assigned weights using the linear stacking model. Specifically, $P_d(x)$ of all decision clusters for all samples in the training set are used as the input; sample labels are used as the learning objective.

Identification of Disease-Specific Gene-ROI Associations

To capture abnormal gene-ROI pairs related to AD, the weight of each decision cluster is assigned to its BPNNs, and then the weight of each BPNN is further assigned to fusion features that are used for training. The weights of each fusion feature are summed, and hence, the final ranking of fusion features is given.

3 Results

3.1 Selection of the Base Classifier

Figure 2(a) reports the classification accuracies and training times among variant combination of base classifier and correlation coefficients, and Fig. 2(b) shows the receiver operating characteristic (ROC) curves. It is clearly seen that the combination of BPNN and Spearman's correlation coefficient achieves the highest accuracy, the largest AUC value, and the smallest standard deviation.

3.2 Parameter Settings

To find optimal parameter settings, experiments under different combinations of parameter settings were performed. Figure 3 and Fig. 4 show the classification accuracies and training times under different combinations of parameter settings, respectively.

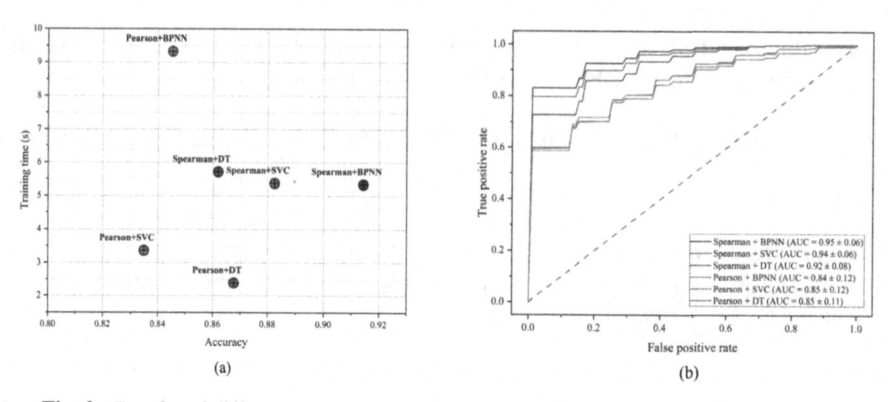

Fig. 2. Results of different combinations of base classifiers and correlation coefficients.

In summary, to balance the classification accuracy and the training time, 450 BPNNs were used in the ensemble, η and τ were set to 8 and 12, respectively. Under this parameter setting, the classification accuracy of the ensemble is 0.914.

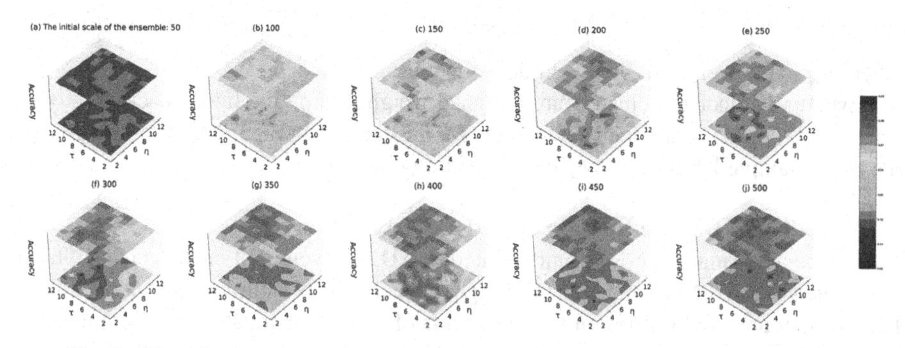

Fig. 3. Classification accuracies under different combinations of parameter settings.

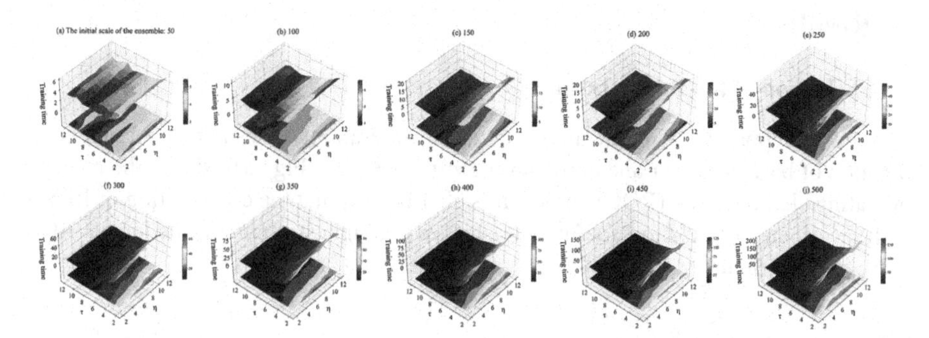

Fig. 4. Training times under different combinations of parameter settings.

3.3 Classification Performance of Sorted Fusion Features

To demonstrate that top fusion features sorted by the TDCOSR have better classification performance than top fusion features sorted by other methods and hence are more likely to be abnormal gene-ROI pairs related to AD. The comparative experiments were carried out using DT, Random Forest, Extra Trees and Ridge. Similar to Sect. 3.1, Spearman's correlation coefficient and Pearson's correlation coefficient were also compared. An SVC is trained by the top fusion features sorted by each compared method, where the numbers of top fusion features are gradually increased from 50 to 1000 with a step of 50.

Table 1 records the highest classification accuracy and the corresponding number of fusion features of each compared method. Figure 5 also shows their experimental results, where Fig. 5(a) reports classification accuracies and the corresponding numbers of fusion features, and Fig. 5(b) plots the ROC curves. It is seen that the highest classification accuracy of the TDCOSR is significantly superior to the highest classification accuracies of compared methods, reaching up to 0.912. Almost all classification accuracies of the TDCOSR are higher than classification accuracies of other methods under different scales of top fusion features. The TDCOSR has the largest AUC value (0.95) and the smallest standard deviation (0.05). Therefore, top fusion features sorted by the TDCOSR are more likely to be abnormal gene-ROI pairs related to AD.

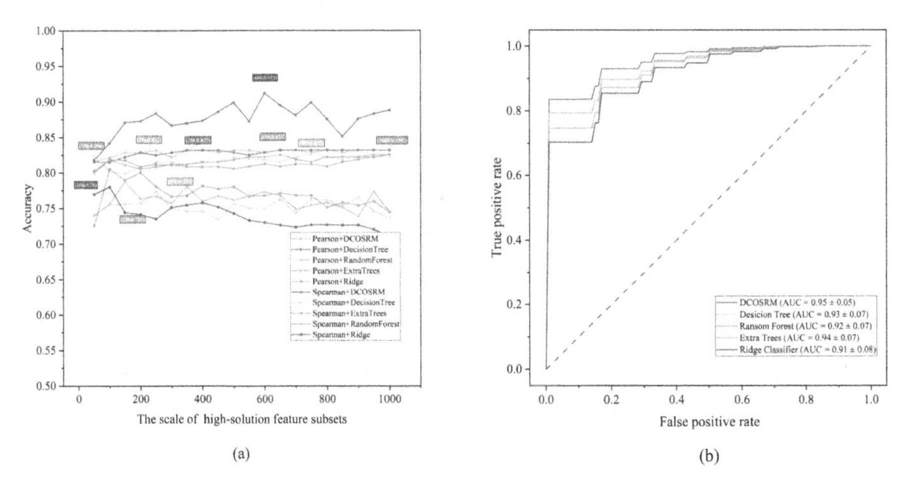

Fig. 5. Classification performance of top fusion features sorted by compared methods.

3.4 Association Analysis of Abnormal Gene-ROI Pairs

To analyze associations between these potential pathogenic genes and ROIs, the gene-ROI network was constructed as shown in Fig. 6(c). In the network, nodes represent ROIs and genes, respectively and each edge corresponds to an abnormal gene-ROI pair. Figure 6(a) and Fig. 6(d) list potential pathogenic ROIs and genes.

Table 1. The highest classification accuracy of each compared method.

Method	Accuracy	Number of fusion features
TDCOSR	**0.912**	600
Spearman + Decision Tree	0.774	250
Spearman + Random Forest	0.788	100
Spearman + Extra Trees	0.806	150
Spearman + Ridge	0.780	100
TDCOSR(Pearson)	0.832	250
Pearson + Decision Tree	0.829	350
Pearson + Random Forest	0.825	750
Pearson + Extra Trees	0.822	650
Pearson + Ridge	0.825	600

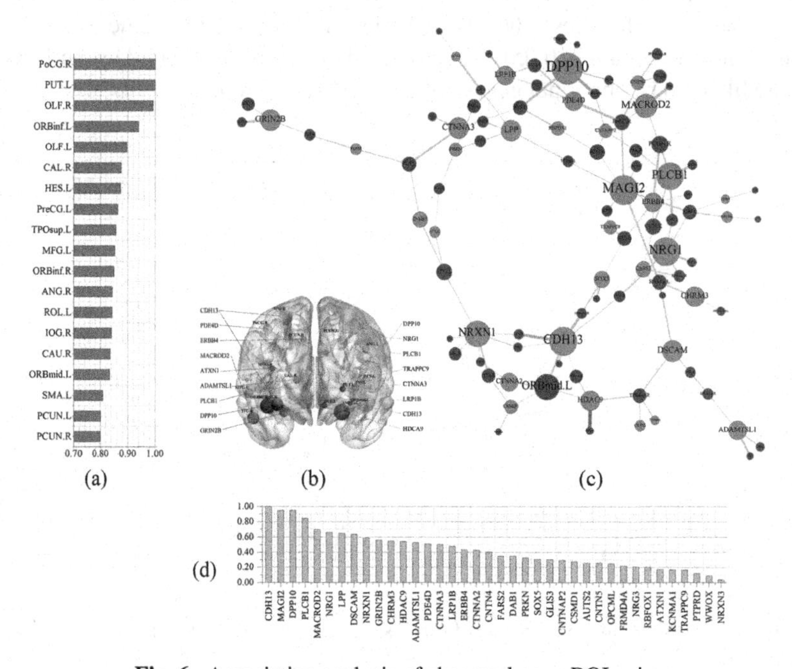

Fig. 6. Association analysis of abnormal gene-ROI pairs

The HDAC9-AMYG. L association pair has the highest edge weight, both of which indeed have an inextricable relationship in many abnormal brain diseases [5]. In the gene-ROI network, some genes, including CDH13 [6], NRG1 [7], ERBB4 [8], CTNNA3 [9], NRXN1 [10], HDAC9 [11], DPP10 [12], MAGI2 [13] and several ROIs, including PoCG [14], PreCG [15], PUT [16], PCUN [17], HIP [18], AMYG [19], STG [20], have been confirmed to be associated with AD.

4 Conclusions

In this study, we proposed the TDCOSR for capturing abnormal gene-ROI associations from multimodality fusion features, which are constructed based on correlation analysis of genes and ROIs. Experiments of the TDCOSR were performed on data from the ADNI database, the results of which showed that the TDCOSR could identify more potential pathogenic genes and ROIs related to AD.

Although the TDCOSR has a higher classification performance, it still has several limitations and needs further improvement, which inspires us to continue working in the future. First, the brain atlas division based on AAL is relatively rough, which might reduce the sensitivity of the TDCOSR to capture pathogenic factors. Further, in addition to genetic data and fMRI data, other multimodality data can be used in combination to deeply identify pathogenic factors of AD.

Acknowledgments. This work was supported by the National Natural Science Foundation of China (61972226, 61902216, 61902430, and 61872220).

Conflict of Interests. The authors declare that there is no conflict of interests regarding the publication of this paper.

References

1. Habib, N., et al.: Disease-associated astrocytes in Alzheimer's disease and aging. Nat. Neurosci. **23**(6), 701–706 (2020). https://www.ncbi.nlm.nih.gov/pubmed/32341542
2. Saykin, A.J., et al.: Genetic studies of quantitative MCI and AD phenotypes in ADNI: progress, opportunities, and plans. Alzheimers Dement **11**(7), 792–814 (2015). https://www.ncbi.nlm.nih.gov/pubmed/26194313
3. Nathoo, F.S., Kong, L., Zhu, H.: A review of statistical methods in imaging genetics. Canadian J. Statist. **47**(1), 108–131 (2019). https://doi.org/10.1002/cjs.11487
4. De Carvalho, L.M., Wiers, C.E., Sun, H., Wang, G.J., Volkow, N.D.: Increased transcription of TSPO, HDAC2, and HDAC6 in the amygdala of males with alcohol use disorder. Brain Behav. **11**(2), e01961 (2021)
5. Liu, F., Zhang, Z., Chen, W., Gu, H., Yan, Q.: Regulatory mechanism of microRNA-377 on CDH13 expression in the cell model of Alzheimer's disease. Eur. Rev. Med. Pharmacol. Sci **22**, 2801–2808 (2018)
6. Jiang, Q., Chen, S., Hu, C., Huang, P., Shen, H., Zhao, W.: Neuregulin-1 (Nrg1) signaling has a preventive role and is altered in the frontal cortex under the pathological conditions of Alzheimer's disease. Mol. Med. Rep. **14**(3), 2614–2624 (2016)
7. Woo, R.-S., Lee, J.-H., Yu, H.-N., Song, D.-Y., Baik, T.-K.: Expression of ErbB4 in the neurons of Alzheimer's disease brain and APP/PS1 mice, a model of Alzheimer's disease. Anatomy Cell Biol. **44**(2), 116 (2011)
8. Giau, V.V., Senanarong, V., Bagyinszky, E., An, S.S.A., Kim, S.: Analysis of 50 neurodegenerative genes in clinically diagnosed early-onset Alzheimer's disease. Int. J. Mol. Sci. **20**(6), 1514 (2019)

9. Swaminathan, S., Kim, S., Shen, L., Risacher, S., Foroud, T., Pankratz, N., Potkin, S., Huentelman, M., Craig, D., Weiner, M., Saykin, A.: Genomic copy number analysis in Alzheimer's disease and mild cognitive impairment: an ADNI study. Int. J. Alzheimer's Disease **2011**, 1–10 (2011). https://doi.org/10.4061/2011/729478
10. Karch, C.M., Goate, A.M.: Alzheimer's disease risk genes and mechanisms of disease pathogenesis. Biol. Psychiat. **77**(1), 43–51 (2015)
11. Kitazawa, M., Kubo, Y., Nakajo, K.: Kv4. 2 and accessory dipeptidyl peptidase-like protein 10 (DPP10) subunit preferentially form a 4: 2 (Kv4. 2: DPP10) channel complex. J. Biol. Chem. **290**(37), 22724–22733 (2015)
12. Ochi, S., et al.: Identifying blood transcriptome biomarkers of Alzheimer's disease using transgenic mice. Mol. Neurobiol. **57**(12), 4941–4951 (2020)
13. Amanzio, M., et al.: Unawareness of deficits in Alzheimer's disease: role of the cingulate cortex. Brain **134**(4), 1061–1076 (2011)
14. Valera-Bermejo, J.M., De Marco, M., Mitolo, M., McGeown, W.J., Venneri, A.: Neuroanatomical and cognitive correlates of domain-specific anosognosia in early Alzheimer's disease. Cortex **129**, 236–246 (2020)
15. Viñas-Guasch, N., Wu, Y.J.: The role of the putamen in language: a meta-analytic connectivity modeling study. Brain Struct. Funct. 1–14 (2017). https://doi.org/10.1007/s00429-017-1450-y
16. Koch, G., et al.: Transcranial magnetic stimulation of the precuneus enhances memory and neural activity in prodromal Alzheimer's disease. Neuroimage **169**, 302–311 (2018)
17. Hyman, B.T., Van Hoesen, G.W., Damasio, A.R., Barnes, C.L.: Alzheimer's disease: cell-specific pathology isolates the hippocampal formation. Science **225**(4667), 1168–1170 (1984)
18. Vogt, L.K., Hyman, B., Van Hoesen, G., Damasio, A.: Pathological alterations in the amygdala in Alzheimer's disease. Neuroscience **37**(2), 377–385 (1990)
19. Galton, C.J., et al.: Differing patterns of temporal atrophy in Alzheimer's disease and semantic dementia. Neurology **57**(2), 216–225 (2001)

Policy-Based Hypertension Monitoring Using Formal Runtime Verification Monitors

Abhinandan Panda[1](✉), Srinivas Pinisetty[1], and Partha Roop[2]

[1] Indian Institute of Technology Bhubaneswar, Bhubaneswar, India
{ap53,spinisetty}@iitbbs.ac.in
[2] The University of Auckland, Auckland, New Zealand
p.roop@auckland.ac.nz

Abstract. Hypertension is a global health threat, manifesting severe complications in vital organs of the human body. The rising prevalence of hypertension urges a novel non-invasive continuous monitoring technique for blood pressure. One of the solutions is to monitor the body's physiological signals, such as an electrocardiogram (ECG), which carries vital information regarding the cardiac system. In earlier works, many deep learning models have been proposed for non-invasive hypertension monitoring based on ECG; however, such models are black-boxes and lack interpretability. In this work, we propose a policy-based formal runtime monitoring framework for the first time for hypertension monitoring using ECG sensing. A decision tree model is implemented from the ECG features to infer ECG patterns/policies related to hypertension, and based on these policies; a formal verification monitor is synthesized. Using the ECG-BP CHARIS dataset, we evaluate the verification monitor's performance.

Keywords: Runtime verification · Hypertension · ECG and Data mining

1 Introduction

Blood pressure (BP) is the vital sign representing a person's cardiovascular health. Blood pressure monitoring is an essential aspect of preventative healthcare practiced in clinical risk and disease management over the years. Hypertension (high blood pressure) poses a high-risk factor for cardiovascular diseases (CVDs) [20], which is the leading cause of mortality around the globe.

For measuring blood pressure, the invasive measurement technique offers the best accuracy among several other methods but is limited in use because of its difficulties and associated risks. Blood pressure measured using auscultation and oscillometric techniques in health centers necessitates the assistance of a trained expert and lacks continuous BP measurement.

This work has been partially supported by The Ministry of Human Resource Development, Government of India (SPARC P#701), IIT Bhubaneswar Seed Grant (SP093).

M. S. Bansal et al. (Eds.): ISBRA 2022, LNBI 13760, pp. 169–179, 2022.
https://doi.org/10.1007/978-3-031-23198-8_16

With the availability of wearable physiological sensors, several non-invasive BP monitoring techniques have been studied using physiological signals such as ECG and PPG. Several machine learning models such as support vector machine, random forest, artificial neural networks (ANN), convolutional neural networks (CNN), recurrent neural network (RNN), Gated Recurrent Unit (GRU), and long short-term memory network (LSTM) have been proposed to monitor hypertension [6,10,15].

The deep learning-based models have proven their effectiveness in terms of hypertension classification tasks; however, these models are considered "black-box." Black-box systems are ones in which the internal process is unknown and so can only be evaluated in terms of input and output. This means that there is no explanation for why any hypertension was classified in a specific way based on ECG. This restricts clinical interpretation, which is essential for informing a diagnosis and increasing confidence in an evaluation. Considering this, there is an urge for explainable monitoring models in high-risk domains such as healthcare [7,19]. There is a need to develop robust and clinically interpretable hypertension analysis tools that are "white-box."

This work proposes a policy-based formal runtime verification framework for hypertension monitoring using ECG. Formal methods-based runtime verification [4,16,17] has been proposed as a dynamic monitoring approach that verifies the execution of a system online/off-line against the set of properties/policies of the system. In this approach, a formally verified runtime monitor is synthesized from the given policies, which are formalized as high-level specifications. During execution, the monitor determines whether the system (under observation) satisfies a specified policy (φ).

1.1 Overview of the Proposed Approach

As presented in Fig. 1, a runtime verification monitor is developed from the ECG timed policies. The ECG policies are extracted from a data mining model (decision tree) implemented using the timed features of ECG. The policies are then formally specified as timed automata for synthesizing an RV monitor. The monitor continuously verifies the ECG policies defined for hypertension monitoring using ECG events and provides a verdict if hypertension is present or not.

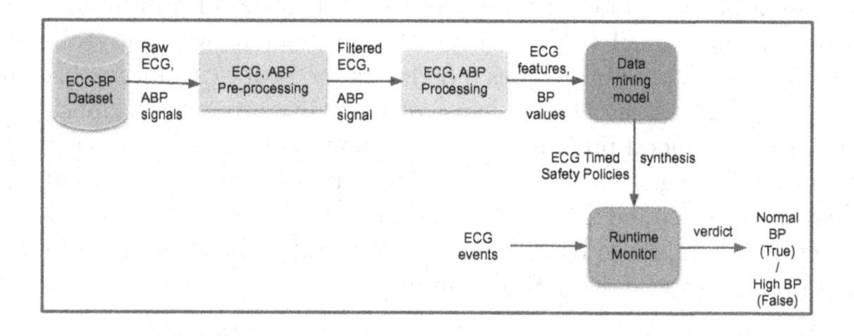

Fig. 1. Proposed BP monitoring framework using RV monitor

The key contributions of the paper are:

- We propose a formal runtime monitor for hypertension monitoring using ECG.
- The ECG features patterns/policies that are closely related to hypertension are presented.
- The suggested method facilitates to create a wearable monitor for non-invasive hypertension monitoring using ECG.

The paper is organized as follows: in Sect. 2, we briefly present an overview of ECG and BP. Section 3 discusses the mining of ECG policies for monitoring hypertension, and in Sect. 4, we explain the formalization of policies and RV monitors with examples. In Sect. 5, we illustrate the behavior of the RV monitor through an example. The performance metrics are discussed in Sect. 6, and conclusions are drawn in Sect. 7.

2 Overview of ECG, BP, ECG-BP Dataset and Feature Extraction

This section presents an overview of typical ECG features and BP, the ECG-BP dataset, and the processing of ECG and ABP signals.

2.1 A Typical ECG and BP

A typical ECG signal represents cardiac activity, such as different phases of blood flow in the body: a diastole phase known as the filling phase and a systole phase known as the pumping phase. As shown in Fig. 2, an ECG contains prominent waves such as the P-wave, QRS-complex, and T-wave. The standard ECG intervals, such as PR, RR, and QT, correspond to the systole and diastole of the cardiac cycle. The PR interval is the time between the start of the P-wave and the start of the QRS-complex. The QT interval is the time from the beginning of the QRS-complex to the end of the T-wave. The RR interval is the duration between two consecutive R-peaks of ECG.

Blood pressure is defined as the force of the blood pushing against the walls of the arteries as the heart pumps blood and is measured in millimeters of mercury (mmHg). During one heartbeat, the maximum blood pressure is systolic blood pressure (SBP), and the minimum pressure between two heartbeats is diastolic blood pressure (DBP). A systolic blood pressure (SBP) of between 140 and 159 mmHg or diastolic blood pressure (DBP) of between 90 and 99 mmHg is defined as the first stage of hypertension, while the second stage is when SBP is higher than 159 mmHg, or DBP is higher than 99 mmHg.

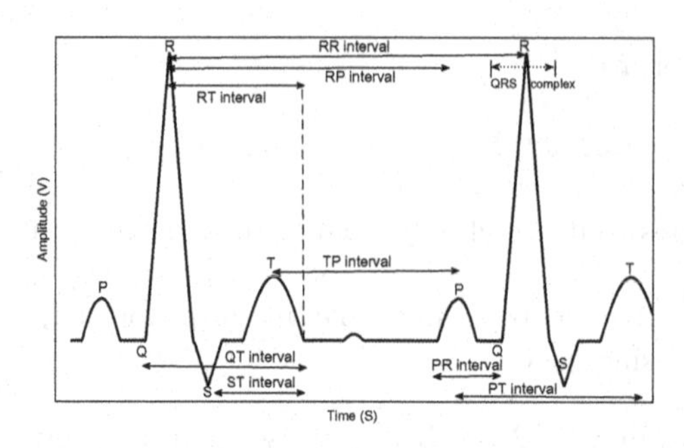

Fig. 2. A typical ECG signal

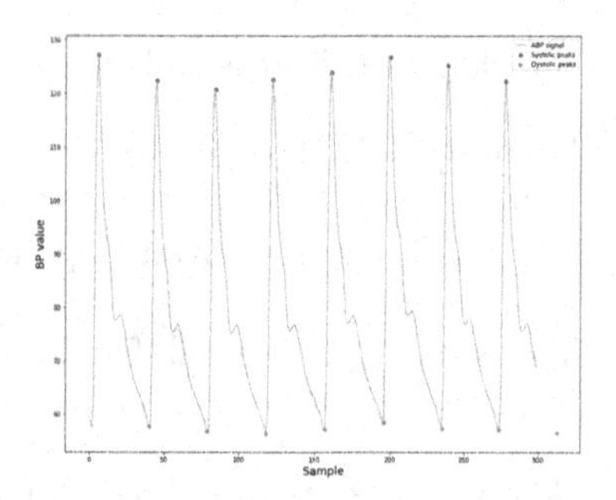

Fig. 3. A processed ABP signal.

2.2 Related Research on the Relationship Between BP and ECG

It is reported that there is a change in the morphological features in an ECG of a hypertensive patient [2]. According to [9], amplitude, dispersion, and other features of the P-wave of ECG are related to changes in blood pressure. In [9], it is studied that there is an increase in the PR interval of ECG with the rise of blood pressure. The QT duration is found to be associated with BP changes [1]. Patients with high blood pressure have a more fragmented QRS-complex and a greater R-wave, and S-wave amplitude [5].

In this work, we study the relationship between ECG systole intervals and systolic blood pressure and thereby extract patterns from the ECG intervals if there are any that strongly determine hypertension in a person.

2.3 ECG-BP Clinical Database and Feature Extraction

Dataset: In this work, we refer to the CHARIS dataset [13] from Physionet [8]. The dataset contains simultaneous ECG and ABP signals recorded 50 Hz by physicians using routinely employed clinical monitors at Rutgers University. We studied the ECG and arterial BP recordings of thirteen patients between the ages of 20 and 74.

Features Considered: In this work, we have considered systole intervals along with other timed features of ECG to study its association with blood pressure. We extract the following ECG timed features:

- PR interval (time interval between P-peak to R-peak)
- RP interval (duration between R-peak to P-peak)
- RR interval (duration between two consecutive R-peaks)
- QT interval (time interval between Q-peak to T-peak)
- RT interval (time interval between R-peak to T-peak)
- ST interval (duration between S-peak to T-peak)
- TP interval (duration between T-peak to P-peak)
- PT interval (time interval between P-peak to T-peak).

Extraction of ECG Features: An ECG recording may be affected by noises such as baseline wander, powerline interference, and motion artefacts. We employ the Neurokit2 [11], a Python-based tool for ECG cleaning and processing. To eliminate baseline drift and high-frequency noise from the ECG signal, we use a high-pass Butterworth filter and a low-pass filter. The R-peaks in an ECG are computed using the Pan-Tompkins approach [14]. We use the Neurokit2 tool's discrete wavelet analysis to compute the P-peak, Q-peak, S-peak, and T-peak events and their times of occurrence.

Extraction of BP Values: A typical arterial blood pressure (ABP) waveform is shown in Fig. 3. The peak values of the ABP signal represent systolic blood pressure (SBP), and the onset indicates diastolic blood pressure (DBP). The SBP and DBP values from the ABP signal are extracted using the Neurokit2 tool.

3 ECG Policy Mining

Since the proposed framework is policy-based, we need to mine the patterns/policies from the ECG features since it is not straightforward to monitor hypertension using ECG sensing. A decision tree [18] is a data mining model widely used for learning patterns from a given feature set. We implemented a decision tree with extracted ECG features such as the PR, RR, RT, QT, TP, and PT intervals, along with class labels as input. Each feature set is associated with a class, such as hypertension (high) or healthy (normal). The decision tree model is input with these features and the class labels. The Scikit-learn module in Python is used to implement the model. The ECG policies are inferred by traversing the generated decision tree's root to its leaves (following the approach in [18]). The decision tree generated from the considered ECG features can be found on GitHub[1].

We inferred the following ECG patterns/policies from the decision tree for hypertension monitoring.

- φ_{ECG1}: if RT $>$ 170 ms and RT $<=$ 219 ms and RR $>$ 540.5 ms and PR $>$ 100.5 ms, then it's hypertension.
- φ_{ECG2}: if RT $<=$ 219 ms and RR $>$ 540.5 ms and PR $<=$ 100.5 ms and QT $>$ 250 ms, then it indicates hypertension.
- φ_{ECG3}: When RT $<=$ 219 ms and RR $<=$ 540.5 ms and QT $>$ 400 ms, then hypertension is present.
- φ_{ECG4}: if RT $>$ 219 ms and TP $>$ 310 ms and PT $>$ 430 ms and ST $<=$ 180 ms, then it's hypertension.

An ECG policy may be a combination of multiple sub-policies; for example, the policy φ_{ECG1} comprises the following sub-policies.

[1] https://github.com/abhinandanpanda/ISBRA-Conference-2022. As the resulting decision tree is large, it's not presented here.

- φ_{ECG1_1}: The RT interval of ECG should not be greater than 170 ms and less than or equal to 219 ms.
- φ_{ECG1_2}: The RR interval of ECG should not be greater than 540.5 ms.
- φ_{ECG1_3}: The PR interval of ECG should not be greater than 100.5 ms.

4 Formalising Policies Using Timed Automata and the Runtime Verification Monitor

In this section, we discuss how to formalize a given ECG policy using timed automaton [3] and the behavior of a runtime verification monitor for a policy specified as timed automata through an example.

4.1 Timed Automata (TA)

Definition 1 (Timed Automata). *A timed automaton* $\mathcal{A} = (Q, q_0, X, \Sigma, \Delta, F)$ *is a tuple, where:* Q *is a finite set of* locations, $q_0 \in Q$ *is the* initial location, X *is a finite set of* clocks, Σ *is a finite set of* events, $\Delta \subseteq L \times \mathcal{G}(X), \Sigma \times 2^X \times L$ *is the* transition relation, $F \subseteq Q$ *is a set of* accepting locations.

Consider the ECG policy φ_{ECG1_1} presented above: The RT interval of the ECG should not be greater than 170 ms and less than or equal to 219 ms. In other words, the policy can be specified as follows: The T event should not be greater than 170 ms and less than or equal to 219 ms after the R event of ECG.

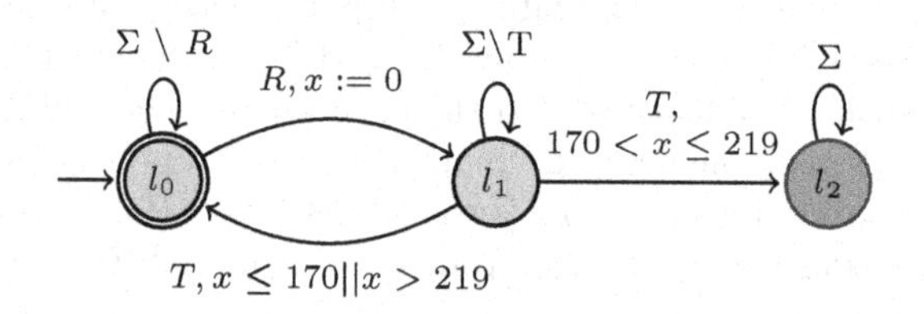

Fig. 4. Timed automata representing ECG policies φ_{ECG1_1}

The timed automaton in Fig. 4 represent the policy φ_{ECG1_1}. The set of locations in the TA is $L = \{l_0, l_1, l_2\}$, where l_0 is the initial location and accepting location (location marked as a double circle with green color. A similar pattern is followed in other automata), l_2 is a non-accepting location (red color). We have considered the set of events, $\Sigma = \{P, Q, R, S, T\}$. Here x is the clock variable that keeps track of the time between the events R and T of ECG. Clocks associated with constraints are known as guards (in this case, $x \leq 170$), and also the clocks can be reset to 0 (in this case, x :=0) and can be associated with constraints known as guards (here $x \leq 170$ is a guard). Here, $X = \{x\}$ is the set of real-valued clocks.

A timed word/trace σ is the sequence of events along with time defined over the alphabet Σ, e.g. $\sigma = (a_1, t_1) \cdot (a_2, t_2) \cdots (a_n, t_n)$ where a_i is an event and t_i is the time of occurrence of the event. $tw(\Sigma)$ is the set of all timed words over Σ.

4.2 Runtime Verification (RV) Monitor

Definition 2 (RV Monitor). *Given a monitoring policy $\varphi \subseteq tw(\Sigma)$ specified as timed automata $\mathcal{A}_\varphi$, the runtime verification monitor is a function $M_\varphi : tw(\Sigma) \rightarrow \mathcal{D}$ for φ, where $\mathcal{D} = \{\mathsf{T}, \mathsf{F}, \mathsf{CT}, \mathsf{CF}\}$ and is defined as follows, with $\sigma \in tw(\Sigma)$ denoting the current observation (a finite timed word over the alphabet Σ):*

$$M_\varphi(\sigma) = \begin{cases} \mathsf{T} & \text{if } \forall \sigma' \in tw(\Sigma) : \sigma \cdot \sigma' \in \varphi \\ \mathsf{F} & \text{if } \forall \sigma' \in tw(\Sigma) : \sigma \cdot \sigma' \notin \varphi \\ \mathsf{CT} & \text{if } \sigma \in \varphi \wedge \exists \sigma' \in tw(\Sigma) : \sigma \cdot \sigma' \notin \varphi \\ \mathsf{CF} & \text{if } \sigma \notin \varphi \wedge \exists \sigma' \in tw(\Sigma) : \sigma \cdot \sigma' \in \varphi \end{cases}$$

The RV monitor reads an input finite timed word (σ) defined over the alphabet Σ. The input is processed by the RV monitor event by event and the monitor may output one of the following after each processing step: T (*true*), F (*false*), CT (*currently true*) and CF (*currently false*).

After reading a timed word σ, the monitor M_φ verdicts T if it is at an accepting location and would remain there for any extension of σ (i.e., for any continuation of σ, φ is satisfied). The monitor M_φ returns the output F if the TA moves to a violating location after reading the current input and remains at a violating location for any further extension of the current input (σ violates φ).

If the policy φ is satisfied by the present observation σ, the monitor M_φ outputs verdict CT; however, all extensions of σ may not meet φ. Otherwise, if the input σ violates φ but there is an extension of σ to meet φ, the monitor will emit CF.

5 Monitoring Policies Using ECG Sensing

Once the ECG policies are formalized as timed automata we follow the approaches in [4,16,17] to develop the RV monitor M_φ. The RV monitor module is implemented in Python 2.7, using UPPAAL DBM libraries [22]. The timed automaton that defines the policy (φ) and input stream (σ) is input into the monitoring algorithm.

The timed automata shown in Fig. 5 model the sub-policies φ_{ECG1_2} and φ_{ECG1_3} of the ECG policy φ_{ECG1}. We compute the product of all TAs presenting the sub-policies that model the TA for φ_{ECG1} since the ECG policy φ_{ECG1} is the intersection of policies φ_{ECG1_1}, φ_{ECG1_2}, & φ_{ECG1_3}.

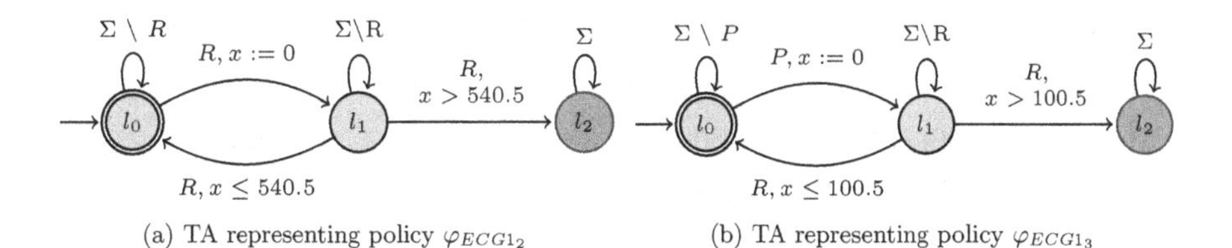

(a) TA representing policy φ_{ECG1_2} (b) TA representing policy φ_{ECG1_3}

Fig. 5. ECG policies φ_{ECG1_2}, φ_{ECG1_3} modelled as Timed automata

All the ECG policies ($\varphi_{ECG1}, \cdots, \varphi_{ECG4}$) are monitored for hypertension. The runtime monitors $\text{M}\varphi_{ECG1}, \cdots \text{M}\varphi_{ECG4}$ corresponding to the policies. The RV monitors read the ECG events, verify the given policy, and provide verdicts. At each step, the verdicts of all the monitors are composed to compute the final verdict. Whenever any policy is violated, the monitor will alarm the user, indicating that the ECG trace shows hypertension signs. The behavior of the RV monitor $\text{M}\varphi_{ECG1}$ for policy φ_{ECG1} is illustrated in the following example for a sample ECG trace (other RV monitors show similar behavior).

Example 1 (Illustrations of RV monitor's ($\text{M}\varphi_{ECG1}$). sequential behaviour for policy $\varphi_{ECG1} = \varphi_{ECG1_1} \wedge \varphi_{ECG1_2} \wedge \varphi_{ECG1_3}$)

Consider the policy φ_{ECG1} presented above to monitor: if RT > 170 ms and RT ≤ 219 ms and RR > 540.5 ms and PR > 100.5 ms then it indicates hypertension. We formalize the policy not to hold.

Let a sample ECG event sequence be: $(P, 30) \cdot (Q, 100) \cdot (R, 120) \cdot (S, 300) \cdot (T, 350) \cdot (P, 500) \cdot (R, 700)$, where P, Q, R, S and T are events of ECG. Each event has a time associated with it, and the RV monitor reads the events in order as they come in. The time associated with an event would be the delay with the prior event or system initialization when feeding the trace to the RV monitor. Table 1 presents the behavior of the RV monitor step-wise.

The monitor senses the first event P at time $t = 30$ and emits the signal CT, signifying that the policy has not been violated. The monitor emits CT for the event Q at time $t = 100$. The event R occurs at $t = 120$, Since the PR interval is less than 100.5 ms, the policy is not broken, and the RV monitor emits CT. The monitor's decision for the S event at $t = 300$ is CT because the policy has not been violated. The event T is sensed at $t = 350$, and the monitor returns CT because the policy is satisfied. The RV monitor emits CT when the next event P is seen at $t = 500$, indicating that the policy is satisfied. When the event R is observed at $t = 700$, the policy gets violated because the RR interval should be less than 540.5 ms, and the PR interval should be less than or equal to 100.5 ms. As a result, the monitor will report the verdict CF, indicating that the presently observed trace violates policy.

Table 1. Step-wise behaviour of RV monitor for the policy φ_{ECG1}

σ	$\text{M}\varphi_{ECG1}(\sigma)$
$(P, 30)$	CT
$(P, 30) \cdot (Q, 100)$	CT
$(P, 30) \cdot (Q, 100) \cdot (R, 120)$	CT
$(P, 30) \cdot (Q, 100) \cdot (R, 120) \cdot (S, 300)$	CT
$(P, 30) \cdot (Q, 100) \cdot (R, 120) \cdot (S, 300) \cdot (T, 350)$	CT
$(P, 30) \cdot (Q, 100) \cdot (R, 120) \cdot (S, 300) \cdot (T, 350) \cdot (P, 500)$	CT
$(P, 30) \cdot (Q, 100) \cdot (R, 120) \cdot (S, 300) \cdot (T, 350) \cdot (P, 500) \cdot (R, 700)$	CF

Remark 1. Here we show how an ECG policy is formalized as timed automata. Once the TA reaches the violating location, it will raise the alarm to the user indicating the policy is violated, and the monitor resets automatically (Fig. 6).

6 Performance Analysis

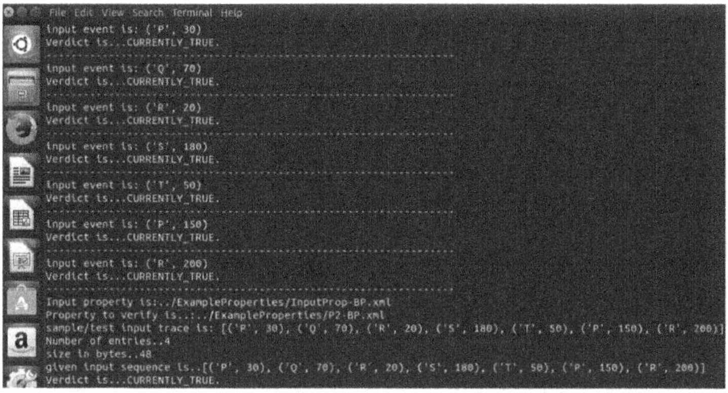

Fig. 6. Execution of a sample ECG trace on RV monitor

We computed accuracy, sensitivity, and specificity to assess the proposed RV monitor's performance using the following equations.

$$Accuracy(\%) = \frac{TruePositives + TrueNegatives}{TruePositives + TrueNegatives + FalsePositives + FalseNegatives} \times 100$$

$$Sensitivity(\%) = \frac{TruePositives}{TruePositives + FalseNegatives} \times 100$$

$$Specificity(\%) = \frac{TrueNegatives}{TrueNegatives + FalsePositives} \times 100$$

The accuracy of the RV framework denotes how well the RV framework classifies normal and hypertension cases, whereas sensitivity and specificity signify the accurate prediction of hypertension and healthy samples out of total samples, respectively. We computed the accuracy, sensitivity, and specificity of the RV framework to be 98.5%, 99.1%, and 97.4%, respectively. In Table 2, we present the RV monitor's performance compared with earlier proposed classification models.

Remark 2. We recorded a few ECG samples of healthy individuals in our lab using a BioRadio device 500 Hz. We found that the ECG policies φ_{ECG2}, φ_{ECG3}, and φ_{ECG4} presented in Sect. 2 are most effective for monitoring showing an accuracy of 99.99% for the two ECG records.

Table 2. Comparison with other works

Author	Dataset	Signal	Features	Model	Accuracy
Shabaan et al., 2020 [21]	CHARIS	ECG	Signal features, age	SVM	98.18%
				ANN	96.5%
				Naïve Bayes	96.08%
Maqsood et al., 2022 [12]	PPG-BP	PPG	Signal features	Logistic Regression	75.13% 0.03
				Linear SVM	75.18% 0.04
				Nearest neighbors	70.92% 1.67
				Gradient boosting classifier	73.65% 0.50
				Decision tree	72.87% 0.48
				Random forest	72.58% 0.48
Our RV framework	CHARIS	ECG	Timed features	Policy based	98.5%

7 Conclusion and Future Work

Hypertension is a global health issue that needs continuous monitoring. In this work, we propose an explainable framework for health monitoring using ECG. We developed a runtime monitor from the patterns/policies of ECG for hypertension monitoring. The proposed approach demonstrates good accuracy and is comparable to other proposed models.

Future work: Other ECG policies may be added to the RV framework to improve hypertension prediction accuracy. The proposed monitoring approach may be tested with other datasets and can be extended for online monitoring. We are exploring the feasibility of deploying the RV monitor on an embedded system.

References

1. Akintunde, A.A., Ayodele, O.E., Opadijo, O.G., Oyedeji, A.T., Familoni, O.B.: QT interval prolongation and dispersion: epidemiology and clinical correlates in subjects with newly diagnosed systemic hypertension in nigeria. J. Cardiovascul. Dis. Res. **3**(4), 290–295 (2012)
2. Aksoy, S., et al.: The effects of blood pressure lowering on p-wave dispersion in patients with hypertensive crisis in emergency setting. Clin. Exper. Hypertens. **32**(7), 486–489 (2010)
3. Alur, R., Dill, D.L.: A theory of timed automata. Theor. Comput. Sci. **126**(2), 183–235 (1994)
4. Bauer, A., Leucker, M., Schallhart, C.: Runtime verification for LTL and TLTL. ACM Trans. Softw. Eng. Methodol. **20**(4), 14:1–14:64 (2011)
5. Bekar, L., et al.: Presence of fragmented QRS may be associated with complex ventricular arrhythmias in patients with essential hypertension. J. Electrocardiol. **55**, 20–25 (2019)
6. Chen, Y., Zhang, D., Karimi, H.R., Deng, C., Yin, W.: A new deep learning framework based on blood pressure range constraint for continuous cuffless BP estimation. Neural Netw. **152**, 181–190 (2022)
7. Gastounioti, A., Kontos, D.: Is it time to get rid of black boxes and cultivate trust in AI? Radiol. Artif. Intell. **2**(3) (2020)

8. Goldberger, A.: PhysioBank, PhysioToolkit, and PhysioNet: components of a new research resource for complex physiologic signals. Circulation **101**(23), e215–e220 (2000). https://doi.org/10.1161/01.CIR.101.23.e215,https://circ.ahajournals.org/content/101/23/e215
9. Hassing, G.J., et al.: Blood pressure-related electrocardiographic findings in healthy young individuals. Blood Press. **29**(2), 113–122 (2020)
10. Jiang, L., Hu, R., Wang, X., Tu, W., Zhang, M.: Nonlinear prediction with deep recurrent neural networks for non-blind audio bandwidth extension. China Commun. **15**(1), 72–85 (2018)
11. Makowski, D., et al.: Neurokit2: a python toolbox for neurophysiological signal processing. Behav. Res. Meth. **53**(4), 1689–1696 (2021)
12. Maqsood, S., et al.: A survey: from shallow to deep machine learning approaches for blood pressure estimation using biosensors. Expert Syst. Appl. 116788 (2022)
13. Kim, N.: Trending autoregulatory indices during treatment for traumatic brain injury. J. Clin. Monitor. Comput. **30**(6), 821–831 (2016)
14. Pan, J., Tompkins, W.J.: A real-time QRS detection algorithm. IEEE Trans. Biomed. Eng. **3**, 230–236 (1985)
15. Paviglianiti, A., Randazzo, V., Villata, S., Cirrincione, G., Pasero, E.: A comparison of deep learning techniques for arterial blood pressure prediction. Cogn. Comput. **14**(5), 1689–1710 (2022)
16. Pinisetty, S., Jéron, T., Tripakis, S., Falcone, Y., Marchand, H., Preoteasa, V.: Predictive runtime verification of timed properties. J. Syst. Softw. **132**, 353–365 (2017)
17. Pinisetty, S., Roop, P.S., Sawant, V., Schneider, G.: Security of pacemakers using runtime verification. In: 2018 16th ACM/IEEE International Conference on Formal Methods and Models for System Design (MEMOCODE), pp. 1–11. IEEE (2018)
18. Quinlan, J.R.: Generating production rules from decision trees. In: ijcai, vol. 87, pp. 304–307. Citeseer (1987)
19. Reyes, M., et al.: On the interpretability of artificial intelligence in radiology: challenges and opportunities. Radiol. Artif. Intell. **2**(3) (2020)
20. Rosendo rff, C., et al.: Treatment of hypertension in patients with coronary artery disease: a scientific statement from the American heart association, american college of cardiology, and american society of hypertension. Circulation **131**(19), e435–e470 (2015)
21. Shabaan, A., Sharawi, A.: Machine learning for blood pressure classification using only the ECG signal. J. Eng. Appl. Sci. **67**(1), 257–274 (2020)
22. UPPAAL DBM Library: The library used to manipulate dbms in uppaal (2020). https://people.cs.aau.dk/~adavid/UDBM/. Accessed 18 June 2020

Deep Learning-Enhanced MHC-II Presentation Prediction and Peptidome Deconvolution

Juntao Deng and Min Liu(✉)

Department of Automation, Tsinghua University, Beijing, China
Lium@tsinghua.edu.cn

Abstract. Antigen-presenting cells can elicit a $CD4^+$ T cell response by displaying foreign peptides on the surface. Identifying such peptides requires robust prediction of the binding and presentation corresponding to peptides and major histocompatibility complexes class II (MHC-II) molecules. However, numerous experimental data suffer from inexact supervision, and the open conformation of MHC-II molecules leads to a complex peptide binding pattern. Though current prediction methods have significantly pushed the development of cancer vaccines and immunotherapies, an urgent desire for better approaches still exists. We practice the powerful multi-head self-attention technique for MHC-II-restricted peptidome deconvolution and antigen presentation prediction problems. According to binding motifs reflected by eluted ligands, the novel expert voting-based deconvolution strategy ensures a reliable MHC-II assignment. Driven by massive trusty annotated peptidome data, our method overwhelms the start-of-the-art MHC-II presentation prediction method, NetMHCIIpan4.0, on two independent single allelic datasets. All these results have demonstrated that our method can boost the performance of MHC-II presentation prediction and peptidome deconvolution.

Keywords: Antigen presentation · $CD4^+$ T cells · Deep learning · Deconvolution

1 Introduction

Antigen presentation based on major histocompatibility complexes (MHC) is an important way for T cells to recognize antigens. As one variant, MHC class II (MHC-II) molecules are expressed by professional antigen-presenting cells, including dendritic cells and macrophages. Peptides derived from extracellular proteins may be bound to MHC-II molecules and then presented on the cell surface to activate $CD4^+$ T cells that play an import role in the immune response [1]. Identifying such peptides is valuable for vaccine design and cancer immunotherapies.

M. S. Bansal et al. (Eds.): ISBRA 2022, LNBI 13760, pp. 180–191, 2022.
https://doi.org/10.1007/978-3-031-23198-8_17

MHC-II molecules in humans are called class II human leukocyte antigen (HLA-II) proteins and are encoded by three polymorphic genes (HLA-DR, HLA-DQ, and HLA-DP) [1]. Different HLA-II proteins have different amino acid sequences and binding grooves and hence can specifically bind peptides with certain patterns and present them to the cell surface. Notably, the binding groove of HLA-II proteins has an open conformation, which leads to a larger length range of peptides that can become ligands for HLA-II proteins [2]. However, it is generally believed that the actual binding core has a fixed length of 9 amino acids [3]. Exploring the MHC-II associated presentation mechanism is paramount for screening MHC-II neoantigens, which have a crucial function in the anti-tumor response [4].

Numerous assay experiments have accumulated a large amount of data to investigate the mechanism of MHC-mediated antigen presentation. For example, vitro assays of MHC protein binding to peptides have generated many binding affinity (BA) data. Based on these data, computational methods can be constructed for the binding affinity prediction of peptides and MHC proteins. In addition, the use of antibodies to elute MHC ligands from the cell surface has also accumulated massive data. Such eluted ligands (EL) data are divided into single allelic eluted ligands (SAEL) and multi allelic eluted ligands (MAEL). Among them, MAEL data are consequences of adopting pan-specific antibodies and cells expressing multiple MHC molecules. This means that for MAEL data, we can only confirm that the peptide was presented on the cell's surface, but we do not know which of the multiple MHC molecules presented the peptide. To address this problem, the motif deconvolution was performed in several studies [5–8], assigning each peptide to its most likely presenting MHC protein. Driven by SAEL and deconvoluted MAEL data, many machine learning methods were implied to establish excellent antigen presentation prediction models. For instance, NetMHCIIpan4.0 employs a conventional feed-forward neural network with only three layers to predict MHC-II associated presentation and binding at the same time [5]. BERTMHC exploits a transformer neural network that leverages self-supervised pretraining from plenty of protein sequences [7]. MARIA adopts a multimodal recurrent neural network, whose training information incorporates not only peptides and MHC-II alleles but also tissue-specific gene-expression levels [9]. Despite the progress made so far, a more robust and accurate predictive model for MHC-II presentation is still needed today.

In this article, we propose a deep learning-based model to predict the MHC-II associated antigen presentation. During the initial training process, we used a hard-parameter sharing architecture to learn both BA and SAEL data. The deep learning network utilizes a transformer encoder with only one multi-headed self-attention layer to extract information from input sequences of peptides and MHC proteins. The output part is three fully connected layers. Based on a novel strategy, we deconvolute the MAEL data, which cannot be used directly for training, using the naive antigen presentation prediction model obtained from the initial training. Driven by SAEL and deconvoluted MAEL data, we retrained the antigen presentation prediction model and updated the deconvolution results on the MAEL data. By iterating this process, the resulting antigen presentation prediction model significantly outperformed NetMHCIIpan4.0, one of the most commonly used MHC-II presentation prediction algorithms, in two independent HLA-DR testing datasets. In addition, the final deconvolution results of our method for MAEL data are remarkably reliable. The binding motif corresponding to deconvoluted MAEL data exhibits very high similarity compared to the binding motif displayed by SAEL data. In particular, for HLA-DRB1*01:01, the Pearson correlation coefficient (PCC) of the two binding motifs reaches 0.979. Overall, by applying powerful deep learning methods and reasonable deconvolution strategies, our method can achieve more accurate MHC-II presentation prediction and motif deconvolution.

2 Methods

2.1 Dataset

We employed the training data of NetMHCIIpan4.0 [5], which include 108959 binding affinity (BA) instances, 651660 SAEL instances and 311949 positive MAEL instances. SAEL data cover 19 MHCs, while MAEL data cover 114 MHCs. We used ninety percent of SAEL data to train our models, leaving the rest for validation. These datasets can be freely downloaded at https://services.healthtech.dtu.dk/suppl/immunology/NAR_NetMHCpan_NetMHCIIpan/.

In order to compare our method with NetMHCIIpan4.0, we obtained two independent testing datasets from supplementary table 5 and 6 of MARIA [9], which are available at https://www.nature.com/articles/s41587-019-0280-2. Both datasets are HLA-DR ligand sets eluted from monoallelic K562 cell lines, one of which contains 2286 ligands for HLA-DRB1*01:01 and the other consist of 1948 ligands for HLA-DRB1*04:04. Both datasets are filtered to include only peptides of length 13–21 following NetMHCIIpan4.0. We enriched the testing data with five times negative instances of equal length sampled from UniProt database [10]. The detailed data enrichment process involves randomly selecting protein sequences in the UniProt database and cutting them out of peptides of the same length as the positive samples. We filtered out instances in the testing dataset that also appeared in the SAEL training data. The final testing data includes 9046 instances for HLA-DRB1*01:01 and 8911 instances for HLA-DRB1*04:04.

2.2 Deep Learning Network for MHC II Presentation Prediction

The popular method, NetMHCIIpan4.1, adopts a conventional feed-forward network with only three dense layers and two output neurons. Simple neural network architectures could limit the predictive performance. It is well known that deep learning techniques have become very powerful and have shown great application values in many fields. In the MHC-II presentation prediction problem, the inputs and outputs are well defined, and a sufficient amount of data has been accumulated to support deep learning.

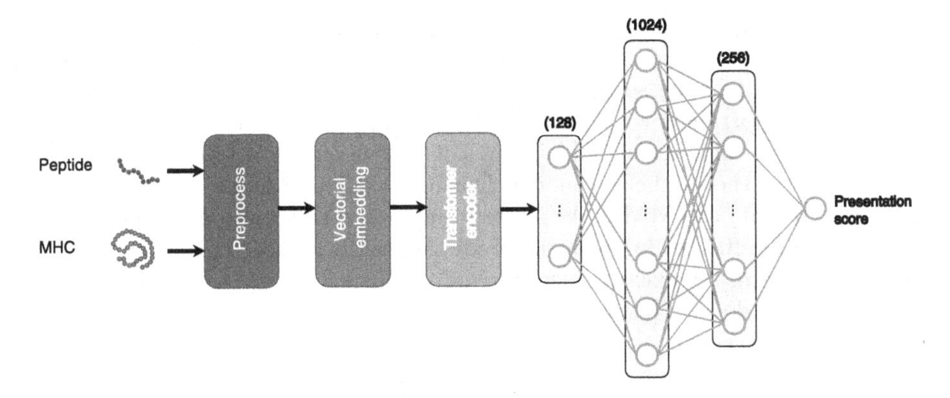

Fig. 1. Structure of the deep learning network. The model inputs the amino acid sequence of the peptide with the pseudo-sequence of the MHC protein and outputs the presentation score of both. The number of neurons included in the three fully connected layers is 128, 1024, and 256, respectively.

We propose a deep learning model to predict the antigen presentation of MHC II, as shown in Fig. 1. The deep learning model's inputs are amino acid sequences of peptides and MHC proteins. Among them, the peptide sequence is padded to a maximum length of 21 with 'X', and the MHC sequence is a pseudo sequence consisting of 34 essential residues related to peptide binding. First, we concatenate the sequences of peptides and MHCs and tokenize input sequences in the preprocessing step. Second, each character in the input sequence is transferred into a vector through word embedding and positional encoding. Word embedding is performed through a trainable embedding layer, which has the identical goal as the one-hot embedding. In order to keep location information for words, positional encoding is achieved through plusing an adjustable position-specific vector to each embedded word. Third, we employ one multi-head self-attention layer to encode the input information. Fourth, three fully-connected layers are adopted to predict the presentation score of input peptide and MHC pairs. We add the BatchNorm layer after the first two fully connected layers to speed up network convergence and control overfitting.

Considering interaction of eluted ligands and binding affinity data can boost the prediction model's performance [11], we fused BA and SAEL training data in the initial training process. BA and SAEL data are combined to train two networks (the BA network and the EL network) with identical shallow layers for extracting features. These shallow layers contain the preprocessing, vectorial embedding, and transformer encoding steps. After that, we can obtain a better EL network than training with single EL data. The EL network is employed as an initial model in the following inexact supervised training process based on MAEL data.

2.3 Inexact Supervised Training of the Deep Learning Model

In the iterative training session, the EL network continuously annotates MAEL data and gets retrained with the merged MAEL and SAEL data (Fig. 2), which is similar to the pipeline of NNAlign_MA [6]. However, deep learning raises the upper limit of the prediction model. During iterative training, more powerful models can annotate MAEL data more accurately, and more accurate deconvolution results further improve the prediction performance. This promotes the model surpassing the existing method NetMHCIIpan.

During the deconvolution step of MAEL data, we employ a novel strategy based on voting. For a positive MAEL instance, we first adopt our trained model to give presentation scores of the peptide to every candidate MHC protein. If our method is confident for the presentation of one of the peptide-MHC protein pairs (score over 0.9), the MHC protein with the highest score is selected as the deconvolution result corresponding to that instance. Otherwise, a committee organized by our method and two experts starts to vote on the pairing situation, and the vote result is used as the deconvolution result for that instance. Here, we assigned the expert 1 as NetMHCIIpan4.0 [5] and the expert 2 as MixMHC2pred [8]. Each labeled positive MAEL instance is paired with five negative MAEL instances sampled from original MAEL data of NetMHCIIpan4.0. Therefore, 1871694 instances are added to the SAEL training set during each iteration to readjust the parameters of the model.

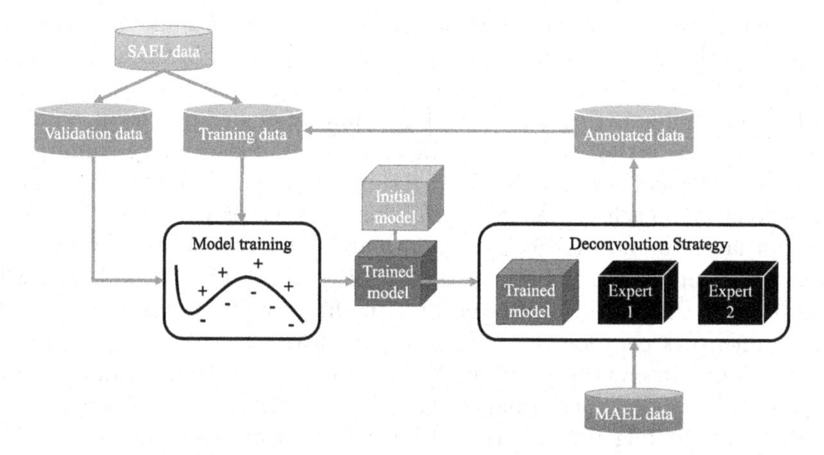

Fig. 2. Pipeline of inexact supervised training process. The initial parameters of the model are derived from the optimal model saved in the previous iteration, except for the first training, which is derived from the initial model.

2.4 Loss Function

For the EL network, we utilize cross entropy (CE) loss to measure the distance between the ground truth and the model's prediction. For BA network, considering binding affinity prediction is a regression problem, we choose mean square error (MSE) to calculate the loss.

$$CE(p, g) = -\frac{1}{N} \sum_{i=1}^{N} (g_i \cdot \log(p_i) + (1 - g_i) \cdot \log(1 - p_i)) \tag{1}$$

$$\mathrm{MSE}(p, g) = \frac{1}{N} \sum_{i=1}^{N} (g_i - p_i)^2 \tag{2}$$

where g is the ground truth and p is the model's prediction.

2.5 Evaluation Metrics

The following criteria are adopted to evaluate the predictive performance of models:

$$\text{Accuracy} = \frac{TP + TN}{TP + TN + FP + FN} \tag{3}$$

$$\text{Recall} = \frac{TP}{TP + FN} \tag{4}$$

$$\text{Precision} = \frac{TP}{TP + FP} \tag{5}$$

$$F1 \text{ score} = \frac{2 \cdot \text{Precision} \cdot \text{Recall}}{\text{Precision} + \text{Recall}} \tag{6}$$

where TP is true positive, FP is false positive, TN is true negative, FN is false negative. Moreover, we employ area under receiver operating characteristics (AUC ROC) as the additional metric to further compare the performance of models.

3 Results

3.1 Iterative Optimization of the Deep Learning Model

MAEL data only indicate whether the peptide can be presented in the cell line without determining precisely which subtype of HLA-II proteins achieve presentation. In this situation, supervision information is not as exact as that of SAEL data. The inexact supervision problem is one branch of weakly supervised learning [12]. In order to take advantage of the massive MAEL data on developing accurate prediction methods, we employed an iterative optimization method similar to pseudo labelling [13].

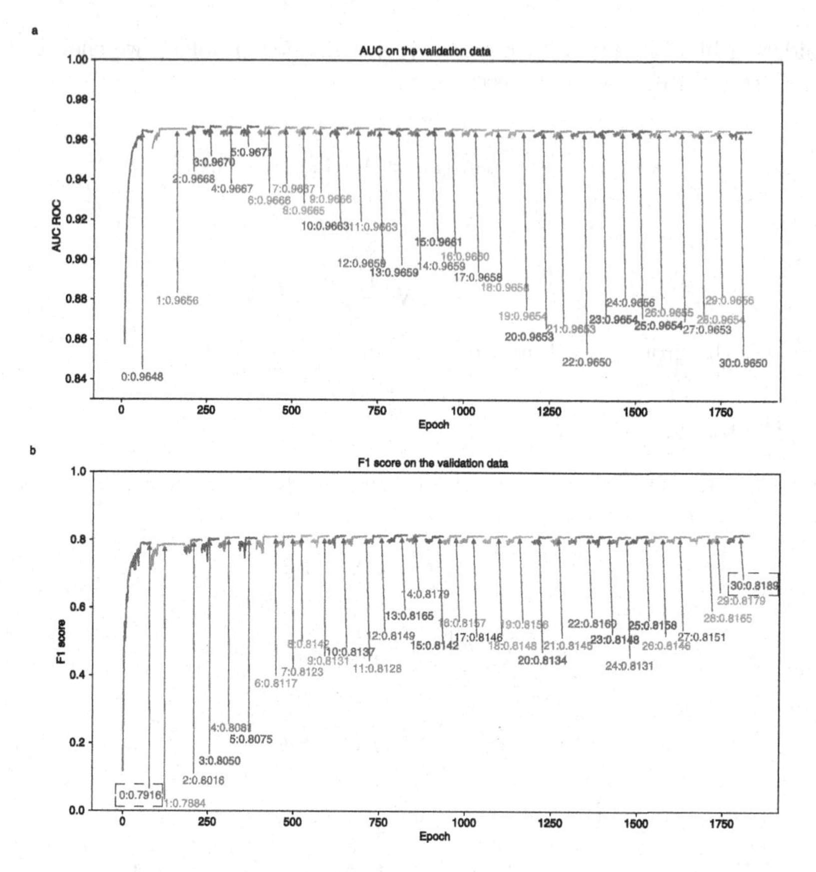

Fig. 3. Model performance on the validation data during the training process. The current number of iterations (CNOI) and the best performance (BP) in each iteration are marked in the form of CNOI:BP. Starting from the 1st iteration, the model saved from the previous iteration deconvolutes MAEL data and adds them to the current training data. The models framed by dashed lines are used for the subsequent comparisons.

During each round of iteration, we computed the performance of the model predictions on the validation set, as shown in Fig. 3. During the initial (round 0) iteration, the BA data and SAEL data were combined to train the model. The initial model achieved an AUC of 0.9648 and an F1 score of 0.7916 on the validation set, which indicates that deep learning can tap into the antigen presentation pattern implied inside the database.

In the following iterative training process, annotated MAEL data and SAEL data were merged to finetune the latest model. We froze the shallow network of the model, which includes preprocessing, vectorial embedding, and transformer encoding steps. In each iteration round, if the AUC ROC of the model on the validation set does not improve for more than 30 epochs, the model optimization is terminated early for the current round. Before the next iteration started,

MAEL data are re-labeled by the latest model and utilized for the next round of training.

In Fig. 3a, we can observe that our model obtained the highest AUC ROC at the fifth round of iteration. However, the optimal F1 score is obtained in the last iteration, as shown in the Fig. 3b. This is possible because the F1 score only considers the model's prediction performance under a specific threshold (e.g., 0.5 in this article), whereas the AUC ROC considers the overall performance of the model under all possible thresholds. Considering that we finally adopted 0.5 as the discriminant threshold, we chose the optimal model saved in the last iteration round for the subsequent performance comparison experiments on independent testing datasets.

3.2 Performance Comparison of MHC-II Presentation Prediction

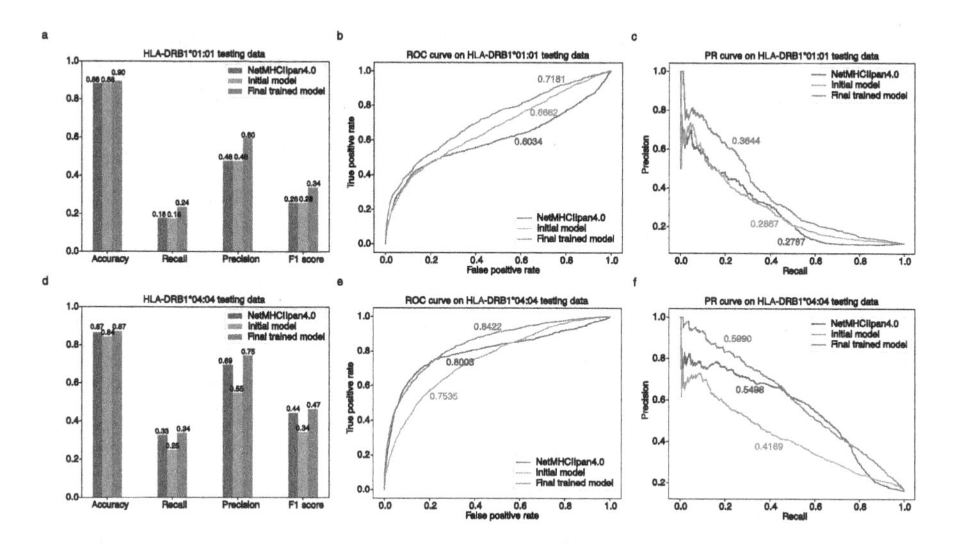

Fig. 4. Benchmarking our methods on MHC-II presentation prediction. (a, b, c), Comparison of our methods with NetMHCIIpan4.0 on HLA-DRB1*01:01 testing data. (d, e, f), Comparison of our methods with NetMHCIIpan4.0 on HLA-DRB1*04:04 testing data. In b and e, AUC ROC is indicated. In c and f, AUC PR is indicated.

To evaluate the ability of our models to predict MHC-II presentation, we compared our models with NetMHCIIpan4.0 on two independent testing datasets of HLA-DR typing. Inspecting all evaluation metrics, the initial model obtained in the round 0 iteration surpasses NetMHCIIpan across the board in the HLA-DRB1*01:01 testing data but lags in the HLA-DRB1*04:04 testing data (Fig. 4). According to the analysis of SAEL training data, positive instances for HLA-DRB1*01:01 have 15661 entries, while positive instances for HLA-DRB1*04:04 only have 1532 entries. The data imbalance between the two HLA-DR subtypes in the SAEL data may explain the difference in prediction performance of the initial model for different HLA-DR subtypes.

As shown in Fig. 4, our final model obtained in the last round iteration improves in all evaluation metrics against the initial model. Especially for HLA-DRB1*04:04, where the initial model performs poorly, the final trained model gains a 38% improvement in F1 score. Furthermore, compared to NetMHCIIpan 4.0, our final model makes better presentation predictions for both types of HLA-DR. These results illustrate that our method can achieve reliable deconvolution of MAEL data and improve the predictive model's performance. The analysis of deconvoluted MAEL data shows that 23275 HLA-DRB1*01:01 positive entries and 10379 HLA-DRB1*04:04 entries are augmented into training data. Richer training data may be the primary reason for the improved model performance.

3.3 Evaluation of Deconvoluted Binding Motifs

To further show the reliability of the deconvolution results, we compared binding motifs exhibited in the final deconvoluted MAEL data and corresponding motifs obtained from SAEL data and NetMHCIIpan4.0, as shown in Fig. 5. We deliberately selected the HLA-DR subtypes in two independent antigen presentation testing datasets for further analysis. GibbsCluster2.0 [14] was adopted here for the alignment of peptide sequences and visualization of binding motifs. Amino acids are divided into four groups, where red indicates negative charge, blue indicates positive charge, green indicates polarity, and black indicates hydrophobicity.

From Fig. 5, We can observe that many amino acid patterns exhibited by SAEL data and NetMHCIIpan are captured by deconvoluted MAEL data, such as hydrophobic Phe(F) and polar Tyr(Y) at position 1(P1) for HLA-DRB1*01:01. Besides, hydrophobic Ile(I) and Leu(L) at P1 for DRB1*04:04 are found by deconvoluted MAEL data. PCC between the binding motifs from SAEL data and deconvoluted MAEL data was calculated to estimate the similarity. For HLA-DRB1*01:01, a very high correlation was observed with a PCC of 0.979. For DRB1*04:04, a PCC of 0.911 also implied the excellent agreement.

4 Discussion

Unlike the more extensively studied MHC class I (MHC-I) molecules, both BA and EL data are relatively less available for MHC-II. Therefore, current computational methods for MHC-II are significantly less than those for MHC class I, especially in antigen presentation prediction. In addition, because of the open conformation of MHC-II molecules, their maximum length of bound peptides is significantly longer than that of MHC-I molecules. The core binding region in MHC-II ligands is difficult to identify, making the peptides' binding process to MHC-II molecules more complex and hence more challenging to model and predict. Besides, current EL data contain more multi-allele data than single-allele data, and training with multi-allele data requires a motif deconvolution operation. The reliability of deconvolution results directly affects the performance of data-driven machine learning models. These obstacles in MHC-II presentation prediction have led to the desire for more robust and accurate computational models.

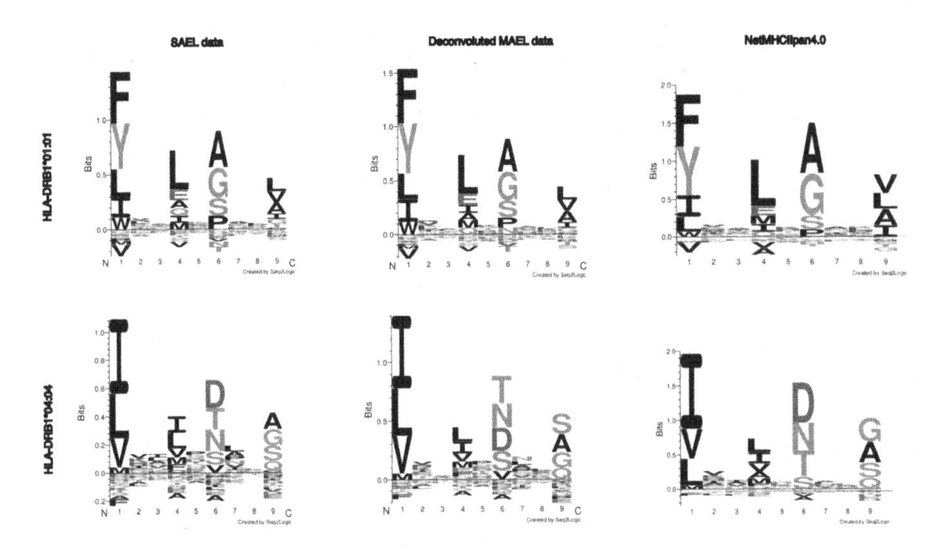

Fig. 5. Comparison of deconvoluted MAEL binding motifs with motifs derived from SAEL data and NetMHCIIpan4.0. For each HLA-II allele, binding motifs of SAEL data, deconvoluted MAEL data, and NetMHCIIpan4.0 are shown in columns 1, 2 and 3 in turn. Binding motifs of NetMHCIIpan can be obtained from https://services.healthtech.dtu.dk/service.php?NetMHCIIpan-4.0. Amino acids are divided into four categories here: hydrophobic (black), polar (green), negatively charged (red), and positively charged(blue). Besides, the letters' size reflects each residue's relative contribution in each position. (Color figure online)

We apply powerful deep learning methods to antigen presentation prediction of MHC-II and deconvolution of MAEL data. The final model achieves an F1 score of 0.34 and an AUC ROC of 0.7181 on the HLA-DRB1*01:01 independent testing dataset, which significantly exceeds the F1 score (0.26) and AUC ROC (0.6034) of NetMHCIIpan4.0. In addition, on the HLA-DRB1*04:04 independent test dataset, our method achieved good performance with an accuracy of 87%, recall of 34%, and precision of 75%. It is worth noting that our model uses exclusively training data derived from NetMHCIIpan 4.0, which ensures a fair comparison. Moreover, our method demonstrates reliable performance in terms of deconvolution. Since there lacks ground truth for the deconvolution of MAEL data, we follow NNAlign_MA to verify the correctness of deconvolution. If deconvolved MAEL data possess similar binding motifs to SAEL data, the deconvolution result is reliable. In our deconvolution result, deconvoluted

MAEL data and SAEL data have highly similar binding motifs, where the PCC for HLA-DRB1*01:01 is 0.979, and the PCC for HLA-DRB1*04:04 is 0.911. The superior performance of our model also benefits from a large amount of reliable deconvoluted MAEL data.

It should be noted that this study has examined the model's MHC-II presentation prediction performance on only two HLA-DR subtypes. A publicly available antigen presentation benchmark with richer MHC-II subtypes can make model comparisons more convenient and convincing, but we failed to find such a benchmark dataset. Therefore, the collection and collation of decentralized MHC-II related antigen presentation data will facilitate model comparisons and algorithmic progress. Similar works have been done for MHC-II binding prediction [15] and MHC-I epitopes prediction [16]. Besides, peptides presented by MHC molecules do not necessarily activate T cells [17]. MHC-II-related immunogenicity prediction is essential to enhance the algorithm's benefit for treating autoimmune, transplantation, and tumor-related diseases.

Acknowledgements. This work was supported in part by funding from the National Science Foundation of China(Grant No. 62173204).

References

1. Rock, K.L., Reits, E., Neefjes, J.: Present yourself! By MHC class I and MHC class II molecules. Trends Immunol. **37**, 724–737 (2016)
2. Unanue, E.R., Turk, V., Neefjes, J.: Variations in MHC class II antigen processing and presentation in health and disease. Annu. Rev. Immunol. **34**, 265–297 (2016)
3. Barra, C., et al.: Footprints of antigen processing boost MHC class II natural ligand predictions. Genome Med. **10**, 84 (2018)
4. Alspach, E., et al.: MHC-II neoantigens shape tumour immunity and response to immunotherapy. Nature **574**, 696-+ (2019)
5. Reynisson, B., Alvarez, B., Paul, S., Peters, B., Nielsen, M.: NetMHCpan-4.1 and NetMHCIIpan-4.0: improved predictions of MHC antigen presentation by concurrent motif deconvolution and integration of MS MHC eluted ligand data. Nucleic Acids Res **48**, W449–W454 (2020)
6. Alvarez, B., et al.: NNAlign_MA; MHC peptidome deconvolution for accurate MHC binding Motif characterization and improved T-cell epitope predictions. Mol. Cell Proteomics **18**, 2459–2477 (2019)
7. Cheng, J., Bendjama, K., Rittner, K., Malone, B.: BERTMHC: improved MHC-peptide class II interaction prediction with transformer and multiple instance learning. Bioinformatics **37**(22), 4172–4179 (2021)
8. Racle, J., et al.: Robust prediction of HLA class II epitopes by deep motif deconvolution of immunopeptidomes. Nat. Biotechnol. **37**, 1283–1286 (2019)
9. Chen, B., et al.: Predicting HLA class II antigen presentation through integrated deep learning. Nat. Biotechnol. **37**, 1332–1343 (2019)
10. UniProt, C.: UniProt: the universal protein knowledgebase in 2021. Nucleic Acids Res. **49**, D480–D489 (2021)
11. Garde, C., et al.: Improved peptide-MHC class II interaction prediction through integration of eluted ligand and peptide affinity data. Immunogenetics **71**(7), 445–454 (2019). https://doi.org/10.1007/s00251-019-01122-z

12. Zhou, Z.-H.: A brief introduction to weakly supervised learning. Natl. Sci. Rev. **5**, 44–53 (2018)
13. Lee, D.H.: Pseudo-label: the simple and efficient semi-supervised learning method for deep neural networks. In: Workshop on challenges in representation learning, ICML, vol. 3, no. 2, p. 896 (2013)
14. Andreatta, M., Alvarez, B., Nielsen, M.: GibbsCluster: unsupervised clustering and alignment of peptide sequences. Nucleic Acids Res. **45**, W458–W463 (2017)
15. Andreatta, M., et al.: An automated benchmarking platform for MHC class II binding prediction methods. Bioinformatics **34**, 1522–1528 (2018)
16. Paul, S., et al.: Benchmarking predictions of MHC class I restricted T cell epitopes in a comprehensively studied model system. Plos Comput. Biol. **16**, e1007757 (2020)
17. Wells, D.K., et al.: Key parameters of tumor epitope immunogenicity revealed through a consortium approach improve neoantigen prediction. Cell **183**, 818-+ (2020)

MMLN: Leveraging Domain Knowledge for Multimodal Diagnosis

Haodi Zhang[1], Chenyu Xu[1,2(✉)], Peirou Liang[1], Ke Duan[1], Hao Ren[2], Weibin Cheng[2(✉)], and Kaishun Wu[1(✉)]

[1] College of Computer Science and Software Engineering, Shenzhen University, Shenzhen, China
2060271017@email.szu.edu.cn, wu@szu.edu.cn
[2] Institute for Healthcare Artificial Intelligence Application, Guangdong Second Provincial General Hospital, Guangzhou, China
chwb817@gmail.com

Abstract. Deep learning models have been widely studied and have achieved expert-level performance in medical imaging tasks such as diagnosis. Recent research also considers integrating data from various sources, for instance, chest X-rays (CXR) radiographs and electronic medical records (EMR), to further improve the performance. However, most existing methods ignore the intrinsic relations among different sources of data, thereby lack interpretability. In this paper, we propose a framework for pulmonary disease diagnosis that combines deep learning and domain-knowledge reasoning. We first formalize the standard medical guidelines into formal-logic rules, and then learn the weights of the rules from medical data, integrating multimodal data for pulmonary disease diagnosis. We verify our method on a real dataset collected from a hospital, and the experimental results show that the proposed method outperforms the previous state-of-the-art multi-modal baselines.

Keywords: Domain knowledge · Multimodal · Interpretability · Disease diagnosis

1 Introduction

In the past few years, advancements in deep learning techniques and the release of serveral large open-source chest X-ray (CXR) datasets have achieved significant contributions to medical images analysis for diagnosis decisions. As one of the most common radiological examinations and the first approach to discovering pulmonary diseases, radiographs of the CXR are utilized in a number of radiology applications. However, the current state-of-the-art deep learning models usually rely only on CXR without clinical context [3]. Among the models, multimodality is a novel way of incorporating various sources of data such as texts and images.

In natural language processing, the Transformer architecture [22] has been successful in the multimodal task involving vision and text. It takes as input images and pieces of text and embeds them to output a prediction. Since text

M. S. Bansal et al. (Eds.): ISBRA 2022, LNBI 13760, pp. 192–203, 2022.
https://doi.org/10.1007/978-3-031-23198-8_18

and image modalities are complementary, these data can contribute to improving diagnostic accuracy. Medical domain knowledge is potentially useful for learning-based models. In medical domains, interpretability based on domain knowledge is in particular important, since the doctors and the patients potentially need to understand how and why the diagnosis result is produced. To achieve the interpretability, the domain knowledge such as authoritative clinical guidelines are quite useful. Proper formalization and application of the domain knowledge could improve the model interpretability, and increase model robustness to uncertain data.

To leverage domain knowledge with multimodal data, we propose a Multimodal Markov Logic Network (MMLN) for pulmonary disease diagnosis. In the MMLN, we first formulate diagnostic rules in the form of Markov logic according to the authoritative clinical medical guidelines. Then, we learn each weight for each rule from a variety of textual data, such as EMRs, laboratory reports, and radiology reports. A pre-trained model is fine-tuned to extract common lung pathologies from CXR images. Finally, we compute the marginal probability of lung disease from the text and image data. We carry out experiments on a real dataset obtained from Guangdong Second Provincial General Hospital (GD2H). The results show that MMLN can accurately diagnose lung disease with good interpretability and outperforms state-of-the-art multimodal baselines. Our contributions can be summarized as follows:

- We formalize the medical diagnosis knowledge into first-order logic (FOL) according to authoritative clinical guidelines, improving the accuracy and interpretability of the model.
- We propose a data-driven and knowledge-driven framework MMLN that combine domain knowledge and deep learning to support lung disease diagnosis.
- We conduct experiments on a real-world dataset collected from a hospital to demonstrate the effectiveness of our model, and the proposal outperforms the previous state-of-the-art baselines.

2 Related Work

Deep learning technique has made a tremendous advancement in the field of medical imaging analysis, which can greatly enhance the capabilities of image-level prediction, segmentation, localization, etc. With the release of several large publicly CXR datasets, including CheXpert [10], chestXray8 [24], PadChest [2], the development of CXR diagnosis is accelerating. A series of advanced CNNs for disease classification are proposed. Singh et al. [20] compare the performance of different architectures with various depths on a given task. Sirazitdinov et al. [21] evaluate the effect of various data augmentation and input pre-processing methods. During the COVID-19, Wang et al. [23] proposed a deep learning pipeline for the diagnosis and discrimination of viral, non-viral and COVID-19 pneumonia from chest X-ray images with AUC of above 0.87. Hou et al. [8] developed an diagnosis model based on deep convolutional neural network based on CXR for COVID-19 pneumonia detection. It can also select instances of CXR to explain the prediction of model. Although deep neural networks have a good performance, they are black box models and are not interpretable.

There are some work integrating knowledge graph (KG) or ontology as complement to knowledge. Li et al. [14] propose a 8-step pipeline to build a medical knowledge graph (KG) with a novel quadruplet structure from EMRs. The KG is used to diagnosis with the top 10 recall of 88.76%. Yin et al. [25] propose a new model by directly introducing domain knowledge from the medical knowledge graph into an RNN architecture, as well as taking the irregular time intervals into account. Results on heart failure risk prediction tasks show that the model associates individual medical events with heart failure onset. Shen et al. [19] construct a clinical Bayesian network with complete causal relationship and probability distribution of ontology from EMRs data to enhance diagnostic inference capability. Jiang et al. [11] propose a Markov logic network (MLN) with disease and symptom nodes from KG constructing from EMRs, which can learn and inference for medical diagnosis. Although domain knowledge is incorporated, it lacks theoretical support and does not explore the intrinsic relationship between domain knowledge and multimodal data.

The most similar work to ours are in [17] which combine neural networks and Bayesian Network with interpretability to classify between pneumonia cases and normal cases. Different from the previous work, the novelty of this paper is to leverage medical domain knowledge from authoritative clinical guidelines and classify between the pneumonia cases and the differential diseases related to pneumonia.

3 Methodology

We propose a data-driven and knowledge-driven framework that integrates knowledge representation and inference to diagnose disease with multimodal data. Figure 1 shows the architechure of the proposed framework. Firstly, we use first-order logic (FOL) to formulate diagnostic rules from the clinical guidelines. Then, the medical entities are extracted from the text and image data to build a multimodal evidence database. Next, we use the evidence database to learn weights of diagnostic rules and perform disease inference based on the weighted rules.

3.1 Formulate Diagnostic Rules

We use first-order logic (FOL) to represent knowledge about clinical guidelines. Figure 2 is the clinical diagnostic criteria for community-acquired pneumonia (CAP). Although we cannot exactly express every detail in the clinical diagnostic criteria, we still leverage the knowledge behind it to improve accuracy and interpretability.

For the first item, we define a query predicate called *Pneumonia* representing CAP. For the second item, we define evidence predicates for each clinical manifestation. For example, we define a predicate called *Fever* corresponding to 2 (2)

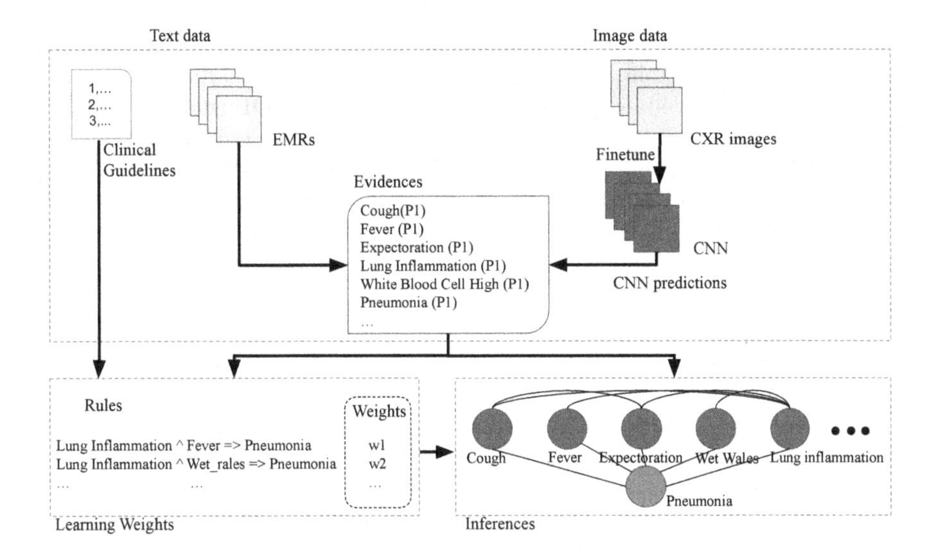

Fig. 1. Multimodal Markov logic network framework.

CAP 的临床诊断标准：
1. 社区发病。
2. 肺炎相关临床表现：
(1) 新近出现的咳嗽、咳痰或原有呼吸道疾病症状加重，伴或不伴脓痰/胸痛/呼吸困难/咯血；
(2) 发热；
(3) 肺实变体征和(或)闻及湿性啰音；
(4) 外周血白细胞>10×10^9/L 或 <4×10^9/L，伴或不伴细胞核左移。
3. 胸部影像学检查显示新出现的斑片状浸润影、叶/段实变影、磨玻璃影或间质性改变，伴或不伴胸腔积液。
符合1、3及2中任何1项，并除外肺结核、肺部肿瘤、非感染性肺间质性疾病、肺水肿、肺不张、肺栓塞、肺嗜酸性粒细胞浸润症及肺血管炎等后，可建立临床诊断。

Clinical diagnostic criteria for CAP:
1. Community acquired.
2. Clinical manifestations related to pneumonia:
(1) Newly developed cough, expectoration or exacerbation of existing respiratory symptoms with or without pus sputum / chest pain / dyspnea / hemoptysis;
(2) Fever;
(3) Solid pulmonary signs and/or wet rales;
(4) peripheral white blood cells >10×10^9/L or <4×10^9/L with or without left shift of nuclei.
3. Chest imaging shows new patchy infiltrative shadow, lobe/segmental solid shadow, ground glass shadow or interstitial changes with or without pleural effusion.
The clinical diagnosis can be established after any one of 1, 3 and 2 is met, and tuberculosis, lung tumor, non-infectious interstitial lung disease, pulmonary edema, pulmonary atelectasis, pulmonary embolism, pulmonary eosinophilic infiltration and pulmonary vasculitis are excluded.

Fig. 2. Clinical diagnostic criteria for CAP (original and its translation).

in Fig. 2. For the third item, we define evidence predicates about chest imaging examination. For example, we define predicates called *Lung_inflammation*.

We then use the defined predicates to construct the FOL rules according to the logical relation described in the last paragraph of the guideline. We select cases that have pneumonia in the admission diagnosis as CAP cases, so the first item is satisfied by default. For the second and third items, we define one of the rules as follows:

$$Fever(x) \wedge Lung\_inflammation(x) \Rightarrow Pneumonia(x): \quad w \tag{1}$$

Since the disease diagnosis in real world is complex and uncertain. For example, if a patient has fever and lung inflammation, it does not mean the patient must have pneumonia. Therefore, we attach a weight w to each rule, which becomes the representation of a Markov Logic Network (MLN). A MLN is a first-order knowledge base with a weight attached to each formula (or clause) [18]. If a patient satisfies one rule, it becomes more probable, while if a patient violates one rule, it becomes less probable but not impossible.

3.2 Building Multimodal Evidence Database

After formulating the diagnostic rules, our goal is to extract each possible grounding predicate appearing in diagnostic rules to build an evidence database from text and image data.

Text Data. We first convert the EMRs from unstructured text to semi-structured JSON format using the regular expression to match the significant fields, such as chief complaint, present illness history, examination, etc. We then extract various clinical manifestations, such as symptoms, physical signs, laboratory items, etc. from these fields. We finally use modified version of NegEx [4], NegBio [16] and named entity recognition (NER) to extract positive symptoms and physical signs.

Figure 3 is an example of JSON format EMRs, from which we extract grounding predicates. In detail, we extract positive symptoms from chief complaints and present illness history, physical signs from physical examination, laboratory items from structured laboratory information system (LIS), etc. For example, we extract the following grounding predicates from a sample case shown in Fig. 3:

$$\begin{aligned} &Cough(P450945) \\ &Expectoration(P450945) \\ &Wet\_rales(P450945) \\ &White\_blood\_cells\_high(P450945) \\ &Pneumonia(P450945) \end{aligned} \tag{2}$$

$P450945$ refers to patient ID. The above grounding predicates show that the patient suffers from cough, expectoration and wet rales. The White Blood Cells (WBC) concentration is higher than the normal level. The patient has been diagnosed with pneumonia.

```
{"ID": "P450945",
"主诉": "咳嗽1周余，加重伴喘息2天",
"现病史": "患儿缘于1周余前无明显诱因出现咳嗽，呈阵发性连声咳，伴咳痰、痰液难咳出，无鼻塞、流涕，无气促、喘息，无发热，无呕吐、
"入院诊断": "肺炎",
"出院诊断": "肺炎",
"一般情况": "无",
"专科情况": "无",
"辅助检查": "2017-12-07我院门诊胸片肺炎。",
"病例特点": "1、幼儿女性，急性病程。2、病史特点患儿缘于1周余前无明显诱因出现咳嗽，呈阵发性连声咳，伴咳痰、痰液难咳出，无鼻塞、
"诊断、诊断依据及鉴别诊断": "1、初步诊断1.肺炎2.急性喘息性支气管炎2、诊断依据幼儿，女性，以“咳嗽”为主要症状。伴有喘息。查体吸
"查体": "T36.6℃,P110次/分,R28次/分。神志清，精神可，全身皮肤无皮疹及出血点，咽部充血，扁桃体I度肿大，双肺呼吸音对称，吸气:

{"ID": "P450945",
"Chief Complaint": "Cough for more than 1 week, aggravated with wheezing for 2 days",
"Present Illness History": "The child developed a cough with no obvious incentive more than 1 week ago , with ...
"Admission Diagnosis": "Pneumonia",
"Discharge Diagnosis": "Pneumonia",
"General": "None",
"Specialty": "None",
"Auxiliary examination": "2017-12-07 Outpatient chest X-ray Pneumonia in our hospital.",
"Characteristics of the case": "1. Young children, female, acute course of disease. 2. The characteristics of ...
"Diagnosis, Diagnosis Basis and Differential Diagnosis": "1. Preliminary diagnosis 1. Pneumonia 2. Acute asthmatic
"Examination": "T36.6℃, P110 times/min, R28 times/min. Consciousness, good spirit, no rash and bleeding spots on .
```

Fig. 3. A sample of EMR (original and its translation).

Image Data. We first collected the CXR images of the selected patients in the hospital. A modified version of Chexpert-labeler [9] was used to extract pathologies as the ground truth labels from radiology reports.

We then use the CXR images and their corresponding labels to finetune a TorchXrayVision [5] CNN model, which has been trained on several large publicly available CXR datasets. The accuracy of the finetuned model is over 92% in an evaluation conducted on 600 CXR images labeled by physicians of the hospital. We regard the CNN predictions as positive grounding predicates about chest imaging examination.

For example, we pass a CXR image through finetuned models and output positive pathology as follows:

$$Lung\_inflammation(P450945) \tag{3}$$

Combining the grounding predicates extracted from the EMRs and CXR images, we obtain a large scale real-world multimodal evidence database.

3.3 Learning Weights and Inference

As mentioned at the end of Sect. 3.1, we construct a multimodal Markov logic network (MMLN) based on diagnostic rules for disease diagnosis. We first initialize one node for each predicate. When the patient has specified clinical manifestation and chest imaging examination, the corresponding grounding predicate is assigned to 1; otherwise, 0.

Next, we split the evidence database into train set and test set in an 80/20 ratio. The train set is used to learn weights, while the test set is used for inference. We use the MLN tool Alchemy [13] and pracmln [15] to learn each weight of rules and infer.

Learning Weights. Given diagnostic rules with the defined predicates and the train set, we learn each weight of rules using MLN learning method in [18]. For example, we learn each weight of rules as Eq. (4) shows. The greater the weight of one rule, the more likely that rule is to happen.

$$\begin{aligned}
&Lung_inflammation(x) \wedge Cough(x) \wedge Expectoration(x) \Rightarrow Pneumonia(x) : 5.12142\\
&Lung_inflammation(x) \wedge Fever(x) \Rightarrow Pneumonia(x) : 6.03559\\
&Lung_inflammation(x) \wedge Wet_rales(x) \Rightarrow Pneumonia(x) : 5.9235\\
&Lung_inflammation(x) \wedge White_blood_cells_high(x) \Rightarrow Pneumonia(x) : 4.23462\\
&Lung_inflammation(x) \wedge White_blood_cells_low(x) \Rightarrow Pneumonia(x) : 2.37089
\end{aligned} \tag{4}$$

Inference. The goal of inference is to compute the marginal probability of a specific illness to make a disease diagnosis. Given weighted rules with defined predicates and the test set, we infer disease probability using MLN inference method in [18]. As for the example mentioned in Sect. 3.2, we get the result as follows:

$$Pneumonia(P450945) \quad 0.99895 \tag{5}$$

The probability represents the illness probability, which shows that the case is highly pneumonia.

4 Experiments

In this section, we introduce the datasets we experiment with and the evaluation results. Besides, we have some discussions about MMLN.

4.1 Datasets

The experiments are conducted on the real-world datasets. We collaborate with the Guangdong Second Provincial General Hospital and are authorized to collect EMRs from the EMR system and CXR images from Picture Archiving and Communication System (PACS). We collect the common lung diseases including Pneumonia, Tuberculosis, Lung cancer and Pulmonary embolism. (see Table 1) In our experiment, disease diagnosis is regarded as a binary classification task of Pneumonia and other differential diseases related to Pneumonia. We treat Pneumonia cases as positive samples and other lung diseases as negative samples.

We first find out all cases with both EMR and CXR. We then balance the number of positive and negative samples so that they are approximately equal in

Table 1. The statistics of the datasets. #cases is the number of cases with both EMR and CXR.

Disease	#cases
Pneumonia	1514
Tuberculosis	344
Lung cancer	1145
Pulmonary embolism	106

Table 2. Performance of our model against multimodal baselines. Best results are shown in bold.

Model	Accuracy	AUC	F1	Sensitivity	Specificity
MMBT (Bert-base-uncased)	0.897	0.883	0.916	0.963	0.803
MMBT (Bert-large-uncased)	0.950	0.946	0.958	0.970	0.923
ConcatBert	0.936	0.929	0.946	0.970	0.889
Ours (Multimodal)	**0.965**	**0.983**	**0.976**	0.954	1.000

our experiments. Next, the main diagnosis in each EMR document is extracted as its disease label. Finally, we split the datasets into train set and test set in an 80/20 ratio. Due to the privacy concern, the original dataset is not allowed to be published according to relevant regulation. Source code and samples are available at https://github.com/charliecaffe/MMLN.

4.2 Comparison of MMLN and Multimodal Baselines

In this section, We compare three multimodal baselines with MMLN in this experiment.

MultiModal BiTransformers (MMBT) [12]: the underlying pre-trained models are BERT [6] and ResNet [7]. The text data is passed through BERT to generate word embedding, while the image data is passed through ResNet to generate CNN embedding. Then the two embeddings are concatenated and passed through a classifier for disease classification. We use two models of different sizes (base and large) to compare.

ConcatBert [1]: it is another model for multimodal classification with text and images, whose text representation is obtained from the pre-trained BERT base model and image representation obtained from the VGG16 pre-trained model.

For the multimodal baselines, we extract the significant text fields, such as chief complaint, present illness history, physical signs, laboratory items, etc. from JSON format EMRs and concatenate them as the text data input. Besides, we take CXR images as the image data input. All baseline are implemented with default settings and run on a Linux server with two Tesla V100 GPU.

As shown in Table 2, our multimodal method achieve the highest performance in accuracy, area under curve (AUC), and F1 score among the multimodal

Table 3. Performance effects of data source and dataset size on MMLN.

Model	Accuracy	AUC	F1	Sensitivity	Specificity
Ours (Multimodal)	0.965	0.983	0.976	0.954	1.000
Ours (Unimodal, EMR)	0.905	0.948	0.929	0.931	0.853
Ours (Multimodal, small size)	0.924	0.967	0.948	0.911	0.965

methods. In particular, three deep learning-based baselines achieve accuracies of 0.89 to 0.95. With large uncased Bert, the MMBT has an improvement in accuracy, but it is still lower than our multimodal model reaching an accuracy near 0.97. The same trend can also be observed in the comparison results of F1 score and AUC. Besides accuracy, F1 score and AUC, sensitivity and specificity are usually used in disease diagnosis. The baselines perform better in sensitivity, while our model performs better in specificity. MMLN based on diagnostic rules inference makes sure we do not misdiagnose but has a little sacrifice of missed diagnosis.

The results show that the proposed method outperforms the other competing methods based on deep learning in accuracy and also has interpretability.

4.3 Discussion

In this section, We discuss the effect of model performance from two perspectives: data source and dataset size. Besides, we also discuss the interpretability of the model.

Compare to Unimodal Model. In this section, we compare the performance of unimodal and multimodal models. Unimodal means that we build an evidence database purely on text data, while multimodal means that we build the evidence database on text and image data. As for the chest imaging examination, we extract grounding predicate from EMRs instead of CXR images by CNN model. We use regular expressions to match the related description from EMRs' fields such as physical examination, auxiliary examination, case characteristics, general and speciality situation, etc. Then we use the evidence database to learn weights and inference mentioned in Sect. 3.3.

Table 3 (Unimodal and Multimodal) shows the performance of unimodal and multimodal models respectively. All evaluation metrics of the multimodal model are better than those of the unimodal model. In particular, our unimodal model reaches an accuracy above 0.90. Incorporating image data, our multimodal model has a significant improvement in accuracy, which is near 0.97.

We also draw the receiver operating characteristic (ROC) curve of the unimodal and the multimodal model, as shown in Fig. 4a. We can see that both the unimodal and multimodal curve is closest to the left top corner and the multimodal model performs better than the unimodal one. Its corresponding AUC score is 0.965, which is higher than the unimodal.

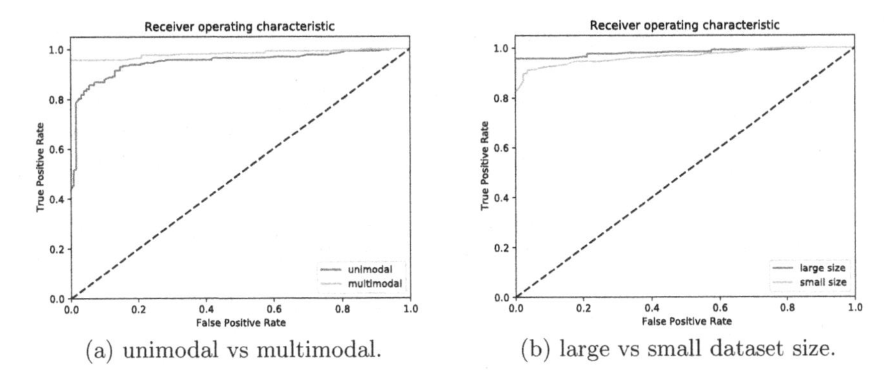

(a) unimodal vs multimodal. (b) large vs small dataset size.

Fig. 4. Receiver operating characteristic (ROC) curve

Generally, the results show that integrating the multimodal data can drastically improve the performance of disease diagnosis.

Compare to Small Dataset Size. When we collect data in the hospital, we find it quite challenging to obtain multimodal data as there is a general problem of missing modalities. In this section, we explore the effect of dataset size to MMLN performance to illustrate that the model is robust regardless of dataset size.

Our original size of cases, including pneumonia cases and other differential diagnosis cases, is about 2400, which is a medium-sized dataset. In this experiment, we only take 1/10 of each disease case and experiment on the data subset.

We find that although the learned weights are smaller than those in Eq. (4), the performance do not drop too much. Table 3 (ours and its small size) lists the results of original size and small size multimodal model performance. Even if the size is only 1/10 of the original, the accuracy reaches 0.924, and the F1 score reaches 0.948, which is even better than MMBT (bert-base-uncased).

In Fig. 4b, two ROC curves are very close and AUC decreases a bit from 0.983 to 0.967. The same trend can also be observed in the comparison results of sensitivity and specificity.

The experimental results show that dataset size has little impact on MMLN. Even with a small dataset, we can build a disease diagnosis system in the startup phase and obtain a decent performance. As we have more and more data over time, it can continuously help the system to improve performance.

The Interpretability of MMLN. One of the major contributions of this work is to bring interpretability into disease diagnosis by integrating multimodal data in a MLN. We illustrate how the diagnosis result is explained, i.e. interpretability, by MLN with bottom right corner in Fig. 1. In the visualization of MLN, each node represents a predicate defined before and the arc lines represent the logic relation of the connected nodes.

If only fever is extracted from a patient's EMR, then pneumonia will be predicted with a low probability because there are many situations causing fever. If fever is extracted from a patient's EMR and lung inflammation is extracted from the same one's CXR, then pneumonia will be predicted with a high probability. This is because fever and lung inflammation exist in one diagnostic rule inferring the pneumonia.

The proposed method can explain the prediction of pneumonia with positive clinical manifestations and chest imaging examination.

5 Conclusion

In this paper, we propose a data-driven and knowledge-driven framework called MMLN for lung disease diagnosis. We formulate diagnosis rules in form of Markov logic according to authoritative clinical guidelines, which improves the accuracy and interpretability of the model. Moreover, MMLN effectively fuse multimodal data and reveal intrinsic relations between different data sources by leveraging the medical domain knowledge. The experimental results show that knowledge and data is fully complementary to better the downstream diagnosis task and MMLN outperforms the state-of-the-art multimodal models for disease classification.

References

1. Arevalo, J., Solorio, T., Montes-y Gómez, M., González, F.A.: Gated multimodal units for information fusion. arXiv preprint arXiv:1702.01992 (2017)
2. Bustos, A., Pertusa, A., Salinas, J.M., de la Iglesia-Vayá, M.: PadChest: a large chest X-ray image dataset with multi-label annotated reports. Med. Image Anal. **66**, 101797 (2020)
3. Çallı, E., Sogancioglu, E., van Ginneken, B., van Leeuwen, K.G., Murphy, K.: Deep learning for chest X-ray analysis: a survey. Med. Image Anal. **72**, 102125 (2021)
4. Chapman, W.W., Bridewell, W., Hanbury, P., Cooper, G.F., Buchanan, B.G.: A simple algorithm for identifying negated findings and diseases in discharge summaries. J. Biomed. Inform. **34**(5), 301–310 (2001)
5. Cohen, J.P., et al.: TorchXRayVision: a library of chest X-ray datasets and models (2020). https://github.com/mlmed/torchxrayvision
6. Devlin, J., Chang, M.W., Lee, K., Toutanova, K.: BERT: pre-training of deep bidirectional transformers for language understanding. arXiv preprint arXiv:1810.04805 (2018)
7. He, K., Zhang, X., Ren, S., Sun, J.: Deep residual learning for image recognition. arXiv preprint arXiv:1512.03385 (2015)
8. Hou, J., Gao, T.: Explainable DCNN based chest X-ray image analysis and classification for COVID-19 pneumonia detection. Sci. Rep. **11**(1), 1–15 (2021)
9. Irvin, J., et al.: CheXpert: a large chest radiograph dataset with uncertainty labels and expert comparison. In: Thirty-Third AAAI Conference on Artificial Intelligence (2019)
10. Irvin, J., et al.: CheXpert: a large chest radiograph dataset with uncertainty labels and expert comparison. CORR abs/1901.07031, 1901 (2019)

11. Jiang, J., Li, X., Zhao, C., Guan, Y., Yu, Q.: Learning and inference in knowledge-based probabilistic model for medical diagnosis. Knowl. Based Syst. **138**, 58–68 (2017)
12. Kiela, D., Bhooshan, S., Firooz, H., Testuggine, D.: Supervised multimodal bitransformers for classifying images and text. arXiv preprint arXiv:1909.02950 (2019)
13. Kok, S., et al.: The alchemy system for statistical relational AI. Technical report, Department of Computer Science and Engineering, University of Washington, Seattle, WA (2005)
14. Li, L., et al.: Real-world data medical knowledge graph: construction and applications. Artif. Intell. Med. **103**, 101817 (2020)
15. Nyga, D., Picklum, M., Beetz, M., et al.: PracMLN - Markov logic networks in Python (2013). https://www.pracmln.org/
16. Peng, Y., Wang, X., Lu, L., Bagheri, M., Summers, R., Lu, Z.: NegBio: a high-performance tool for negation and uncertainty detection in radiology reports. AMIA Summits Transl. Sci. Proc. **2018**, 188 (2018)
17. Ren, H., et al.: Interpretable pneumonia detection by combining deep learning and explainable models with multisource data. IEEE Access **9**, 95872–95883 (2021). https://doi.org/10.1109/ACCESS.2021.3090215
18. Richardson, M., Domingos, P.: Markov logic networks. Mach. Learn. **62**(1), 107–136 (2006)
19. Shen, Y., et al.: CBN: constructing a clinical Bayesian network based on data from the electronic medical record. J. Biomed. Inform. **88**, 1–10 (2018)
20. Singh, R., et al.: Deep learning in chest radiography: detection of findings and presence of change. PloS One **13**(10), e0204155 (2018)
21. Sirazitdinov, I., Kholiavchenko, M., Kuleev, R., Ibragimov, B.: Data augmentation for chest pathologies classification. In: 2019 IEEE 16th International Symposium on Biomedical Imaging (ISBI 2019), pp. 1216–1219. IEEE (2019)
22. Vaswani, A., et al.: Attention is all you need. In: Guyon, I., et al. (eds.) Advances in Neural Information Processing Systems, vol. 30. Curran Associates, Inc. (2017). https://proceedings.neurips.cc/paper/2017/file/3f5ee243547dee91fbd053c1c4a845aa-Paper.pdf
23. Wang, G., et al.: A deep-learning pipeline for the diagnosis and discrimination of viral, non-viral and COVID-19 pneumonia from chest X-ray images. Nat. Biomed. Eng. **5**(6), 509–521 (2021)
24. Wang, X., Peng, Y., Lu, L., Lu, Z., Bagheri, M., Summers, R.M.: ChestX-ray8: hospital-scale chest X-ray database and benchmarks on weakly-supervised classification and localization of common thorax diseases. In: Proceedings of the IEEE Conference on Computer Vision and Pattern Recognition, pp. 2097–2106 (2017)
25. Yin, C., Zhao, R., Qian, B., Lv, X., Zhang, P.: Domain knowledge guided deep learning with electronic health records. In: 2019 IEEE International Conference on Data Mining (ICDM), pp. 738–747. IEEE (2019)

Optimal Sequence Alignment to ED-Strings

Njagi Moses Mwaniki and Nadia Pisanti(✉)

Department of Computer Science, University of Pisa, Pisa, Italy
njagi.mwaniki@di.unipi.it, nadia.pisanti@unipi.it

Abstract. Partial Order Alignment (POA) was introduced by Lee et al. in 2002 to allow the alignment of a string to a graph-like structure representing a set of aligned strings (a Multiple Sequence Alignment, MSA). However, the POA edit transcript (the sequence of edit operations that describe the alignment) does not reflect the possible elasticity of the MSA (different gaps sizes in the aligned string), leaving room for possible misalignment and its propagation in progressive MSA. Elastic-Degenerate Strings (ED-strings) are strings that can represent the outcome of an MSA by highlighting gaps and variants as a list of strings that can differ in size and that can possibly include the empty string. In this paper, we define a method that optimally aligns a string to an ED-string, the latter compactly representing an MSA, overcoming the ambiguity in the POA edit transcript while maintaining its time and space complexity.

1 Introduction

In genomic data analyses, as well as in other applications of sequence alignment, insertions and deletions can involve fragments of different lengths, raising issues on how, and how much, to account for their costs. For example, a gap cost purely proportional to the gap length can erroneously discourage long gaps, whereas a low cost for gaps can fragment the alignment into too many short gaps. This has motivated the definitions of gap penalty functions such as the affine gap penalty score, introduced for genomic sequences [28], that we will use in this paper.

Multiple Sequence Alignments (MSA) are typically performed in a progressive manner by iteratively adding a new linear sequence. To this purpose, the MSA can be linearized into a lossy representation of the actual similarities and differences of the aligned strings, and then the iterative, crucial, step of progressive MSA is brought down to a pairwise alignment of the new string against the MSA. This alignment involves two strings of different type, one linear and a more complex one, breaking up the traditional symmetry of pairwise alignments, and therefore suggesting new challenges that include re-opening the issue of how to account for gaps. Indeed, whereas in pairwise alignments an insertion is just the dual of a deletion and their cost is the same, here the insertion can mean two events that are substantially different: wherever the MSA has already located, and already accounted, a gap, then this insertion cost should not be accounted again and again to each new string; on the other hand, when the gap

M. S. Bansal et al. (Eds.): ISBRA 2022, LNBI 13760, pp. 204–216, 2022.
https://doi.org/10.1007/978-3-031-23198-8_19

is an insertion that had not occurred yet in the MSA, or a deletion in the new string, then this is a new event, and it has to be penalized as such [3,10,11].

An Elastic Degenerate String (ED-String), denoted $\widetilde{T}$, is a way to compactly represent the outcome of a MSA by collapsing common fragments into linear strings, and highlighting gaps and variants as list of strings that can have different size and that can include the empty string [14,15,25]. For example, for the following MSA of three closely-related sequences

AC TAG A TA GGGGCA
AC CA - A AA GGGGCA
AC - - - A AT GGGGCA

its ED-string representation is as follows, where ϵ denotes the empty string, and those in curly brackets are called *degenerate positions*:

$$\widetilde{T} = AC \begin{Bmatrix} TAG \\ CA \\ \epsilon \end{Bmatrix} A \begin{Bmatrix} TA \\ AA \\ AT \end{Bmatrix} GGGGCA$$

In the literature, there is currently a lot of attention on graphs representations of variants [13,19] that are more suitable than ED-strings to represent structural variants and repeats, but lack an explicit skip expression. The variants represented by an ED-String introduce a partial order in its letters (letters are ordered from left to right as usual, but as opposed to the total order of letters in a linear string, here letters that belong to distinct strings of the same degenerate positions are not comparable). In this case, its edit distance can be computed using Partial Order Alignment (POA) [7] as shown in [23] for the simpler case of D-strings. However, POA alone does not yield the optimal edit transcript as the empty string is not properly managed. We here solve this problem by adding a new edit event in the edit transcript: the *skip operation*. In this paper we suggest a dynamic programming algorithm to optimally align a linear string P and an ED-String $\widetilde{T}$ with (free) *skip operations*.

For example, the optimal alignment of $P = ACATGGGGAAA$ to $\widetilde{T}$ above involves the letters of $\widetilde{T}$ highlighted with colors with (i) red letters being matches, (ii) a skip operation in that the ϵ represent an existing gap locus in the MSA, and hence a free insertion, also highlighted in red (ii) the blue A is instead a new insertion in P and therefore it has a cost, (iii) the green letter C is a mismatch (at its cost), and (iv) the last letter of P has been deleted (at a cost).

Our aim is to compute the correct *edit distance with skips* (the lowest number of operations needed to convert P into one of the strings represented by $\widetilde{T}$ with free cost of skips) and an unambiguous optimal *edit transcript* [12]. For example, assuming unitary cost for all edit operations and no cost for matches and skips, then the distance between P and $\widetilde{T}$ above is 3 (1 insertion, 1 mismatch, 1 deletion), and the edit transcript representing an optimal alignment is `MMSMMIMMMMXMD`, where `M` denotes a match, `S` is a skip, `X` is a mismatch, and `D` denotes a deletion.

The notion of ED-strings and, over them, the exact (that is, no edit operations other than exact match are allowed) matching problem *elastic-degenerate string matching* (EDSM) problem has attracted some attention in the combinatorial pattern matching community in recent years. Since its introduction in [9], a series of results have been published [4,16,17,25], and a few tools produced [1,2]. In [14,15], the approximate version of the problem has been addressed both with Hamming and edit distance admitted between the pattern and its occurrence inside the ED-string. In particular, the current state of the art for computing the edit distance between a pattern P of length m and an ED-string $\widetilde{T}$ of size[1] N takes time in $\mathcal{O}(k \cdot m \cdot G + k \cdot N)$, where $G \in \mathcal{O}(N)$ is the total number of strings in $\widetilde{T}$ and k is the computed edit distance [14,15].

In this paper we extend the traditional dynamic programming alignment algorithm to work with ED-strings, by adding a skip (S) operation that inserts in the pattern an empty character of the ED-string at no cost. This new edit operation, together with a suitably designed $\widetilde{T}'$ (rewriting of the ED-string $\widetilde{T}$ that conserves its size and its edit distance with any linear string), and the consequent recurrent formula we introduce for the dynamic programming computation of the edit distance between $\widetilde{T}'$ and a string P, lead to our final result: a dynamic programming algorithm that computes an optimal alignment of $\widetilde{T}$ and P in time and space complexity in $\mathcal{O}(m \cdot N)$ (where $m = |P|$ and N is the size of $\widetilde{T}$), independently from the actual distance.

The algorithm we describe here is for a global alignment (the natural choice for the progressive MSA application mentioned above), but its extension to local and semiglobal alignment is straightforward. A semiglobal alignment could be used, for example, in computational pan-genomics, with the ED-string representing a pan-genome [8,20,21,29] and the pattern being a read to be mapped therein: this is currently one of the most relevant problems in bioinformatics [1,2,5,18,19,22].

In Sect. 3 we explain how we propose to manage skips using as an example the basic case of an elastic but not degenerate strings (which we define in Sect. 2 with other preliminary definitions), and in Sect. 4 we show how to extend the dynamic programming framework to ED-strings.

2 Preliminary Definitions

A *string* X is a sequence of elements on an alphabet Σ, where the *alphabet* Σ is a non-empty finite set of letters of size $|\Sigma|$: we say that $X \in \Sigma^*$, and specifically that $X \in \Sigma^n$ if n is the length of X ($n = |X|$). We denote with ε the empty character (a character that does not change the string it is in), and with ϵ the *empty string*: a string made up of empty characters only. A string Υ containing one or more empty characters is an *elastic string* (E-string) ($\Upsilon \in (\Sigma \cup \{\varepsilon\})^*$) where ε is the *skip*. The length $|\Upsilon|$ of Υ is computed including its skips.

For any string X, we denote by $X[i, j]$ the *substring* of X that *starts* at position i and *ends* at position j. In particular, $X[0, j]$ is the *prefix* of X that

[1] In [14,15] the size of the ED-string $\widetilde{T}$ is defined in a slightly different way as the actual number of letters or empty strings that appear in $\widetilde{T}$.

ends at position j (it is the empty prefix if $j = 0$), and $X[i, |X|]$ is the *suffix* of X that starts at position i (being the empty suffix if $i = 0$).

Definition 1. *An* elastic degenerate string (ED-string) $\widetilde{S} = \widetilde{S}\{1\}\widetilde{S}\{2\}\ldots\widetilde{S}\{n\}$ *of* length n *over an alphabet Σ is a finite sequence of n positions $\widetilde{S}\{i\}$'s. Each position $\widetilde{S}\{i\}$ is a finite non-empty set of s_i strings (also called variants) that can possibly properly include the empty string. We denote with ℓ_i the length of the longest string in $\widetilde{S}\{i\}$.*

When $s_i = 1$, then $\ell_i = 1$ as well, and in that case $\widetilde{S}\{i\}$ is just a simple letter of Σ and we say that $\widetilde{S}\{i\}$ is *solid*, while if $s_i > 1$ then we say that it is a *degenerate position*. The following parameters measure the degree of degeneracy of an ED-string: the *total size* N and *total width* W of an ED-string $\widetilde{S}$ are respectively defined as $N = \sum_{i=1}^{n} s_i \cdot \ell_i$ and $W = \sum_{i=1}^{n} \ell_i$.

For example, $\widetilde{T}$ above has length $n = 11$, $W = 14$, and $N = 24$. Moreover, there we have that all positions are solid except for 3 and 5, and hence all ℓ_i's are equal to 1 except $\ell_3 = 3$ and $\ell_5 = 2$, and therein we have $s_3 = s_5 = 3$.

3 Edit Distance with Skips

An *alignment* between two strings is described by an *edit transcript*: a sequence of edit events among `M,X,I,D,S` (where `M` denotes a match, `X` is a mismatch, `I` is an insertion, `D` is a deletion, and `S` denotes a skip), each one having its own cost, that transform a string into the other. The alignment is *optimal* when such cost is minimized. The affine gap penalty functions assigns a unique cost $o+k \cdot e$ to a gap of length k, that is a maximal run of k consecutive insertions `I` (resp. deletions `D`), with o being for gap opening and e for gap extension.

In this section we describe a dynamic programming algorithm to optimally compute the edit distance with affine gap penalty cost with skips between a string $P \in \Sigma^m$ and an E-string $\Upsilon \in (\Sigma \cup \{\varepsilon\})^n$. A *skip* is an insertion of an empty letter ε of Υ whose cost s will be 0 in the rest of this paper[2].

Example 1. Let Υ = GεTCCCATεGA and P = GTCAAATGC, and let $a = 0$ be the cost of a match[3], $x = 1$ for a mismatch, and $o = 2$ (resp. $e = 1$) for gap opening (resp. extension). The edit transcript `MSMIIMMDDMSMX` is an alignment of Υ and P with cost 9 (twice $o + 2e = 4$ for the two gaps of size two, and 1 for the final mismatch), while `MSMMXXMMSMX` is an optimal alignment of cost 3.

---

[2] The method can generalize to the case of cost $s > 0$ for skips.

[3] In an edit distance computation framework, the match has typically null cost in order to fulfill the metric requirement that a string has zero-distance to itself; for this reason in our examples we assume $a = 0$. However, since the the dynamic programming method we design also works when one wants to compute a similarity score rather than a distance (it suffices to adapt the penalty scores and seek the maximum instead of the minimum), then in our problem statement as well as in the recurrence formula that describe our algorithm, we parametrize the score of a match with a.

The formal statement of the problem StoES we solve here is the following:

String Alignment to E-string (StoES)
Input: An E-string Υ of length n, a pattern P of length m, and penalty scores a, x, o, e.
Output: An optimal alignment between P and Υ using scores: a for match, x for mismatch, o for gap opening, and e for gap extension.

Like in the traditional dynamic programming (DP) framework, we incrementally compute the optimal alignments of prefix $P[0,j]$ of P against prefix ($\Upsilon[0,i]$) of Υ starting with i and/or j equal to 0 up to the final lengths n and m. As it is well known [12], the problem of computing the optimal alignment with such gap penalty function does not directly fulfill the requirement of having optimal substructure (needed to correctly apply dynamic programming), and to overcome this limitation, the alignment is conceptually split in blocks grouping runs of edit events. To this purpose, the traditional computation of an optimal alignment with affine gap penalty function requires to use three matrices: (I) for insertions, (D) for deletions, (M) for (mis)matches. Since the skip operation we want to implement raises a similar issue, we here add an additional matrix (S) for skips. We compute the DP matrices M, S, I, D of size $(n+1) \times (m+1)$ whose entries for $0 \leq i \leq n$ and $0 \leq j \leq m$ have the following meaning:

$M[i,j] =$ score of best alignment of $P[0,j]$ and $\Upsilon[0,i]$ ending with a match or a mismatch of $P[j]$ and $\Upsilon[i]$.

$S[i,j] =$ score of best alignment of $P[0,j]$ and $\Upsilon[0,i]$ ending with a skip at $\Upsilon[i]$

$I[i,j] =$ score of best alignment of $P[0,j]$ and $\Upsilon[0,i]$ ending with an insertion of $\Upsilon[i]$ in P

$D[i,j] =$ score of best alignment of $P[0,j]$ and $\Upsilon[0,i]$ ending with a deletion of $P[j]$ after $\Upsilon[i]$

The initialization of the matrices is as follows:

M: Like in traditional dynamic programming with affine gap penalty, we set $M[0,0] = 0$ and $M[i,0] = M[0,j] = \infty$ for $0 < i \leq n$ and $0 < j \leq m$ because no alignment with an empty prefix can end with match or mismatch.

D: Again like in traditional dynamic programming with affine gap penalty, we set $D[0,0] = 0$, while for $0 < i \leq n$ and $0 < j \leq m$ we have $D[0,j] = o + e \cdot j$ because we align an empty prefix of Υ with $P[0,j]$, while $D[i,0] = \infty$ because no alignment of $\Upsilon[0,i]$ and the empty prefix of P ends with a deletion in P.

I: Dually, $I[0,0] = 0$ and $I[0,j] = \infty$ because no alignment of $P[0,j]$ and an empty prefix of Υ can end with an insertion of a letter of Υ. In order to set the values $I[i,0]$ of the first column, instead, we need to take into account whether $\Upsilon[i] = \varepsilon$ and, if not, then we need to pay the gap opening only the first time this is the case, and pay the gap extension only for successive insertions: let i' be the smallest i from 1 to n such that $\Upsilon[i] \neq \varepsilon$. Then we have:

$$I[i,0] = min \begin{cases} \infty & \Upsilon[i] = \varepsilon \\ I[i-1,0] + o + e & \Upsilon[i] \neq \varepsilon \ \& \ i = i' \\ I[i-1,0] + e & \Upsilon[i] \neq \varepsilon \ \& \ i \neq i' \\ S[i-1,0] + e & \Upsilon[i] \neq \varepsilon \ \& \ i \neq i' \end{cases} \tag{1}$$

because we need to align $\Upsilon[0,i]$ and the empty prefix of P ending with an insertion of $\Upsilon[i]$, and this is only possible if $\Upsilon[i] \neq \varepsilon$, and else it either opens a new gap (when $i = i'$), or it extends an existing gap that was already open at row $i-1$ whose cost is either in the matrix I itself, or in S if $\Upsilon[i-1] = \varepsilon$, in which case the gap has included a skip at no cost.

S: We have that $S[0,j] = \infty$ because no alignment of an empty prefix of Υ and $P[0,j]$ can end with a skip of Υ, while:

$$S[i,0] = \begin{cases} min\{S[i-1,0], I[i-1,0]\} & \Upsilon[i] = \varepsilon \\ \infty & \Upsilon[i] \neq \varepsilon \end{cases} \tag{2}$$

because we need to align $\Upsilon[0,i]$ and the empty prefix of P ending with a skip and this latter is not possible if $\Upsilon[i] \neq \varepsilon$. On the other hand, when $\Upsilon[i] = \varepsilon$, then either a skip had occurred already at the previous row (in which case we inherit its score), or we have to inherit the score of an insertion gap.

We remark that, besides the apparent circular definition of the first columns of I and S, they can just both filled in a non ambiguous way once the index i' has been located (in linear time).

We can then progressively fill in the matrices for $i, j > 0$ as follows:

$$M[i,j] = min \begin{cases} \infty & \Upsilon[i] = \varepsilon \\ f(i,j) + M[i-1,j-1] & \Upsilon[i] \neq \varepsilon \\ f(i,j) + S[i-1,j-1] & \Upsilon[i] \neq \varepsilon \\ f(i,j) + D[i-1,j-1] & \Upsilon[i] \neq \varepsilon \\ f(i,j) + I[i-1,j-1] & \Upsilon[i] \neq \varepsilon \end{cases} \tag{3}$$

where

$$f(i,j) = \begin{cases} a \text{ if } \Upsilon[i] = P[i] \\ x \text{ otherwise} \end{cases} \tag{4}$$

as in traditional dynamic programming with affine gap penalty, except that no (mis)match is defined in rows for which $\Upsilon[i] = \varepsilon$ because in that case no alignment between $P[0,j]$ and $\Upsilon[0,i]$ can end with a match or a mismatch.

The deletion matrix D is the very same as in the traditional alignment:

$$D[i,j] = min \begin{cases} D[i,j-1] + e \\ M[i,j-1] + o + e \\ S[i,j-1] + o + e \\ I[i,j-1] + o + e \end{cases} \tag{5}$$

The DP recurrence relations to fill in the insertion matrix I is as follows:

$$I[i,j] = min \begin{cases} \infty & \Upsilon[i] = \varepsilon \\ D[i-1,j] + o + e & \Upsilon[i] \neq \varepsilon \\ M[i-1,j] + o + e & \Upsilon[i] \neq \varepsilon \\ S[i-1,j] + e & \Upsilon[i] \neq \varepsilon \\ I[i-1,j] + e & \Upsilon[i] \neq \varepsilon \end{cases} \quad (6)$$

because when a skip is encountered at row i, then no costly insertion can take place and hence no alignment of $P[0,j]$ and $\Upsilon[0,i]$ can end with an insertion when $\Upsilon[i] = \varepsilon$, and finally:

$$S[i,j] = min \begin{cases} \infty & \Upsilon[i] \neq \varepsilon \\ S[i-1,j] & \Upsilon[i] = \varepsilon \\ M[i-1,j] & \Upsilon[i] = \varepsilon \\ I[i-1,j] & \Upsilon[i] = \varepsilon \\ D[i-1,j] & \Upsilon[i] = \varepsilon \end{cases} \quad (7)$$

because no skip operation can be performed if $\Upsilon[i] \neq \varepsilon$, while if $\Upsilon[i] = \varepsilon$, then we can copy in $S[i,j]$ the best score in the cell above from any of the four matrices.

The computation can end when all matrices have been filled in up to the last row for $i = n$ and the last column for $j = m$, and then we have that the final result of StoES is $min(M[n,m], I[n,m], D[n,m], S[n,m])$, and an optimal edit transcript can be recovered by tracing back the computation of the optimal cost suitably jumping among the four matrices like in the traditional dynamic programming algorithm with affine gap score [12,24,28], except that here we have matrix S as well. We remark that matrices I and S could possibly be merged into one to save space, but we maintain the distinction for the sake of clarity of their distinct meaning.

The correctness of the algorithm derives from the observation that any alignment either ends with a match, or a mismatch, or an insertion, or a deletions, or a skip, and that for each pair (i,j) all these options have been correctly taken into account in the matrices as we defined them, and with the incremental computation of optimal alignments of prefixes of Υ and P using the recurrence relations that, inductively, correctly refer to the optimal solution of all sub-problems of the possible sub-alignments that can precede them and according to the cost definition. In particular, our method has shown how to realize a null cost for skips operations. The algorithm we just described optimally solves the problem StoES in time and space $\mathcal{O}(|\Upsilon| \cdot |P|)$ as this is the size of the matrices and each entry can be computed in constant time.

4 Edit Distance with Skips on ED-Strings

In this section we shown how to compute an optimal alignment with skips for a string P and an ED-string $\widetilde{T}$ with affine gap penalty score using the free skips rules we introduced in Sect. 3. We can see an ED-string as the complete set of the linear strings with skips it represents: let us denote such set with $S(\widetilde{T})$. Seeking an optimal alignment between a linear string P and $\widetilde{T}$ means to find the E-string $P' \in S(\widetilde{T})$ that minimizes the edit distance with skips between P and P'. In the example shown in Sect. 1, P' is the coloured string in $\widetilde{T}$.

The problem we want to solve is therefore defined as follows:

STRING ALIGNMENT TO ED-STRING (STOEDS)
Input: An ED-string $\widetilde{T}$ of length n, size N, and width W, a pattern P of length m, and penalty scores a, x, o, e.
Output: An optimal alignment between P and $\widetilde{T}$ using scores: a for match, x for mismatch, o for gap opening, and e for gap extension.

The first step of our algorithm is a rewriting of the input ED-string $\widetilde{T}$ that modifies degenerate positions $\widetilde{T}\{i\}$'s in order to let all strings in $\widetilde{T}\{i\}$ reach length ℓ_i by padding shorter strings with ε's (we recall that ℓ_i is the length of the longest string in $\widetilde{T}\{i\}$). For example, the ED-string $\widetilde{T}\{i\}$ of Sect. 1 becomes:

$$\widetilde{T}' = A\ C \left\{ \begin{matrix} T\ A\ G \\ C\ A\ \varepsilon \\ \varepsilon\ \varepsilon\ \varepsilon \end{matrix} \right\} A \left\{ \begin{matrix} TA \\ AA \\ AT \end{matrix} \right\} G\ G\ G\ G\ C\ A$$

$$i' \quad 1\ 2 \quad 3\ 4\ 5 \quad 6 \quad 7\ 8 \quad 9\ 10\ 11\ 12\ 13\ 14$$

Observe that the width W of $\widetilde{T}'$ is the same as that of $\widetilde{T}$. We can then build dynamic programming matrices with $W+1$ rows and $m+1$ columns, associating matrices rows i' with the columns of $\widetilde{T}'$: each one is one base wide, but some have several variants, and therefore refer to *tuples* of letters. In $\widetilde{T}'$ above $W = 14$ and the numbers below $\widetilde{T}'$ show the (set of) letters that will be associated with rows of the dynamic programming; for example, $\widetilde{T}[1] = A$, $\widetilde{T}[2] = C$, $\widetilde{T}[3]$ is the tuple $<\varepsilon, C, T>$, $\widetilde{T}[4] <\varepsilon, A, A>$, etc. We denote with $\widetilde{T}[i'][h]$ the hth variant of the tuple $\widetilde{T}[i']$.

The cells of the DP matrices within degenerate positions will thus contain tuples of values: one possible different edit distance per variant. Namely, we have now matrices M', S', D', I' whose entries are in general tuples. We denote with $M'[i', j][h]$ (with $1 \leq h \leq s$) the hth element of a tuple of size s in $M'[i', j]$ (resp. in S', I', D'), and we denote with $\widetilde{T}[a, b][h]$ the substring of $\widetilde{T}$ that starts at $\widetilde{T}[a]$ and ends at $\widetilde{T}[b]$ choosing the hth variant of $\widetilde{T}[b]$. Then the entries of the matrices are defined as follows:

$M'[i', j][h]$ = score of best alignment of $P[0, j]$ and $\widetilde{T}[0, i'][h]$ ending with a match or a mismatch of $P[j]$ and $\widetilde{T}[i'][h]$.

$S'[i', j][h]$ = score of best alignment of $P[0, j]$ and $\widetilde{T}[0, i'][h]$ ending with a skip at $\widetilde{T}[i'][h]$.

$I'[i', j][h]$ = score of best alignment of $P[0, j]$ and $\widetilde{T}[0, i'][h]$ ending with an insertion of $\widetilde{T}[i'][h]$ in P

$D'[i', j][h]$ = score of best alignment of $P[0, j]$ and $\widetilde{T}[0, i'][h]$ ending with a deletion of $P[j]$ after $\widetilde{T}[i'][h]$

We remark that we must ensure that, when a partial alignment has chosen a variant h in a degenerate position, then as long as rows involve variants of the same degenerate position, then the hth entry of all tuples must always refer to the corresponding h^{th} entry at previous rows and/or columns of the matrices. The function $f'(i,j)$ that checks whether we have a match or a mismatch must then be modified from Eq. (4) as follows:

$$f'(i',j)[h] = \begin{cases} a \text{ if } \widetilde{T'}[i'][h] = P[j] \\ x \text{ otherwise} \end{cases} \tag{8}$$

and we have that the generic entry of, say, M' is filled in as follows

$$M'[i',j][h] = min \begin{cases} \infty & \widetilde{T'}[i'][h] = \varepsilon \\ f'(i',j)[h] + M'[i'-1,j-1][h] & \widetilde{T'}[i'][h] \neq \varepsilon \\ f'(i',j)[h] + S'[i'-1,j-1][h] & \widetilde{T'}[i'][h] \neq \varepsilon \\ f'(i',j)[h] + D'[i'-1,j-1][h] & \widetilde{T'}[i'][h] \neq \varepsilon \\ f'(i',j)[h] + I'[i'-1,j-1][h] & \widetilde{T'}[i'][h] \neq \varepsilon \end{cases} \tag{9}$$

when both $\widetilde{T'}[i']$ and $\widetilde{T'}[i'-1]$ belong to the same degenerate position (and hence their tuples have the same size and letters of $\widetilde{T'}[i'][h]$ and $\widetilde{T'}[i'-1][h]$ correspond to the very same variant). Similarly, if $\widetilde{T'}[i']$ and $\widetilde{T'}[i'-1]$ are both solid positions, then the formula (9) applies just ignoring h because none of the involved entries is a tuple (basically formula (3) of the previous section can be used). More in general, we must take into account also the cases in which $\widetilde{T'}[i']$ and $\widetilde{T'}[i'-1]$ are such that:

1. $\widetilde{T'}[i'-1]$ is solid and $\widetilde{T'}[i']$ is degenerate
2. $\widetilde{T'}[i'-1]$ is degenerate and $\widetilde{T'}[i']$ is solid
3. $\widetilde{T'}[i'-1]$ and $\widetilde{T'}[i']$ are both degenerate but for two distinct consecutive degenerate positions.

In case 1. above, we have that $\widetilde{T'}[i'-1]$ is a letter of Σ and $M'[i'-1,j]$ is a single entry, while $\widetilde{T'}[i']$ is a tuple of ℓ letters and $M'[i',j]$ is also a tuple of size ℓ. In this case, whenever the computation of $M'[i',j]$ needs to refer to the previous row, then all variants h of the tuple have to consider the same entry at row $i'-1$, that is:

$$M'[i',j][h] = min \begin{cases} \infty & \widetilde{T'}[i'][h] = \varepsilon \\ f'(i',j)[h] + M'[i'-1,j-1] & \widetilde{T'}[i'][h] \neq \varepsilon \\ f'(i',j)[h] + S'[i'-1,j-1] & \widetilde{T'}[i'][h] \neq \varepsilon \\ f'(i',j)[h] + D'[i'-1,j-1] & \widetilde{T'}[i'][h] \neq \varepsilon \\ f'(i',j)[h] + I'[i'-1,j-1] & \widetilde{T'}[i'][h] \neq \varepsilon \end{cases} \tag{10}$$

In case 2, we have the opposite situation: $\widetilde{T'}[i'-1]$ is a tuple of letters and $M'[i'-1,j]$ is tuple of values, while $\widetilde{T'}[i']$ is a letter of Σ and $M'[i',j]$ is a single entry. In this case, whenever the computation of $M'[i',j]$ needs to refer to the previous row, then all variants h of the tuple at row $i'-1$ have to be taken into account, and among them the dynamic programming framework has to pick the value that minimizes the distance. That is:

$$M'[i',j] = min \begin{cases} \infty & \widetilde{T'}[i'] = \varepsilon \\ f'(i',j) + \min_h\{M'[i'-1,j-1][h]\} & \widetilde{T'}[i'] \neq \varepsilon \\ f'(i',j) + \min_h\{S'[i'-1,j-1][h]\} & \widetilde{T'}[i'] \neq \varepsilon \\ f'(i',j) + \min_h\{D'[i'-1,j-1][h]\} & \widetilde{T'}[i'] \neq \varepsilon \\ f'(i',j) + \min_h\{I'[i'-1,j-1][h]\} & \widetilde{T'}[i'] \neq \varepsilon \end{cases} \tag{11}$$

Finally, In case 3. both $\widetilde{T'}[i'-1]$ and $\widetilde{T'}[i']$ are tuples of letters (say of size ℓ_1 and ℓ_2, respectively), and consequently $M'[i'-1,j]$ is a tuple of ℓ_1 values, while $M'[i',j]$ is a tuple of ℓ_2 values. Since they belong to distinct degenerate positions, whenever the computation of $M'[i',j]$ needs to refer to the previous row, there are no restrictions on which preceding alignment must be picked, and therefore it is again the minimum that must drive the choice. The recurrence formula is then as follows, where h_2 is the index on the tuple of $M'[i',j]$, and h_1 is that of $M'[i'-1,j]$:

$$M'[i',j][h_2] = min \begin{cases} \infty & \widetilde{T'}[i'][h_2] = \varepsilon \\ f'(i',j)[h_2] + \min_h\{M'[i'-1,j-1][h_1]\} & \widetilde{T'}[i'][h_2] \neq \varepsilon \\ f'(i',j)[h_2] + \min_h\{S'[i'-1,j-1][h_1]\} & \widetilde{T'}[i'][h_2] \neq \varepsilon \\ f'(i',j)[h_2] + \min_h\{D'[i'-1,j-1][h_1]\} & \widetilde{T'}[i'][h_2] \neq \varepsilon \\ f'(i',j)[h_2] + \min_h\{I'[i'-1,j-1][h_1]\} & \widetilde{T'}[i'][h_2] \neq \varepsilon \end{cases} \tag{12}$$

Also matrices D', I', S' must be filled in using recurrences formula obtained modifying formulae (5), (6), (7) in the same way as formulae (9), (10), (11), (12) have replaced formula (3). To avoid clutter, we do not show all such recurrence relations, as well as we omit the initialization of M', D', I' and S'. The computation ends when all matrices have been filled in up to the last row for $i' = W$ and the last column with $j = m$, and then have that the final result is $min(M'[W,m], I'[W,m], D'[W,m], S'[W,m])$, and an optimal edit transcript can be recovered by tracing back the computation of the optimal cost suitably jumping among the four matrices.

The algorithm above correctly solves problem StoEDS in time and space in $\mathcal{O}(m \cdot N)$ where $m = |P|$ and N is the size of $\widetilde{T}$, and its trace-back reconstruct an explicit edit transcript with skips.

5 Progressive MSA with ED-Strings and StoEDS

Progressive multiple alignment can be improved by using our optimal solution for StoEDS by performing the following steps, (assuming the order in which the sequences are aligned is known, e.g. guided by a tree that is given in input or that can be computed as a pre-processing step [6,10,11]:

1. Performing a first pairwise global alignment that produces a partial MSA that is turned into an ED-string.
2. Progressively adding a new string to the MSA.
3. Producing a new richer MSA as a result of Step 2. and hence a new ED-string.
4. Iterate step 2. and 3. until all input sequences have been included.

Step 1. can be performed suitably turning an edit transcript of the pairwise alignment into an ED-string as follows: runs of matches will correspond to solid positions; a mismatch can be expressed with a degenerate position with two variants and unitary width; gaps can be represented by degenerate positions that include an empty string among its options. Step 2. can be performed with our optimal solution for StoEDS. Step 3. can be done similarly to Step 1, with new degenerate positions that include the empty string being added when insertions and deletions appear in the edit transcript, while skips in the ED-string will just *confirm* existing ones in the MSA.

6 Conclusions and Further Work

The state of the art for progressive MSA has been for decades reducing the MSA to a linear profile and then iterating a pairwise alignment. Partial Order Alignment (POA) [7] overcomes problems from earlier progressive methods by not reducing partial MSA to a linear profile, but rather aligning against a graph-like text and correctly computing the edit distance. However, POA does not yield the optimal edit transcript because the traceback does not reflect the distance introduced by variants with gaps of different lengths. In this paper this problem is solved with the skip operation. Our alignment can also be sped up in implementation using SIMD such as is done in spoa [27] and abPOA [30]. As a future direction, we might move the paradigm to circular strings [26].

Acknowledgment. This work is part of the ALPACA project that has received funding from the European Union's Horizon 2020 research and innovation programme under the Marie Skłodowska-Curie grant agreement No 956229.

References

1. Cisłak, A., Grabowski, S.: SOPanG2: online searching over a pan-genome without false positives. arXiv:2004.03033 [cs] (2020)
2. Cisłak, A., Grabowski, S., Holub, J.: SOPanG: online text searching over a pan-genome. Bioinformatics **34**(24), 4290–4292 (2018)
3. Loytynoja, A.L., Goldman, N.: An algorithm for progressive multiple alignment of sequences with insertions. Proc. Natl. Acad. Sci. **102**(30), 10557–10562 (2005)
4. Aoyama, K., Nakashima, Y., I, T., Inenaga, S., Bannai, H., Takeda, M.: Faster online elastic degenerate string matching. In: 29th Annual Symposium on Combinatorial Pattern Matching (CPM). LIPIcs, vol. 105 (2018)

5. Darby, C.A., Gaddipati, R., Schatz, M.C., Langmead, B.: Vargas: heuristic-free alignment for assessing linear and graph read aligners. Bioinformatics **36**(12), 3712–3718 (2020)
6. Grasso, C., Lee, C.: Combining partial order alignment and progressive multiple sequence alignment increases alignment speed and scalability to very large alignment problems. Bioinformatics **20**(10), 1546–1556 (2004)
7. Lee, C., Grasso, C., Sharlow, M.F.: Multiple sequence alignment using partial order graphs. Bioinformatics **18**(3), 452–464 (2002)
8. The Computational Pan-Genomics Consortium: Computational Pan-Genomics: Status, Promises and Challenges. Brief. Bioinform. **19**(1), 118–135 (2018)
9. Iliopoulos, C.S., Kundu, R., Pissis, S.P.: Efficient pattern matching in elastic-degenerate texts. In: Drewes, F., Martín-Vide, C., Truthe, B. (eds.) LATA 2017. LNCS, vol. 10168, pp. 131–142. Springer, Cham (2017). https://doi.org/10.1007/978-3-319-53733-7_9
10. Feng, D.-F., Doolittle, R.F.: Progressive sequence alignment as a prerequisitet to correct phylogenetic trees. J. Mol. Evol. **25**(4), 351–360 (1987)
11. Higgins. D.G., Sharp, P.M.: CLUSTAL: a package for performing multiple sequence alignment on a microcomputer. Gene **73**(1), 237–244 (1988)
12. Gusfield, D.: Algorithms on Strings, Trees, and Sequences: Computer Science and Computational Biology. Cambridge University Press, Cambridge (1997)
13. Birmelé, E., et al.: Efficient bubble enumeration in directed graphs. In: Calderón-Benavides, L., González-Caro, C., Chávez, E., Ziviani, N. (eds.) SPIRE 2012. LNCS, vol. 7608, pp. 118–129. Springer, Heidelberg (2012). https://doi.org/10.1007/978-3-642-34109-0_13
14. Bernardini, G., Pisanti, N., Pissis, S.P., Rosone, G.: Pattern matching on elastic-degenerate text with errors. In: Fici, G., Sciortino, M., Venturini, R. (eds.) SPIRE 2017. LNCS, vol. 10508, pp. 74–90. Springer, Cham (2017). https://doi.org/10.1007/978-3-319-67428-5_7
15. Bernardini, G., Pisanti, N., Pissis, S.P., Rosone, G.: Approximate pattern matching on elastic-degenerate text. Theor. Comput. Sci. **812**, 109–122 (2020)
16. Bernardini, G,. Gawrychowski, P., Pisanti, N., Pissis, S.P., Rosone, G.: Even faster elastic-degenerate string matching via fast matrix multiplication. In: 46th International Colloquium on Automata, Languages, and Programming (ICALP). LIPIcs, vol. 132, pp. 21:1–21:15 (2019)
17. Bernardini, G., Gawrychowski, P., Pisanti, N., Pissis, S.P., Rosone, G.: Elastic-degenerate string matching via fast matrix multiplication. SIAM J. Comput. **51**(3), 549–576 (2022)
18. Li, H., Feng, X., Chu, C.: The design and construction of reference pangenome graphs with minigraph. Genome Biol. **21**, 265 (2020)
19. Eizenga, J.M., et al.: Efficient dynamic variation graphs. Bioinformatics **36**(21), 5139–5144 (2021)
20. Alzamel, M., et al.: Degenerate string comparison and applications. In: 18th International Workshop on Algorithms in Bioinformatics (WABI). LIPIcs, vol. 113, pp. 21:1–21:14 (2018)
21. Alzamel, M., et al.: Comparing degenerate strings. Fundamenta Informaticae **175**(1–4), 41–58 (2020)
22. Rautiainen, M., Marschall, T.: GraphAligner: rapid and versatile sequence-to-graph alignment. Genome Biol. **21**, 253 (2020)
23. Mwaniki, N.M. Garrison, E. Pisanti, N.: Fast exact string to d-texts alignments. CoRR, abs/2206.03242 (2022)

24. Gotoh, O.: An improved algorithm for matching biological sequences. J. Mol. Biol. **162**(3), 705–708 (1982)
25. Grossi, R., et al.: On-line pattern matching on similar texts. In: 28th Annual Symposium on Combinatorial Pattern Matching (CPM). LIPIcs, vol. 78, pp. 9:1–9:14 (2017)
26. Grossi, R., et al.: Circular sequence comparison: algorithms and applications. Algorithms Mol. Biol. **11**, 12 (2016)
27. Vaser, R., Sović, I., Nagarajan, N., Šikić, M.: Fast and accurate de novo genome assembly from long uncorrected reads. Genome Res. **27**(5), 737–746 (2017)
28. Smith, T.F., Waterman, M.S.: Identification of common molecular subsequences. J. Mol. Biol. **147**(1), 195–197 (1981)
29. Carletti, V., Foggia, P., Garrison, E., Greco, L., Ritrovato, P., Vento, M.: Graph-based representations for supporting genome data analysis and visualization: opportunities and challenges. In: Conte, D., Ramel, J.-Y., Foggia, P. (eds.) GbRPR 2019. LNCS, vol. 11510, pp. 237–246. Springer, Cham (2019). https://doi.org/10.1007/978-3-030-20081-7_23
30. Gao, Y., Liu, Y., Ma, Y., Liu, B., Wang, Y., Xing, Y.: abPOA: an SIMD-based C library for fast partial order alignment using adaptive band. bioRxiv (2020)

Heterogeneous PPI Network Representation Learning for Protein Complex Identification

Peixuan Zhou, Yijia Zhang(✉), Fei Chen, Kuo Pang, and Mingyu Lu

School of Information Science and Technology, Dalian Maritime University, Dalian 116024, Liaoning, China
zhangyijia@dlmu.edu.cn

Abstract. Protein complexes are critical units for studying a cell system. How to accurately identify protein complexes has always been the focus of research. Most of the existing methods are based on the topological structure of the Protein-Protein Interaction (PPI) network and introduce some biological information to analyze the correlation between proteins to identify protein complex. However, these methods only comprise a homogenous network of biological information and protein nodes. Most of them ignore that different types of nodes have different importance for protein complex identification. Therefore, there is an urgent need for a method to integrate different types of biological information. This paper proposes a new protein complex identification method GHAE based on heterogeneous network representation learning. Firstly, GHAE combines Gene Ontology (GO) attribute information and PPI data to construct a heterogeneous PPI network. Secondly, based on the constructed network, we use the heterogeneous representation learning method to obtain the vector representation of protein nodes. Finally, we propose a complex identification method based on a heterogeneous network to identify protein complexes. Extensive experiments show that our method achieves state-of-the-art performance in most cases.

Keywords: Protein complexes identification · Heterogeneous PPI network · Network representation learning · Attention mechanism

1 Introduction

All life activities are based on proteins. However, in the life system, the function of cells cannot be realized by a single protein. On the contrary, cell function is realized by aggregating many protein nodes. These aggregated protein nodes are called protein complexes. Because most biological processes are based on protein complexes, it is essential to understand the formation mechanism of protein complexes. Accurate and efficient identification of protein complexes can help us better understand cells' tissue structure and function [1]. In recent years, identifying protein complexes has been a research focus of bioinformatics researchers. The traditional method depends on the topology of the PPI network and uses some classical clustering algorithms to find closely related subgraphs. In addition, many studies have proved that adding some additional biological

M. S. Bansal et al. (Eds.): ISBRA 2022, LNBI 13760, pp. 217–228, 2022.
https://doi.org/10.1007/978-3-031-23198-8_20

information can effectively assist the traditional clustering algorithm in improving the performance of protein complex identification.

We roughly summarize the existing protein complex identification methods. The first group methods are based on the PPI network. These methods search dense subgraphs from the PPI network as identified protein complexes [2]. They merge, grow or split strategies to identify protein complexes. Here, we list some typical methods in this category. Such as CFinder [3], MCODE [4], CMC [5], Cluster One [6], MCL [7] and PEWCC [8]. The second group methods add some additional information, such as biological information, core-attachment structure, biological evolution information and so on. Core [9], coach [10] and Hunter [11] identify protein complexes according to the principle that each complex is composed of a core and its attachment nodes. RNSC [12] and Decaff [13] combine topological and GO information, they use GO annotation as attribute information to identify protein complexes.

Recently, the network embedding method has proved to be effective in many downstream tasks. The heterogeneous information network has more abundant information than the traditional homogeneous network. In the heterogeneous network, the biological attribute information is added. The importance of different nodes is also considered. We can learn the structure and biological information of the PPI network better. So we explore using heterogeneous PPI network embedding to identify protein complexes.

We propose a method called **G**O attribute **H**eterogeneous **A**ttention network **E**mbedding (GHAE) to identify protein complexes. The PPI network is represented as a heterogeneous attributed network in which the protein nodes are associated with GO slims. We summarize the contributions of this paper as follows:

1. We construct a heterogeneous PPI network, which fused the PPI data with the GO attribute information of the protein. Integrating GO information can provide valuable biological domain information for protein complex prediction and effectively improve the performance of protein complex identification.
2. We use the heterogeneous representation learning method to mine the structure information and the global representation of protein nodes in PPI network. Our method can reduce the noise in the traditional PPI network and improve the accuracy of complex identification.
3. We propose a new protein complex identification method based on GO heterogeneous PPI network. The experimental results show that our method achieves state-of-the-art performance in most cases.

2 Method

This paper proposes a novel method based on a heterogeneous graph neural network for protein complexes identification. Our method consists of the following steps. To alleviate the sparsity of PPI networks and enrich biological information, we propose a heterogeneous network to model proteins and GO attribute information. Then, we propose a node & type attention mechanism to learn the representation of protein nodes. Our method can assign weights to MF and BP GO attribute nodes. Finally, we propose a protein complex identification method based on the weighted PPI network. Figure 1 introduces the flow of our model.

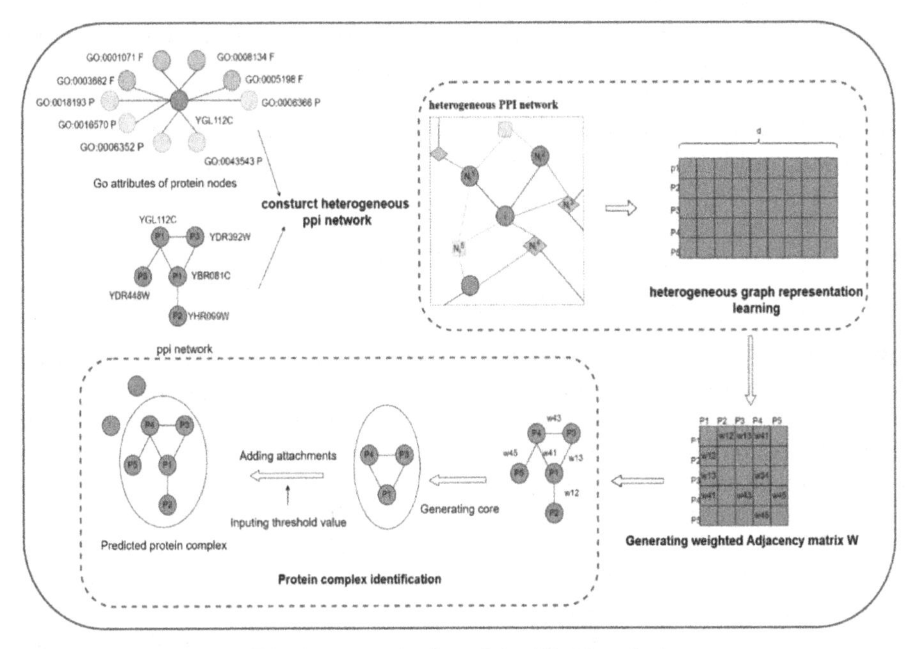

Fig. 1. Basic pipeline of the GHAE method

2.1 Heterogeneous PPI Network Construction

Gene Ontology is widely used in the field of bioinformatics [14]. It covers three aspects of biology: cell components (CC), molecular functions (MF) and biological processes (BP). GO slims offer an overview of the GO contents but do not provide detailed information about specific fine-grained terms. Since GO slims of CC include some protein complexes information, we only select BP and MF into our model as GO attributes. Firstly, we need to integrate the original PPI network with additional GO attributes to build a heterogeneous PPI network. Compared with the homogeneous PPI network, it can contain more types of nodes and analyze the relationship between additional GO attributes and protein nodes. As shown in Fig. 2, we construct the heterogeneous PPI network $G = (V, E)$ contains the protein nodes $N = \{n_1, \ldots, n_m\}$, BP attribute nodes $P = \{p_1, \ldots, p_k\}$ and MF attribute nodes $F = \{f_1, \ldots, f_n\}$ as nodes V, V = N ∪ P ∪ F.

We establish the connection between protein nodes and two types of attribute nodes according to the GO attribute library. We use word2vec to get initial embeddings of attribute nodes based on the short text description in the GO attribute library. By combining two types attribute nodes and protein nodes, the semantics of the PPI network is enriched. It will help us to learn protein nodes embedding better.

2.2 Heterogeneous Graph Convolution Network

GCN [15] is a classic graph neural network (GNN) model. Its function is similar to CNN, but the object it processes is graph data. It can aggregate the neighbor-hood information of nodes to obtain the embedding of target nodes. Suppose we have a batch of PPI network data $\mathcal{G} = (\mathcal{V}, \mathcal{E})$. $\mathcal{V}$ and $\mathcal{E}$ denote the collection of protein nodes and all the interactions between them. Each node has its own features, we assume that the features of these nodes form a matrix $X \in R^{|\mathbb{V}|\times q}$. Each row in matrix X is a feature vector for a node $v(x_v \in R^q)$. The relationship between protein nodes also forms an N × N dimensional matrix A. The way it propagates from layer to layer is as follows.

$$H^{(l+1)} = \sigma\left(D^{-\frac{1}{2}}A'D^{-\frac{1}{2}} \cdot H^{(l)} \cdot W^{(l)}\right) \tag{1}$$

$W^{(l)}$ is a trainable transformation matrix. σ denotes an activation function such as ReLU. $H^{(l)} \in R^{|\mathbb{V}|\times q}$ is characteristic of each layer protein nodes in the lth layer. For the input layer, $H^{(0)} = X$. Our heterogeneous PPI network has three types of nodes: protein nodes, MF attribute nodes and BP attribute nodes. For protein nodes, we use the topology information in the PPI network, the row vector in the adjacency matrix as the initial feature X_n. According to the description in GO slim database, we divide these protein nodes into eight categories and assign labels for them.

GCN encoder cannot aggregate the information of different kinds of neighbor nodes in the PPI network. Previous models have combined GCN with the heterogeneous network, but they have some limitations. These models splice feature spaces of different types of nodes to form a new large feature space. It is easy to ignore the heterogeneity of different types of nodes, resulting in poor features learned. To address this problem, we propose the heterogeneous graph convolution, which considers the difference between various types of information and projects them into an implicit common space with their respective transformation matrices.

$$\mathrm{H}^{(l+1)} = \sigma\left(\sum_{\tau\in T} \tilde{\mathrm{A}}_\tau \cdot \mathrm{H}_\tau^{(l)} \cdot \mathrm{W}_\tau^{(l)}\right) \tag{2}$$

where $\widetilde{A_\tau} \in R^{|\mathbb{V}|\times|\mathbb{V}_\tau|}$ represents the adjacency matrix of all τ type nodes, the rows of the matrix represents the number of nodes in our heterogeneous graph and the columns of the matrix represent all τ type neighbor nodes. We use three types of adjacency matrix $\widetilde{A_\tau}$ based on different feature transformation matrices to aggregate the features of their different types of neighbor nodes $H_\tau^{(l)}$ respectively. Then we can obtain the representation of the target protein node $H(l+1)$.

2.3 Node and Type Attention Mechanism

Different adjacent nodes may have different effects on target protein nodes. The neighboring nodes of the same type may carry more useful information and different neighboring nodes of the same type could also have different importance. Therefore, we design a node & type attention mechanism to capture the different effects on both node level and type level.

The type-level attention assigns weights to different types of nodes around the target protein node. We calculate the attention scores of different types according to the target node embedding h_v and the type embedding h_τ. The calculation formula is as follows:

$$a_\tau = Leaky\ ReLU\left(\mu_\tau^T \cdot [h_v||h_\tau]\right) \tag{3}$$

$h_\tau = \sum_{v'} \widetilde{A_{vv'}}\, h_{v'}$ is the embedding of the type τ, $h_{v'}$ is the sum of neighbor nodes embedding. Then we put the attention score a_τ into the softmax classifier to get the type level attention:

$$\alpha_\tau = \frac{exp(a_\tau)}{\sum_{\tau' \in \mathcal{T}} exp\left(a_{\tau'}\right)} \quad (4)$$

Node-level attention is calculated based on type-level attention, it can reduce the weight of noisy connections in PPI networks. The formula is as follows.

$$e_{vv'} = Leaky\ ReLU\left(v^T \cdot \alpha_{\tau'}\left[h_v || h_{v'}\right]\right) \quad (5)$$

Taking the target node of type τ as an example, node-level attention needs to use node v to calculate the attention weights $a_{\tau'}$ of different types of nodes. Then we put the attention score into the softmax layer, and the calculation formula is as follows:

$$\beta_{vv'} = \frac{exp\left(e_{vv'}\right)}{\sum_{i \in N_v} exp\left(e_{vi}\right)} \quad (6)$$

Finally, we apply type-level and node-level attention to our network through equation replacement. The operation process is as follows:

$$H^{(l+1)} = \sigma\left(\sum_{\tau \in \mathcal{T}} \mathcal{B}_\tau \cdot H_\tau^{(l)} \cdot W_\tau^{(l)}\right) \quad (7)$$

Here, $\mathcal{B}_\tau$ represents the attention matrix, we can get the embedding of protein nodes $H^{(L)}$. We use the final characterization of all protein nodes to weight our PPI network. Here, we use cos_{sim} method. cos_{sim} is also known as cosine distance. It uses cosine between two vectors to measure the difference between two individuals. The cosine value is closer to 1, the more similar the two vectors are. Our network weighting method is shown as follows:

$$w_{ij} = \begin{cases} cos\_sim(\varphi_i, \varphi_j) & a_{ij} = 1 \\ 0 & a_{ij} = 0 \end{cases} \quad (8)$$

where φ_i and φ_j represent the embedding of two protein nodes i and j with interaction in the PPI network.

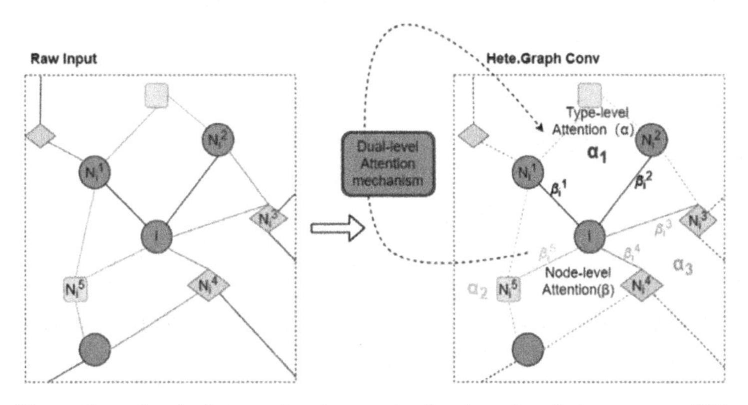

Fig. 2. Illustration of node & type attention mechanism based on heterogeneous PPI network

2.4 Protein Complex Identification Based on GHAE

Protein complexes are usually composed of core and attachment parts. Based on this principle, our protein complex identification method can be described as follows. First, we need to generate a set of seed cores. The group mining algorithm can find closely related proteins in the weighted PPI network. We screen and eliminate the cliques with less than three protein nodes. The remaining cliques are considered as the candidate cores and we will put them into the *Alternative_core* set. We prune the *Alternative_core* set to get the *Seed_core* set based on the following steps:

1. Cliques in *Alternative_core* set are sorted in descending order by *density_score*. The *density_score* can consider the inside connective density and biological information at the same time:

$$density\_score(Clique_q) = \sum_{i,j\in\ Clique_q} w_{ij} \tag{9}$$

2. At this time, $Clique_1$ has the largest *Density_score* in the *Alternative_core* set. Remove $Clique_1$ from the *Alternative_core* set and put it to the *Seed_core* set.
3. For $Clique_i$ in the *Alternative_core* set that has an overlap with $Clique_1$, $Clique_1$ is updated with $Clique_i - Clique_1$. If $|\ Clique_i\ | < 3$, we remove $Clique_i$ from the *Alternative_core* set.

This process repeats until the *Alternative_core* set is empty. The cliques in the *Seed_core* set are regarded as the core proteins in protein complexes.

The internal adjacency strength of the core includes topology and biology. The correlation score between cliques and candidate attachment proteins is calculated as below:

$$correlation\_score\left(p_i,\ Clique_j\right) = \frac{\sum_{k\in\ Clique_j} w_{ik}}{|\ Clique_j|} \tag{10}$$

where protein p_i is one of the neighbors of the corresponding core $Clique_j$. If the correlation score between protein p_i and $Clique_j$ is larger than a threshold value θ, $p_i$is considered as one attachment of the corresponding clique. We combine the protein core with its accessory nodes to form the final identification protein complexes.

3 Experiments and Results

3.1 Datasets and Evaluation Metrics

We use five yeast PPI networks in our experiments: DIP, Krogan-core, Krogan14k, Biogrid and Collins. The detailed information of these datasets is shown in Table 1. We selected MIPs, cyc2008, SGD, alloy, and tap06 as the standard complexes. There are 789 protein complexes in the standard complex set.

For the fairness and reliability of our experimental results, we use the evaluation indexes commonly used in the field of protein complex identification. P denotes the set

Table 1. The PPI datasets used in the experiment

PPI networks	Number of proteins	Number of interactions	Average clustering coefficient	Average number of neighbors
DIP	4928	17201	0.095	6.981
Krogan-core	2708	7123	0.188	5.261
Krogan14K	3581	14076	0.122	7.861
Biogrid	5640	59748	0.246	21.187
Collins	1622	9074	0.555	11.189

of identified protein complexes from one method. We suppose that B and P represent the standard and predicted protein complex sets, respectively.

If we select a real protein complex $b \in B$ and an identification protein complex $p \in P$, we can calculate their similarity. Neighborhood affinity score is as follow:

$$NA(p, b) = \frac{|V_p \cap V_b|^2}{|V_p| \times |V_b|} \tag{11}$$

where V_p and V_b represent the collection of protein molecules in complexes b and p respectively. $|V_p \cap V_b|$ represents the number of proteins shared in the two protein complexes. Generally speaking, if NAp, b is larger than 0.25, the two protein complexes are considered to be matched.

N_{cp} represents protein complexes that match at least one actual complex, N_{cb} is the number of real complexes that match at least one predicted complex.

$$N_{cp} = |\{p|p \in P, \exists b \in B, NA(p, b) \geq \omega\}|, \text{Precision} = \frac{N_{cp}}{|P|} \tag{12}$$

$$N_{cb} = |\{b|b \in B, \exists p \in P, NA(p, b) \geq \omega\}|, \text{Precision} = \frac{N_{cb}}{|B|} \tag{13}$$

ω is the threshold parameter, its reasonable value is usually 0.25. First three measures used in the experiments for evaluating the performance of different methods are Precision, Recall and F-score. Precision is the proportion of identified protein complexes that match at least one reference complex. The recall is the proportion of reference protein complexes that match at least one predicted complex. F-score is the harmonic mean of Precision and Recall.

$$F - \text{score} = \frac{2 \times \text{Precision} \times \text{Recall}}{\text{Precision} + \text{Recall}} \tag{14}$$

We also use geometric accuracy (Acc) as our indicator. F – score and Acc can effectively evaluate the ability of complex identification methods to rediscover known complexes from PPI networks. Their value range is between 0 and 1.

3.2 Performance Comparison

We compare our algorithm with the six most advanced protein complex identification methods: COACH, CMC, MCODE, ClusterOne [10], GANE [16] and PEWCC. All experimental results are listed in Table 2.

According to the data in Table 2, we can observe that GHAE obtains the highest Acc scores in all datasets and get the highest F-score in two datasets: Biogrid and Krogan14k. GHAE does not achieve the highest Recall. It is probably because the number of identified protein complexes is small. It illustrates that the overall accuracy of protein complexes identified by GHAE is better than prevalent algorithms.

GANE method also has no bad F-score and Acc values. The F-score on the DIP dataset and Krogan-core dataset is better than GHAE. However, our method is superior to GANE in Acc index. The GANE method has a similar idea as our method. We both learn the vector representation of each protein from the PPI network of GO attributes and construct a weighted adjacency matrix based on the similarity of the vector representation. The difference between them is no graph neural network is used on GANE. So GANE cannot learn the global topology information of PPI network very well. GHAE uses the representation learning method of heterogeneous graph neural network and combines it with the dual attention mechanism. Compared with GANE, it can generate a representation of the target node by aggregating the features of the node itself and the features of its attributes. This makes the protein node embedding more accurate. It can also distinguish the importance of different types of GO attribute nodes. That is why our proposed GHAE method achieves the highest Acc on all datasets.

3.3 Parameter Sensitivity Experiment

Our method can be summarized as two steps. First, we learn the expression of protein nodes based on heterogeneous PPI network. Then, we use the protein complexes clustering method based on the weighted PPI network. Therefore, the embedding dimension of protein nodes D and threshold value θ affect the final performances directly. In this part, we study the representation D and threshold θ. In the experiment, the dimension D of the embedding vector varies from 32 to 224. Figure 3 shows that the fluctuation of our experimental data with the change of dimension is not obvious. When the dimension is 128, the performance is the best. The threshold θ determines the neighborhood range of the core. When the value θ is higher, the protein nodes near the central node will be more difficult to become affiliated nodes. On the other hand, the internal connections of protein complexes identified in this way are also more reliable. When θ is less than 0.1, the performance is very poor. This is because when θ is small, it will integrate many neighbor protein nodes to form complexes. The performance reaches its peak when θ is 0.3, so we set 0.3 as its default value.

3.4 Influence of Different Network Representation Learning Methods

To verify the impact of the heterogeneous network representation learning method in GHAE. We chose four network representation learning methods: DeepWalk [17], HOPE [18], LINE [19] and SDNE [20] for experiment. We compare the experimental results with GHAE. The dimension of the network embedding is set to 128. The comparison between GHAE and the other four network representation learning methods on the DIP datasets is shown in Fig. 4. Among the remaining four network representation learning methods, the result of the DeepWalk method is slightly lower than GHAE. The results obtained using the other three network representation learning methods are much

Table 2. Performance comparison based on five evaluation metrics on the five datasets

Datasets	Methods	Precision	Recall	F-score	Acc
DIP	COACH	0.450	0.620	0.521	0.243
	CMC	0.603	0.394	0.477	0.219
	MCODE	0.542	0.118	0.194	0.149
	ClusterOne	0.390	0.343	0.365	0.227
	GANE	0.623	0.550	**0.584**	0.294
	PEWCC	0.620	0.469	0.534	0.230
	GHAE	0.610	0.513	0.564	**0.319**
Krogan-core	COACH	0.592	0.460	0.518	0.320
	CMC	0.672	0.304	0.419	0.442
	MCODE	0.732	0.198	0.311	0.364
	ClusterOne	0.364	0.464	0.408	0.489
	GANE	0.774	0.436	**0.558**	0.229
	PEWCC	0.675	0.406	0.507	0.387
	GHAE	0.654	0.446	0.531	**0.535**
Krogan14k	COACH	0.461	0.465	0.463	0.342
	CMC	0.472	0.440	0.455	0.421
	MCODE	0.612	0.112	0.189	0.270
	ClusterOne	0.467	0.302	0.366	0.473
	GANE	0.448	0.442	0.445	0.234
	PEWCC	0.535	0.418	0.470	0.367
	GHAE	0.537	0.437	**0.482**	**0.505**
Biogrid	COACH	0.311	0.657	0.422	0.376
	CMC	0.157	0.553	0.245	0.365
	MCODE	0.276	0.043	0.075	0.181
	ClusterOne	0.393	0.497	0.439	0.486
	GANE	0.345	0.464	0.395	0.310
	PEWCC	0.375	0.677	0.338	0.347
	GHAE	0.443	0.489	**0.465**	**0.509**
Collins	COACH	0.749	0.522	**0.615**	0.362
	CMC	0.680	0.390	0.496	0.493
	MCODE	0.847	0.400	0.540	0.362
	ClusterOne	0.733	0.511	0.602	0.561

(continued)

Table 2. *(continued)*

Datasets	Methods	Precision	Recall	F-score	Acc
	GANE	0.819	0.491	0.615	0.520
	PEWCC	0.837	0.426	0.564	0.449
	GHAE	0.609	0.571	0.590	**0.619**

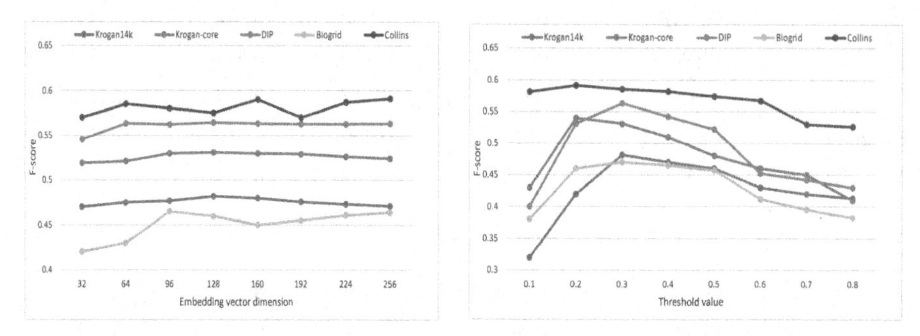

Fig. 3. Performance comparison on different embedding dimension D and threshold value θ

lower than GHAE. Therefore, when obtaining protein vector representations in the PPI network, the GHAE method can be used to get better performance. This shows that the GHAE method not only learns the topological features of the protein network but also incorporates the GO attribute information of the protein node. It reflects the superiority of GHAE compared to traditional homogeneous network representation learning methods for protein complexes identification.

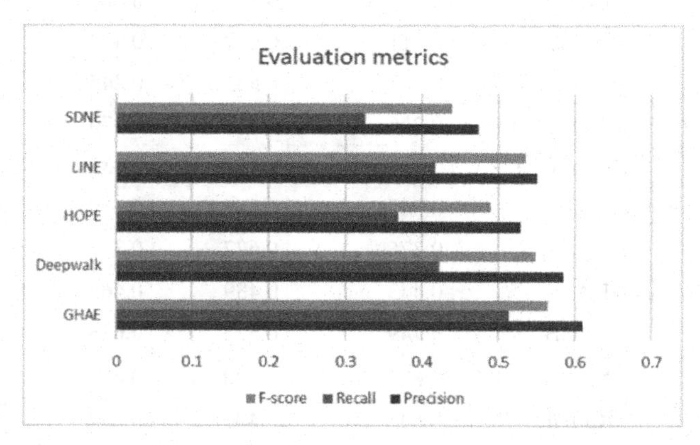

Fig. 4. The impact of different network representation learning methods on the experimental performance of the DIP network

3.5 Biological Significance of the Identification Protein Complex

To assess whether protein complexes identified by our method have biological significance, we calculate the $MinP - value$. $P - value$ is defined as follows:

$$P - value = 1 - \sum_{i=0}^{k-1} \frac{\binom{|F|}{i}\binom{|V|-|F|}{|C|-i}}{\binom{V}{C}} \quad (15)$$

where an identified complex C contains k proteins in the functional group F and the whole PPI network contains $|V|$ protein. The functional homogeneity of a predicted the complex is the $MinP - value$ overall the possible functional groups. An identified complex with a low functional homogeneity indicates it is enriched by proteins from the same function group.

We can know that the simultaneous appearance of these proteins in the complex is not a coincidence but can be explained. Table 3 lists the results of GO analysis in the Krogan-core network. There are some identified complexes that cannot match any reference with high biological significance actually. These complexes stand a good chance of the new complexes that have not been discovered. Thus, these identified complexes have a high probability of being real and have great biological significance.

Table 3. Examples of predicted complexed on the Krogan-core dataset

Protein complex	Size	mr	Min P-value	GO-Term	GO-Description
YEL005C TGL079W YNL086W YLR408C	4	1.0	5.96E-10	0031083	BLOC-1 complex
YHR069C YOR001W YHR081W YOL021C	4	1.0	6.56E-09	0000178	Exosome
YBR231C YDR334W TGR002C YLR385C YLR399C	5	1.0	2.04E-11	0000812	Swr1 complex
YLR071C YGL127C YHR058C YOL051W YBR193C YPR070W	6	1.0	3.54E-13	0070847	Core mediator complex

4 Conclusion

In this paper, we propose a protein complex identification method called GHAE. The heterogeneous PPI network we build has richer semantic information than the traditional PPI network. By introducing a dual attention mechanism, we can assign weights to different nodes and types to improve our complex identification accuracy. The experimental performance proves that the GHAE algorithm can obtain the best results in most scenarios. The introduction of heterogeneous network with GO attributes can be combined with many other algorithms to improve the performance of protein complex identification in the future.

Acknowledgment. This work is supported by grant from the Natural Science Foundation of China (No. 62072070).

References

1. Hanna, E.M., Zaki, N.: Dynamic protein-protein interaction networks and the detection of protein complexes: an overview. In: Proceedings of the International Conference on Bioinformatics and Computational Biology, p. 1 (2014)
2. Xu, Y., Zhou, J., Zhou, S., Guan, J.: CPredictor3.0: Detecting protein complexes from PPI networks with expression data and functional annotations. BMC Syst. Biol. **11**(S7), 45–56 (2017)
3. Adamcsek, B., Palla, G., Farkas, I.J., et al.: CFinder: locating cliques and overlapping modules in biological networks. Bioinformatics **22**(8), 1021–1023 (2006)
4. Bader, G.D., Hogue, C.W.V.: An automated method for finding molecular complexes in large protein interaction networks. BMC Bioinform. **4**(1), 1–27 (2003)
5. Liu, G., Wong, L., Chua, H.N.: Complex discovery from weighted PPI networks. Bioinformatics **25**(15), 1891–1897 (2009)
6. Nepusz, T., Yu, H., Paccanaro, A.: Detecting overlapping protein complexes in protein-protein interaction networks. Nat. Methods **9**(5), 471–472 (2012)
7. Asur, S., Ucar, D., Parthasarathy, S.: An ensemble framework for clustering protein–protein interaction networks. Bioinformatics **23**(13), i29–i40 (2007)
8. Zaki, N., Efimov, D., Berengueres, J.: Protein complex detection using interaction reliability assessment and weighted clustering coefficient. BMC Bioinformatics **14**(1), 1–9 (2013)
9. Leung, H.C.M., Xiang, Q., Yiu, S.M., et al.: Predicting protein complexes from PPI data: a core-attachment approach. J. Comput. Biol. **16**(2), 133–144 (2009)
10. Wu, M., Li, X., Kwoh, C.K., et al.: A core-attachment based method to detect protein complexes in PPI networks. BMC Bioinform. **10**(1), 1–16 (2009)
11. Chin, C.H., Chen, S.H., Ho, C.W., et al.: A hub-attachment based method to detect functional modules from confidence-scored protein interactions and expression profiles. BMC Bioinform. **11**(1), 1–9 (2010)
12. King, A.D., Pržulj, N., Jurisica, I.: Protein complex prediction via cost-based clustering. Bioinformatics **20**(17), 3013–3020 (2004)
13. Li, X.L., Foo, C.S., Ng, S.K.: Discovering protein complexes in dense reliable neighborhoods of protein interaction networks. Comput. Syst. Bioinform. **6**, 157–168 (2007)
14. Lambrix, P., Habbouche, M., Perez, M.: Evaluation of ontology development tools for bioinformatics. Bioinformatics **19**(12), 1564–1571 (2003)
15. Kipf, T.N., Welling, M.: Semi-supervised classification with graph convolutional networks. In: ICLR (2017)
16. Xu, B., et al.: Protein complexes identification based on go attributed network embedding. BMC Bioinform. **19** (2018). https://doi.org/10.1186/s12859-018-2555-x
17. Perozzi, B., Al-Rfou, R., Deepwalk, S.S.: Online learning of social representations. In: Proceedings of the 20th ACM SIGKDD International Conference on Knowledge Discovery and Data Mining, pp. 701–710. ACM (2014)
18. Ou, M., Cui, P., Pei, J., et al.: Asymmetric transitivity preserving graph embedding. In: Proceedings of the 22nd ACM SIGKDD International Conference on Knowledge Discovery and Data Mining, pp. 1105–1114. San Francisco, USA (2016)
19. Tang, J., Qu, M., Wang, M., et al.: Line: large-scale information network embedding. In: Proceedings of the 24th International Conference on World Wide Web, pp. 1067–1077 (2015)
20. Heimann, M., Koutra, D.: On generalizing neural node embedding methods to multi-network problems. KDD MLG Workshop (2017)

A Clonal Evolution Simulator for Planning Somatic Evolution Studies

Arjun Srivatsa[1], Haoyun Lei[1], and Russell Schwartz[1,2(✉)]

[1] Computational Biology Department, Carnegie Mellon University, Pittsburgh, PA 15213, USA
russells@andrew.cmu.edu

[2] Department of Biological Sciences, Carnegie Mellon University, Pittsburgh, PA 15213, USA

Abstract. Somatic evolution plays a key role in development, cell differentiation, and normal aging, but also diseases such as cancer. Understanding mechanisms of somatic mutability and how they can vary between cell lineages will likely play a crucial role in biological discovery and medical applications. This need has led to a proliferation of new technologies for profiling single-cell variation, each with distinctive capabilities and limitations that can be leveraged alone or in combination with other technologies. The enormous space of options for assaying somatic variation, however, presents unsolved informatics problems with regards to selecting optimal combinations of technologies for designing appropriate studies for any particular scientific questions. Versatile simulation tools are needed to explore and optimize potential study designs if researchers are to deploy multiomic technologies most effectively. In this paper, we present a simulator allowing for the generation of synthetic data from a wide range of clonal lineages, variant classes, and sequencing technology choices, intended to provide a platform for effective study design in somatic lineage analysis. Users can input various properties of the somatic evolutionary system, mutation classes, and biotechnology options and then generate samples of synthetic sequence reads and their corresponding ground-truth parameters for a given study design. We demonstrate the utility of the simulator for testing and optimizing study designs for various experimental queries.

Keywords: Somatic evolution · Study design · Cancer genomics · Simulation

1 Introduction

Advanced sequencing technologies have made it possible to profile genetic variation at the single-cell level on population scales, revealing in part that the human body is a continuously evolving genetic mosaic [1,9]. Genetic and epigenetic modifications in somatic cells over many generations of cell growth and replication result in heterogeneity between cells, tissues, and organs in normal aging

M. S. Bansal et al. (Eds.): ISBRA 2022, LNBI 13760, pp. 229–242, 2022.
https://doi.org/10.1007/978-3-031-23198-8_21

and development, as well as in disease conditions such as neurodegeneration and, most notably, cancer [21]. Accumulating genomic data has made it apparent that somatic mutability is much more complicated than early models of tumor clonal evolution first suggested [5] and far more extensive in even healthy tissues (c.f., [4]). Somatic variation produces complex patterns of "mutational signatures" [2] reflecting different endogenous and exogenous mechanisms of mutability. In cancers and other precancerous conditions, high levels of somatic mutability are frequently observed due to damage to cell replication or error-correction machinery [24]. They further may include not just single nucleotide variations (SNVs) but potentially extensive copy number alterations (CNAs) and structural varations (SVs), including complex chromosomal rearrangement patterns and genome duplication events [18].

As we have come to understand the extent and importance of somatic evolution, enormous effort has been put into developing biotechnological tools for profiling somatic variability at ever greater scales, precision, and accuracy [7]. No one technology is able to comprehensively characterize somatic variability across a complex tissue and do so with precision and accuracy and at low cost. Current work increasingly depends on multiomic biotechnology combinations(e.g., long and short read, single cell, whole genome and exome, etc.), along with various other involved study design choices (e.g., number of biopsy replicates), with uncertain knowledge of how these choices together with analysis software will influence one's ability to quantify any particular feature of the somatic evolution process [14]. There is currently little empirical or theoretical basis on which an investigator planning a study can select a combination of technologies and study design well suited for any particular investigation.

Simulation presents a viable solution to these issues by allowing for efficient tests of various study designs with direct knowledge of most biological parameters of interest. The popular BAMSurgeon simulator [8] has been valuable in testing tumor variant calling algorithms. However, current simulators fail to capture the broad range of hypermutability processes that occur in cancer cell populations, and often focus on one particular aspect of cancer evolution (e.g., copy number or spatial analysis). Furthermore, no simulator currently exists that allows for the exploration of widely varying study designs and multiomic technologies. For a comparison between features available in our simulator versus others, see [25].

Here, we seek to meet the needs of sequencing study design for somatic variation studies through a new clonal evolution simulator. Our simulator links a coalescent model of clonal evolution to a versatile model of read generation with user-configurable variant classes, mutation rates, evolutionary models, sequencing setups, and study design decisions. Our framework focuses on general properties of sequencing that allow for the design of better experiments and future sequencing technologies. It also introduces a wide variety of features important to somatic variation studies that are not, to our knowledge, found in any other current simulator, such as capturing broad classes of complex structural variation that have been implicated in certain cancers. We demonstrate utility of this simulator through application to a series of hypothetical questions in testing and optimizing study design for somatic evolution studies.

2 Materials and Methods

The complete simulator consists of four main modules: 1) sampling an evolutionary lineage tree for the clonal evolution, 2) sampling mutation events on the lineage tree, 3) simulating sequence reads, and 4) sampling reads based on experimental design decisions. Below, we describe each module in turn. Each module has a number of user-tunable parameters to control different biological parameters of the presumed cell lineages as well as experimental parameters of the sequencing strategy. Additional information on mutation types and module implementations are provided in [25]. The major tunable parameters are summarized in Table 1.

2.1 Lineage Simulation and Mutation Events

For each simulation, we generate a cell lineage assuming that mutations are selectively neutral and generally follow the assumptions of the standard coalescent model [20]. User-definable parameters include a total population size (N_e), as well as a number of clones (k) to be sampled. Coalescent sampling of tree topologies and edge lengths is implemented using `msprime` [12]. The unit of time we use is a generation, i.e., the time of a single cell division. A single sample from this process represents a single-cell lineage tree, each cell of which is taken to be representative of some 'clone' in the tree.

The simulator supports commonly discovered types of somatic variations, particularly those implicated in the development of cancer [6]. These currently include the following: SNVs, CNAs, insertions, deletions, kataegis, chromothripsis, chromoplexy, aneuploidy, translocations, inversions, and breakage fusion bridge cycles. Each mutation type is implemented by sampling from various probability distributions for location and length while simultaneously maintaining constraints encapsulating our knowledge of the mechanism for each mutation type. The scale of many of these forms of variations can be tuned, but with default values set based on estimated distributions of sizes found in current studies of cancer genomes [18]. Size distributions for structural variants are modeled as a truncated mixture of negative binomials to represent small, medium, and large scale events. Each mutation type also has a rate at which that mutation appears which we assume may differ between tumor stages and in healthy tissues per the "mutator phenotype" hypothesis [19]. The simulator also currently supports simulating mutations drawn from single base substitution signatures derived from the COSMIC dataset [26]. Distributions over mutation size and location are also flexible.

Once a lineage is simulated, we apply mutations to this lineage going forward from the most recent common ancestor of all the clones to be sampled. Mutation rates for each class of variation are defined as a uniform discrete distribution of potential rates, M_i. For each edge of the lineage, a specific rate is generated via $r_i \sim M_i$ and mutation times are generated via a Poisson process, with rate r_i. This is done for each edge of the phylogeny independently for each class of mutation. The end result of this process is a list of times for each mutation type and its occurrence on each edge of the phylogeny. Next, the simulator imposes each of these mutations on a reference genome to establish the sequences of clones at all nodes of the lineage tree. Given that mutations were independently simulated for each mutational class, we first merge and sort the mutational events by time. We first compute all root-to-leaf paths in the tree, which allows us to generate all potential clonal genomes. We start with the root node as the reference and impose sorted mutations for each root-to-leaf path. The end result of this process is a stored genomic sequence for each clone, including those at internal nodes, that is later sampled in the sequencing step.

2.2 Sequencing Implementation and Experimental Decisions

Sequencing procedures differ depending on the type of sequencing chosen (e.g., whole genome sequencing or targeted sequencing). The general strategy, however, is similar. First, a clone from the tree is sampled from a Dirichlet process with distribution $g(k)$ and concentration parameter α, i.e. $k \sim G \sim DP(g(k), \alpha)$. This allows for a flexible, and potentially skewed, distribution of the samples clones, as may occur in a biopsy sample. The sampled clone may be either a leaf node or an internal node, but not the root node. For this selected clone, stochastic fragment lengths are drawn according to the read length, i.e., $fl \sim TruncatedNegBin(rl)$. Using these fragment lengths, operations are performed on fragmented clonal reads defined by the user-parameters. Specifically, clone k is loaded, "chopped" according to fragment length, subsetted based on read length and paired sequencing type, seeded with errors according to the error rate, and written to a FASTQ file. After this particular clone is sequenced, the coverage of the simulator is updated according to the fraction of the genome covered depending on the read and fragment length. This process repeats with independent repeats of each stochastic process until the desired coverage is reached.

The process for exome and targeted sequencing requires additionally identifying reads that align to exon or targeted gene sequences. This approximate string matching problem is computationally infeasible over every possible read so we make a few simplifying assumptions. First, we create k-mer sets for each of our target sequences and use locality sensitive min-hashing to index these sequences for fast lookup [22]. Then, for genome locations surrounding our target sequence intervals, we calculate whether reads originating from these genome locations match some target sequence in the hashing index above a certain threshold. If this location matches some sequence, then we add it to a list of locations. We repeat this sampling according to the user set parameter G_N times, and use this list combined with the original sequence locations to generate a discrete probability density to sample read locations. We sample reads from this generated list with the given read length while simultaneously sampling different clones until our desired coverage is reached.

Modifications also occur for single cell sequencing, where we do not continually re-sample cell clones, but instead sample only once and use the given clone until the desired coverage is reached. Finally, for liquid biopsy, we perform a similar iterative clonal sampling procedure, but do not chop the sequence uniformly. Rather, we draw random fragments from the genome strings of each clone and mix them at some frequency with reference DNA then return these in a read file.

The full simulation is defined by looping the lineage, mutation, and sequence sampling over tumors and samples. Specifically, we independently define and execute a separate lineage sampling and mutational frequency for the number of user-defined tumors generated. Similarly, we configure parameters related to sequencing decisions, and execute the procedures listed above according to the number of user-defined samples requested. The final result of the simulator is a labeled directory with sub-directories corresponding to reference reads, tumor reads, and sample reads. Each of these directories holds ground truth parameters with information about the tumor and sequencing parameters.

2.3 Runtime, Space, and Resource Analysis

Defining c_{ij} as the sequencing coverage of the jth sample on tumor t_i, n_{ij} as the number of single cell samples drawn on the on the jth sample of tumor t_i, and s_i as the number of samples drawn on tumor t_i. The run time is approximately proportional to $O(\sum_{i=1}^{t}\sum_{j=1}^{s_i} c_{ij}(1+n_{ij}))$, which approximately calculates the amount of times we traverse and sequence the genome.

The maximum memory usage is constrained by the genome size, read length, and sequencing type. Generally, this factors to around 5–10x the size of the genome in standard sequencing settings. If the user wants to minimize memory footprint, then he or she should set the batch and subblock size to 1, or avoid generating long-read exome sequenced data. The amount of disk storage of the program is bounded by the storage of the clonal genomes during the mutational process as well as the sequencing FASTQ files. This is approximately proportional to $O([\sum_{i=1}^{t} k_i + \sum_{i=1}^{t}\sum_{j=1}^{s_i} c_{ij}(1+n_{ij})] \times d)$, where d is the genome size, k_i denotes the number of clones in tumor t_i and c_{ij} and n_{ij} are defined as above.

A single run of the simulator only uses one core, allowing for parallelization across cores. As a reference point, three 30x-paired WGS samples take around 3.5 h to generate on our system and use around 250-300GB of disk storage and less than 40G of memory.

2.4 Simulator Usage

The simulation process described above has a number of tunable biological and experimental parameters (detailed in Table 1). The main biological parameters include mutation rate, number of clones, and number of tumors. The main experimental parameters include the number of samples and various parameters describing the sequencing modalities. Each major tunable parameter is encoded as a list, which is randomly sampled, allowing for the user to vary both the biology of the tissue as well as the different experimental setups. The intended usage of the simulator revolves around testing the limitations of various study design paradigms against different somatic evolution instances. For example, if one were interested in finding a study design for testing presumably "healthy" tissue for mutation burden, he or she might fix the mutation rate list to a low rate and replicate a large number of tumors as well as search over a broad experimental design space. On the other hand, if one were interested in the limitations of a particular study design (for instance, 30x WGS), he or she might replicate a broad range of tumors with various mutation rates and clone numbers while fixing the experimental strategy. A description of the directory structure and output of a sample simulator run is shown in Fig. 1.

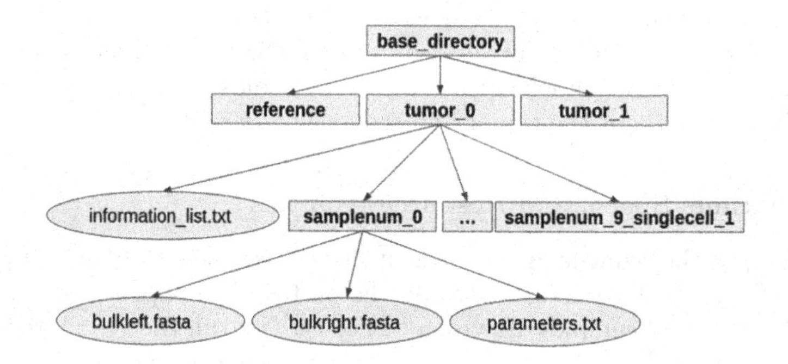

Fig. 1. Example simulator output and directory structure for a single run.

3 Experimental Results

In this section, we demonstrate the utility of our simulator in evaluating or optimizing sequencing study designs for profiling clonal evolution. A primary motivation of this simulator is to plan study designs for evaluating questions about differences in somatic mutability between subsets of samples. The questions for study design are to test whether a difference can be detected between two subsets of samples under a given study design or to find a study design optimizing power to detect such differences. Here, study design might include changes in the types of sequencing applied, the informatics software used to evaluate it, and features such as the number of tumors and tumor sites or regions to be examined. We would then assume that the study is being used to test for differences in biological parameters between subsets of samples. Such biological parameters might include mutation rate differences, presence of rare variations, variation in mutational signatures, phylogeny structures, or clonal frequencies.

3.1 Notation and Performance Measures

For the analyses presented here, we take the sequencing read outputs from our simulator and perform alignment to the hg38 reference genome, after which we

Table 1. Summary of the main tunable parameters

Parameter	Symbol	Units	Description
Effective population size	N_e	Cells	Total cellular population of the region to be sampled. Impacts coalescent times
Number of clones	k	Clones	Number of distinct genetic somatic cell populations to be sampled
Mutation rate lists	M_i	Mutation events per locus per cell division	Lists for each variant class, defining rates per locus per cell division
Mutation size and location distributions	$S, z(x)$	Number of bases, none	Each mutation type can be tuned over size distributions and single base substitutions can be tuned over signature distributions
Number of tumors	t	Tumors	Number of distinct sites of somatic evolution to be sampled
Number of samples	s	Samples	Number of distinct tissue biopsies (regions) to be drawn from each "tumor" site
Read length	rl	Bases	Size of reads to be generated from sequencer
Fragment length	fl	Bases	Defines a superstring from which reads are derived during the sequencing process
Depth/coverage	c	Reads per base	Average number of times each nucleotide of the genome is sequenced
Error rate	e	Fraction of incorrectly sequenced bases	Rate of Incorrectly sequenced nucleotides
Dirichlet concentration, clonal frequency distribution	α, $g(k)$	None, none	Parameter of a Dirichlet process which is used to derive the concentration of the baseline distribution in a sample. A high value will lead to approximately uniform sampling of clones during sequencing, whereas a low value would favor very uneven clonal frequencies. The clonal frequency distribution is the baseline distribution at high α
Number of single cells	n	Cells	Number of individually sequenced cells per sample
Paired-end/single-end	$Paired$	Boolean	Binary parameter describing whether the reads are paired end or single end. Paired-end reads have two related reads derived from the same fragment.
Whole genome/whole exome/targeted sequencing	$Genome$	Boolean	Binary parameter describing whether the sequencer extracts genes from the entire genome or only a subset of the genome
Liquid biopsy	$LiquidBiopsy$	Boolean	True or false parameter to produce liquid biopsy sequenced reads
Sampling number	G_N	Genome positions	Number of random positions on the genome to sample to build a hash table for approximate string matching for exon/targeted sequencing

call several forms of variation for analysis. The aligners used were minimap2 [17], Bowtie [15], and bwa-mem [16]; the callers used were Strelka [13] and Delly [23]. Throughout the tests, we reference our study design, which we formally define as a collection of matrices $\mathbf{X} = \{X_1, ..., X_t\}$, where t denotes the number of tumors. The matrices X_i denote the sequencing decisions taken on tumor i and encapsulate every sample. Namely, each matrix X_i is of dimension $7 \times s_i$, where s_i denotes the number of samples taken on tumor i. The columns of X_i denote sequencing and informatics choices on a single sample s_i. That is, the matrix is of the form:

$$\mathbf{x_i} = \begin{pmatrix} rl_i/fl_i \\ c_i \\ 1-e_i \\ n_i \\ Paired_i \\ Genome_i \\ Informatics_i \end{pmatrix} X_i = \begin{bmatrix} | & | & | \\ ... & \mathbf{x_i} & ... \\ | & | & | \end{bmatrix} \tag{1}$$

The vector $\mathbf{x_i}$ contains the variables defined in Table 1, along with a variable to encode informatics options. In our experimental tests, we consider the single tumor, single sample case. With this instance, we can collapse the collection of matrices down to a single vector $\mathbf{x}$.

To judge a study design's utility, we require performance measures that assess whether the design is recovering a signal related to the intended hypothesis. We use recall as the metric for highly mutated SNV samples and F1 score for less heavily mutated SNV samples. For structural variants, we use the measure described following. Call the variant output locations for chromosome i, $C_i = \{(a, b)\}$, and call our ground truth set of structural variant locations, $D_i = \{(c, d)\}$. The measure is then defined as follows: $J = \sum_i \frac{|C_i+D_i|}{\sum_i |C_i+D_i|} \sum_{j \in D_i} \frac{\mathbb{1}(|j \cap C_i|>0)}{|D_i|}$, i.e., we weight the fraction of ground truth events that overlap with called events using the total mutation count over chromosomes.

3.2 Statistical Test for Mutation Rate Variation

We first evaluate whether a given study design would be able to detect a difference in SNV mutation rates between tumors. A motivating hypothesis for these tests is the idea that cancerous and precancerous tissues should typically exhibit hypermutability phenotypes, that is elevated rates of particular kinds of variation that lead to genetic heterogeneity across cells. We would then wish to detect whether a specific study design would be powered to detect a hypothetical difference in mutation rate between two samples indicative of a hypermutability phenotype specific to one sample.

We assume that we have two independently sampled tissues, and we wish to determine whether the mutation rates of various mutation classes is different between these tissues. To create a specific scenario, we generated two sets of SNV data, one where the average rate was high and the other where it was comparatively low (denote the rates as $\lambda_1 \approx 10^{-8}$, $\lambda_2 \approx 10^{-10}$ mutations per nucleotide per generation). We specifically tested whether a 30x WGS screen on both tissues would be statistically powered to detect the mutation rate difference. All other parameters were kept equal at reasonable values (0 error rate, paired-end sequencing, 0 single cells, with 1 sample per tissue). We assume that the first tissue has mutation count generated as $M_1 \sim Poisson(\lambda_1 t_1)$ and the second tissue has mutation count generated as $M_2 \sim Poisson(\lambda_2 t_2)$. We can then test for a rate difference using the following null and alternative hypotheses: $H_0 : \frac{\lambda_1}{\lambda_2} \leq 1$ and $H_1 : \frac{\lambda_1}{\lambda_2} > 1$. In our setting, we have an estimate of the mutation counts M_1 and M_2 from the output of our variant calling software, and we estimate t_1 and t_2 as biologically plausible tumor formation times as described below. Several statistics can be used to establish p-values for a two-sample Poisson rate test (c.f., [10]), but here we favor a conditional test based on the fact that the conditional distribution of M_1 given $M_1 + M_2$ is binomial. Under the assumptions of the problem, this expression, which we denote $p(t_1, t_2)$ to emphasize its dependence on each time, is: $2min(P(X_1 \geq M_1 | n = M1 + M2, p = \frac{t_1/t_2}{1+\frac{t_1}{t_2}}), P(X_1 \leq M_1 | n = M1 + M2, p = \frac{t_1/t_2}{1+\frac{t_1}{t_2}}))$. These values are both computed easily using binomial cdf packages.

In our empirical test case, 244 mutations were called in the higher SNV rate dataset and 2 were called in the lower SNV rate dataset. Times t_1 and t_2 are estimated, with uncertainty, as depths of somatic lineages which could plausibly lead to a tumor; we assign them random variables $t_1, t_2 \overset{ind.}{\sim} Unif[1, 30]$ years. Current evidence suggests that the time from a normal cell to clinical cancer sequencing could be as high as many decades, though studies are in their nascence for somatic charting in healthy and precancerous tissues [11]. To generate a p-value, we approximate the expected p-value with respect to the time via a bootstrap procedure, i.e. $\mathbb{E}_{t_1,t_2}[p(t_1, t_2)] = \int_1^{30} \int_1^{30} p(t_1, t_2) dt_1 dt_2 \approx \frac{1}{N} \sum_{i \in [N]} p(t_{i1}, t_{i2})$. In our test, case we took 500 random draws of times t_1, t_2 and computed an average p-value of 6.9×10^{-7} with variance 1.4×10^{-10}. Therefore, we can conclude in this test case that the given study design should be powered to provide a strongly significant detection of the given rate variation.

3.3 Optimizing a Study Design

The most involved intended use of the simulator is to optimize a study design to evaluate a particular hypothesis about somatic variability. The presumed goal is to design a study that is optimally powered to detect the hypothesized signal within available resource constraints. Here, we demonstrate the use of our simulator to answer such a study design question to find an approximately optimal study design for a particular hypothesis. The prior premise can be framed as the following optimization:

$$argmin_{\mathbf{x}} \quad \mathbb{E}_{q \sim Q}[\mathbf{c}(\mathbf{x}) + \lambda L_q(\mathbf{x})] \tag{2}$$

$$\text{s.t.} \quad \mathbf{d}(\mathbf{x}) \leq \mathbf{b} \tag{3}$$

$$\mathbf{x} \geq \mathbf{0} \tag{4}$$

Here $\mathbf{x}$ is the study design vector defined in Eq. (1). $\mathbf{c}(\mathbf{x})$ is a cost function for a particular study design, $\mathbf{d}(\mathbf{x})$ is a budget function that assigns a resource usage to a study design, and $\mathbf{b}$ is a maximal budget vector. We also define a loss

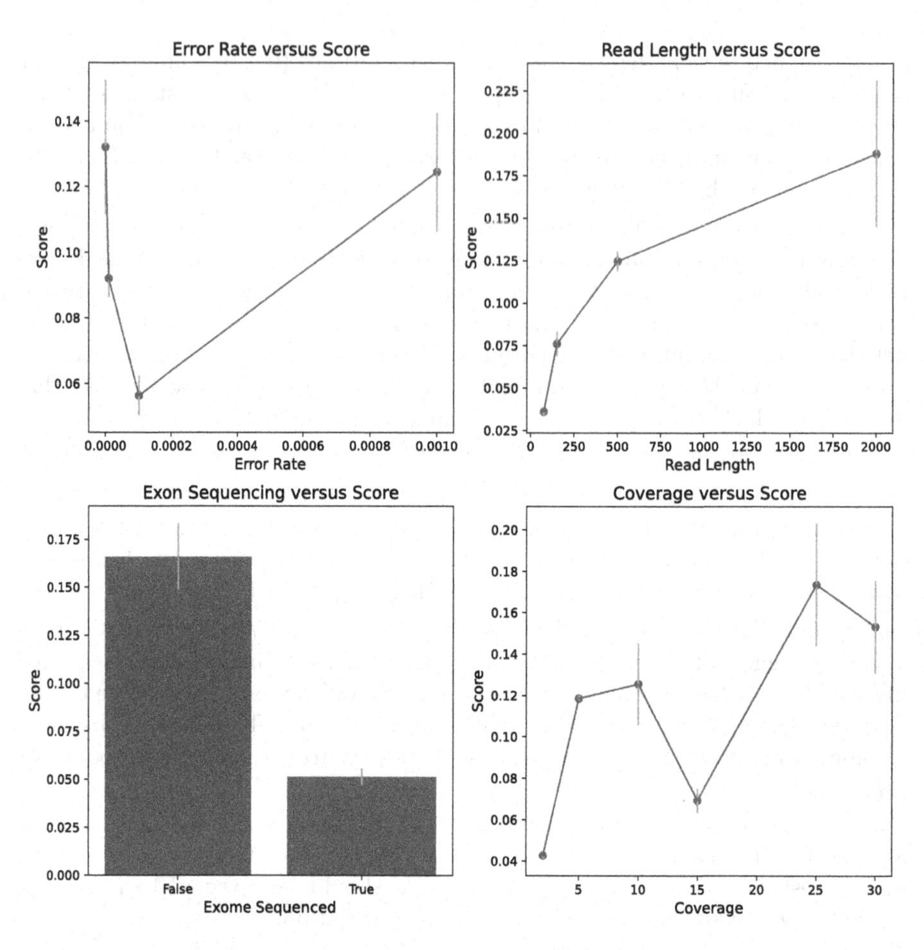

Fig. 2. Visualizations of various parameters and their impact on the efficacy of a study design. Read length showed the strongest positive correlation with study design score, and similarly whole genome sequenced data had a higher score than exome sequenced data. Higher coverage had a generally positive, but not entirely consistent, effect on the study design score. Higher error rates also generally produced lower scoring study designs, but high error rates did not preclude a study from having a high score. Error bars are depicted in orange which assign confidence to our results and depend on the sample sizes from our experiment. (Color figure online)

function $L_q(\mathbf{x})$, which describes the error with which the study design answers the question we pose. The subscript q and the expectation term over the distribution Q are used to emphasize that instances of our simulator are evaluated on a biological parameter vector q drawn from a stochastic high-dimensional biological parameter distribution Q. The expectation can be approximated by Monte-Carlo methods, where a singular study design, $\mathbf{x}$ would be evaluated on multiple instances q_i, and the result averaged, i.e. $\frac{1}{N}\sum_{i=1}^{N}\mathbf{c}(\mathbf{x}) + L_{q_i}(\mathbf{x})$.

We assume that our study design variable $\mathbf{x}$ can vary only in numerical read length, coverage, error rate, and a binary decision of whole genome versus whole exome sequencing. We assume the study is meant to recover inversions, deletions, and SNVs in a sample, so we define a scoring function intended to provide a balanced measure of performance at these tasks:

$$Score = 1 - L_q(\mathbf{x}) = 0.4 \times [\frac{3Recall_{snv} + F1_{snv}}{4}] + 0.4 \times [\frac{J_{del} + J_{inv}}{2}] + 0.2 * \mathbb{1}(Recall_{snv} > 0.1 \wedge J_{del} > 0.1 \wedge J_{inv} > 0.1).$$

We fix our callers as `strelka` and `delly`, fix most biological parameters, and fix our number of samples at 1. We assume that all study designs have fixed cost $\mathbf{c}(\mathbf{x}) = 0$ and constrain our study design as follows: $rl \in [75, 2000]$; $c \in [2, 30]$, $e \in [0, 0.001]$, $\mathbb{1}(Paired) \in \{True\}$, $\mathbb{1}(Genome) \in \{True, False\}$, $n \in \{0\}$. We replicated tumors independently with fixed biological parameters. For evaluation, approximately 75 study design vectors $\mathbf{x}$ were generated and evaluated from around 15 tumors. With additional computational time and resources, a user might consider generating a larger number of tumor replicates, or exploring a larger study design space. The score function was averaged across all tumors with respect to study design to generate a final score for each study design.

The best study design had a read length of 2000, a coverage of 25x, a 0.0 error rate, and was whole-genome sequenced. Further study design scores can be found in [25], as well as the github repository. As expected, the best study designs were mostly whole genome sequenced, which allowed the recovery of more variants across non-coding regions. Similarly, high coverage did seem to boost the power of the design, but there did not appear to be a large difference between 10x and 30x coverage. Increasing the read length boosted our ability to detect structural variants in samples, in both exome and genome sequenced samples. The error rate parameter did impact the rate of false calls and the ability to detect SNVs, but the overall scores were not heavily affected by the error rate even though many more false SNV calls were generated. However, with respect to structural variation, the larger error rate samples did not appear to do significantly worse. This meant that long-read high error-rate designs did well with respect to our scoring measure. There were also some sample size effects, as shown by the variance bars in Fig. 2. For instance, the high error-rate figure point had a relatively high variance, which might be due to its low sample size in conjunction with association with other advantageous study design parameters. We might expect more confident trends with an increase in the number of replicated tumors. In particular, we might expect the "dip" seen in the coverage plot to revert to a smooth logistic-type curve, and the error rate plot to have a more clear negative trend. The worst study designs were often exome-sequenced, had poor coverage, or completely failed to recover a particular variant type due to noise. Figure 2 visualizes tradeoffs between score and the model design parameters in our exploration of the search space. Different conclusions may have been reached if we imposed budget or cost constraints for various parameters.

Aside from identifying efficacious study design vectors and study design parameters, our simulations yielded a number of tangentially interesting results. High frequency structural variation introduced a substantial amount of noise in our SNV caller report. As expected, the task of calling layered structural variations at various frequencies was more challenging than that of single nucleotide variation. With regards to analysis, the patterns that large scale hypermutability, high error, and imbalanced clonal samples can induce on sequencing reads should be considered when developing the next generation of cancer informatics tools. A final observation generated by the simulator was that ensembled and merged study designs often tended to perform better than single-sample designs with respect to profiling the variant spectrum of a tumor, though this often came at the cost of increased false positive mutations.

3.4 Comparison to Real Cancer Reads

We compared reads generated by our simulator with an Ion Torrent targeted cancer sequencing panel in a colorectal cancer patient, found in the sequencing read archive (SRA Access Key: SRX9731615). This particular case used single-end reads in the read-length range 50–300 base pairs. For a comparison case, we generated 150 base pair exome-sequenced reads. The GC content distributions as well as the GC percentage of the reads in the real reads seemed to closely mirror that in our simulated reads, with a real GC content of 46 percent versus simulated GC content of 44 percent. GC bias is likely platform-specific [3] and may be less pronounced for this system than others. The read length distribution for the Ion Torrent reads had a range of $[25, 354]$, whereas ours ranged within $[0, 150]$ by a user-set distribution. The final noteworthy difference was with respect to quality scores. In our simulator, we placed a uniform quality score on all bases. The real sequencing case had fairly uniform sequencing quality scores except for extremely long reads, after which point the quality dropped. One main conclusion is that it appears possible to mimic arbitrary sequencer properties by altering distributions within our simulator.

4 Discussion and Conclusion

In this paper, we introduced a new simulation toolkit to generate sequencing reads from somatic variation processes under a wide range of biological and technological parameters. We demonstrated the utility of our simulator for several hypothetical questions in evaluating and optimizing study designs for profiling somatic variability in cell lineages. While we incorporated many features of the somatic evolution process, some simplifying assumptions were made that can be improved upon in future work. In particular, our evolutionary model assumes selective neutrality and does not explicitly model the spatial structure of the tumor. We aim to expand the evolutionary modeling incorporated in this simulator beyond the neutral coalescent model to incorporate various selection pressures, clonal dynamics, spatial distributions, and bottleneck effects. There is also room for improvement in modeling the mutation process, where the simulator could implement non-independent mutation processes or arbitrary genome rearrangement patterns. A major future direction will likely involve multiparameter optimization algorithms intended to provide low-cost study designs for experimental queries.

Acknowledgements. Portions of this work have been funded by Pennsylvania Dept. of Health award FP00003273. Research reported in this publication was supported by the National Human Genome Research Institute of the National Institutes of Health under award number R01HG010589. The content is solely the responsibility of the authors and does not necessarily represent the official views of the National Institutes of Health. The Pennsylvania Department of Health specifically disclaims responsibility for any analyses, interpretations or conclusions.

Availability. https://github.com/CMUSchwartzLab/MosaicSim. Additional pseudocodes and tests can be found in [25].

References

1. Abascal, F., et al.: Somatic mutation landscapes at single-molecule resolution. Nature **593**(7859), 405–410 (2021)
2. Alexandrov, L.B., et al.: The repertoire of mutational signatures in human cancer. Nature **578**(7793), 94–101 (2020)
3. Benjamini, Y., Speed, T.P.: Summarizing and correcting the gc content bias in high-throughput sequencing. Nucleic Acids Res. **40**(10), e72–e72 (2012)
4. Colom, B., et al.: Mutant clones in normal epithelium outcompete and eliminate emerging tumours. Nature **598**, 510–514 (2021)
5. Coorens, T.H., et al.: Extensive phylogenies of human development inferred from somatic mutations. Nature **597**(7876), 387–392 (2021)
6. Dentro, S.C., et al.: Characterizing genetic intra-tumor heterogeneity across 2,658 human cancer genomes. Cell **184**(8), 2239–2254 (2021)
7. Ellis, P., et al.: Reliable detection of somatic mutations in solid tissues by laser-capture microdissection and low-input DNA sequencing. Nat. Protoc. **16**(2), 841–871 (2021)
8. Ewing, A.D., et al.: Combining tumor genome simulation with crowdsourcing to benchmark somatic single-nucleotide-variant detection. Nat. Methods **12**(7), 623–630 (2015)
9. García-Nieto, P.E., Morrison, A.J., Fraser, H.B.: The somatic mutation landscape of the human body. Genom. Biol. **20**(1), 1–20 (2019)
10. Gu, K., Ng, H.K.T., Tang, M.L., Schucany, W.R.: Testing the ratio of two Poisson rates. Biom. J. J. Math. Methods Biosci. **50**(2), 283–298 (2008)
11. Jolly, C., Van Loo, P.: Timing somatic events in the evolution of cancer. Genome Biol. **19**(1), 1–9 (2018)
12. Kelleher, J., Etheridge, A.M., McVean, G.: Efficient coalescent simulation and genealogical analysis for large sample sizes. PLoS Comput. Biol. **12**(5), e1004842 (2016)
13. Kim, S., et al.: Strelka2: fast and accurate calling of germline and somatic variants. Nat. Methods **15**(8), 591–594 (2018)

14. Koboldt, D.C.: Best practices for variant calling in clinical sequencing. Genome Med. **12**(1), 1–13 (2020)
15. Langmead, B., Salzberg, S.L.: Fast gapped-read alignment with Bowtie 2. Nat. Methods **9**(4), 357–359 (2012)
16. Li, H.: Aligning sequence reads, clone sequences and assembly contigs with bwa-mem. arXiv preprint arXiv:1303.3997 (2013)
17. Li, H.: Minimap2: pairwise alignment for nucleotide sequences. Bioinformatics **34**(18), 3094–3100 (2018)
18. Li, Y., et al.: Patterns of somatic structural variation in human cancer genomes. Nature **578**(7793), 112–121 (2020)
19. Loeb, L.A.: A mutator phenotype in cancer. Cancer Res. **61**(8), 3230–3239 (2001)
20. Nordborg, M.: Coalescent theory. In: Handbook of Statistical Genomics: Two Volume Set, pp. 145–177 (2019)
21. Olafsson, S., Anderson, C.A.: Somatic mutations provide important and unique insights into the biology of complex diseases. Trends Genet. **37**, 872–881 (2021)
22. Rajaraman, A., Ullman, J.D.: Mining of Massive Datasets. Cambridge University Press, Cambridge (2011)
23. Rausch, T., Zichner, T., Schlattl, A., Stütz, A.M., Benes, V., Korbel, J.O.: DELLY: structural variant discovery by integrated paired-end and split-read analysis. Bioinformatics **28**(18), i333–i339 (2012)
24. Salk, J.J., Fox, E.J., Loeb, L.A.: Mutational heterogeneity in human cancers: origin and consequences. Annu. Rev. Pathol. Mech. Dis. **5**, 51–75 (2010)
25. Srivatsa, A., Lei, H., Schwartz, R.: A simulator for somatic evolution study design. bioRxiv (2022)
26. Tate, J.G., et al.: Cosmic: the catalogue of somatic mutations in cancer. Nucleic Acids Res. **47**(D1), D941–D947 (2019)

Prediction of Drug-Disease Relationship on Heterogeneous Networks Based on Graph Convolution

Jiancheng Zhong, Pan Cui, Zuohang Qu, Liuping Wang, Qiu Xiao(✉), and Yihong Zhu

School of Information Science and Engineering, Hunan Normal University, Changsha 410081, China
jczhongcs@gmail.com, xiaoqiu@hunnu.edu.cn

Abstract. Drug-disease association prediction is essential in drug development and repositioning. At present, the proposed drug-disease association prediction models based on graph convolution usually learn the characterization of the entire drug-disease heterogeneous network. However, the obtained characterization information come more from the characteristics of neighboring nodes in the homogeneous network, it lacks attribute information of nodes in the heterogeneous network, thus affecting the model's predictive performance. In this paper, an end-to-end model named DAHNGC based on graph convolutional neural networks is proposed to predict drug-disease association, which divides the characteristic learning of drugs and disease nodes into two parts. The proposed model uses the graph convolutional network to learn the attribute characteristics of drugs and disease nodes in the homogeneous network. Based on the known relationship between drugs and diseases, we design a method to automatically learn the characteristic information of drugs and disease nodes in heterogeneous networks. Subsequently, the drug-disease association matrix is reconstructed using a bilinear decoder to obtain a potential drug-disease association. In addition, we also adopt the DropEdge method to alleviate the over-smoothing problem of graph convolution. The experimental results show that the average AUC of the DAHNGC is 0.9113 through five-fold cross-verification, which is superior to that of the comparative method.

Keywords: Graph convolutional neural network · Drug-disease association prediction · Heterogeneous networks

1 Introduction

Drug discovery is significant research, and the vast market demand has led to continuous additional investment in the pharmaceutical industry. New drug development is usually divided into three stages: drug discovery, clinical trials and marketing approval [1], of which more than 90% of the candidate drugs are eliminated at the clinical stage, and less than 10% are finally approved for marketing [2]. To meet this challenge, drug repositioning has emerged [3]. Drug repositioning refers to the identification and development new

M. S. Bansal et al. (Eds.): ISBRA 2022, LNBI 13760, pp. 243–254, 2022.
https://doi.org/10.1007/978-3-031-23198-8_22

uses for existing drugs [4]. Therefore, drug repositioning has become one of the current auxiliary means of drug research and development, attracting widespread attention from the pharmaceutical and research communities.

With the development of computer technology and the continuous deepening of research in the field of bioinformatics, many computationally based drug repositioning methods have been proposed, most of which discover new drug efficacy by looking for drug redirection paths or new drug-disease associations. For example, Wang et al. [5] proposed a computational model based on heterogeneous networks, on which the strength of the association between drugs and diseases was obtained using iterative computing strategies on a heterogeneous map composed of disease, drug and target information. Dai et al. [6] derived potential drug-disease associations by using matrix decomposition on known drug-disease associations, which include drug-gene interactions, disease-gene interactions and gene-gene interactions. At present, the drug-disease association prediction method based on computation is mainly to build heterogeneous networks to get drug and disease characteristics. Still, this type of method relies too much on the edges between nodes in heterogeneous networks, ignores the attribute information of the nodes, separates the network from the attributes and does not consider the network from the perspective of graph data [7]. From the graph perspective, network and attribute information can be learned simultaneously, and it is also suitable for studying the association between drug molecules, which is indispensable for further improving the performance of the drug-disease association prediction model.

In recent years, graph convolutional neural networks have received much attention in processing graph structure data, which can solve data problems in non-Euclidean spaces very well. Studies have shown that graph convolutional networks also play an important role in biomedicine. For example, Wang et al. [8] proposed a method for predicting drug-disease association based on graph neural networks. Yu et al. [9] proposed a method to introduce a layer attention mechanism into graph convolutional neural networks, and predict potential drug-disease associations by obtaining embedded information about drugs and diseases in heterogeneous networks. However, the use of graph convolutional neural networks to extract characteristics on the overall heterogeneous network usually only considers the neighborhood closer to the node [10]. The information obtained by the node comes more from the characteristics of neighboring nodes in the homogeneous network and the characteristic information of other nodes in the heterogeneous network is missing. For this problem, we propose a new model for predicting drug-disease associations on heterogeneous networks based on graph convolution by learning features in homogeneous and heterogeneous networks, respectively.

In this paper, a drug-disease association prediction model on the heterogeneous network based on graph convolution (DAHNGC) is proposed. By integrating the interactions of drug-disease, drug-protein and disease-protein, the proposed model constructs a drug-protein-disease heterogeneous network and calculates drug and disease similarity based on this heterogeneous network. In addition, two feature extraction methods are designed to obtain the characteristics between the homogeneous and heterogeneous networks of drugs and diseases. The graph convolutional neural network is used in the homogeneous network to obtain the characteristics of the drug or disease, and the DropEdge method is introduced to alleviate the GCN's over-smoothing. We also design a method based on known drug-disease associations to automatically learn the characteristics of drugs and diseases in heterogeneous networks. The DAHNGC model can obtain not only the attribute information of the nodes in the graph, but also the structure information of the network.

2 Materials and Methods

2.1 Dataset

In this study, the standard dataset was derived from Wu et al. [11], and only known drug-protein, drug-disease, and disease-protein associations were used to construct the model. Among them, the drug-disease associations are derived from the Gottlieb dataset [12] (2011), containing 1827 experimentally validated drug-disease associations. The drug and drug-protein associations are derived from the DrugBank database [13] (DrugBank 5.0, 2017), including 1186 drug small molecules and 4642 known drug-protein associations. The diseases and disease-gene (protein) associations come from the OMIM database [14] (2005), 449 diseases and 1365 known disease-protein associations can be acquired from the OMIM database by associating with 1467 proteins.

2.2 Method

The proposed model DAHNGC: a drug-disease association prediction model on the heterogeneous network based on graph convolution neural networks can be mainly divided into three steps:

(1) Measuring the similarities of drugs and diseases. In the three-layer heterogeneous network of drug-disease-protein composition, the similarities of drugs and diseases are evaluated according to the "shared similarity" hypothesis. The hypothesis means that drugs with common proteins or diseases may be more similar.
(2) Encoding. The first is to construct a drug-disease heterogeneous network, and the independent drug similarity network diagram and the disease similarity network diagram are input into the graph convolutional neural network to obtain the characteristic information of the drug and disease in the homogeneous network. The homogeneous network contains only nodes of the same type. The DropEdge method is introduced in the graph convolutional neural network to avoid over-fitting or over-smoothing problems. The second is to design a method to get the characteristic information of drugs and disease nodes in heterogeneous networks through drug-disease associations. The third is to fuse the above two kinds of characteristic information and input them into the fully connected network. The final embedding vector can be obtained through nonlinear transformation. These learned embedding vectors of drugs and disease are used as the input of the downstream scoring structure.
(3) Decoding. The drug-disease association matrix is reconstructed by bilinear decoding to obtain the potential drug-disease association relationship. The overall workflow of the model DAHNGC proposed in this article is shown in Fig. 1.

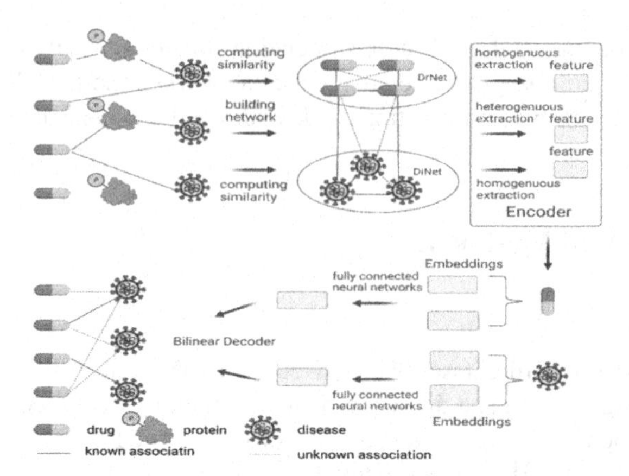

Fig. 1. Framework for the DAHNGC model.

Similarity Measurements. A three-layer drug-protein-disease network is constructed using known drug-disease, drug-protein, and disease-protein associations between drug nodes. The drug-protein and disease topological similarities are calculated on the heterogeneous network [11]. The node topology is obtained by calculating the number of common neighbors of nodes, and cosine similarity is applied to measure the topological similarities. According to the similarity matrix, two independent similarity networks were constructed. Based on the known drug-disease association, the two independent networks were mapped to a large network space to form a unified drug-disease heterogeneous network.

Encode. In order to obtain complete characteristics information and topology of the nodes in the graph, we use the autoencoder to learn the node characteristics in homogeneous networks, and the node characteristics and topologies in heterogeneous networks.

Graph convolutional neural networks (GCN) can generate characterization vectors for each node in the graph by aggregating the neighborhood of the nodes. In the process of coding, the drug and the disease similarity network diagram can be learned by GCN respectively to obtain the characteristic information of the drug and disease nodes [15]. In this paper, let *Gr* represent the drug similarity network graph, and *m* represent the number of drug nodes in the graph. *Gd* represents the disease similarity network graph, and *n* represents the number of disease nodes in the graph. *Ar* represents *Gr*'s adjacency matrix and *Ad* represents *Gd*'s adjacency matrix. *X* and *Y* represent drug characteristics and disease characteristics, respectively. GCN is a layer propagation mechanism, in which each layer gathers neighbor information through the direct link of the graph to reconstruct the embedding and serve as the input of the next layer. The formula of drug node learning embedding vector from *Gr* is:

$$X^{(l+1)} = \widetilde{D}_r^{-0.5}\widetilde{A}_r\widetilde{D}_r^{-0.5}X^l W^l \tag{1}$$

where $\widetilde{A}_r$ is based on the adjacency matrix add its own ring, namely $\widetilde{A}_r = A_r + I$. $\widetilde{D}_r$ is the $\widetilde{A}_r$'s degree matrix. *W* is the weight parameter, which is automatically updated during training. Formula (2) can be regarded as a linear layer feedforward neural network, X^l is the output of layer L, as well as the input to layer $(L + 1)$. And formula (2) can be extended to add a nonlinear activation function to enhance the performance and presentation of the model. The formula is expressed as:

$$X^{(l+1)} = \delta\left(\widetilde{D_r}^{-0.5}\widetilde{A_r}\widetilde{D_r}^{-0.5}X^l W^l\right) \quad (2)$$

The activation function is usually a RELU function, and the initial input of the function is characteristics of the drug node.

Studies show that the effect of the GCN model is usually excellent when its number of layers is 2–3. Too many layers will lead to over-fitting and over-smoothing of the GCN, which makes the performance of the model show a downward trend [16]. Over-smoothing is a phenomenon unique to graph neural networks. To alleviate over-smoothing, the DropEdge algorithm is proposed [17]. Information transmission relies on the edges between adjacent nodes in GCN. The DropEdge method randomly deletes some edges in the graph, making the connection of the nodes in the graph loose and reducing the convergence speed. Thus, the over-smoothing problem caused by layer increasing can be alleviated.

The DropEdge method is added in layer-to-layer propagation, which brings more randomness to the raw data, obtains more sufficient neighbor information, and improves the performance of characteristics extraction in homogeneous network. The DropEdge is expressed as follows:

$$X^{(l+1)} = dropout(X^{(l+1)}) \quad (3)$$

let $\widetilde{L_r}$ represents $\widetilde{D_r}^{-0.5}\widetilde{Ar}\widetilde{D_r}^{-0.5}$, the $(L + 1)$ layer of GCN extraction drug node characteristics can be expressed as:

$$X^{(L+1)} = relu\left(dropout\left(\widetilde{L_r}relu\left(..dropout\left(relu\left(\widetilde{L_r}XW_r\right)\right)..\right)W_r^l\right)\right) \quad (4)$$

Node characteristics are transformed between layers by GCN. Similarly, the characterization of disease nodes is obtained by GCN in the disease similarity network diagram.

$$Y^{(L+1)} = relu\left(dropout\left(\widetilde{L_d}relu\left(..dropout\left(relu\left(\widetilde{L_d}YW_d\right)\right)..\right)W_d^l\right)\right) \quad (5)$$

where $\widetilde{L_d} = \widetilde{D_d}^{-0.5}\widetilde{A_d}\widetilde{D_d}^{-0.5}$, $\widetilde{A_d} = A_d + I$, $\widetilde{D_d}$ is the degree matrix of $\widetilde{A_d}$.

The characterization information of drug nodes and disease nodes in the homogeneous network is learned by calculating the formula (4) and formula (5). At this time, the characterization information only contains the information of the same kind of node. To obtain more adequate information about nodes, we design a method to learn the characterization in a heterogeneous network. First, the adjacency matrix about drug-disease association and disease similarity is obtained by using drug-disease association and disease similarity information. Then, the disease node associated with the drug can be obtained based on two adjacency matrices. The related disease nodes are aggregated, and the characterization information of the drug nodes can be learned by normalizing the aggregated information:

$$X_{ri} = \frac{1}{Dri} AGG(Y_{d1}, Y_{d2}, \cdots) \quad (6)$$

where X_{ri} represents the characteristics of the drug node *ri*, Y_{d1} represents the characteristics of the disease node *d1*, *AGG* represents the function of aggregating the disease node information, D_{ri} represents the degree of node *ri*. Normalization of the aggregated information can learn the characterization information of drug nodes. The characteristic information of disease node *dj* can be expressed as:

$$Y_{dj} = \frac{1}{Ddj} AGG(X_{r1}, X_{r2}, \cdots) \quad (7)$$

Two characteristic information obtained by the drug node through learning can be fused to obtain the final characteristics or embedding vector of the drug node. The two characteristic vectors obtained by the disease node after learning can be integrated to get the final embedding vector of the disease node. Figure 2 shows the process of getting the embedding vector.

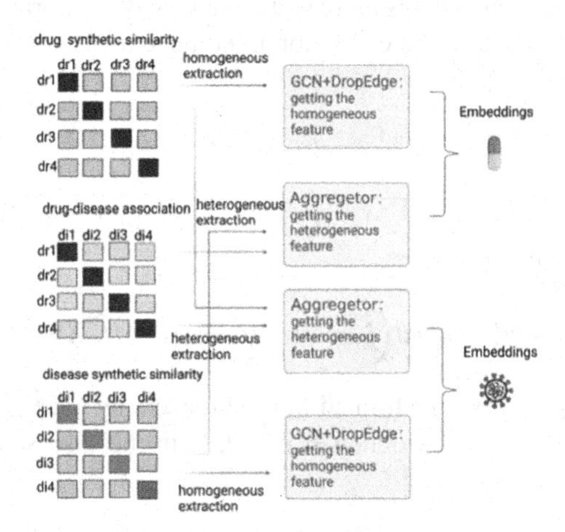

Fig. 2. The process of extracting homogeneous and heterogeneous features.

In order to make the characterization information learned better applied to the downstream learning task, the above-mentioned embedded vectors are input into the fully connected network respectively. The embedded vectors required for the decoding process are obtained through nonlinear transformation, which can be represented as:

$$X'_{ri} = LeakyRELU\left(h\left(X^{(L+1)}, X_{ri}\right)\right) \quad (8)$$

where *h*(.) is the function to combine $X^{(L+1)}$ and X_{ri}.$X^{(L+1)}$ represents the characteristic information of the drug node obtained after passing through the *L* layer GCN in the homogeneous network. X_{ri} represents the characteristic information obtained by the drug node in the heterogeneous network. The final embedding vector of the disease node is as follows:

$$Y'_{dj} = LeakyRELU\left(h\left(Y^{(L+1)}, Y_{dj}\right)\right) \tag{9}$$

Decode. The drug-disease prediction is regarded as a link prediction problem in this paper. The decoding process refers to reconstructing the drug-disease associations. The reconstructed drug-disease associations are compared with the original ones to obtain a potential drug-disease associations pair. The bilinear decoder is used to predict the probability of a link between drug nodes and disease nodes, the formula is as follows:

$$\widehat{y_{ij}} = \sigma\left(X'_{ri} W \left(Y'_{dj}\right)^T\right) \tag{10}$$

where X'_{ri} is the embedding vector obtained by the drug node ri by encoding, and Y'_{dj} is the embedding vector obtained by the disease node dj by encoding. Since this model can also be regarded as a dichotomous model, and the sigmoid function has its own unique advantages in dealing with the dichotomous problems, so the activation function in formula (10) can use the sigmoid function. W is the parameter matrix, which is constantly updated during training.

The cross-entropy loss function is used to optimize the parameters in the model during training. The expression of the function is as follows:

$$\text{LOSS} = -\sum\nolimits_{i,j \in R} \left(y_{ij} \log \widehat{y_{ij}} + \left(1 - y_{ij}\right) \log\left(1 - \widehat{y_{ij}}\right)\right) \tag{11}$$

where R is the set of all nodes in the training set and y_{ij} is the true label.

3 Result

3.1 Model Evaluation Metrics

In this paper, the final result for predicting the disease-drug association is a drug-disease interaction probability score after decoding. We set a threshold, if the score is greater than the threshold value, the drug-disease pair is considered to have associations, otherwise it is not. The above problem can be reduced to dichotomies, so the index of the classifier classification performance is used as the evaluation index in this paper. For instance, we adopt the classic evaluation index ROC curve and its AUC. The ROC curve is a curve drawn with the true positive rate (TPR) as the ordinate and the false positive rate (FPR) as the abscissa.

The PR curve is also an important indicator for predictive performance evaluation as the ROC curve. The PR curve is drawn with precision as the ordinate and recall as the abscissa.

3.2 Analysis of Experimental Results

In experiments, drug-disease pairs with known relationships were taken as positive samples, while drug-disease pairs with unknown relationships or no associations were taken as negative samples. In order to balance the positive and negative samples, we randomly select negative samples with the same number of positive samples, and fuse them as the input of the model. Five-fold cross-validation is used to evaluate the predictive performance of the model.

The model DAHNGC is implemented by the Deep Graph Library with an MXNET backend. The model's random initial parameters are initialized using Xavier during training. The optimization of the model's parameters is achieved by the Adam algorithm. The model also uses five common evaluation criteria of AUC, ACC, RE, PRE and F1 to measure the performance. We take the average of the five results as the final experimental result. The experimental results are shown in Table 1, and the ROC curve is shown in Fig. 3. The AUC of the DAHNGC model can reach 0.9113 in five-fold cross-validation. At the same time, other evaluation indicators of the DAHNGC can obtain excellent results, suggesting that our proposed model has favorable stability.

Table 1. DAHNGC 5-fold cross-validation results.

	ACC	Pre	RE	F1	AUC
Fold 1	0.8434	0.8272	0.8541	0.8404	0.8968
Fold 2	0.8769	0.8649	0.8889	0.8767	0.9126
Fold 3	0.8529	0.8684	0.8315	0.8495	0.9039
Fold 4	0.8714	0.8792	0.8792	0.8792	0.9189
Fold 5	0.8705	0.8622	0.8778	0.8699	0.9241
AVE	0.8630	0.8604	0.8663	0.8631	0.9113

To further analyze the advantages of DAHNGC, we compare the DAHNGC with three other drug repositioning methods. The performance of the proposed model is comprehensively proved by a variety of evaluation indicators, including the TPR and FPR values of DAHNGC and GADTI [10], NeoDTI [18] and SCMFDD [19] under multiple thresholds, as well as the Precision value and the Recall value. The corresponding ROC curve is plotted for each evaluation indicator, which can be seen in Fig. 4. DAHNGC achieve the best performance with an average AUC of 0.911 among 1186 drugs. The results show that the DAHNGC has better classification performance than comparative methods. Therefore, the DAHNGC can provide more accurate recommendations than other methods for predicting potential drug-disease associations. As can be seen from Table 2, DAHNGC has excellent results for other indicators, suggesting that the model proposed in this article has favorable stability.

There are two reasons for the superiority of the DAHNGC model. One is that the model has unique advantages in characteristic extraction. In the drug and disease similarity network diagram, the attribute information of the drug and the disease node is

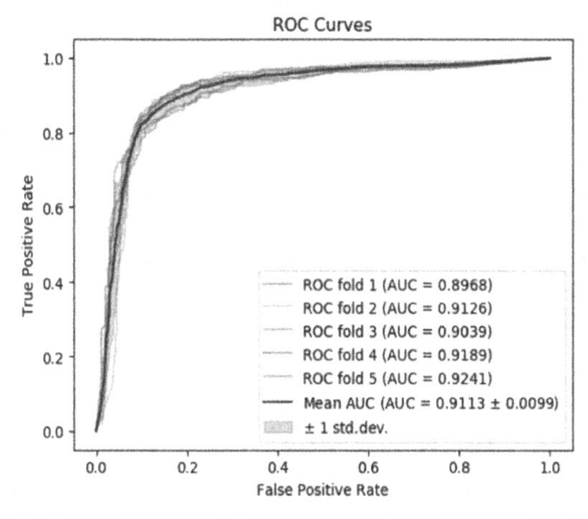

Fig. 3. ROC curve under DAHNGC 5-fold cross-validation.

learned. The drug-disease association relationship can be used in the heterogeneous network to learn the disease or drug node with a known relationship, other diseases similar to this disease or drug node, attribute information of the drug node and topology information of networks. The DropEdge method is added to alleviate the over-smoothing of graph convolution. Another reason is that the other models do not reflect their own unique advantages in the benchmark dataset. All comparison methods in this paper use balanced data sets without considering the influence of unbalanced data sets. The performance of the DAHNGC model is superior to other methods. The experimental results fully prove that the model can discover more hidden characteristics of drugs and diseases, and significantly improve the prediction ability of nodes.

Table 2. Comparison results of DAHNGC with other three methods.

	ACC	Pre	RE	F1	AUC
NeoDTI	0.836	0.635	0.683	0.674	0.901
GADTI	0.843	0.646	0.711	0.693	0.909
SCMFDD	0.804	0.769	0.763	0.764	0.872
DAHNGC	0.865	0.859	0.873	0.866	0.911

3.3 Case Studies

In order to prove the effectiveness of the model in predicting the drug-disease association relationship, some drugs and diseases are selected as the research objects. The reliability of the model is further verified by predicting the relevant drug candidates for a certain disease. First, the relationship between drugs and diseases was predicted using the DAHNGC method on a complete dataset. The dataset includes 449 diseases,

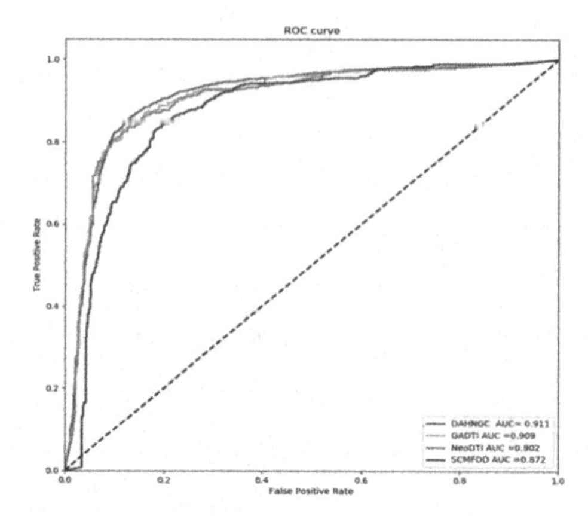

Fig. 4. ROC curves of DAHNGC and three other methods.

1186 drug small molecules and 1827 known drug-disease pairs; Then, looking for drugs predicted to have associations with the disease; Finally, drugs known to be associated with the disease are deleted, and the top 15 related drugs are collected in order of the predicted value of drugs with the disease. CTD databases and literature were used to verify that there was an association between the drug candidates and the disease (Table 3).

Table 3. DAHNGC algorithm ranks the top 30 drug candidates associated with hypertension.

Rank	DrugBank ID	Drug	Evidence
1	DB00490	Buspirone	CTD
2	DB00413	Pramipexole	Refs. [20]
3	DB00472	Fluoxetine	CTD
4	DB00302	Tranexamic acid	CTD
5	DB00268	Ropinirole	CTD
6	DB00388	Phenylephrine	CTD
7	DB00458	Imipramine	CTD
8	DB00502	Haloperidol	
9	DB00715	Paroxetine	CTD
10	DB00163	Vitamin E	CTD
11	DB00915	Amantadine	CTD
12	DB00626	Bacitracin	CTD
13	DB00216	Eletriptan	CTD
14	DB00419	Miglustat	Refs. [21]
15	DB00437	Allopurinol	CTD

4 Conclusion

A drug-disease association prediction model on the heterogeneous network based on graph convolution (DAHNGC) is proposed in this study. This model integrates different association information to obtain the drug and disease similarity matrix. Two characteristic learning methods are designed using graph convolution on the constructed drug-disease heterogeneous network. When learning attribute characterizations in homogeneous networks, the DropEdge method is introduced to alleviate the over-smoothing of graph convolution. The DAHNGC model proposed in this paper enhances the characteristic information of nodes through the deep integration of two parts of drugs and disease characterization learning. Based on the five-fold cross-validation, the unique advantages of DAHNGC are demonstrated by comparing it with other three drug repositioning methods. Experimental results show that DAHNGC has high accuracy and balanced performance, and effectively explore the deep characteristics of the drug-disease network. In addition, the case study indicates that the DAHNGC model can achieve high-precision prediction of unobserved drug-disease associations, which can provide great help for biomedical researchers.

Acknowledgment. This work was supported by the China Scholarship Council (201906725017), the Collaborative Education Project of Industry University cooperation of the Chinese Ministry of Education (201902098015), the Teaching Reform Project of Hunan Normal University (82).

References

1. Paul, S.M., et al.: How to improve R&D productivity: the pharmaceutical industry's grand challenge. Nat. Rev. Drug Discov. **9**(3), 203–214 (2010)
2. Adams, C.P., Brantner, V.V.: Estimating the cost of new drug development: is it really $802 million? Health Aff. **25**(2), 420–428 (2006)
3. Li, J., Zheng, S., Chen, B., Butte, A.J., Swamidass, S.J., Lu, Z.: A survey of current trends in computational drug repositioning. Brief. Bioinform. **17**(1), 2–12 (2016)
4. Ashburn, T.T., Thor, K.B.: Drug repositioning: identifying and developing new uses for existing drugs. Nat. Rev. Drug Discov. **3**(8), 673–683 (2004)
5. Wang, W., Yang, S., Zhang, X., Li, J.: Drug repositioning by integrating target information through a heterogeneous network model. Bioinformatics **30**(20), 2923–2930 (2014)
6. Dai, W., et al.: Matrix factorization-based prediction of novel drug indications by integrating genomic space. Comput. Math. Methods Med. **2015** (2015)
7. Zhao, B.-W., You, Z.-H., Hu, L., Wong, L., Ji, B.-Y., Zhang, P.: A multi-graph deep learning model for predicting drug-disease associations. In: Huang, D.-S., Jo, K.-H., Li, J., Gribova, V., Premaratne, P. (eds.) ICIC 2021. LNCS (LNAI), vol. 12838, pp. 580–590. Springer, Cham (2021). https://doi.org/10.1007/978-3-030-84532-2_52
8. Wang, B., Lyu, X., Qu, J., Sun, H., Pan, Z., Tang, Z.: GNDD: a graph neural network-based method for drug-disease association prediction. In: 2019 IEEE International Conference on Bioinformatics and Biomedicine (BIBM), pp. 1253–1255, November 2019
9. Yu, Z., Huang, F., Zhao, X., Xiao, W., Zhang, W.: Predicting drug–disease associations through layer attention graph convolutional network. Brief. Bioinform. **22**(4), bbaa243 (2021)
10. Liu, Z., Chen, Q., Lan, W., Pan, H., Hao, X., Pan, S.: GADTI: graph autoencoder approach for DTI prediction from heterogeneous network. Front. Genet. **12**, 650821 (2021)

11. Wu, G., Liu, J.: Predicting drug-disease treatment associations based on topological similarity and singular value decomposition. In: 2019 IEEE International Conference on Bioinformatics and Biomedicine (BIBM), pp. 153–158, November 2019
12. Gottlieb, A., Stein, G.Y., Ruppin, E., Sharan, R.: PREDICT: a method for inferring novel drug indications with application to personalized medicine. Mol. Syst. Biol. **7**(1), 496 (2011)
13. Wishart, D.S., et al.: DrugBank 5.0: a major update to the DrugBank database for 2018. Nucleic Acids Res. **46**(D1), D1074–D1082 (2018)
14. Hamosh, A., Scott, A.F., Amberger, J.S., Bocchini, C.A., McKusick, V.A.: Online mendelian inheritance in man (OMIM), a knowledge base of human genes and genetic disorders. Nucleic Acids Res. **33**(suppl._1), D514–D517 (2005)
15. Li, J., Zhang, S., Liu, T., Ning, C., Zhang, Z., Zhou, W.: Neural inductive matrix completion with graph convolutional networks for miRNA-disease association prediction. Bioinformatics **36**(8), 2538–2546 (2020)
16. Huang, W., Rong, Y., Xu, T., Sun, F., Huang, J.: Tackling over-smoothing for general graph convolutional networks. arXiv preprint arXiv: 2008.09864 (2020)
17. Rong, Y., Huang, W., Xu, T., Huang, J.: Dropedge: Towards deep graph convolutional networks on node classification. arXiv preprint arXiv: arXiv:1907.10903 (2019)
18. Wan, F., Hong, L., Xiao, A., Jiang, T., Zeng, J.: NeoDTI: neural integration of neighbor information from a heterogeneous network for discovering new drug–target interactions. Bioinformatics **35**(1), 104–111 (2019)
19. Zhang, W., et al.: Predicting drug-disease associations by using similarity constrained matrix factorization. BMC Bioinformatics **19**(1), 1–12 (2018)
20. Li, X.X., et al.: Adrenergic and endothelin B receptor-dependent hypertension in dopamine receptor type-2 knockout mice. Hypertension **38**(3), 303–308 (2001)
21. Buemi, M., et al.: Reduced bcl-2 concentrations in hypertensive patients after lisinopril or nifedipine administration. Am. J. Hypertens. **12**(1), 73–75 (1999)

t-SNE Highlights Phylogenetic and Temporal Patterns of SARS-CoV-2 Spike and Nucleocapsid Protein Evolution

Gaik Tamazian[1], Andrey B. Komissarov[2], Dmitry Kobak[3], Dmitry Polyakov[4], Evgeny Andronov[5], Sergei Nechaev[6], Sergey Kryzhevich[7](✉), Yuri Porozov[8], and Eugene Stepanov[4,9,10,11]

[1] Centre for Computational Biology, Peter the Great Saint Petersburg Polytechnic University, 195251 St. Petersburg, Russia
mail@gtamazian.com
[2] Smorodintsev Research Institute of Influenza, St. Petersburg, Russia
[3] University of Tübingen, Tübingen, Germany
[4] HSE University, Moscow, Russia
[5] All-Russia Research Institute for Agricultural Microbiology, Russian Academy of Sciences, Pushkin-8, St. Petersburg 196608, Russia
[6] LPTMS, Universitè Paris Saclay, 91405 Orsay Cedex, France
[7] Institute of Applied Mathematics, Faculty of Applied Physics and Mathematics and BioTechMed Center, Gdańsk University of Technology, ul. Narutowicza 11/12, 80-233 Gdańsk, Poland
kryzhevicz@gmail.com
[8] The Center of Bio- and Chemoinformatics, I.M. Sechenov First Moscow State Medical University, 119435 Moscow, Russia
[9] Scuola Normale Superiore, Pisa, Italy
[10] St. Petersburg Branch of the Steklov Mathematical Institute of the Russian Academy of Sciences, Fontanka 27, 191023 St. Petersburg, Russia
[11] Department of Mathematical Physics, Faculty of Mathematics and Mechanics, St. Petersburg State University, St. Petersburg, Russia

1 Introduction

Since the beginning of the COVID-19 pandemic, whole-genome sequences of SARS-CoV-2 have been continuously added to public databases, such as NCBI Virus [4] and GISAID [3]. As of July 2022, the SARS-CoV-2 Data Hub of the NCBI Virus database stored more than one million complete whole-genome sequences of the coronavirus. For navigating the SARS-CoV-2 genome sequences, the Pango nomenclature [18] and the Pangolin software [14] were developed. The nomenclature and the software have been also extensively used for tracking the coronavirus evolution and rapidly classifying new genomes. To supplement the Pango nomenclature, we propose applying modern manifold learning techniques [2,12,19,20] to protein sequences of SARS-CoV-2 to construct, visualize and study the global evolutionary space [16] of the coronavirus. The basic idea is to explore the COVID-19 evolution space by trying to find geometric structure

M. S. Bansal et al. (Eds.): ISBRA 2022, LNBI 13760, pp. 255–262, 2022.
https://doi.org/10.1007/978-3-031-23198-8_23

hidden in the evolutionary distances between variants. The adequate visualization of such a geometric structure might provide more information on evolution compared to the conventional phylogeny methods. For instance, phylogenetic trees do not provide any information on possible evolutionary patterns, nor they may reveal the "empty spaces" in the global evolutionary space left by extinguished specimens. Moreover, phylogenetic trees, especially very large ones, may be not well adapted for cluster analysis of the species diversity, which was also the motivation under the application of manifold learning techniques, among all, to COVID-19 genomic data, in [6,11].

Evolutionary distances have been calculated based on the structures of the nucleocapsid (N) and spike (S) proteins. For the manifold learning technique to apply to these distances we propose to use the t-distributed stochastic neighbor embedding (t-SNE) [19]. In fact, many existing global methods of hidden manifold reconstruction from big data, such as the prototypical MDS (or its dimension-reduction counterpart PCA) are well adapted to Euclidean distances, that is, to the case when the data are supposed to belong to a linear subspace of some Euclidean space, and may distort a lot the geometry of the data, e.g., when the latter belongs to some curved manifold and the distances are geodesic inside this manifold [1,9,10]. On the contrary modern methods like t-SNE and UMAP rely more on local structure of distances and were shown to be more adapted to the study of data originating from some nonlinear manifolds. In the particular case of COVID-19 evolutionary data we have studied, the t-SNE revealed to give the clearest picture well-adapted for further analysis, in particular for finding intrinsic clusters or possible evolutionary patterns.

2 Methods

We implemented the t-SNE visualization of coronavirus sequences in the form of a pipeline with the following steps.

1. We sampled 100,000 coronavirus records with whole-genome sequences from the NCBI Virus database. In the original database all known sequences have been inserted once they were discovered which naturally resulted in an over-representation of most recent variants (e.g. delta and omicron strains). The sampling has been made to override this problem partially equalizing the number of representatives of older variants.
2. For the selected records, amino-acid sequences of N and S proteins were obtained. The information on amino-acid sequences was deliberately used instead of that on nucleotide sequences encoding the respective proteins to explore potentially more stable and evolutionary significant data or, in other words, to ignore a big part of "noise" generated by random single-nucleotide mutations not affecting any change in protein structure.
3. The selected protein sequences were deduplicated by retaining a single representative for each series of duplicates and keeping original duplicate counts. The deduplication resulted in 6,280 and 18,953 unique sequences of the N and S proteins, respectively.

4. N and S protein sequences from the SARS-CoV-19 reference sequence [15] were added to the dataset. NCBI Protein accessions for the reference proteins sequences were `YP_009724397.2` and `YP_009724390.1` for N and S proteins, respectively.
5. Multiple sequence alignments of the N and S proteins were computed using the MAFFT software [7] with its default options.
6. For each protein, pairwise distance matrices were obtained from multiple sequence alignments. The distance between a pair of proteins was calculated as the ratio of the sum of weighted amino acid substitutions in mutual non-gap positions to the total number of the mutual non-gap positions. The substitution weights were based on the BLOSUM62 matrix [5]. Note that the word "distance" is an obvious abuse of the language since triangle inequality needs not to be satisfied.
7. The obtained pairwise distance matrices were passed to routines from the R package `Rtsne` [8] to apply t-SNE, thus deriving the mapping of protein variants into Cartesian coordinates of the protein points in a two-dimensional plane.
8. The resulting mapping has been visualized using meta-information about the original coronavirus samples: the collection date, the Pango lineage [13], and the WHO-named group. The visualization was implemented in `R` [17] using routines from packages `tidyverse` [22] and `ggplot2` [21].

Following our previous study [16], we also identified groups of "reference diversity representatives" for each gene by iterating triplets of protein sequences originating from three Pango lineages B.1.617.2, P.1, and BA.5. They have been obtained as follows: after adding the reference sequence to the triplets, we checked that pairwise evolutionary distances between the respective quadruples of points be such that they could be embedded isometrically into a three-dimensional Euclidean space (in particular, this implies also that the triangle inequality held). Finally, we ranked the resulting triplets by the minimal pairwise distance between the points and selected the one with the greatest value. The diversity representative quadruple selected for the N protein gene was `YP_009724397.2` (reference), `QZE91429.1` (B.1.617.2), `QYG34624.1` (P.1), and `USF49821.1` (BA.5); the quadruple for the S protein gene was `YP_009724390.1` (reference), `UKF20319.1` (B.1.617.2), `UCX76691.1` (P.1), and `USJ64862.1` (BA.5).

Software implementing the computational analysis is available at https://gitlab.com/gtamazian/t-sne-sars-cov-2-patterns.

3 Results

Figures 1 to 3 show t-SNE visualizations of the N and S proteins with reference diversity representatives denoted by crosses (non-reference sequences) or asterisks (reference sequences). Though t-SNE is based mainly on the local structure of distances, we are aimed at reconstructing the whole (global) picture, so the diversity representative sequences provide an insight on how adequately the global structure of distances has been reconstructed.

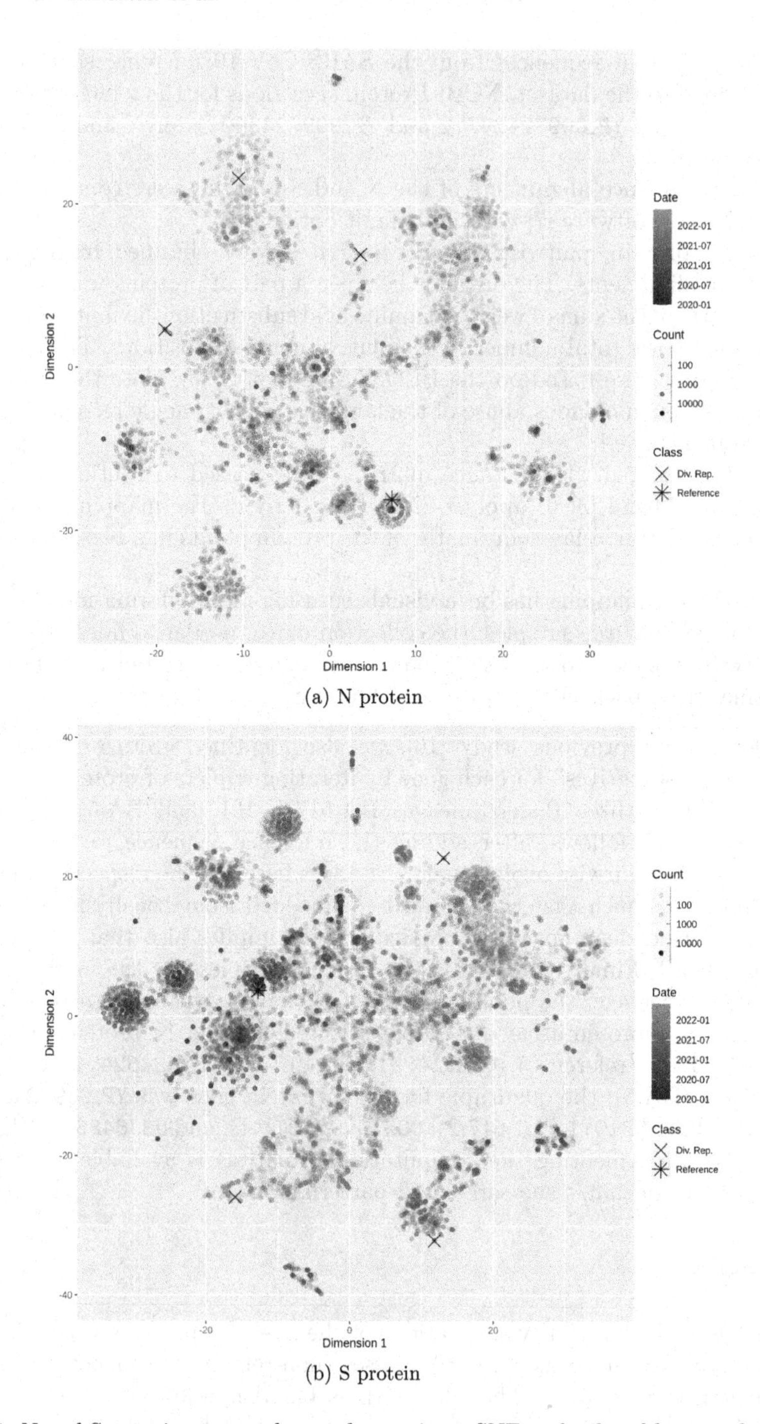

Fig. 1. N and S proteins mapped to a plane using t-SNE and colored by sample collection dates. Each dot corresponds to a deduplicated sequence. The dot size designates the count (multiplicity) of original sequences in the COVID database sample.

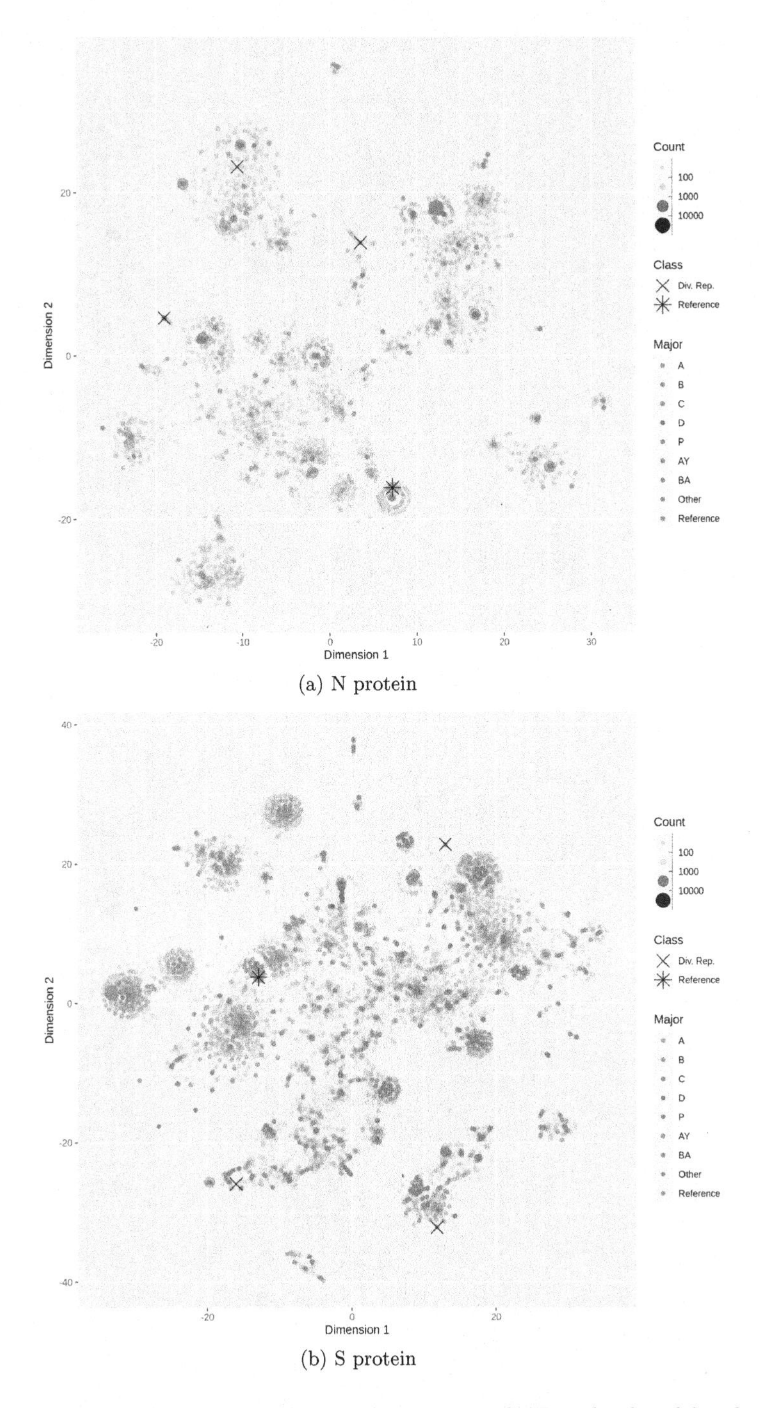

(a) N protein

(b) S protein

Fig. 2. N and S proteins mapped to a plane using t-SNE and colored by the Pango lineage major levels. Each dot corresponds to a deduplicated sequence. The dot size designates the count (multiplicity) of original sequences in the COVID database sample.

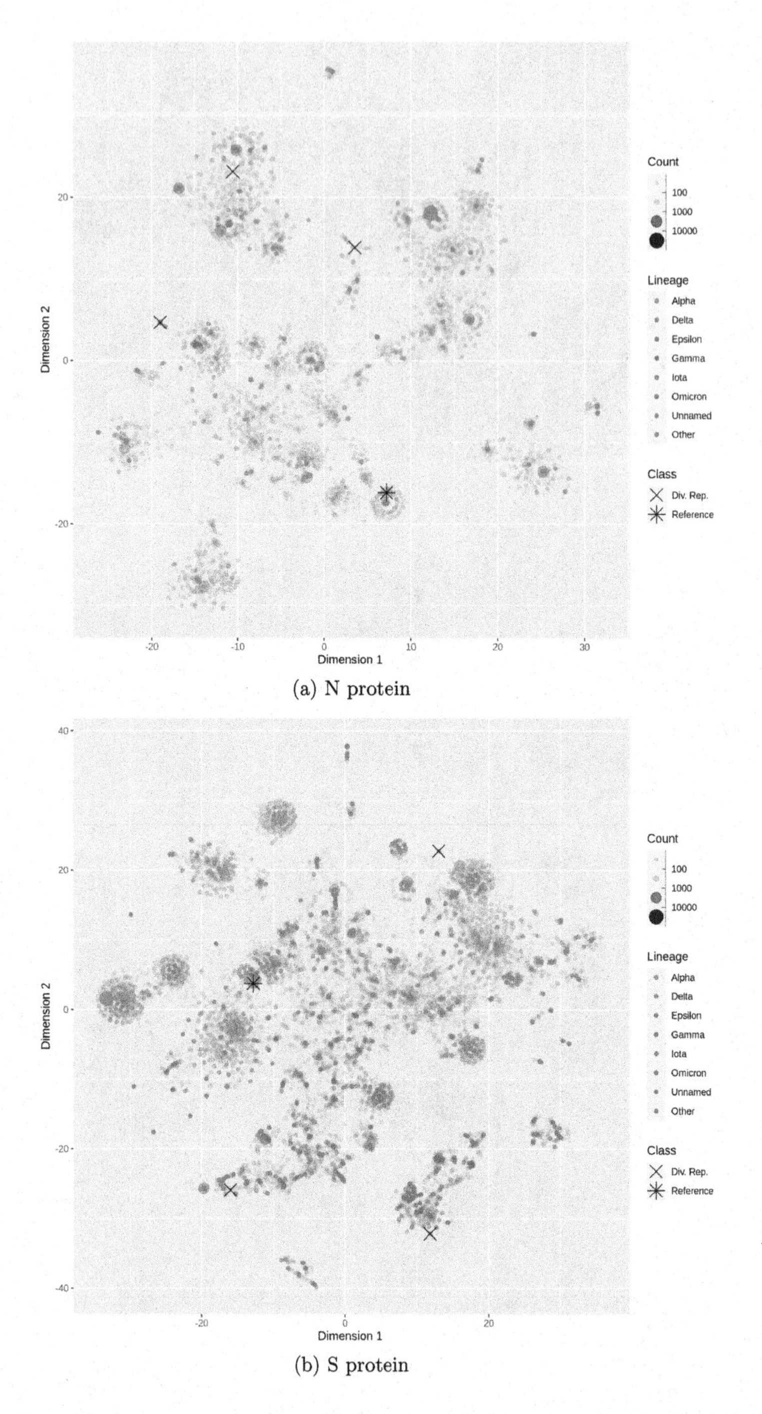

(a) N protein

(b) S protein

Fig. 3. N and S proteins mapped to a plane using t-SNE and colored by WHO-assigned names. Each dot corresponds to a deduplicated sequence. The dot size designates the count (multiplicity) of original sequences in the COVID database sample.

Figure 1 highlights the temporal distribution of the selected coronavirus samples while Figs. 2 and 3 show that the t-SNE protein coordinates are consistent with the Pango lineage and conventional WHO-named coronavirus groups. The Pearson correlation coefficients between original pairwise distances and Euclidean distances calculated from the t-SNE restored Cartesian coordinates were 0.571 and 0.508 for the N and S proteins, respectively.

4 Discussion

It is quite clear from Figs. 1 and 3 that t-SNE highlights represent the evolutionary space quite agreeing with basic biological knowledge and in particular with known phylogeny. However, t-SNE reveals evolutionary patterns which are not visible on classical phylogenetic trees. In particular, it suggests that although "locally" the evolution of COVID-19 seems to be driven by a random process, globally it is far from being purely random (i.e. on Fig. 1 the distribution of colors corresponding to the evolution of COVID specimens in time, is not chaotic at all), and in fact, the overall evolutionary trend is visible. However to interpret in biological terms the trend visible on the picture, one needs to have more metadata related to the COVID genomic data, such as lethality, contagiousness etc. The available COVID databases unfortunately lack such information. The information of that kind could also contribute to other published COVID studies, such as [6,11], and enhance their reported findings.

Acknowledgment. G. Tamazian was supported by Peter the Great St. Petersburg Polytechnic University in the framework of the Russian Federation's Priority 2030 Strategic Academic Leadership Programme (Agreement 075-15-2021-1333).

S. Kryzhevich was supported by Gdańsk University of Technology by the DEC 14/2021/IDUB/I.1 grant under the Nobelium - 'Excellence Initiative - Research University' program.

References

1. Adams, H., Blumstein, M., Kassab, L.: Multidimensional scaling on metric measure spaces. Rocky Mt. J. Math. **50**(2), 397–413 (2020)
2. Belkin, M., Niyogi, P.: Laplacian eigenmaps for dimensionality reduction and data representation. Neural Comput. **15**(6), 1373–1396 (2003)
3. Elbe, S., Buckland-Merrett, G.: Data, disease and diplomacy: GISAID's innovative contribution to global health. Glob. Challenges **1**(1), 33–46 (2017). https://doi.org/10.1002/gch2.1018
4. Hatcher, E.L., et al.: Virus variation resource - improved response to emergent viral outbreaks. Nucleic Acids Res. **45**(D1), D482–D490 (2016). https://doi.org/10.1093/nar/gkw1065
5. Henikoff, S., Henikoff, J.G.: Amino acid substitution matrices from protein blocks. Proc. Natl. Acad. Sci. **89**(22), 10915–10919 (1992). https://doi.org/10.1073/pnas.89.22.10915

6. Hozumi, Y., Wang, R., Yin, C., Wei, G.-W.: UMAP-assisted K-means clustering of large-scale SARS-CoV-2 mutation datasets. Comput. Biol. Med. **131**, 104264 (2021)
7. Katoh, K., Standley, D.M.: MAFFT multiple sequence alignment software version 7: improvements in performance and usability. Mol. Biol. Evol. **30**(4), 772–780 (2013). https://doi.org/10.1093/molbev/mst010
8. Krijthe, J.H.: Rtsne: T-Distributed Stochastic Neighbor Embedding using Barnes-Hut Implementation. R package version 0.16 (2015). https://github.com/jkrijthe/Rtsne
9. Kroshnin, A., Stepanov, E., Trevisan, D.: Infinite multidimensional scaling for metric measure spaces. In: ESAIM: COCV, pp. 28: 58 (2022). https://doi.org/10.1051/cocv/2022053
10. Lim, S., Memoli, F.: Classical MDS on metric measure spaces. arXiv preprint arXiv:2201.09385 (2022)
11. Lin, Q., Huang, Y., Jiang, Z., Feng, W., Ma, L.: Deciphering the subtype differentiation history of SARS-CoV-2 based on a new breadth-first searching optimized alignment method over a global data set of 24,768 sequences. Front. Genet. **11**, 591833 (2021)
12. McInnes, L., Healy, J., Melville, J.: Umap: uniform manifold approximation and projection for dimension reduction. arXiv preprint arXiv:1802.03426 (2018)
13. O'Toole, Á., et al.: Tracking the international spread of SARS-CoV-2 lineages B.1.1.7 and B.1.351/501Y-V2 with grinch [version 2; peer review: 3 approved]. Wellcome Open Res. **6**(121) (2021). https://doi.org/10.12688/wellcomeopenres.16661.2
14. O'Toole, Á., et al.: Assignment of epidemiological lineages in an emerging pandemic using the pangolin tool. Virus Evol. **7**(2) (2021). https://doi.org/10.1093/ve/veab064.veab064
15. Okada, P., et al.: Early transmission patterns of coronavirus disease 2019 (COVID-19) in travellers from Wuhan to Thailand, January 2020. Eurosurveillance **25**(8), 2000097 (2020). https://doi.org/10.2807/1560-7917.ES.2020.25.8.2000097
16. Pershina, E.V., et al.: The evolutionary space model to be used for the metagenomic analysis of molecular and adaptive evolution in the bacterial communities. In: Pontarotti, P. (eds) Evolutionary Biology: Genome Evolution, Speciation, Coevolution and Origin of Life, pp. 339–355. Springer, Cham (2014). https://doi.org/10.1007/978-3-319-07623-2_16
17. Team, R.C.: A language and environment for statistical computing. R Foundation for Statistical Computing, Vienna, Austria (2018). https://www.R-project.org/
18. Rambaut, A., et al.: A dynamic nomenclature proposal for SARS-CoV-2 lineages to assist genomic epidemiology. Nat. Microbiol. **5**(11), 1403–1407 (2020). https://doi.org/10.1038/s41564-020-0770-5
19. van der Maaten, L.J.P., Hinton, G.E.: Visualizing high-dimensional data using t-SNE. J. Mach. Learn. Res. **9**, 2579–2605 (2008)
20. Wang, J.: Geometric Structure of High-Dimensional Data and Dimensionality Reduction, vol. 13. Springer, Cham (2012)
21. Wickham, H.: ggplot2: Elegant Graphics for Data Analysis. Springer, New York (2016)
22. Wickham, H., et al.: Welcome to the tidyverse. J. Open Source Softw. **4**(43), 1686 (2019). https://doi.org/10.21105/joss.01686

MPCDDI: A Secure Multiparty Computation-Based Deep Learning Framework for Drug-Drug Interaction Predictions

Xia Xiao[1], Xiaoqi Wang[1], Shengyun Liu[2], and Shaoliang Peng[1,3](✉)

[1] College of Computer Science and Electronic Engineering, Hunan University, Changsha 410082, China
{xiaoxia,xqw,slpeng}@hnu.edu.cn

[2] Shanghai Jiao Tong University, Shanghai 200240, China
shengyun.liu@sjtu.edu.cn

[3] The State Key Laboratory of Chemo/Biosensing and Chemometrics, Hunan University, Changsha 410082, China

Abstract. Drug-drug interaction (DDI) is a key concern in drug development and pharmacovigilance. It is important to improve DDI predictions by integrating multi-source data from various pharmaceutical companies. Unfortunately, the data privacy and financial interest issues seriously influence the inter-institutional collaborations for DDI predictions. We propose MPCDDI, a secure multiparty computation-based deep learning framework for drug-drug interaction predictions. MPCDDI leverages the secret sharing technologies to incorporate the drug-related feature data from multiple institutions and develops a deep learning model for DDI predictions. In MPCDDI, all data transmission and deep learning operations are integrated into secure multiparty computation (MPC) frameworks to enable high-quality collaboration among pharmaceutical institutions without divulging private drug-related information. The results suggest that MPCDDI is superior to other five baselines and achieves the similar performance to that of the corresponding plaintext collaborations. More interestingly, MPCDDI significantly outperforms methods that use private data from the single institution. In summary, MPCDDI is an effective framework for promoting collaborative and privacy-preserving drug discovery.

Keywords: Secure multiparty computation · Deep learning · Drug-drug interaction predictions · Privacy-preserving

1 Introduction

Drug-drug interactions (DDIs) can occur frequently when taking more than two drugs concurrently [12]. The drugs, which change the activity of other drugs [10,11], will cause adverse drug reactions (ADRs) that may lead to diseases or deaths [32]. Furthermore, with the growth of the drug number taken by patients,

M. S. Bansal et al. (Eds.): ISBRA 2022, LNBI 13760, pp. 263–274, 2022.
https://doi.org/10.1007/978-3-031-23198-8_24

DDI events increase rapidly. Therefore, it is important to effectively recognize potential DDIs that can maximize synergistic benefits and reduce ADRs [10] when treating a disease.

To improve DDI predictions, a number of computational techniques have been created. In particular, various deep learning models have achieved excellent performance and emerged as a promising paradigm for DDI predictions. Ma *et al.* proposed an attentive multi-view graph auto-encoders-based drug similarity integration for DDI predictions in which the weights of each view are determined by an attentive mechanism [25]. Zitnik *et al.* proposed Decagon for predicting ADRs, which creates a brand-new multimodal graph convolutional neural network [35]. Deac *et al.* predicted drug-drug adverse effects by utilizing a co-attentional mechanism, which was able to get advanced results [7]. These deep learning-based DDI prediction methods have achieved good prediction performance. However, most previous models excessively rely on the lager-scale of drug data. Unfortunately, the high-quality drug data is time-consuming and expensive for real drug discovery. Hence, these deep learning-based DDI prediction models may not be satisfactory to real drug development scenarios [33].

With the rapid development of biomedical research, there are various publicly available bioinformatics and pharmacogenomics databases. For instance, DrugBank [22] is a comprehensive database for drug discovery, which provides exhaustive drug information and target data. The therapeutic target database (TTD) [26] provides a number of drug information, including nucleic acid targets, therapeutic protein, pathways, and the targeted diseases. The SuperTarget database [14] is a one-stop data warehouse that is a resource for exploring drug-target interactions. More importantly, there are a lot of valuable and private data from the various pharmaceutical companies and academic institutions. Simultaneously, existing researches shown that integrating multi-source data from various public databases and institutions can improve the performance of DDI predictions. However, pharmaceutical companies are reluctant to disclose their data due to drug data privacy concerns.

To overcome the difficult issues of data sharing, there are a few attempts using the secure multiparty computation (MPC) techniques to facilitate data sharing and collaboration in drug discovery and other biomedical fields [3,15,17,24]. MPC is a modern cryptographic technique that can realize the fusion of data among various institutions without revealing private information. While preventing their private data from being disclosed to other participating institutions, it enables many participating institutions to work together to execute private calculations on their secret data. Cho *et al.* leveraged MPC for large-scale genome-wide analysis under the premise of ensuring that the private data of underlying genotypes and phenotypes are not leaked [3]. Hie *et al.* proposed a predictive model for securely training drug-target interactions (DTI) [15]. Jagadeesh *et al.* used MPC for genomic diagnostics [17]. Ma et al. proposed DTIMPC and QSARMPC to improve the privacy-preserving drug discovery [24]. Unfortunately, MPC protocol-based deep learning methods for drug discovery are still relatively limited. Especially, it is challenging to develop MPC-based deep learning for high-precision DDI predictions.

In this study, we propose a secure multiparty computation-based deep learning framework for drug-drug interaction predictions, named MPCDDI. First, the drug-related data from various pharmaceutical companies and institutions are integrated via secret sharing technologies. Second, MPCDDI develops a deep neural network that includes a projection model and multilayer perceptrons for DDI predictions. In MPCDDI, all deep learning operations are executed under MPC frameworks, thus improving collaboration among pharmaceutical institutions and privacy-preserving in drug discovery. The experiments suggest that MPCDDI outperforms five existing methods and achieves the similar performance to that of the corresponding plaintext collaborations. In summary, MPCDDI is a useful framework for promoting collaborative and privacy-preserving drug discovery.

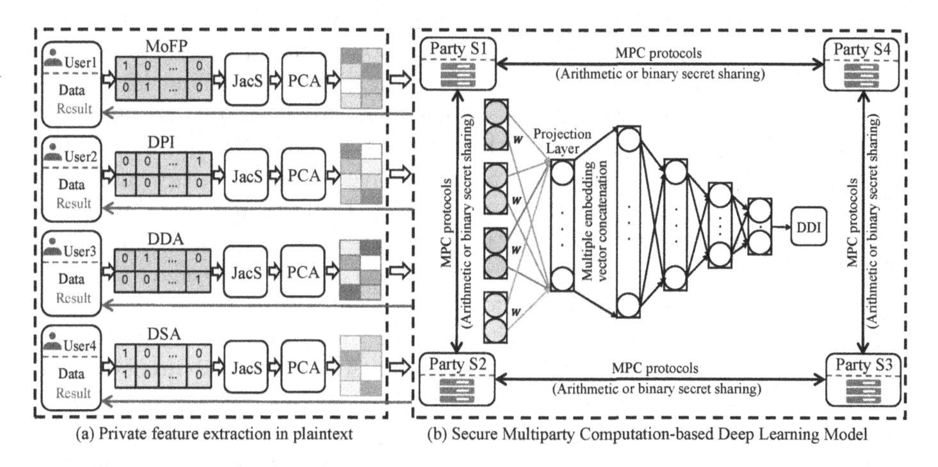

Fig. 1. The schematic workflow of MPCDDI. MoFP, DPI, DDA, and DSA denote the molecular fingerprints, drug-protein interactions, drug-disease associations, and drug-side effect associations, respectively. JacS and PCA represent the Jaccard similarity and Principal Component Analysis, respectively. (a) Under plaintext state, each user performs JacS calculation and PCA dimensionality reduction on its private drug data to obtain the initial features and then sends the secret shares of initial features to four parties S1, S2, S3, and S4. (b) Four parties initiate the MPC protocol together for joint training deep learning model and DDI predictions via the secure multiparty computation frameworks. Finally, the parties respectively send the prediction results to the corresponding users.

2 Methods

The schematic workflow of MPCDDI as shown in Fig. 1, which depicts the application scenario where four pharmaceutical institutions have different feature data of drugs. MPCDDI aims to integrate these feature data and make four pharmaceutical institutions to collaboratively improve DDI predictions without revealing sensitive data.

2.1 Secure Multiparty Computation Protocols

In this study, secure multiparty computation (MPC) is treated as a user-server model, where each pharmaceutical institution with private data is represented by a also named user. There are four users (e.g., pharmaceutical institutions) that is represented by $User_1, User_2, User_3, User_4$, and each user $User_i$ owns private drug feature data X_i. There are four pharmaceutical institutions that want to jointly train a deep learning model to improve DDI predictions without divulging the privacy of drug features and the parameters of the deep learning model. In addition, there are four key assumes. (1) There are four parties that are based on semi-honest protocol. In real drug discovery scenario, there can be any number of semi-honest parties. In a semi-honest protocol, each party abides by the established rules and refrains from sending fictitious or incorrect information to other parties. (2) Every two parties do not collude with each other. (3) The channels, that are used for inter-node communication, are secure. (4) Adversaries are not able to see or modify any data in the channels.

Based on the above assumptions, we carry out MPC by two key steps. In the first stage, the private data of each user is divided into four parts and then is sent to four servers S_1, S_2, S_3, and S_4, respectively. The above operations are conducted by the secret sharing technology [29]. In particular, our MPC uses 4-out-of-4 additive secret sharing to perform this task [18]. To be specific, it supposes that there is a private value x in one party. In order to share the value x, the party divides the value x into four secret shares (i.e., x_1, x_2, x_3, and x_4), and then distributes the four secret shares to the four servers S_1, S_2, S_3, and S_4 by generating a pseudorandom zero-share [4] with four random numbers that sum to 0. No one server can know anything about x, but the sum of all secret shares in S_1, S_2, S_3, and S_4 can reconstruct the original value of x. Therefore, the private information of each user is not disclosed to the four semi-honest parties. In addition, each party does not know anything about the data of other parties.

In the second stage, we break down the deep learning algorithm into a set of fundamental operations including private addition, private multiplication, comparison, and a few linear and nonlinear functions. Each party fails to capture the information about input data when executing a basic private operation. The four parties carry out a series of private actions sequentially to finish the assignments. To increase efficiency, some private activities can be run concurrently. The confidentiality of input is preserved because neither side can discover the input, intermediate values, or final results from either secret shares or private operations. All the basic private operations are implemented by using the protocols available in CRYPTEN that adopts binary secret sharing [13], arithmetic secret sharing [5], and conversions between these two types of sharing to facilitate secure computations. Most of basic operations, such as bilinear operations and multiplications, adopt arithmetic secret sharing mechanisms. Similarly, binary secret sharing techniques are used for the evaluation of logical expressions, such as the activation function rectified linear units (ReLU). Some other operations involve the conversion between arithmetic secret sharing and binary secret sharing, which can be achieved by the ABY share-conversion techniques [8]. At the cryptographic level, Beaver multiplication triples [1] can help implement basic arithmetic operations, while the GMW protocol [13] can help implement binary circuit evaluation. The four parties acquire the secret results after finishing all the private activities. Finally, S_1, S_2, S_3, and S_4 respectively send their secret result to the corresponding users, that is, these secret results are revealed to corresponding users.

Private Feature Extraction. In this work, we assume that there are four pharmaceutical institutions that respectively have different drug data, i.e., molecular fingerprints which are derived by RDKit [21], drug-side effect associations, drug-disease associations, and drug-protein interactions. Four binary feature vectors can be used to represent a given drug. The values (1 or 0) in feature vectors denote the presence or absence of the relevant descriptors. Based on these different drug descriptors, four kinds of drug-drug similarity are computed by the Jaccard similarity metric and treated as four kinds of drug features, respectively.

In real drug discovery, there are a great number of drugs or molecules, thus leading to high-dimension and sparse drug similarity features. To overcome this problem, all similarity features are fed into principal component analysis for reducing the dimensionality of each drug. The low-dimensional features of drugs in each user are secretly shared with various parties S_1, S_2, S_3, and S_4 in MPC. All parties then initiate 4-out-of-4 additive secret sharing together. Finally, the privacy-preserving deep learning algorithm is executed by the four MPC parties.

2.2 MPC-Based Deep Learning

Each party obtain four secret features $F_{MoFP}, F_{DPI}, F_{DDA}$, and F_{DSA} that are derived from molecular fingerprints, drug-disease associations, drug-side effect associations, and drug-protein interactions. These features are fed into the privacy-preserving deep learning model to predict DDIs. The deep learning model includes two key modules: feature projection modules and DDI prediction modules. The projection module aims to map the different types of drug features into the same space. The projection module is defined as:

$$h_0 = BNLayer(h_{MoFP}||h_{DPI}||h_{DDA}||h_{DSA}) \quad (1)$$

where $||$ denotes the concatenation operation, $BNlayer$ denotes batch normalization layers [16], and $h_l \in \{h_{MoFP}, h_{DPI}, h_{DDA}, h_{DSA}\}$ can be calculated by:

$$h_l = WF_l + b \quad (2)$$

where W is the weight matrix of linear transformations, b is a bias vector, and $F_l \in \{F_{MoFP}, F_{DPI}, F_{DDA}, F_{DSA}\}$. It is noted that four secret features share a weight matrix and bias vector to convert diverse types of features into the same space. Finally, h_0 is fed into for DDI predictions.

The DDI prediction module adopts a multilayer perceptron (MLP) with three fully connected layers to predict DDIs. The two hidden layers and output layer employ rectified linear unit (ReLU) and sigmoid activation function, respectively. The batch normalization is used to increase the speed of convergence. Furthermore, we add the dropout layers [30] to improve generalization performance and reduce over-fitting. By fully utilizing the drug feature data from all users to forecast DDIs, MPCDDI trains the deep learning models using private operation protocols, while only disclosing the predicted actions to the corresponding users. After running MPCDDI, each user will obtain the predicted DDIs and trained deep learning models.

The proposed MPCDDI is implemented based on CRYPTEN framework [18] to realize DDI predictions from private drug data under MPC protocol. The CRYPTEN framework provides a secure computational backend for PyTorch, while still retaining the PyTorch front-end APIs that enable experimentation with deep neural networks. The framework is similar to PyTorch while all values are secretly shared among the modes of multiple parties.

3 Experiments and Results

3.1 Datasets

In this work, we used a heterogeneous dataset [23] in which include the drug-drug interactions (DDI) and the drug-protein interactions (DPI) network derived from DrugBank 3.0 [19]. The drug-disease association (DDA) network was derived from the Comparative Toxicogenomics Database [6]. We extracted the drug-side effect (DSA) associations from Comparative Toxicogenomics Database and SIDER [20]. In addition, for each drug, its simplified molecular input line entry system was transformed into the Molecular fingerprints (MoFP) in the shape of a 1024-bit vector by using RDKit program [21]. The information about drug structure was included in the molecular fingerprints. The sizes of heterogeneous dataset was shown in Table 1. In MPCDDI, we assumed that MoFP, DPI, DDA, and DSA were the private feature of four users.

Table 1. The sizes of heterogeneous dataset.

Name	Feature shape	Edge size
Drug-Drug	708×708	10,036
Drug fingerprint	$708 \times 1,024$	30,661
Drug-protein	$708 \times 1,512$	1,923
Drug-disease	$708 \times 5,603$	199,214
Drug-side effect	$708 \times 4,192$	80,164

3.2 Experiments Setting

To evaluate the performance of MPCDDI, it was compared with five methods, including DDIMDL [9], DeepDDI [28], LINE [31], GraRep [2], and struc2vec [27] models. DDIMDL and DeepDDI are deep learning-based DDI prediction methods and have been tested on various datasets. The LINE, GraRep, and struc2vec generated great performance in various biomedical applications. Every known DDI was a positive sample. From the unidentified interactions, we used random ways to choose negative samples. There were the same number of negative samples and positive samples that were split into 10% testing set and 90% training set. In this work, DDI prediction is a binary classification task. Therefore, the area under the receiver operating characteristic curve (AUROC), Recall, and precision-recall curve (AUPR) were treated as evaluation metrics.

In this work, the representation vectors generated by LINE, GraRep, and struc2vec were fed into Logistic Regression models for DDI predictions. For hyperparameters of MPCDDI, we set 256 neurons in the projection layer. The MLP includes three hidden layers for DDI predictions, where there are 128, 64, and 2 neurons in each layer, respectively. We used Adam optimizer with a 1e-4 learning rate, and set the dropout rate at 0.3. The hyperparameters of the LINE, GraRep, and struc2vec models were set as the default values in [34], because Yue *et al.* carefully optimized them by grid search. The parameters of DDIMDL and DeepDDI were selected according to the suggestions in respective articles [9,28].

DDIMDL: It is a multimodal deep learning framework that uses a joint mechanism to integrate various drug features for DDI predictions. In DDIMDL, the number of units in current layer was set to half of that in former layer, the batch size and dropout rate were set to 256 and 0.3, respectively.

DeepDDI: It first calculates the structure similarity profiles among drugs by using molecular fingerprints. Second, these structural profiles are fed into the deep neural networks to predict DDIs. DeepDDI includes nine hidden layers where there are 1,024 nodes in each hidden layer. The batch size and learning rate were set to 256 and 0.0001, respectively.

LINE: The method obtains local and global network structures by jointly training the objective functions of first-order similarity and second-order similarity. The algorithm is applicable to various types of information networks. In LINE, we set the negative ratio as 5 and the order as 2.

GraRep: It designs a more accurate loss function that is able to capture rich local and global structural properties of the biomedical network. In GraRep, we set to k-step = 4, learning rate = 0.01.

struc2vec: The method mainly focuses on the structure of nodes and uses a hierarchical structure to measure the similarity of nodes at different scales. In struc2vec, we set the length of random walk as 64, the window size as 10, and the number of random walks as 32.

3.3 Comparisons Between MPCDDI and Baselines

The results of MPCDDI and baselines were shown in Fig. 2. We found that MPCDDI (AUROC = 0.938, AUPR = 0.942, Recall = 0.870) was superior to baselines on all evaluation metrics, with 4.82% higher AUROC, 6.8% higher AUPR, and 3.48% higher Recall than the average performance of the five baselines. In particular, MPCDDI outperformed GraRep with 6.8%, 7.8%, and 10.1% improvements in terms of Recall, AUROC, and AUPR for DDI predictions, respectively. These results suggested that MPCDDI can achieve better performance than other methods.

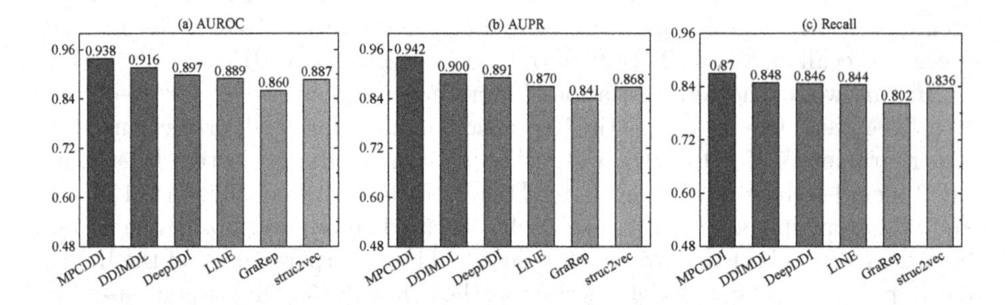

Fig. 2. The results of MPCDDI and baseline methods for DDI predictions.

Table 2. The performances of MPCDDI and PlainDDI.

Method	AUROC	AUPR	Recall
MPCDDI	**0.938**	**0.942**	0.870
PlainDDI	0.932	0.919	**0.886**

3.4 Comparison Between Public and MPC Collaborations

We looked at whether our MPC cooperation protocols caused a reduction in prediction accuracy when compared to the corresponding plaintext learning method (named PlainDDI) under the open collaboration using all shared information. To be specific, PlainDDI used four plaintext features to drive the proposed deep learning model for DDI predictions. The performances of MPCDDI and PlainDDI were summarized in Table 2 where the best results were marked in boldface. We observed that MPCDDI and PlainDDI achieved the similar performance. For example, there was 0.6% difference between MPCDDI and PlainDDI in terms of AUROC. Interestingly, although MPCDDI generated the poor results in terms of Recall, MPCDDI was significantly higher than PlainDDI in terms of AUPR. These results demonstrated that MPCDDI helped pharmaceutical institutions to obtain high-quality collaboration and achieve almost the same performance as the corresponding models for plaintext predictions on publicly shared information. In other words, MPCDDI not only achieved high-quality results, but also protected the data privacy on pharmaceutical institutions.

Table 3. The results of MPCDDI and other methods using private data owned by a single institution.

Method	AUROC	AUPR	Recall
MPCDDI	**0.938**	**0.942**	**0.870**
Fing-DDI	0.885	0.865	0.827
Pro-DDI	0.801	0.797	0.664
Dis-DDI	0.850	0.827	0.798
Se-DDI	0.883	0.860	0.838

3.5 Comparison Between Predictions Using More Data with MPC Collaboration and Using Private Data of a Single Institution

In a perfect collaborative situation, medication feature data from several pharmaceutical institutions are combined to create a multi-modal feature space that is conducive to training better deep learning models. In MPCDDI, we imitated the application scenario where four different data (i.e., MoFP, DPI, DDA, and DSA) were randomly distributed to four pharmaceutical companies and treated as their private feature data. These private features from different institutions were integrated via MPC protocols and then used to jointly training a deep learning model for DDI predictions. Therefore, MPCDDI was converted to Fing-DDI without MPC protocols, that is, only using the molecular fingerprints data owned by a single institution. Similarly, MPCDDI was transformed to Pro-DDI, Dis-DDI, and Se-DDI. In Table 3, we described the results of MPCDDI and other methods using only a single feature (i.e., MoFP, DPI, DDA, and DSA). The best results were marked in boldface. We find that Fing-DDI achieved relatively high results than other three methods (i.e., Pro-DDI, Dis-DDI, and Se-DDI). For example, Fing-DDI obtained 8.4% higher AUROC, 6.8% higher AUPR, and 16.3% higher Recall than Pro-DDI for DDI predictions. Furthermore, we observed that MPCDDI consistently achieved higher performance than other methods. Interestingly, MPCDDI generates 5.5%–13.7% AUROC, 7.7%–14.5% AUPR, and 3.2%–20.6% Recall improvements than other methods for DDI predictions. In particularly, MPCDDI is superior to Pro-DDI, with AUROC, AUPR, and Recall improvements of 13.7%, 14.5%, and 20.6%, respectively. These results suggested that MPCDDI can help pharmaceutical institutions or organizations to achieve better prediction performance without divulging private information.

3.6 Feature Evaluation

In this part, we assessed the impact of several pharmacological properties on DDI predictions under the encrypted state. MPCDDI is converted to FPD-MPCDDI when we use only three feature including MoFP, DPI, and DDA. Similarly, MPCDDI is converted to PDS-MPCDDI, FDS-MPCDDI, and FPS-MPCDDI. The results of MPCDDI and different feature models are shown in Fig. 3. We observe that PDS-MPCDDI obtained relatively poorer results than other three methods (i.e., FDS-MPCDDI, FPS-MPCDDI, and FPD-MPCDDI). In particularly, FDS-MPCDDI was 2.9% higher AUROC, 3.4% higher AUPR, and 5.3% higher Recall than PDS-MPCDDI for DDI predictions. However, we found that FDS-MPCDDI (AUROC = 0.928, AUPR = 0.934, and Recall = 0.863) obtained the most similar results to MPCDDI. Furthermore, we observed that MPCDDI achieved higher performance than other four methods. These results suggested that the combinations of four features are contribute to DDI predictions, and that molecular fingerprints have the greater impact on DDI predictions relative to the drug-protein interactions.

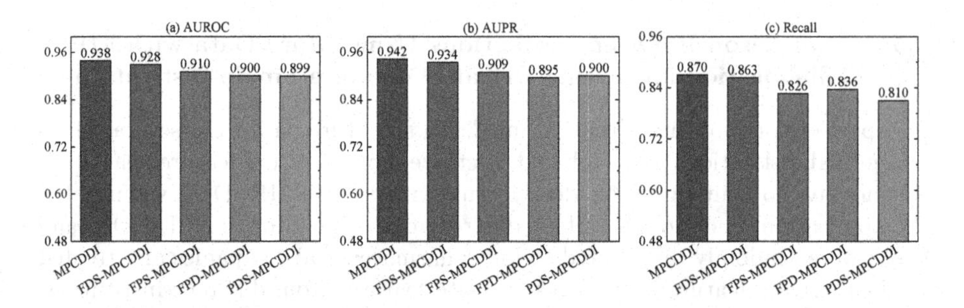

Fig. 3. The performance of MPCDDI with different feature combinations.

4 Conclusion

In this work, we propose a secure multiparty computation-based deep learning framework for drug-drug interaction predictions. This work integrates multiple feature data from various pharmaceutical companies and executes the deep learning operations via the secure multiparty computation frameworks to enable high-quality collaboration among pharmaceutical institutions without divulging private information. The experiments suggest that the proposed method can achieve great performance under different scenarios. In summary, the proposed method is an effective framework for promoting collaborative and privacy-preserving drug discovery. In the future, we will further verify the secure multiparty computation-based deep learning framework for more biomedical problems, for example, electronic medical records, and medical image analysis. In addition, secure multiparty computation-based deep learning calculations still have a large time overhead in an encrypted state, resulting in low efficiency. Therefore, we will pay attention to improving the computation efficiency.

Acknowledgements. This work was supported by NSFC Grants U19A2067; National Key R&D Program of China 2022YFC3400404; Science Foundation for Distinguished Young Scholars of Hunan Province (2020JJ2009); Science Foundation of Changsha Z202069420652, kq2004010; JZ20195242029, JH20199142034; The Funds of State Key Laboratory of Chemo/Biosensing and Chemometrics, the National Supercomputing Center in Changsha (http://nscc.hnu.edu.cn/), and Peng Cheng Lab.

References

1. Beaver, D.: Efficient multiparty protocols using circuit randomization. In: Feigenbaum, J. (ed.) CRYPTO 1991. LNCS, vol. 576, pp. 420–432. Springer, Heidelberg (1992). https://doi.org/10.1007/3-540-46766-1_34
2. Cao, S., Lu, W., Xu, Q.: GraRep: learning graph representations with global structural information. In: Proceedings of the 24th ACM International on Conference on Information and Knowledge Management, pp. 891–900 (2015)
3. Cho, H., Wu, D.J., Berger, B.: Secure genome-wide association analysis using multiparty computation. Nat. Biotechnol. **36**(6), 547–551 (2018)
4. Cramer, R., Damgård, I., Ishai, Y.: Share conversion, pseudorandom secret-sharing and applications to secure computation. In: Kilian, J. (ed.) TCC 2005. LNCS, vol. 3378, pp. 342–362. Springer, Heidelberg (2005). https://doi.org/10.1007/978-3-540-30576-7_19
5. Damgård, I., Pastro, V., Smart, N., Zakarias, S.: Multiparty computation from somewhat homomorphic encryption. In: Safavi-Naini, R., Canetti, R. (eds.) CRYPTO 2012. LNCS, vol. 7417, pp. 643–662. Springer, Heidelberg (2012). https://doi.org/10.1007/978-3-642-32009-5_38
6. Davis, A.P., et al.: The comparative toxicogenomics database: update 2013. Nucleic Acids Res. **41**(D1), D1104–D1114 (2013)
7. Deac, A., Huang, Y.H., Veličković, P., Liò, P., Tang, J.: Drug-drug adverse effect prediction with graph co-attention. arXiv preprint arXiv:1905.00534 (2019)
8. Demmler, D., Schneider, T., Zohner, M.: ABY-a framework for efficient mixed-protocol secure two-party computation. In: NDSS (2015)
9. Deng, Y., Xu, X., Qiu, Y., Xia, J., Zhang, W., Liu, S.: A multimodal deep learning framework for predicting drug-drug interaction events. Bioinformatics **36**(15), 4316–4322 (2020)
10. Edwards, I.R., Aronson, J.K.: Adverse drug reactions: definitions, diagnosis, and management. Lancet **356**(9237), 1255–1259 (2000)
11. Evans, W.E., McLeod, H.L.: Pharmacogenomics-drug disposition, drug targets, and side effects. N. Engl. J. Med. **348**(6), 538–549 (2003)
12. Giacomini, K.M., Krauss, R.M., Roden, D.M., Eichelbaum, M., Hayden, M.R., Nakamura, Y.: When good drugs go bad. Nature **446**(7139), 975–977 (2007)
13. Goldreich, O., Micali, S., Wigderson, A.: How to play any mental game, or a completeness theorem for protocols with honest majority. In: Providing Sound Foundations for Cryptography: On the Work of Shafi Goldwasser and Silvio Micali, pp. 307–328 (2019)
14. Hecker, N., et al.: SuperTarget goes quantitative: update on drug-target interactions. Nucleic Acids Res. **40**(D1), D1113–D1117 (2012)
15. Hie, B., Cho, H., Berger, B.: Realizing private and practical pharmacological collaboration. Science **362**(6412), 347–350 (2018)
16. Ioffe, S., Szegedy, C.: Batch normalization: accelerating deep network training by reducing internal covariate shift. In: International Conference on Machine Learning, pp. 448–456. PMLR (2015)
17. Jagadeesh, K.A., Wu, D.J., Birgmeier, J.A., Boneh, D., Bejerano, G.: Deriving genomic diagnoses without revealing patient genomes. Science **357**(6352), 692–695 (2017)
18. Knott, B., Venkataraman, S., Hannun, A., Sengupta, S., Ibrahim, M., van der Maaten, L.: CRYPTEN: secure multi-party computation meets machine learning. In: Advances in Neural Information Processing Systems, vol. 34, pp. 4961–4973 (2021)

19. Knox, C., et al.: DrugBank 3.0: a comprehensive resource for 'omics' research on drugs. Nucleic Acids Res. **39**(suppl_1), D1035–D1041 (2010)
20. Kuhn, M., Campillos, M., Letunic, I., Jensen, L.J., Bork, P.: A side effect resource to capture phenotypic effects of drugs. Mol. Syst. Biol. **6**(1), 343 (2010)
21. Landrum, G.: RDKit documentation. Release **1**(1–79), 4 (2013)
22. Law, V., et al.: DrugBank 4.0: shedding new light on drug metabolism. Nucleic Acids Res. **42**(D1), D1091–D1097 (2014)
23. Luo, Y., et al.: A network integration approach for drug-target interaction prediction and computational drug repositioning from heterogeneous information. Nat. Commun. **8**(1), 1–13 (2017)
24. Ma, R., et al.: Secure multiparty computation for privacy-preserving drug discovery. Bioinformatics **36**(9), 2872–2880 (2020)
25. Ma, T., Xiao, C., Zhou, J., Wang, F.: Drug similarity integration through attentive multi-view graph auto-encoders. arXiv preprint arXiv:1804.10850 (2018)
26. Qin, C., et al.: Therapeutic target database update 2014: a resource for targeted therapeutics. Nucleic Acids Res. **42**(D1), D1118–D1123 (2014)
27. Ribeiro, L.F., Saverese, P.H., Figueiredo, D.R.: struc2vec: learning node representations from structural identity. In: Proceedings of the 23rd ACM SIGKDD International Conference on Knowledge Discovery and Data Mining, pp. 385–394 (2017)
28. Ryu, J.Y., Kim, H.U., Lee, S.Y.: Deep learning improves prediction of drug-drug and drug-food interactions. Proc. Natl. Acad. Sci. **115**(18), E4304–E4311 (2018)
29. Shamir, A.: How to share a secret. Commun. ACM **22**(11), 612–613 (1979)
30. Srivastava, N., Hinton, G., Krizhevsky, A., Sutskever, I., Salakhutdinov, R.: Dropout: a simple way to prevent neural networks from overfitting. J. Mach. Learn. Res. **15**(1), 1929–1958 (2014)
31. Tang, J., Qu, M., Wang, M., Zhang, M., Yan, J., Mei, Q.: Line: large-scale information network embedding. In: Proceedings of the 24th International Conference on World Wide Web, pp. 1067–1077 (2015)
32. Vilar, S., et al.: Similarity-based modeling in large-scale prediction of drug-drug interactions. Nat. Protoc. **9**(9), 2147–2163 (2014)
33. Wang, X., Cheng, Y., Yang, Y., Li, F., Peng, S.: Multi-task joint strategies of self-supervised representation learning on biomedical networks for drug discovery. arXiv preprint arXiv:2201.04437 (2022)
34. Yue, X., et al.: Graph embedding on biomedical networks: methods, applications and evaluations. Bioinformatics **36**(4), 1241–1251 (2020)
35. Zitnik, M., Agrawal, M., Leskovec, J.: Modeling polypharmacy side effects with graph convolutional networks. Bioinformatics **34**(13), i457–i466 (2018)

A Multimodal Data Fusion-Based Deep Learning Approach for Drug-Drug Interaction Prediction

An Huang[1], Xiaolan Xie[1,2(✉)], Xiaoqi Wang[3], and Shaoliang Peng[3,4(✉)]

[1] College of Information Science and Engineering, Guilin University of Technology, Guilin 541006, China

[2] Guangxi Key Laboratory of Embedded Technology and Intelligent System, Guilin 541006, Guangxi, China

xie_xiao_lan@foxmail.com

[3] The State Key Laboratory of Chemo/Biosensing and Chemometrics, Hunan University, Changsha 410082, China

[4] College of Computer Science and Electronic Engineering, Hunan University, Changsha 410082, China

slpeng@hnu.edu.cn

Abstract. Prediction of drug-drug interaction (DDI) is one of the vital topics in drug development. Many computational methods have been present for DDI prediction. However, these methods are often limited to exploiting the drug's molecular structure and ignoring other features of modalities, necessary for capturing the complicated DDI patterns. In this study, we proposed HF-DDI, a hybrid fusion-based deep learning framework, for DDI event prediction using various biomedical information about drugs. At first, HF-DDI uses multiple drug similarities based on drug substructure, target, and enzyme as representations of drug-drug interaction events. Afterward, HF-DDI combines two different levels of fusion strategies and utilizes a score calculation module with adaptive weighted averaging to help prediction-making. Experimental results demonstrated that our proposed method outperformed existing approaches for interaction prediction, which provided a high accuracy of 0.948.

Keywords: Drug-drug interaction · Biomedical data fusion · Multimodal deep learning

1 Introduction

Many patients take many prescriptions at once to treat complex or co-existing diseases, and prescribing multiple medications for individual patients has reached become rather widespread. More than four in ten older persons take five or more prescription drugs daily, a number that has tripled over the preceding two decades, according to an analysis from the Lown Institute. However, some drugs

M. S. Bansal et al. (Eds.): ISBRA 2022, LNBI 13760, pp. 275–285, 2022.
https://doi.org/10.1007/978-3-031-23198-8_25

interact with others in the body, causing unexpected drug-drug interactions [5]. The likelihood of drug-drug interactions rises steeply with the number of drugs a patient is taking. It is not possible to evaluate clinically all of the drug interactions that may be associated with taking medications since the combinations of drugs are seemingly infinite. Accurate prediction of drug-drug interactions (DDIs) using computational methods can help clinicians make effective decisions and improve the safety and efficacy of prescribing [14].

There are many computational methods available for predicting drug-drug interactions, and they can be loosely divided into six categories. The similarity-based methods [8,29,34] predict interactions by assuming that similar drugs would interact similarly with other drugs. The network-based methods [18,25,28] construct a network of drug-drug interactions and then infer potential DDIs using techniques such as label propagation and random walks [4]. The matrix factorization-based approaches [32,35,36] take interaction prediction as a matrix completion task where drugs in the interaction space are projected into a low-dimensional space [19]. The machine learning-based models [1,9,15] mainly use logistic regression, support vector machines, and other conventional algorithms to build classifiers for discovering the unknown DDIs. The ensemble learning-based frameworks [3,10] integrate several machine learning models to outperform individual models in terms of accuracy and generalization. The deep learning-based methods [4,6,7,11,12,21] use multiple transformation steps to learning representations to predict novel DDIs, which tend to outperform other prediction methods [31].

Many methods have shown promising results, however, they have not considered heterogeneous features of different modalities in their models. For example, Nyamabo et al. introduced a novel deep learning-based neural network that learns chemical substructures for interaction prediction [17]. The unimodal data is insufficient to comprehensively describe drugs due to the complexity of drug reactions. Generally, the chemical substructure of a small molecule drug determines its physicochemical properties and pharmacological activity [22]. The therapeutic target is inherently linked to a particular disease process when the metabolizing enzyme participates in the absorption, distribution, metabolism, and excretion of many drugs [13].

In this study, we put our full focus on biomedical data fusion. Integrating multimodal data with fusion technologies allows more complementary information to be captured, which can help the prediction model increase its accuracy [20]. For establishing an efficient multimodal deep learning framework, we attempt to predict DDIs based on different fusion strategies: feature-level fusion and decision-level fusion. Different integrating strategies of diverse features generate varied prediction performances [24]. The feature-level fusion [shown in Fig. 1(a)] is referred to as early fusion, which concatenates several feature vectors of different modalities to obtain a single feature vector. Compared to any of the input feature vectors, the single feature vector is more discriminative [23,27]. However, the redundant information in a joint representation can lead to prediction errors. The decision-level fusion [shown in Fig. 1(b)] aggregates the decisions from various classifiers, which are trained on features of different modalities. The decision-level fusion strategy ensures that the training of each modality will not interfere with one another during training, however, it is difficult to achieve high accuracy due to the lack of global information. According to the above-mentioned challenges, we proposed the hybrid fusion framework with better performance of interaction prediction, which combines two different levels of fusion strategies, allowing the model to capture the underlying complex relationships between features of different modalities.

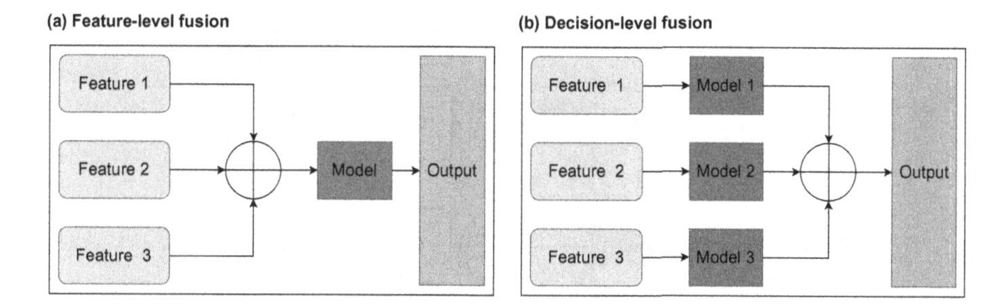

Fig. 1. Two different multimodal data fusion frameworks.

The main contributions are summarized as follows: (1) Different from most existing unimodal prediction methods, we presented a novel framework, which uses multiple drug similarity features to predict drug-drug interaction events. (2) The proposed network framework is designed with the structure of combing two different levels of fusion strategies. (3) The experimental results demonstrated that our proposed strategy outperformed existing approaches in the task of predicting interaction events.

2 Methods

2.1 Data Description

DrugBank is a source that provides comprehensive information about drugs, which has seen a progressive increase in drug quantity and a significant enhancement to the quality of its information over the few decades [30]. The chemical structure, drug target, and enzyme information are important for many pharmaceutical research applications. Therefore, in this study, we collected chemical structures, targets, and enzymes of drugs, and 323,539 pairwise interaction events between the 1258 drugs from DrugBank 5.1.9 released on January 4, 2022.

2.2 Feature Extraction

The dataset includes chemical structures, targets, and enzymes of 1258 small molecule drugs. We need to transform a drug into a mathematical representation

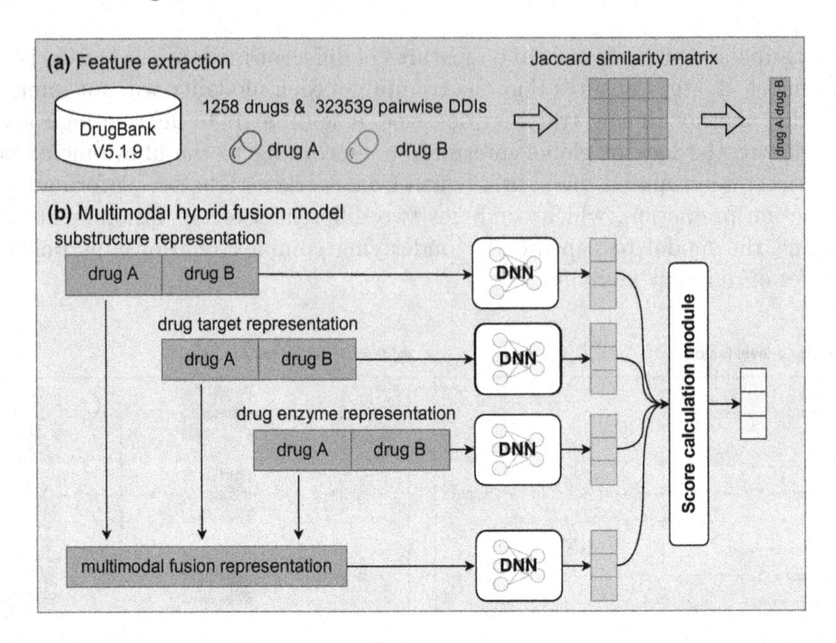

Fig. 2. Overview of the interaction prediction framework.

that can be used for deep learning. In this work, we employed the open-source toolkit RDKit to generate the bit fingerprint vector based on the Morgan fingerprint algorithm and use binary to represent the presence or absence of different targets and enzymes. Morgan fingerprint perceives the presence of specific circular substructures around each atom in a molecule, which is predictive of the biological activities of small organic molecules [2]. The representation of a drug chemical substructure is converted as a 2040-dimensional bit vector and the drug target feature is converted as a 1650-bit vector, whereas the enzyme feature is converted as a 314-bit vector. Based on the assumption that similar drugs may interact with the same drugs, we use these bit vectors of drug features to calculate the Jaccard similarity coefficient with other drugs to obtain the 1258-bit vectors. Jaccard similarity coefficient is calculated by Eq. 1.

$$J(drugA, drugB) = \frac{|drugA \bigcap drugB|}{|drugA| + |drugB| - |drugA \bigcap drugB|} \tag{1}$$

In which, $drugA$ and $drugB$ are the bit vectors of two drugs. We can obtain three 1258×1258 drug-drug similarity matrices and a 1258-dimensional row vector is then used to represent each feature of drugs. The feature vectors of the two drugs are concatenated to represent a drug pair. The order of Drug A and Drug B is important and means interaction.

2.3 Deep Learning Models Based Data Fusion

The proposed hybrid fusion network (HF-DDI) with the structure of combing two different levels of fusion strategies is shown in Fig. 2. It treats the DDI prediction as a multi-classification task. We construct a sub-model for each representation and then merge those sub-models to create the prediction model. The main task of the score calculation module is to fuse these outputs to calculate the final prediction.

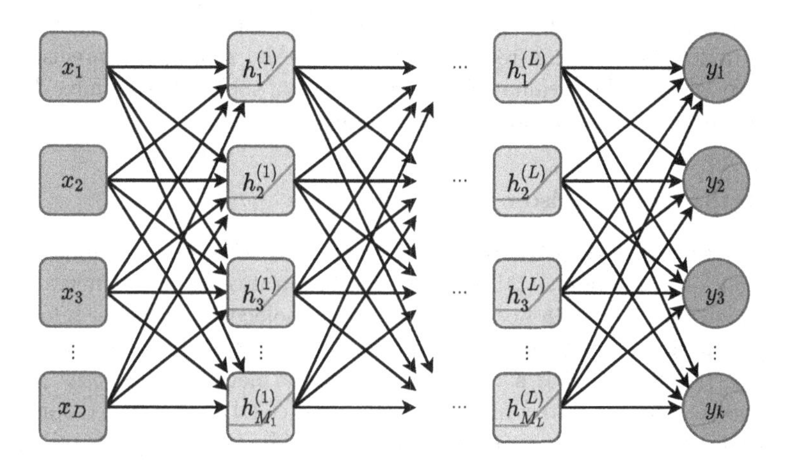

Fig. 3. Illustration of the fundamental structure of the deep neural networks

We used deep neural networks (DNN) to construct sub-models. As shown in Fig. 3, DNN is the simplest type of neural network that propagates signals unidirectionally from the input layer to the output layer and minimizes the loss function to optimize a set of model parameters, which include weights and biases between each layer. To arrive at the final decision fusion, the score calculation module computes a weighted average of the projected probability for each classifier.

Deep neural networks compose computations performed by multiple layers. Denoting the output of hidden layers by $h^{(l)}(x)$, the calculation for a network with L hidden layers is the following formula:

$$f(\boldsymbol{x}) = f\left[a^{(L+1)}\left(h^{(L)}\left(a^{(L)}\left(\cdots\left(a^{(2)}\left(h^{(1)}\left(a^{(1)}(\boldsymbol{x})\right)\right)\right)\right)\right)\right)\right] \quad (2)$$

To achieve superior results and mitigate the gradient vanishing or exploding problem, the rectified linear unit (ReLU) is used as a hidden-layer activation function [16,31]. In addition, we add a batch normalization layer and a rejection layer after each hidden layer to speed up the training process and reduce overfitting [26]. All these techniques can improve the generalization ability of the model. Cross-entropy is used as a loss function when optimizing classification models during training.

3 Experiments and Results

3.1 Hyperparameters Setting and Evaluation Metrics

The different hyperparameter configurations in the experiments are as follows: the batch size is set to 128, and the dropout rate is set to 0.3. We empirically choose 3 hidden layers for neural networks and the number of neurons was 1024, 512, and 256. Each neural network is trained with Adam, an optimizer algorithm based on adaptive estimates. The average values of accuracy classification score(ACC), the area under the receiver operating characteristic curve (AUC), the area under the precision-recall curve (AUPR), F1 score (F1), precision, and recall were calculated to evaluate the performance of the models we have designed utilizing the 5-fold cross-validation technique.

3.2 Method Comparison

We compared HF-DDI with several prediction methods to demonstrate the advantages of our model: (1)DeepDDI: a computational neural network with eight hidden layers of 2048 nodes that uses the molecular structure information of drug pairs as inputs to predict their interaction types between drug combinations [21]. (2)CNN-DDI: a convolutional neural network that uses pathway, target, enzyme, and category information to predict DDIs [33]. Since our dataset lacks pathway and category information, we apply the substructure replicate method to allow for a fair comparison. (3)DDIMDL: a multimodal framework based on deep learning that combines diverse drug features for drug-drug interaction event prediction [5]. We also took into account several well-liked classification techniques, including random forests (RF), K-nearest neighbors algorithm (KNN), and Decision Trees (DT) as baseline methods.

The output results for the evaluated performance metrics are displayed in Table 1, which is evident that the proposed method performs better than the existing methods for all of the performance measures. Although we used simple deep neural networks as classifiers, the results of the 5-fold cross-validation achieved an accuracy of 0.948, which shows the powerful ability of HF-DDI in predicting drug-drug interaction events. To further demonstrate the effectiveness of the framework we designed, the AUC and AUPR scores of several approaches for 100 different types of events were displayed using Fig. 4. It is worth mentioning that these boxplots unequivocally demonstrate that HF-DDI yields superior performances than comparison approaches.

3.3 Performance Evaluation

In this study, the main parameters affecting the performance of the model involve the number of layers and the number of neurons in the deep neural network. When the model is too complex, it can start to learn the irrelevant information within the dataset and become overfitted. However, if the model has too few parameters or excludes too many important features, the model may encounter

Table 1. Performance comparison

	ACC	AUC	AUPR	F1	Precision	Recall
HF-DDI	0.948	1.000	0.986	0.928	0.942	0.918
DDIMDL	0.928	0.999	0.966	0.909	0.929	0.896
DeepDDI	0.918	0.999	0.969	0.894	0.906	0.884
CNN-DDI	0.794	0.998	0.868	0.777	0.795	0.772
RF	0.775	0.997	0.842	0.635	0.831	0.548
KNN	0.762	0.985	0.832	0.701	0.856	0.622
DT	0.725	0.964	0.781	0.590	0.764	0.503

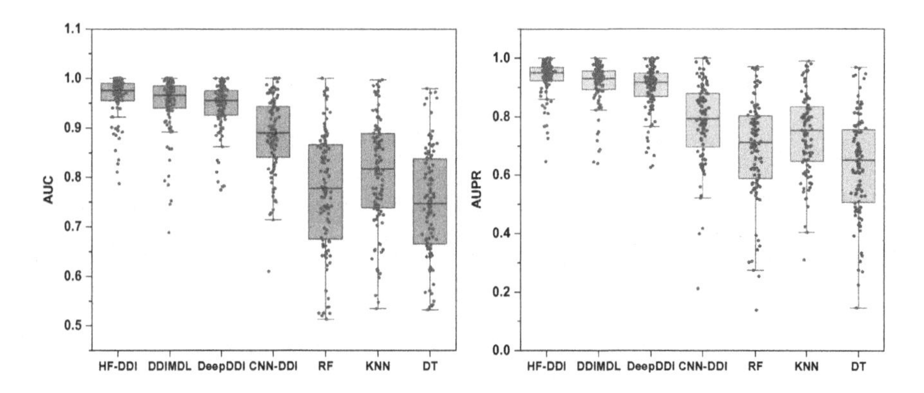

Fig. 4. Boxplots showing the distribution of the AUC and AUPR of for each event

the opposite problem, and instead, underfitting occurs. In order to find the optimal neural network architecture, the simplest strategy is to start with a model that has just one hidden layer between the input layer and the output layer, and then add hidden layers. We set the number of neurons in the last hidden layer to a fixed number of 128 and consider 1, 2, 3, and 4 hidden layers. The number of neurons in the former layer is twice as many as in the latter layer. According to Fig. 5, as the number of hidden layers is changed, the performance of the model does not vary dramatically, and the network architecture with 3 hidden layers can achieve the best accuracy. Furthermore, we evaluated the importance of different drug features and feature combinations for interaction prediction. We trained prediction models based on individual features or combinations of features and the experiment results are shown in Tables 2. We observed that the substructure obtains an accuracy of 0.931 and seems to be the most instructive. Comparing the combination of features to the single features, a significant improvement is provided.

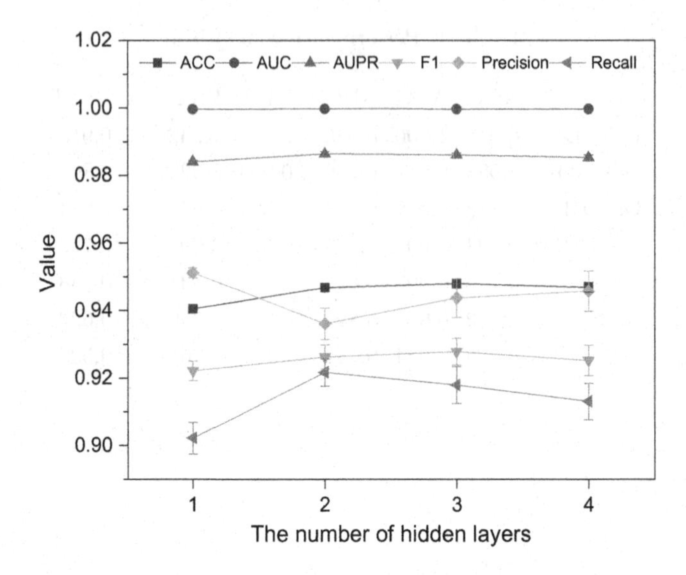

Fig. 5. The effect of different number of hidden layers on model performance

Table 2. The performance of HF-DDI while using various features and feature combinations.

	ACC	AUC	AUPR	F1	Precision	Recall
S	0.931	1.000	0.980	0.908	0.915	0.903
T	0.856	0.999	0.941	0.841	0.841	0.846
E	0.667	0.994	0.765	0.478	0.598	0.430
S+T	0.935	1.000	0.974	0.918	0.928	0.911
S+E	0.933	0.999	0.957	0.907	0.925	0.894
T+E	0.886	0.999	0.933	0.855	0.873	0.847

Note: S, substructure; T, target; E, enzyme

4 Conclusion

This study collected heterogeneous drug information and interaction data from DrugBank with 323,539 pairwise interactions between the 1258 drugs. We investigated and compared different fusion strategies and presented a multimodal deep learning framework with the structure of combing two different levels of fusion strategies, which uses multiple drug similarity features to predict drug-drug interaction events. Our method effectively exploits complementary and cooperative features of several modalities to produce excellent prediction accuracy and reliability. The results of the comparison experiment showed that our proposed method has achieved better performance than the state-of-the-art approaches in the task of predicting interaction events.

There are further research should be carried out to validate the effectiveness of our presented method. As more heterogeneous data becomes available, we will embed more drug feature data and biological domain knowledge to enhance the reliability and interpretability of the method, which is an important issue for future research. Furthermore, We will explore further applications and research on multimodal data fusion strategies, applying such strategies discussed to other tasks in the biomedical field to demonstrate their generalizability.

Acknowledgement. This work was supported by NSFC Grants (62262011, U19A2067, 61762031); Natural Science Foundation of Guangxi (2021JJA170130); Science Foundation for Distinguished Young Scholars of Hunan Province (2020JJ2009); National Key R&D Program of China (2017YFB0202602, 2018YFC0910405, 2017YFC13 11003, 2016YFC1302500); Science Foundation of Changsha (Z202069420652, kq2004010; JZ20195242029, JH20199142034); The Funds of State Key Laboratory of Chemo/Biosen-sing and Chemometrics and Peng Cheng Lab.

References

1. Cami, A., Manzi, S., Arnold, A., Reis, B.Y.: Pharmacointeraction network models predict unknown drug-drug interactions. PloS One **8**(4), e61468 (2013)
2. Capecchi, A., Probst, D., Reymond, J.-L.: One molecular fingerprint to rule them all: drugs, biomolecules, and the metabolome. J. Cheminform. **12**(1), 1–15 (2020). https://doi.org/10.1186/s13321-020-00445-4
3. Cheng, F., Zhao, Z.: Machine learning-based prediction of drug-drug interactions by integrating drug phenotypic, therapeutic, chemical, and genomic properties. J. Am. Med. Inform. Assoc. **21**(e2), e278–e286 (2014)
4. Deng, Y., et al.: Meta-DDIE: predicting drug-drug interaction events with few-shot learning. Briefings Bioinform. **23**(1), bbab514 (2022)
5. Deng, Y., Xu, X., Qiu, Y., Xia, J., Zhang, W., Liu, S.: A multimodal deep learning framework for predicting drug-drug interaction events. Bioinformatics **36**(15), 4316–4322 (2020)
6. Feeney, A., et al.: Relation matters in sampling: a scalable multi-relational graph neural network for drug-drug interaction prediction. arXiv preprint arXiv:2105.13975 (2021)
7. Feng, Y.Y., Yu, H., Feng, Y.H., Shi, J.Y.: Directed graph attention networks for predicting asymmetric drug-drug interactions. Brief. Bioinform. **23**(3) (2022)
8. Ferdousi, R., Safdari, R., Omidi, Y.: Computational prediction of drug-drug interactions based on drugs functional similarities. J. Biomed. Inform. **70**, 54–64 (2017)
9. Gottlieb, A., Stein, G.Y., Oron, Y., Ruppin, E., Sharan, R.: Indi: a computational framework for inferring drug interactions and their associated recommendations. Mol. Syst. Biol. **8**(1), 592 (2012)
10. Kastrin, A., Ferk, P., Leskošek, B.: Predicting potential drug-drug interactions on topological and semantic similarity features using statistical learning. PloS one **13**(5), e0196865 (2018)
11. Kim, E., Nam, H.: Deside-ddi: interpretable prediction of drug-drug interactions using drug-induced gene expressions. J. Cheminform. **14**(1), 1–12 (2022)
12. Lin, X., Quan, Z., Wang, Z.J., Ma, T., Zeng, X.: KGNN: knowledge graph neural network for drug-drug interaction prediction. In: IJCAI, vol. 380, pp. 2739–2745 (2020)

13. Liu, L., et al.: Analysis and prediction of drug-drug interaction by minimum redundancy maximum relevance and incremental feature selection. J. Biomol. Struct. Dyn. **35**(2), 312–329 (2017)
14. Lyu, T., Gao, J., Tian, L., Li, Z., Zhang, P., Zhang, J.: MDNN: a multimodal deep neural network for predicting drug-drug interaction events. In: International Joint Conferences on Artifical Intelligence (2022)
15. Mei, S., Zhang, K.: A machine learning framework for predicting drug-drug interactions. Sci. Rep. **11**(1), 1–12 (2021)
16. Nair, V., Hinton, G.E.: Rectified linear units improve restricted boltzmann machines. In: Icml (2010)
17. Nyamabo, A.K., Yu, H., Liu, Z., Shi, J.Y.: Drug-drug interaction prediction with learnable size-adaptive molecular substructures. Brief. Bioinform. **23**(1), bbab441 (2022)
18. Park, K., Kim, D., Ha, S., Lee, D.: Predicting pharmacodynamic drug-drug interactions through signaling propagation interference on protein-protein interaction networks. PloS one **10**(10), e0140816 (2015)
19. Qiu, Y., Zhang, Y., Deng, Y., Liu, S., Zhang, W.: A comprehensive review of computational methods for drug-drug interaction detection. IEEE/ACM Trans. Comput. Biol. Bioinform. (2021)
20. Ramachandram, D., Taylor, G.W.: Deep multimodal learning: a survey on recent advances and trends. IEEE Sign. Process. Mag. **34**(6), 96–108 (2017)
21. Ryu, J.Y., Kim, H.U., Lee, S.Y.: Deep learning improves prediction of drug-drug and drug-food interactions. Proc. Natl. Acad. Sci. **115**(18), E4304–E4311 (2018)
22. Shen, Y., et al.: Drug2vec: knowledge-aware feature-driven method for drug representation learning. In: 2018 IEEE International Conference on Bioinformatics and Biomedicine (BIBM), pp. 757–800. IEEE (2018)
23. Shi, J.Y., Li, J.X., Gao, K., Lei, P., Yiu, S.M.: Predicting combinative drug pairs towards realistic screening via integrating heterogeneous features. BMC Bioinform. **18**(12), 1–9 (2017)
24. Shi, J.Y., et al.: Predicting combinative drug pairs via multiple classifier system with positive samples only. Comput. Meth. Prog. Biomed. **168**, 1–10 (2019)
25. Sridhar, D., Fakhraei, S., Getoor, L.: A probabilistic approach for collective similarity-based drug-drug interaction prediction. Bioinformatics **32**(20), 3175–3182 (2016)
26. Srivastava, N., Hinton, G., Krizhevsky, A., Sutskever, I., Salakhutdinov, R.: Dropout: a simple way to prevent neural networks from overfitting. J. Mach. Learn. Res. **15**(1), 1929–1958 (2014)
27. Stahlschmidt, S.R., Ulfenborg, B., Synnergren, J.: Multimodal deep learning for biomedical data fusion: a review. Brief. Bioinform. **23**(2), bbab569 (2022)
28. Takarabe, M., Shigemizu, D., Kotera, M., Goto, S., Kanehisa, M.: Network-based analysis and characterization of adverse drug-drug interactions. J. Chem. Inf. Model. **51**(11), 2977–2985 (2011)
29. Vilar, S., et al.: Similarity-based modeling in large-scale prediction of drug-drug interactions. Nat. Protocols **9**(9), 2147–2163 (2014)
30. Wishart, D.S., et al.: Drugbank 5.0: a major update to the drugbank database for 2018. Nucleic Acids Res. **46**(D1), D1074–D1082 (2018)
31. Witten, I.H., Frank, E., Hall, M.A., Pal, C.J.: Chapter 10 - deep learning. In: Data Mining 4th edn., pp. 417–466. Morgan Kaufmann, (2017)
32. Yu, H., et al.: Predicting and understanding comprehensive drug-drug interactions via semi-nonnegative matrix factorization. BMC Syst. Biol. **12**(1), 101–110 (2018)

33. Zhang, C., Zang, T.: CNN-DDI: a novel deep learning method for predicting drug-drug interactions. In: 2020 IEEE International Conference on Bioinformatics and Biomedicine (BIBM), pp. 1708–1713. IEEE (2020)
34. Zhang, P., Wang, F., Hu, J., Sorrentino, R.: Label propagation prediction of drug-drug interactions based on clinical side effects. Sci. Rep. **5**(1), 1–10 (2015)
35. Zhang, W., Chen, Y., Li, D., Yue, X.: Manifold regularized matrix factorization for drug-drug interaction prediction. J. Biomed. Inform. **88**, 90–97 (2018)
36. Zhu, J., Liu, Y., Zhang, Y., Li, D.: Attribute supervised probabilistic dependent matrix tri-factorization model for the prediction of adverse drug-drug interaction. IEEE J. Biomed. Health Inform. **25**(7), 2820–2832 (2020)

GNN-Dom: An Unsupervised Method for Protein Domain Partition via Protein Contact Map

Lei Wang(✉) and Yan Wang

School of Life Science and Technology, Huazhong University of Science and Technology, Wuhan, China
wanglei94@hust.edu.cn

Abstract. Protein domains are the basic building blocks of protein structures, which fold and function independently. Protein domain partition could be used as measure to decompose the modeling of a large, multi-domain protein in to smaller segments and has great significance for protein structure prediction. Although many methods have been proposed for proteins domain partition, there is still room for improvement. In this work, we propose an unsupervised method (GNN-Dom) by applying Graph Convolutional Neural Network on the protein contact map. We show GNN-Dom has competitive performance against state-of-the-art baselines across the CASP 14 test proteins, especially in discontinuous-domain prediction.

Keywords: Protein domain · Unsupervised method

1 Introduction

Protein domains are the basic building blocks of protein structures, which fold and function independently. The domain partition of proteins is an essential step for determining their structural folds and understanding their biological functions. Furthermore, the protein structure prediction tools such as trRosetta [26] and I-TASSER [17] are limited in single domain protein. Even though alphafold2 has made a great breakthrough in the field of protein structure prediction, its prediction effect on the entire protein is not as good as the protein domain level [10]. Protein domain partition could be used as measure to decompose the modeling of a large, multi-domain protein in to smaller segments.

Due to the importance of the protein domain partition, many methods of domain partition have been proposed to determine the domain boundaries of proteins. One class of methods delineates and defines domains directly from the experimental structures of proteins; these methods include PDP [1], DomainParser [6], DDOMAIN [28] and SWORD [15]. Another important class of methods predicts domain boundaries from the protein sequences. These domain prediction methods can typically be categorized into three general groups. The

M. S. Bansal et al. (Eds.): ISBRA 2022, LNBI 13760, pp. 286–294, 2022.
https://doi.org/10.1007/978-3-031-23198-8_26

first group is mainly based on deep learning, including DOMPro [3], DoBo [4], ConDo [8] and DNN-Dom [18]. The second group of methods is based on structural templates detected from the PDB, typically by threading [19]. For example, ThreaDom [25] deduces domain boundary locations based on multiple threading alignments [24]. Because ThreaDom cannot directly detect discontinuous domains beyond the templates, an extended version, ThreaDomEx [23], was further developed to assign discontinuous domains by domain-segment assembly. Here, a discontinuous domain is defined as a domain that contains two or more segments from separate regions of the query sequence. The third group is based on 3D structure prediction [2,5]. Finally, the last group is based on the accurate protein contact map such as FUpred [27].

Despite their successes, each of these methods have their own limitations. The methods, which predicts domains from experimental protein structures, e.g. generally have higher accuracies than the methods that start from sequences but can only be applied to a small portion of proteins that have known experimental structures. For deep learning-based methods, the accuracy of prediction is often not high and could not conduct the discontinuous-domain partition. The threading-based methods generally have higher accuracy when close templates are identified, but the domain partition will be worse for targets lacking homologous templates. 3D model-based methods depends on the prediction quality of the 3D models, which can only be limited to proteins with short lengths. Furthermore, most approaches cannot deal with the prediction of discontinuous multi-domains, except for ThreaDomEx and FUpred.

In this work, we propose a new unsupervised method, named GNN-Dom for protein domain partition from protein sequences based on contact map prediction, partly motivated by the quick progress recently achieved in the field of contact prediction [16]. Although there are some methods that utilize contact information as a machine learning feature to help predict domains, GNN-Dom is the first unsupervised method to deduce protein domain partition from contact map predictions. In particular, the case study demonstrates that GNN-Dom can accurately predict the discontinuous-domain proteins with complex domain structures.

2 Method

In this section, we present our proposed GNN-dom approach. We first develop a graph auto-encoder which effectively integrates both structure and sequence information to learn a latent representation. Based on the representation, self-training module and Conditional Random Fields (CRF) are proposed to guide the protein domain partition algorithm towards better performance.

2.1 Protein Contact Graph

To aggregate inter-residue information from proteins, we construct a protein contact graph based ESM-MSA-1 [16]. In order to get a sparse graph $\mathbf{G}$, we choose a fixed cut-off weight for adjacency matrix $\mathbf{A}$.

Note that **G** preserves the residue contact from the three-dimensional structure. In addition to the three-dimensional space, it also has a position relationship in the sequence. For the target residue, the update information between the target and direct neighbors should consider the position encoding. For node feature information, we encode the protein-specific information into the encoder network's input features from two aspects:

1) Amino-acid sequence one-hot encoding: For an amino-acid sequence of length L, we encode it into a $L * 20$ matrix of one-hot encoding vectors.
2) Protein secondary structure one-hot encoding: The 3-class SS (Helix, Strand, Loop) are predicted using Spider3 [7], and we encode it into $L * 3$ matrix.

2.2 Graph Convolutional Encoding

Here, we firstly encode the position-wise information $\mathbf{H}^0$ by one-layer convolutional network [9]. Then, two-layer graph convolutional network was used to aggregate the residue similarity information in a simple propagation equation [11]. GCN could be considered as message passing across connected nodes where messages are node attribute feature. The GCN model follows this formula:

$$\mathbf{H}^{(l+1)} = \sigma\left(\tilde{\mathbf{D}}^{-\frac{1}{2}}\tilde{\mathbf{A}}\tilde{\mathbf{D}}^{-\frac{1}{2}}\mathbf{H}^{(l)}\mathbf{W}^{(l)}\right) \quad (1)$$

where $\mathbf{H}^{(l)}$ denotes the lth layer in the network, σ is the non-linearity, and $\mathbf{W}^l$ is the weight matrix for this layer. $\tilde{\mathbf{D}}$ and $\tilde{\mathbf{A}}$ are the renormalized degree and adjacency matrices of the protein contact map **G**.

2.3 Inner Product Decoder

In order to ensure that the protein latent embedding can truly represent the original protein contact map, we need to reconstruct protein contact map based on embedding information. As our latent embedding of protein residues already combines both protein sequence and potential structure information, we choose to adopt a simple inner product decoder method [12] to predict the space interaction between protein residues:

$$\hat{A}_{ij} = sigmoid\left(h_i^\top h_j\right) \quad (2)$$

where $\hat{A}$ is the predicted structure matrix of the protein contact graph.

2.4 Self-optimizing Training

Protein domain partitioning is considered a kind of graph clustering. One of the main challenges of graph clustering methods is that there is no true domain guidance. To confront this challenge, we adopt a self-optimizing training algorithm from [22]. Apart from optimizing the reconstruction error, the hidden embedding is fed into a self-optimizing clustering module, which minimizes the following objective:

$$L_c = \sum_i \sum_u p_{iu} \log \frac{p_{iu}}{q_{iu}} \tag{3}$$

where q_{iu} measures the similarity between protein residue embedding h_i and each domain cluster center embedding μ_u. The initial cluster domain embedding centers are obtained through the simple K-means algorithm, but the cluster centers are gradually optimized with training each time. For K values, we chose the silhouette coefficient as the measure. The K corresponding to the largest average silhouette score is the best number of clusters. It is measured with a Student's t-distribution pattern so that it could handle different scaled protein domains with different number of residues and is defined as:

$$q_{iu} = \frac{\left(1 + \|z_i - \mu_u\|^2\right)^{-1}}{\sum_k \left(1 + \|z_i - \mu_k\|^2\right)^{-1}}, \tag{4}$$

it can be seen as a soft clustering assignment distribution of each residue belonging to the corresponding domains. Here, p_{iu} is the target distribution of residues from all domains which is defined as:

$$p_{iu} = \frac{q_{iu}^2 / \sum_i q_{iu}}{\sum_k \left(q_{ik}^2 / \sum_i q_{ik}\right)} \tag{5}$$

2.5 Conditional Random Fields

Considering that the protein partition should be continuous in protein sequence, the residues are assigned by using Conditional Random Fields to strengthen the correlation between protein sequence cluster consistency. Conditional random fields (CRFs) were proposed by [13] for labeling sequence data. Given a protein sequence (length is T) of observations $X = (x_1, x_2, \ldots, x_T)$, the most probable clustering as soft label is $Y = (y_1, y_2, \ldots, y_T)$. By the fundamental theorem of a random field, the joint distribution over label sequence Y given X can be given by the following conditional probability:

$$P(Y|X) = \frac{1}{Z(h)} \prod_{t=1}^{T} exp(h_t) \prod_{t=1}^{T-1} exp\left(\varphi_{y_t,\ y_{t+1}}\right) \tag{6}$$

where $h = (h_1,\ h_2, \ldots, h_T)$ is the embedding of the encoder network directly below the conditional random field, $Z(h)$ is the normalization constant of the distribution $P(Y|X)$, obtains outputs of m clusters and $\varphi_{y_t,\ y_{t+1}}$ is a learnable transition matrix with $m \times m$ parameters which is initialized as a identity matrix. The use of the identity matrix for the transition matrix is equivalent to enhancing an inductive bias to ensure the continuity of the sequence, which can make the model easier to converge.

2.6 Joint Training and Protein Domain Partition Optimization

We jointly optimize the graph auto-encoder embedding reconstruction loss, clustering learning and partition consistency, and define our total objective function as:

$$L = L_r + \gamma L_c + \delta L_s \tag{7}$$

where L_r,L_c and L_s are the reconstruction loss, clustering loss and CRF loss respectively, γ and δ are the coefficient that controls the balance of loss. It is worth mentioning that we could gain our clustering result directly from the most probable clustering $Y^* = argmax_Y P(Y|X)$.

3 Experiment

3.1 Setup

Datasets and Data Processing. We select all 14 multi-domain and discontinuous-domain test proteins from the CASP14 [14] in which the discontinuous-domain prediction of proteins is the most difficult task. All domain definitions for these proteins can be seen from Table 1. In order to construct the protein contact graph, the cut-off of inter-residues weight is 0.01. In these test proteins, the number of discontinuous-domain proteins is 5. The definition of the protein structure domain comes from the official CASP definition.

Table 1. All 14 test protein domain definition of CASP 14

CASP ID	Length	Number of domains	Definition of domains
T1024	408	2	(2-194)(203-406)
T1030	273	2	(1–154)(155-273)
T1052	832	3	(1-539)(540-588;669–832)(589-668)
T1053	580	2	(2-406)(407-577)
T1058	382	2	(1-41;117-238;325-382)(42-116;239-324)
T1070	335	4	(4-79)(80-180)(181-256)(265-332)
T1085	588	3	(173-339)(340-521)(524-580)
T1086	408	2	(1-216)(221-408)
T1092	426	2	(1-245)(246-426)
T1093	631	3	(1-143)(144-401;508-631)(402-507)
T1094	496	2	(1-126;334-484)(127-333)
T1096	464	2	(6-260)(294-464)
T1100	338	2	(1-53;220-337)(54-219)
T1101	318	2	(12-94)(95-318)

Evaluation Metrics. In this paper, we employ the definitions of NDO-score [20] in CASP 6 , DBD-score [21] in CASP 7, Precision-B and Recall-B [18] for evaluating domain partition accuracy. Furthermore, we also compare the number Q value of the correct prediction (boundaries error ± 20) of discontinuous-domain proteins.

Table 2. Performance comparisons of protein domain partition on the 14 test data of CASP 14.

Methods	Pre-B	Rec-B	NDO	DBD	Q
FUpred	0.338	0.396	0.516	0.470	0
ThreaDomEx	0.75	0.208	0.526	0.557	1
GNN-Dom	0.32	0.434	0.456	0.521	2

Baselines. To compare GNN-dom's performance for protein domain partition, especial discontinuous-domain, we choose two state-of-the-art methods: 1) FUpred, 2) ThreaDomEx, which could predict the discontinuous-domain based on protein contact map and protein templates respectively.

Parameter Settings. For our method, we set the clustering coefficient γ to 0.5 and CRF loss coefficient δ to 5. The size of amino-acid and secondary structure embedding layers are 4, 8 respectively. The kernel size of one-layer convolutional network is 32 and the GCN encoder is also 32.

3.2 Result

GNN-Dom Is a Robust, Competitive Protein Domain Partition Algorithm. The prediction result measured by above metrics is reported in Table 2. We see that in Precision-B, ThreaDomEX achieved the highest performance, the performance of FUpred and GNN-dom is close. In Recall-B, GNN-Dom's performance (0.434) is better than FUpred (0.396) and ThreaDomEX (0.208). For the NDO score, GNN-Dom's performance (0.456) is worse than FUpred (0.516) and ThreaDomEX (0.526). For the DBD score, GNN-Dom's performance (0.521) is worse than ThreaDomEX (0.557), better than FUpred (0.470). This suggests GNN-Dom has very good predictive performances, and in often times, better predictions. For the performance of 5 discontinuous-domain test proteins, GNN-Dom successfully predicts 2 proteins (T1058 and T1100), ThreaDomEX only predicts one protein (T1094) and FUpred could not predict any discontinuous-domain test proteins. In sum, GNN-Dom achieves the state-of-the-art performance in discontinuous-domain prediction.

Case Study. We select the most complex discontinuous-domain protein (T1058) in CASP14: the first domain includes three discontinuous segments and the second domain includes two discontinuous segments. T1058 is a membrane protein of Pseudomonas aeruginosa, which includes 4 transmebrane Helix segments in

Table 3. T1058 domain partition at three different methods.

CASP ID	T1058
CASP 14	(1-41,117-238,325-382)(42-116,239-324)
GNN-Dom	(1-36,127-228,338-382)(37-126,229-337)
ThreaDomEX	(1-382)
FUpred	(1-382)

the first domain. The detail definition is in Table 3. In order to find out the predicted domain partition in 3D space, we use different colors to represent the predicted output of two discontinuous domains in Fig. 1. We also report the results of three different methods in Table 2. Within the margin of error (±20), GNN-Dom predicts all discontinuous domains in T1058. Considering that the size of the T1058 protein is small (382 residues) and two domains in 3D space are close, it is difficult to predict the discontinuous domain for ThreaDomEX and FUpred, which regard the T1058 as a single-domain protein.

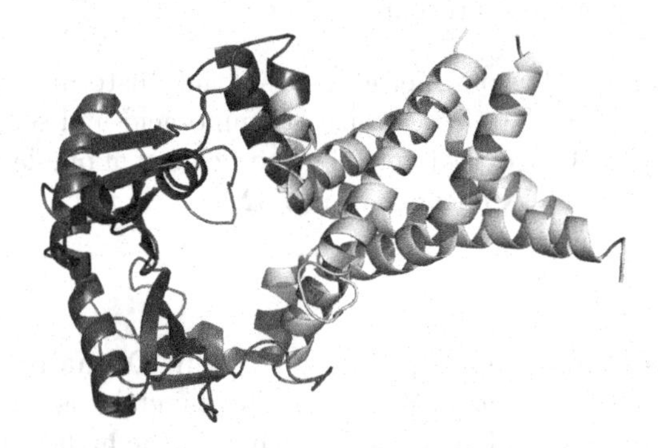

Fig. 1. The PDB structure of T1058 (PDB ID: 7ABW): the green part represents the first discontinuous domain and the blue part represents the second discontinuous domain. (Color figure online)

4 Conclusion

In this work, we present a novel, simple, unsupervised and efficient computational method that unifies protein contact map, secondary structure and sequence and also work well for predicting the discontinuous-domain proteins. GNN-Dom achieves great performance and improves the prediction of discontinuous-domain proteins. Our model adopts predicted protein secondary structures as input feature, and this brings some error for more accurate boundaries. For example, the boundaries of domains is usually located at the linker regions, but some tools give a not accurate boundary prediction. The protein contact graph plays the role of inductive bias, prompting the representation of adjacent residues to be similar, and the use of graph autoencoder is one of the core of the our model. Furthermore, self-training can enhance that our domain center emebedding can better adapt to the representation of graph. In our practice, CRF can indeed guarantee the continuity of the sequence, it is one of the very important components. Although GNN-Dom has achieved some good results, its boundary partition is still not accurate enough. Here, it needs to be emphasized that ThreaDomEx is still excellent if there is a good template and GNN-Dom also has a great advantage over ThreaDomEx in terms of computational efficiency. In the furture, we will consider more accurate protein contact (distance) map and secondary structure feature to reduce the limitations.

References

1. Alexandrov, N., Shindyalov, I.: Pdp: protein domain parser. Bioinformatics **19**(3), 429–430 (2003)
2. Cheng, J.: Domac: an accurate, hybrid protein domain prediction server. Nucleic Acids Res. **35**(suppl_2), W354–W356 (2007)
3. Cheng, J., Sweredoski, M.J., Baldi, P.: Dompro: protein domain prediction using profiles, secondary structure, relative solvent accessibility, and recursive neural networks. Data Min. Knowl. Disc. **13**(1), 1–10 (2006)
4. Eickholt, J., Deng, X., Cheng, J.: Dobo: protein domain boundary prediction by integrating evolutionary signals and machine learning. BMC Bioinform. **12**(1), 1–8 (2011)
5. George, R.A., Heringa, J.: Snapdragon: a method to delineate protein structural domains from sequence data. J. Mol. Biol. **316**(3), 839–851 (2002)
6. Guo, J.t., Xu, D., Kim, D., Xu, Y.: Improving the performance of domainparser for structural domain partition using neural network. Nucleic Acids Res. **31**(3), 944–952 (2003)
7. Heffernan, R., Yang, Y., Paliwal, K., Zhou, Y.: Capturing non-local interactions by long short-term memory bidirectional recurrent neural networks for improving prediction of protein secondary structure, backbone angles, contact numbers and solvent accessibility. Bioinformatics **33**(18), 2842–2849 (2017)
8. Hong, S.H., Joo, K., Lee, J.: Condo: protein domain boundary prediction using coevolutionary information. Bioinformatics **35**(14), 2411–2417 (2019)
9. Islam, M.A., Jia, S., Bruce, N.D.B.: How much position information do convolutional neural networks encode? CoRR abs/2001.08248 (2020)
10. Jumper, J., et al.: Highly accurate protein structure prediction with AlphaFold. Nature **596**(7873), 583–589 (2021). https://doi.org/10.1038/s41586-021-03819-2
11. Kipf, T.N., Welling, M.: Semi-supervised classification with graph convolutional networks. arXiv preprint arXiv:1609.02907 (2016)
12. Kipf, T.N., Welling, M.: Variational graph auto-encoders. arXiv preprint arXiv:1611.07308 (2016)
13. Lafferty, J., McCallum, A., Pereira, F.C.: Conditional random fields: Probabilistic models for segmenting and labeling sequence data. Numerical Analysis and Scientific Computing Commons (2001)
14. Moult, J.: 14th community wide experiment on the critical assessment of techniques for protein structure prediction. CASP 14 (2020)
15. Postic, G., Ghouzam, Y., Chebrek, R., Gelly, J.C.: An ambiguity principle for assigning protein structural domains. Sci. Adv. **3**(1), e1600552 (2017)

16. Rao, R., Liu, J., Verkuil, R., Meier, J., Canny, J.F., Abbeel, P., Sercu, T., Rives, A.: Msa transformer. bioRxiv (2021)
17. Roy, A., Kucukural, A., Zhang, Y.: I-tasser: a unified platform for automated protein structure and function prediction. Nat. Protoc. **5**(4), 725–738 (2010)
18. Shi, Q., et al.: Dnn-dom: predicting protein domain boundary from sequence alone by deep neural network. Bioinformatics **35**(24), 5128–5136 (2019)
19. Söding, J.: Protein homology detection by hmm-hmm comparison. Bioinformatics **21**(7), 951–960 (2005)
20. Tai, C.H., Lee, W.J., Vincent, J.J., Lee, B.: Evaluation of domain prediction in casp6. PROTEINS: Struct. Function Bioinform. **61**(S7), 183–192 (2005)
21. Tress, M., et al.: Assessment of predictions submitted for the casp7 domain prediction category. Proteins Struct. Function Bioinform. **69**(S8), 137–151 (2007)
22. Wang, C., Pan, S., Hu, R., Long, G., Jiang, J., Zhang, C.: Attributed graph clustering: a deep attentional embedding approach. arXiv preprint arXiv:1906.06532 (2019)
23. Wang, Y., Wang, J., Li, R., Shi, Q., Xue, Z., Zhang, Y.: Threadomex: a unified platform for predicting continuous and discontinuous protein domains by multiple-threading and segment assembly. Nucleic Acids Res. **45**(W1), W400–W407 (2017)
24. Wu, S., Zhang, Y.: Lomets: a local meta-threading-server for protein structure prediction. Nucleic Acids Res. **35**(10), 3375–3382 (2007)
25. Xue, Z., Xu, D., Wang, Y., Zhang, Y.: Threadom: extracting protein domain boundary information from multiple threading alignments. Bioinformatics **29**(13), i247–i256 (2013)
26. Yang, J., Anishchenko, I., Park, H., Peng, Z., Ovchinnikov, S., Baker, D.: Improved protein structure prediction using predicted interresidue orientations. Proc. Natl. Acad. Sci. **117**(3), 1496–1503 (2020)
27. Zheng, W., Zhou, X., Wuyun, Q., Pearce, R., Li, Y., Zhang, Y.: Fupred: detecting protein domains through deep-learning-based contact map prediction. Bioinformatics **36**(12), 3749–3757 (2020)
28. Zhou, H., Xue, B., Zhou, Y.: Ddomain: dividing structures into domains using a normalized domain-domain interaction profile. Protein Sci. **16**(5), 947–955 (2007)

A Locality-Constrained Linear Coding-Based Ensemble Learning Framework for Predicting Potentially Disease-Associated MiRNAs

Yi Shen[1], Ying-Lian Gao[2](✉), Shu-Zhen Li[1], Boxin Guan[1], and Jin-Xing Liu[1]

[1] School of Computer Science, Qufu Normal University, Rizhao 276826, Shandong, China
[2] Qufu Normal University Library, Qufu Normal University, Rizhao 276826, Shandong, China
yinliangao@126.com

Abstract. With the development of biology and scientific technologies, miRNAs are gradually revealed to be associated with the generation of complex diseases. Nevertheless, predicting miRNA-disease associations (MDAs) by traditional wet experiments requires significant resource consumption. Therefore, identifying novel MDAs through effective computational models aids in the treatment and intervention of diseases. In this paper, a locality-constrained linear coding-based ensemble learning framework is proposed to predict potentially disease-associated miRNAs (LLCELF). Specifically, LLCELF ensembles multi-source biological information while preserving the local structure. In addition, the hypergraph-regular term is introduced in the learning process to establish complex relationships between samples. As a result, the AUC of LLCELF under five-fold cross validation (CV) reaches 0.9533, which outperforms the state-of-the-art methods. Further, all of the top 20 novel miRNAs associated with Kidney Neoplasms are validated experimentally in HMDD v3.2, miRCancer, and the existing biological studies. It is concluded that LLCELF serves as an effective model for predicting potential MDAs.

Keywords: MicroRNA · Disease · Association prediction · Ensemble learning

1 Introduction

MicroRNAs (miRNAs), which are non-coding RNAs (about 19–25 nucleotides), play an indispensable role in life processes [1]. Since miRNAs control gene expression by complementing the targeted mRNA sequences, abnormal expression of miRNAs can lead to a variety of complex human diseases. In short, identifying miRNA-disease associations (MDAs) makes a significant contribution to the treatment of diseases.

To date, various computational methods have been proposed to predict MDAs. For example, Ding et al. [2] embedded hypergraphs in a bipartite local model to predict associations (HGBLM). To enhance the robustness of the similarity, Qu et al. [3] utilized locality-constrained linear coding (LLC) to reconstruct similarities and obtained related scores based on label propagation (LLCMDA).

M. S. Bansal et al. (Eds.): ISBRA 2022, LNBI 13760, pp. 295–302, 2022.
https://doi.org/10.1007/978-3-031-23198-8_27

In this study, a locality-constrained linear coding-based ensemble learning framework is proposed to predict miRNA-disease associations (LLCELF). The main contributions of LLCELF are as follows: (1) Locality-constrained linear coding (LLC) reconstructs robust similarity networks based on interaction profile (IP). (2) Multi-source biological information in miRNA (disease) space is fed into the ensemble learning to improve the comprehensiveness of prediction. (3) Instead of simple graph structure, hypergraph-regular terms are introduced into the ensemble learning framework to describe higher-order relationships between samples.

2 Materials and Methods

2.1 Human MiRNA-Disease Associations

We obtained 5430 experimentally validated associations between 495 miRNAs and 383 diseases in HMDD v2.0 [4]. Here, the sparse matrix **M** is utilized to represent MDAs. If miRNA m_i is associated with disease d_j, $\mathbf{M}(m_i, d_j)= 1$ otherwise 0.

2.2 MiRNA Space

To better characterize miRNAs from different biological data sources, miRNA features (i.e., IP feature and sequence feature) and similarities (i.e., IP similarity, sequence similarity, and functional similarity) are considered [5].

Firstly, IP is applied to construct feature vector based on the association matrix. Since LLC can retain local information during encoding, IP features are input into LLC to reconstruct IP similarity [3]. The objective function of LLC is expressed as:

$$\begin{gathered}\arg_{z_i}\|x_i - Dz_i\|_2^2 + \mu\|\mathbf{S}_i \odot z_i\|_2^2 \\ s.t.\ I^T z_i = 1,\end{gathered} \tag{1}$$

where x_i denotes the descriptor of the i-th sample and D is a dictionary matrix. Besides, μ represents the regularization parameter and $\odot$ represents the element-wise multiplication. $\mathbf{S}_i$ is the local adapter vector that denotes the distance between the i-th sample and other samples. The equation of $\mathbf{S}_i$ is as follows:

$$\mathbf{S}_{i,j} = \exp\left(\frac{\|\mathbf{M}(i,:) - \mathbf{M}(j,:)\|_2}{\sigma}\right), \tag{2}$$

where $\mathbf{M}(i,:)$ and $\mathbf{M}(j,:)$ correspond to the i-th and j-th feature vectors of **M**, respectively. σ is the kernel bandwidth parameter [3]. By minimizing the Lagrangian function of Eq. (1), the reconstructed IP similarity $\mathbf{MS}_1$ is obtained.

Secondly, miRNA sequence similarity $\mathbf{MS}_2$ is constructed based on the sequence alignment results calculated by the Needleman-Wunsch algorithm [6]. Finally, based on the hypothesis that diseases with similar phenotypes may be related to functionally similar miRNAs, miRNA functional similarity is calculated by Wang et al. [7]. Here, miRNA functional similarity is represented as $\mathbf{MS}_3$.

2.3 Disease Space

In the disease space, disease IP feature and multiple types of similarities (i.e., IP similarity, disease semantic similarity 1, and disease semantic similarity 2) are calculated.

According to the same calculation rules, IP features and IP similarity $\mathbf{DS}_1$ are obtained. Next, the associations between different diseases are downloaded from the MeSH database [8]. Among them, the Directed Acyclic Graph (*DAG*) describing disease *Dis* is obtained.

The semantic contribution value $D1_{Dis}(d)$ of node d to the disease *Dis* is expressed as:

$$D1_{Dis}(d) = \begin{cases} 1, & if\ d = Dis, \\ \max(\Delta \times D1_{Dis}(d')|d' \in children\ of\ d), & if\ d \neq Dis, \end{cases} \tag{3}$$

where Δ is a semantic contribution factor with a value of 0.5 [7]. Since the semantic contribution value of ancestor node d decreases with increasing the number of layers in *DAG*, the semantic score of disease *Dis* is calculated as follows:

$$DV1(Dis) = \sum_{d \in T(Dis)} D1_{Dis}(d). \tag{4}$$

As the similarity between diseases is proportional to the shared parts of diseases in the *DAGs*, the disease semantic similarity 1 between diseases is shown as follows:

$$\mathbf{DS}_2(Dis_i, Dis_j) = \frac{\sum_{t \in T(Dis_i) \cap T(Dis_j)} D1_{Dis_i}(t) + D1_{Dis_j}(t)}{DV1(Dis_i) + DV1(Dis_j)}. \tag{5}$$

At last, considering that different disease terms in the same level of the *DAG* may contribute differently to the semantic value of disease *Dis*, the disease semantic similarity 2 $\mathbf{DS}_3$ is defined as:

$$\mathbf{DS}_3(Dis_i, Dis_j) = \frac{\sum_{t \in T(Dis_i) \cap T(Dis_j)} D2_{Dis_i}(t) + D2_{Dis_j}(t)}{DV2(Dis_i) + DV2(Dis_j)}, \tag{6}$$

$$D2_{Dis}(d) = -\log\left(\frac{the\ number\ of\ DAGs\ including\ Dis}{the\ number\ of\ disease}\right), \tag{7}$$

$$DV2(Dis) = \sum_{d \in T(Dis)} D2_{Dis}(d). \tag{8}$$

2.4 Hypergraph Learning

Complex relationships between objects in large-scale data modeling can be established through hypergraph learning [2]. In hypergraph $G = (V, E, T)$, V is the set of vertices, E is the set of hyperedges, and T denotes the weight assigned to the hyperedges. Among them, the hyperedges are formed by the k-Nearest Neighbor (KNN) algorithm. Here, association matrix $\mathbf{H}_G$ is used to describe the relationship between vertices, where

$\mathbf{H}_G(i, j)$ is equal to 1 if the vertex v_i is contained in the hyperedge e_j otherwise 0. The degree of vertex and the degree of hyperedge are expressed as:

$$g(v) = \sum_{e \in E} \mathbf{H}_G(v, e), \tag{9}$$

$$\psi(e) = \sum_{v \in V} \mathbf{H}_G(v, e). \tag{10}$$

Next, the diagonal matrices of $g(v)$ and $\psi(e)$ are denoted by $\mathbf{D}_v$ and $\mathbf{D}_e$, respectively. Besides, the diagonal matrix of hyperedge weight is represented by $\mathbf{A}$. Thus, the hypergraph Laplacian matrix $\mathbf{L}$ is defined as:

$$\mathbf{L} = \mathbf{I} - \Theta, \tag{11}$$

$$\Theta = \mathbf{D}_v^{-1/2}\mathbf{H}_G\mathbf{A}\mathbf{D}_e^{-1}\mathbf{H}_G^T\mathbf{D}_v^{-1/2}, \tag{12}$$

where $\mathbf{I}$ represents the identity matrix.

2.5 Ensemble Learning for Predicting Novel Associations

By inputting the above-mentioned multi-source features and similarities into the ensemble learning framework, the diversity and reliability of the model are improved [9]. In miRNA space, the feature matrix $\mathbf{X}_i$ is projected onto the prediction matrix $\mathbf{F}_m$ by the projection matrix $\mathbf{P}_i$. The objective function is defined as follows:

$$\begin{gathered} \min_{\mathbf{P}_i} \textstyle\sum_{i=1}^n \left\|\mathbf{X}_i\mathbf{P}_i^T - \mathbf{F}_m\right\|_F^2 \\ s.t.\mathbf{P}_i \geq 0, \end{gathered} \tag{13}$$

where $\|\cdot\|_F^2$ means the Frobenius norm. To ensure the smoothness of $\mathbf{P}_i$, the $l_{1,2}$-norm regularization term is added. In addition, $\mathbf{F}_m$ needs to be approximated to the known association matrix $\mathbf{M}$. Thus, Eq. (13) is transformed into:

$$\begin{gathered} \min_{\mathbf{P}_i, \mathbf{F}_m} \|\mathbf{F}_m - \mathbf{M}\|_F^2 + \alpha \textstyle\sum_{i=1}^n \left\|\mathbf{X}_i\mathbf{P}_i^T - \mathbf{F}_m\right\|_F^2 + \beta \sum_{i=1}^n \|\mathbf{P}_i\|_{1,2}^2 \\ s.t.\mathbf{P}_i \geq 0, \end{gathered} \tag{14}$$

where α denotes the trade-off parameter that measures the error between the predicted value and the projection matrix. β is the regularization coefficient controlling the contribution of $\mathbf{P}_i$.

After introducing the hypergraph-regular term, the objective function of ensemble learning is expressed as:

$$\begin{gathered} \min_{\mathbf{P}_i, \mathbf{F}_m, \omega} \|\mathbf{F}_m - \mathbf{M}\|_F^2 + \alpha \textstyle\sum_{i=1}^n \left\|\mathbf{X}_i\mathbf{P}_i^T - \mathbf{F}_m\right\|_F^2 + \sum_{i=1}^m \omega_i^\lambda tr(\mathbf{F}_m^T\mathbf{L}_i\mathbf{F}_m) + \beta \sum_{i=1}^n \|\mathbf{P}_i\|_{1,2}^2 \\ s.t.\mathbf{P}_i \geq 0, \sum_i \omega_i = 1, \end{gathered} \tag{15}$$

where ω_i is the weight vector controlling the contribution of different hypergraph-regular terms. λ refers to the intensity of the different similarities.

After introducing the Lagrangian function, the above equation is transformed into:

$$Lf = \|\mathbf{F}_m - \mathbf{M}\|_F^2 + \alpha \sum_{i=1}^{n} \left\|\mathbf{X}_i\mathbf{P}_i^T - \mathbf{F}_m\right\|_F^2 + \sum_{i=1}^{m} \omega_i^\lambda tr(\mathbf{F}_m^T\mathbf{L}_i\mathbf{F}_m) + \beta \sum_{i=1}^{n} \|\mathbf{P}_i\|_{1,2}^2$$
$$- \delta(\sum_{i=1}^{m} \omega_i - 1) - \sum_{i=1}^{n} tr(\Gamma_i\mathbf{P}_i). \tag{16}$$

By taking partial derivatives of Eq. (16), $\mathbf{F}_m$, ω_i and $\mathbf{P}_i$ are calculated as:

$$\mathbf{F}_m = (\sum\nolimits_{i=1}^{m} \omega_i^\lambda \mathbf{L}_i + (1 + n\alpha)I)^{-1}(\mathbf{M} + \alpha \sum\nolimits_{i=1}^{n} \mathbf{X}_i\mathbf{P}_i^T), \tag{17}$$

$$\omega_i = \frac{\left(\frac{1}{tr(\mathbf{F}_m^T\mathbf{L}_i\mathbf{F}_m)}\right)^{\frac{1}{\lambda-1}}}{\sum_{i=1}^{m}\left(\frac{1}{tr(\mathbf{F}_m^T\mathbf{L}_i\mathbf{F}_m)}\right)^{\frac{1}{\lambda-1}}}, \tag{18}$$

$$\mathbf{P}_i = \mathbf{P}_i \odot \sqrt{\frac{\mathbf{P}_i(\alpha\mathbf{X}_i^T\mathbf{X}_i + \beta ee^T)^+ + \alpha(\mathbf{F}_m^T\mathbf{X}_i)^-}{\mathbf{P}_i(\alpha\mathbf{X}_i^T\mathbf{X}_i + \beta ee^T)^- + \alpha(\mathbf{F}_m^T\mathbf{X}_i)^+}}, \tag{19}$$

where e is a column vector with element values of 1.

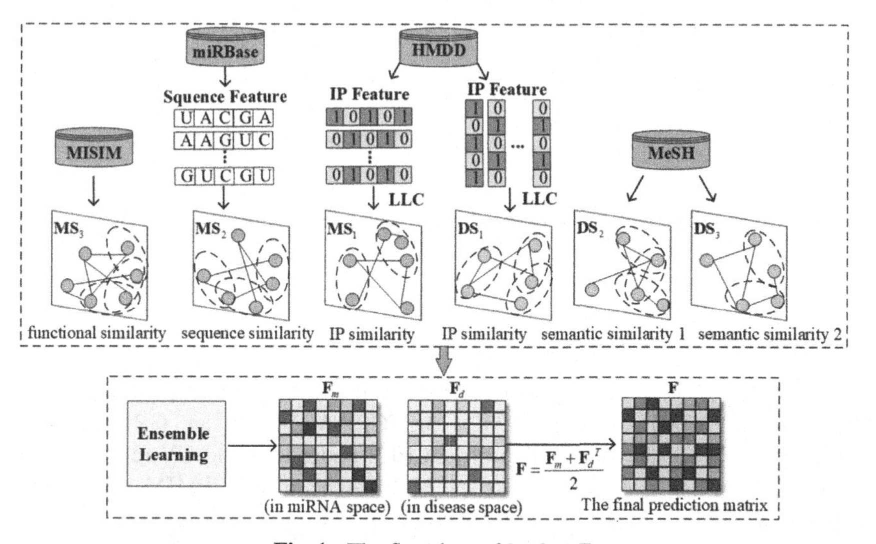

Fig. 1. The flowchart of LLCELF.

Similarly, the prediction matrix $\mathbf{F}_d$ in disease space is calculated. Finally, the association score matrix $\mathbf{F}$ is obtained by integrating association information in miRNA space and disease space. The entire process of LLCELF is shown in Fig. 1.

$$\mathbf{F} = \frac{\mathbf{F}_m + \mathbf{F}_d^T}{2}. \tag{20}$$

3 Results and Discussion

3.1 Performance Evaluation

In this work, five-fold cross validation (CV) is implemented to evaluate the performance of LLCELF. Among them, the evaluation metric Area Under the Curve (AUC) value is the area under the Receiver Operating Characteristic (ROC) curve [10].

3.2 Parameter Selection

In this work, four parameters need to be adjusted, namely the regularization parameter α, β, the exponential parameter λ, and the number of nearest neighbours K in KNN algorithm. As shown in Fig. 2, our method achieves an optimal AUC value when $\alpha = 10^0$, $\beta = 10^{-2}$, $\lambda = 2^4$, and $K = 15$.

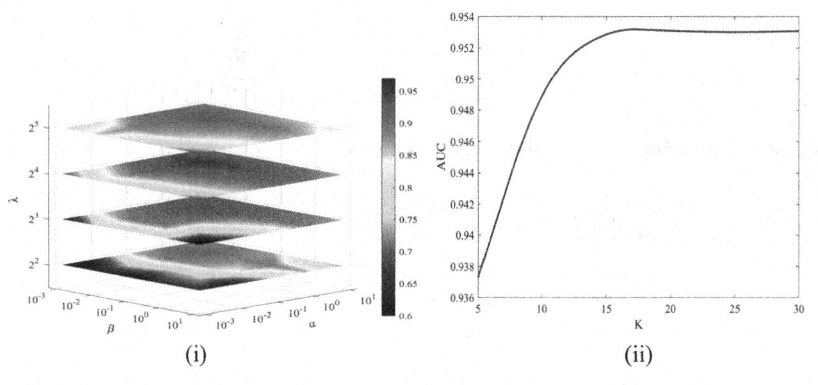

Fig. 2. The effect of (i) parameters α, β, and λ, and (ii) parameter K on AUC values.

3.3 Methods Comparison

In this section, LLCELF is compared with other state-of-the-art models, including NCP-MDA [11], NCMCMDA [12], NMFMC [13], TCRWMDA [14], and GRPAMDA [15]. As shown in Fig. 3, the AUC value of LLCELF under five-fold CV reaches 0.9533, which is 7.55%, 4.48%, 3.68%, 3.24%, and 1.37% higher than NCPMDA, NCMCMDA, NMFMC, TCRWMDA, and GRPAMDA respectively.

3.4 Case Studies

To confirm the performance of LLCELF in practical application, we perform the case analysis on Kidney Neoplasms. In case studies, the top 20 Kidney Neoplasms-associated miRNAs are selected from the prediction matrix sorted by association score. As listed in Table 1, 9 known MDAs are successfully predicted. In addition, the remaining novel associations can be confirmed by HMDD v3.2 [16], miRCancer [17], and relevant literature.

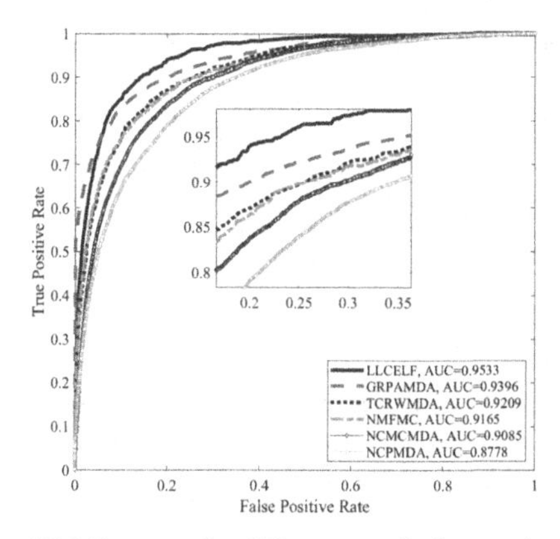

Fig. 3. AUC values and ROC curves for different prediction methods under five-fold CV.

Table 1. Predicted miRNAs for Kidney Neoplasms.

Rank	miRNA	Evidence	Rank	miRNA	Evidence
1	**hsa-mir-1**	known	11	**hsa-mir-23b**	known
2	**hsa-mir-200**	known	12	hsa-mir-9	PMID: 30532596
3	**hsa-mir-141**	known	13	hsa-mir-155	HMDD v3.2
4	**hsa-mir-215**	known	14	hsa-mir-99a	PMID: 23173671
5	**hsa-mir-192**	known	15	hsa-mir-125b	PMID: 25155155
6	**hsa-mir-15a**	known	16	hsa-mir-375	miRCancer
7	**hsa-mir-21**	known	17	hsa-mir-146a	PMID: 21975861
8	**hsa-mir-200c**	known	18	hsa-mir-203	HMDD v3.2, miCancer
9	hsa-mir-200a	HMDD v3.2	19	hsa-mir-20a	PMID: 34360679
10	hsa-mir-429	PMID: 31814979	20	hsa-mir-200b	PMID: 31130475

4 Conclusion

Identification of disease-related miRNAs by efficient computational models has become a popular topic. In this paper, a locality-constrained linear coding-based ensemble learning framework is proposed to predict miRNA-disease associations (LLCELF). First of all, the extracted IP feature descriptors are converted into encoding matrix by LLC to reconstruct the similarity network. Next, features and similarities obtained from different perspectives are input into the ensemble learning framework. Then, the hypergraph-regular terms are introduced to learn the high-order relationships between samples during the projection of the features onto the known association matrix. Eventually, many potential association pairs are discovered. The AUC value of LLCELF is superior to state-of-the-art models. Further, the case study in Kidney Neoplasms confirms the reliability of the model. However, the insufficiency of the validated known MDAs may affect prediction performance. In the future, the richness of the input features needs to be further enhanced.

Acknowledgment. This work was supported by the National Natural Science Foundation of China (Grant Nos. 62172254, and 61872220).

References

1. Cheng, A.M., Byrom, M.W., Shelton, J., Ford, L.P.: Antisense inhibition of human miRNAs and indications for an involvement of miRNA in cell growth and apoptosis. Nucleic Acids Res. **33**(4), 1290–1297 (2005)
2. Ding, Y.J., Jiang, L.M., Tang, J.J., Guo, F.: Identification of human microRNA-disease association via hypergraph embedded bipartite local model. Comput. Biol. Chem. **89**, 107369 (2020)
3. Qu, Y., Zhang, H.X., Lyu, C., Liang, C.: LLCMDA: a novel method for predicting miRNA gene and disease relationship based on locality-constrained linear coding. Front. Genet. **9**, 576 (2018)
4. Li, Y., et al.: HMDD v2. 0: a database for experimentally supported human microRNA and disease associations. Nucleic Acids Res. **42**(D1), D1070–D1074 (2014)
5. Zhou, F., et al.: Predicting miRNA-disease associations through deep autoencoder with multiple kernel learning. IEEE Trans. Neural Netw. Learn. Syst. (2021). https://doi.org/10.1109/TNNLS.2021.3129772
6. Wallin, E., Wettergren, C., Hedman, F., von Heijne, G.: Fast Needleman—Wunsch scanning of sequence databanks on a massively parallel computer. Bioinformatics **9**(1), 117–118 (1993)
7. Wang, D., Wang, J., Lu, M., Song, F., Cui, Q.H.: Inferring the human microRNA functional similarity and functional network based on microRNA-associated diseases. Bioinformatics **26**(13), 1644–1650 (2010)
8. Lowe, H.J., Barnett, G.O.: Understanding and using the medical subject headings (MeSH) vocabulary to perform literature searches. JAMA **271**(14), 1103–1108 (1994)
9. Zhang, W., Yue, X., Tang, G.F., Wu, W.J., Huang, F., Zhang, X.N.: SFPEL-LPI: sequence-based feature projection ensemble learning for predicting LncRNA-protein interactions. PLoS Comput. Biol. **14**(12), e1006616 (2018)
10. Fawcett, T.: An introduction to ROC analysis. Pattern Recogn. Lett. **27**(8), 861–874 (2006)
11. Gu, C.L., Liao, B., Li, X.Y., Li, K.Q.: Network consistency projection for human miRNA-disease associations inference. Sci. Rep. **6**(1), 1–10 (2016)
12. Chen, X., Sun, L.G., Zhao, Y.: NCMCMDA: miRNA-disease association prediction through neighborhood constraint matrix completion. Brief. Bioinform. **22**(1), 485–496 (2021)
13. Zheng, X., Zhang, C.J., Wan, C.: MiRNA-Disease association prediction via non-negative matrix factorization based matrix completion. Signal Process. **190**, 108312 (2022)
14. Yu, L.M., Shen, X.J., Zhong, D., Yang, J.C.: Three-layer heterogeneous network combined with unbalanced random walk for miRNA-disease association prediction. Front. Genet. **10**, 1316 (2020)
15. Zhong, T.B., Li, Z.W., You, Z.H., Nie, R., Zhao, H.: Predicting miRNA-disease associations based on graph random propagation network and attention network. Brief. Bioinform. **23**(2), bbab589 (2022)
16. Huang, Z., et al.: HMDD v3. 0: a database for experimentally supported human microRNA-disease associations. Nucleic Acids Res. **47**(D1), D1013–D1017 (2019)
17. Xie, B.Y., Ding, Q., Han, H.J., Wu, D.: miRCancer: a microRNA–cancer association database constructed by text mining on literature. Bioinformatics **29**(5), 638–644 (2013)

Gaussian-Enhanced Representation Model for Extracting Protein-Protein Interactions Affected by Mutations

Da Liu, Yijia Zhang(✉), Ming Yang, Fei Chen, and Mingyu Lu

School of Information Science and Technology, Dalian Maritime University, Liaoning 116024, China
zhangyijia@dlmu.edu.cn

Abstract. The biomedical literature contains many protein-protein interactions (PPIs) affected by genetic mutations. Automatic extraction of PPIs affected by gene mutations described in biomedical literature can help evaluate the clinical significance of gene variations, which plays a crucial role in the realization of precision medicine. This paper proposes a novel Gaussian-enhanced representation model (GRM) to extract PPI, which uses Gaussian probability distribution to generate target entity representation based on BioBERT pre-trained model. The proposed GRM enhanced the weight of target protein entities and their adjacents, solved the problem of long input text and scattered distribution of target entities in the PPI task, and introduced a supervised contrast learning method to improve the effectiveness and robustness of the model. Experiment results on the BioCreative VI data set show that our proposed model GRM leads to a new state-of-the-art performance.

Keywords: PPI extraction · Gaussian probability distribution · Contrastive Learning

1 Introduction

To help health professionals and precision medical researchers, it is of clinical importance to study the effect of gene mutation on protein-protein interaction. The automatic extraction of PPI described in biomedical literature contains many protein-protein interactions affected by gene mutations, which is an essential step towards the ultimate goal of precision medicine. Recently, BioCreative VI task [1] organizers released the PM task, consisting of two sub-tasks: (1) Identification of relevant PubMed citations describing genetic mutations affecting protein-protein interactions. (2) Extracting PPI affected by genetic mutations. This paper mainly studies the PPI extraction task.

In the previous PPI work, Zhou proposed a knowledge-aware attention network (KAN) [2] to extract PPI by fusing prior knowledge and context information of protein-protein pairs. They use prior knowledge and contextual information to achieve the best performance for the time. However, their approach [2] is no longer competitive with the widespread use of pre-trained model in the NLP field. Wang et al. [3] proposed a

M. S. Bansal et al. (Eds.): ISBRA 2022, LNBI 13760, pp. 303–314, 2022.
https://doi.org/10.1007/978-3-031-23198-8_28

method combining the pre-training model with the auxiliary task for PPI extraction, which improved the accuracy of the PPI extraction and achieved a pretty good result compared to previous work. However, this method only improved model performance through general technology without considering the characteristics of the PPI task itself.

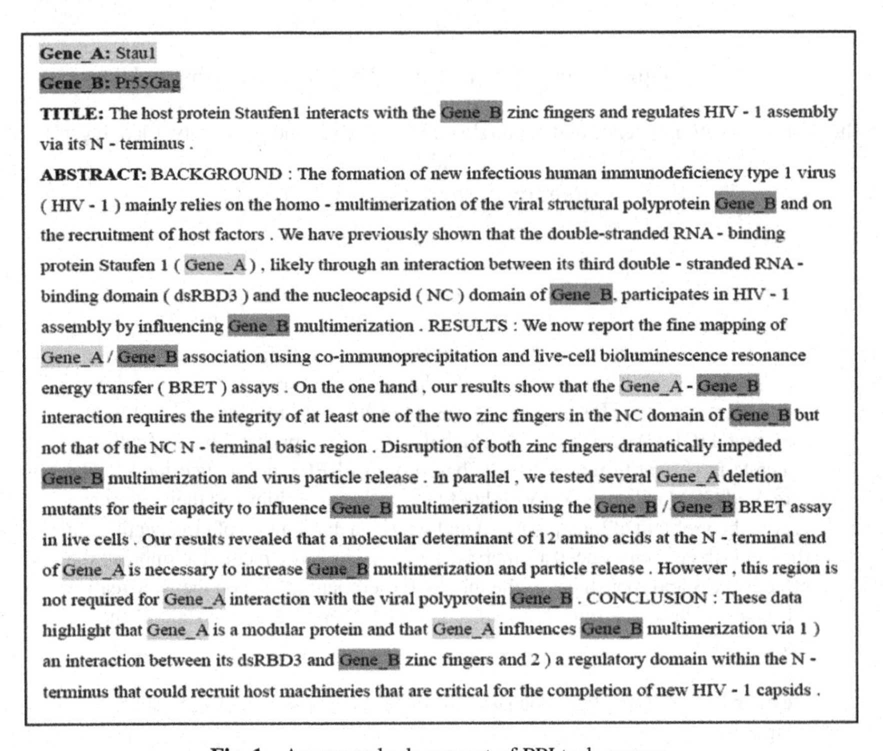

Gene_A: Staul

Gene_B: Pr55Gag

TITLE: The host protein Staufen1 interacts with the Gene_B zinc fingers and regulates HIV - 1 assembly via its N - terminus .

ABSTRACT: BACKGROUND : The formation of new infectious human immunodeficiency type 1 virus (HIV - 1) mainly relies on the homo - multimerization of the viral structural polyprotein Gene_B and on the recruitment of host factors . We have previously shown that the double-stranded RNA - binding protein Staufen 1 (Gene_A) , likely through an interaction between its third double - stranded RNA - binding domain (dsRBD3) and the nucleocapsid (NC) domain of Gene_B, participates in HIV - 1 assembly by influencing Gene_B multimerization . RESULTS : We now report the fine mapping of Gene_A / Gene_B association using co-immunoprecipitation and live-cell bioluminescence resonance energy transfer (BRET) assays . On the one hand , our results show that the Gene_A - Gene_B interaction requires the integrity of at least one of the two zinc fingers in the NC domain of Gene_B but not that of the NC N - terminal basic region . Disruption of both zinc fingers dramatically impeded Gene_B multimerization and virus particle release . In parallel , we tested several Gene_A deletion mutants for their capacity to influence Gene_B multimerization using the Gene_B / Gene_B BRET assay in live cells . Our results revealed that a molecular determinant of 12 amino acids at the N - terminal end of Gene_A is necessary to increase Gene_B multimerization and particle release . However , this region is not required for Gene_A interaction with the viral polyprotein Gene_B . CONCLUSION : These data highlight that Gene_A is a modular protein and that Gene_A influences Gene_B multimerization via 1) an interaction between its dsRBD3 and Gene_B zinc fingers and 2) a regulatory domain within the N - terminus that could recruit host machineries that are critical for the completion of new HIV - 1 capsids .

Fig. 1. An example document of PPI task corpus.

In the dataset of the PPI extraction task, there is a problem that the input text is long, and the target protein entities are scattered. Taking the document in Fig. 1 as an example, target entity 1 and target entity 2 are marked orange and blue, respectively. This example shows the length of the document and the degree of dispersion of the target entity. We generally assume that the semantic information contained in different words in input text is of varying importance. Intuitively, words adjacent to the target entity typically contribute more to the semantics of the target entity than words far away from it (Guo et al. 2019) [4]. Therefore, we introduce Gaussian probability distribution to improve the weight of target protein entity and its adjacent words, so as to better capture the semantic information around target entity and enhance the feature extraction ability of the model. At the same time, supervised contrastive learning is introduced, which makes full use of the tag information and regards the samples of the same class as positive examples and illustrations of different classes as negative examples. The motivation is to facilitate PPI extraction by reducing the intra-class distance in the embedded space and increasing the distance between positive and negative example samples, so as to alleviate the problem of long input text in this task.

This paper proposes a novel Gaussian-enhanced representation model (GRM) for PPI extraction. Specifically, the method starts with a high-quality contextual representation of the sequence of instances generated by Bert. Secondly, the Gaussian probability distribution is introduced to improve the weight of target protein entities and their neighbors in the instance sequence. Then supervised contrast learning is used to integrate in-class and inter-class samples into the embedded space. Finally, related representations are concatenated in the inference phase and entered into the Softmax function to extract PPI. The main contributions of this paper can be summarized as follows:

- We propose a novel model GRM for PPI extraction. By introducing the Gaussian probability distribution, the weight of the target protein entity and its neighboring words is improved, effectively solving the problem of long input text and scattered distribution of the target entity in the biomedical field.
- We use supervised contrast learning to distinguish semantic features in the PPI task, which improves the effectiveness and robustness of the model.
- Experiment results on the BioCreative VI precision medicine track corpus show that the proposed approach leads to a new state-of-the-art performance.

2 Related Work

This section will summarize the related work into two main categories: PPI extraction, and Contrastive Learning.

2.1 PPI Extraction

Biomedical relationship extraction methods can be divided into three main categories: those based on Hand-Write patterns, statistical machine learning, and neural network. PPI extraction, as a subclass of biomedical relational extraction, can also be divided into three kinds of extraction methods.

In the template-based approach [5–8], Chen et al. [8] used a simple rule-based co-occurrence strategy, assuming that if specific heuristic design criteria are met, then two proteins occurring simultaneously within a document scope are considered PPI relations. The template-based method has high precision and is simple. However, its recall rate is low, and because its template is specific to a certain field, it needs expert construction, which is difficult and time-consuming.

Methods based on statistical machine learning [9–11] include feature-based methods and kernel-based methods. They address some of the shortcomings of template-based approaches by eliminating the need for experts to build templates. Phan et al. [10] used a feature-based approach combining different features, such as vocabulary and word context features from sentences, syntactic features from parsing trees, and automatic extraction of PPIs features from articles using existing patterns. Qian et al. [11] proposed a new tree core-based PPI extraction method, in which the tree representation generated by the component parser is further refined, using the shortest dependent path between two proteins from the dependency parser. Statistical machine learning methods do not need to define rules in advance, but need to manually design features or core functions to improve performance, which is time-consuming and laborious.

Deep learning uses neural networks to extract feature information and thus reduces the dependence on feature engineering in statistical machine learning methods. Peng et al. [12] used the multi-channel CNN model to extract PPI. Zhou et al. [2] used the attention network of knowledge perception to extract PPI.

2.2 Contrastive Learning

Contrastive learning, as a popular method of unsupervised representational learning, can be traced back to Hadsell et al., 2006 [13]. The author proposed Dimensionality Reduction by Learning an Invariant Mapping (DrLIM), map similar inputs to approximate points in the representation space. Chen et al. (2020) [14] proposed SIMCLR, which improved contrast learning with the help of data enhancement technology and introduced learnable nonlinear transformation between representation and contrast loss, thus improving the quality of representation. He et al. (2020) [15] proposed momentum contrast (MoCo) in order to obtain more negative samples, and they established a large and consistent dynamic dictionary to promote unsupervised contrast learning.

In addition, the idea of contrastive learning is also applied to supervised learning. Khosla et al. [16] put forward SupCon, which extended batch self-supervised comparison learning to supervised tasks and made full use of label information to make points of the same class closer and points of different classes further in the embedding space. Wang et al. [17] proposed ClusterSCL, which introduced cluster-aware data enhancement strategy on the basis of SupCon to alleviate the negative impact caused by intra-class difference and inter-class similarity.

3 Method

This section will summarize the Methods into two main categories: PPI extraction, and Gaussian-enhanced representation model (GRM).

3.1 PPI Extraction

PPI extraction is a task to detect the existence of PPI relationships between target protein-protein pairs in a set of documents. If there is a PPI relation, the interacting protein pair (and the corresponding gene Entrez ID) is returned. So, we define PPI extraction as a two-class classification problem (yes or no). The classification problem is defined as follows: given the input document $\{s_1, ..., s_i, ..., s_j, \ldots s_n\}$, the goal is to classify the relation r for protein pair s_i, s_j. Actually, GRM is to estimate the probability $P(r|s_i, s_j)$, where $R = \{\text{PPI and non} - \text{PPI}\}$, $r \in R$, $1 \leq i \leq n$.

3.2 Gaussian-Enhanced Representation Model (GRM)

The Fig. 2 shows the overall structure of GRM. The input of the model is the sequence of document-level instances. Firstly, we extract the features of the input instances through

BioBERT to generate a high-quality context representation of instances. Secondly, Gaussian probability distribution is introduced to improve the weight of the target entity and its adjacencies in the illustration, and the enhanced representation of related entities is generated. Thirdly, the input instance representation extracted by BioBERT is concatenated with the enhanced representation of the first target entity and the second target entity obtained by the Gaussian enhancement module to get the final fusion representation. Finally, the supervised contrast loss and cross entropy loss of the fusion representation are calculated in the training phase, and the fusion representation is input into the Softmax function to extract the PPI in the inference phase. Below we introduce each part of the model in detail.

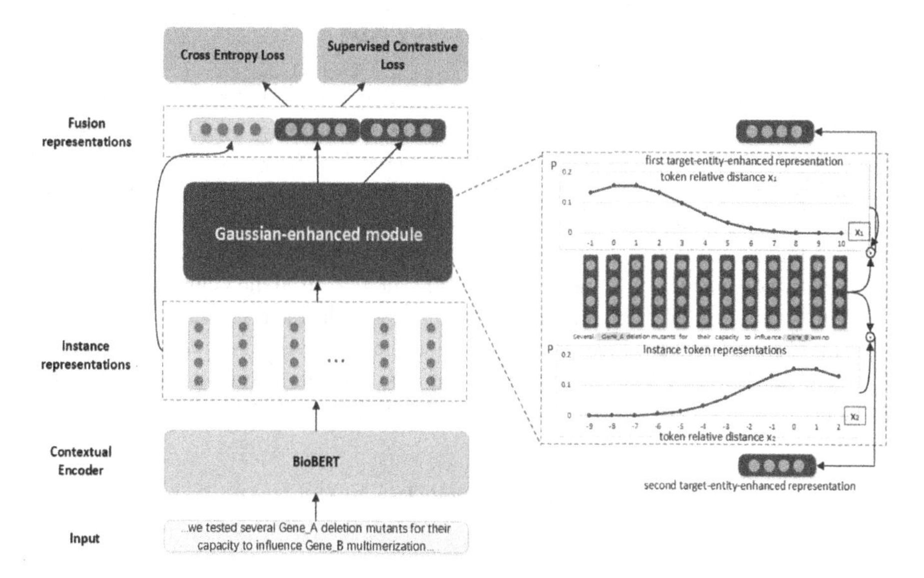

Fig. 2. Gaussian-enhanced representation model(GRM)

BioBERT Representation

The extensive application of pre-training models in NLP has proved its effectiveness in NLP tasks. In addition, studies [18] have shown that BioBERT trained on a biomedical corpus performs better than BERT in most biomedical text mining tasks. Therefore, in the PPI extraction task, we use BioBERT [18] to obtain the representation of the input document and the input document representation is denoted as R_{ins}. We segment the instance sequence of documents through the WordPiece classifier [19] to alleviate the problem of insufficient vocabulary in the word list, and then construct the instance sequence as "[CLS] Document [SEP]".

Gaussian-Enhanced Module

In this paper, we introduce Gaussian probability distribution to improve the weight of the target protein entity and its adjacent words, so as to the model can learn the semantic information near the target entity. The Gaussian probability density function is:

$$\mathrm{f}(x) = \frac{1}{\sqrt{2\pi}\delta}\exp\left(-\frac{(x-\mu)^2}{2\delta^2}\right) \tag{1}$$

the Gaussian cumulative distribution function is:

$$\mathrm{F}(x) = \int_{-\infty}^{x} f(x)dx \tag{2}$$

the Gaussian probability distribution function is:

$$\mathrm{P}(x) = \mathrm{F}(x) - \mathrm{F}(x - \mathrm{w}) \tag{3}$$

where x is the actual number, μ is the mean of the distribution, δ is the standard deviation, and w is the token window. During the experiment, we set the token window to 1 to represent the distance of each token itself, and set μ to 0. Then we experimented with different values of δ and finally determined that the model works best when δ is 4. The remaining parameters will be adjusted on the development set. We first numbered the tokens according to the two target entities in the instance, thus obtaining two token relative distance lists (x_1 and x_2 in Fig. 2). Then we used the Gaussian probability distribution function to calculate the probability of each token in the instance. Afterward, these token probabilities were performed element-wise multiplication with the token representations. Finally, we obtained the first and second target-entity-enhanced representations, respectively. The formula is as follows:

$$\mathrm{R}_{tar1} = \sum_{1}^{M} tanh(P(x_1)u_i) \tag{4}$$

$$\mathrm{R}_{tar2} = \sum_{1}^{M} tanh(P(x_2)u_i)$$

where M is the sequence length, x_1 and x_2 are token numbers, u_i denote the instance token representations, R_{tar1} and R_{tar2} denote the first and second target-entity-enhanced representations, respectively.

Contrastive Learning

The Contrastive method distinguishes semantic features by comparing the input of the same class and the input of different categories. The information of the same type is a positive sample, and the input of other type is a negative sample. In the training process, representations of the same kind in the embedding space will be closer together, and different representations will be further apart. Our model uses the cross-entropy function as the primary loss function to deal with the classification problem. At the same time, the supervised contrast loss function proposed by Khosla et al. [16] is used to distinguish semantic features and promote the extraction of PPI. The positive sample consists of the same category of samples, and the negative sample consists of other types of examples in the batch. The standard definition of the multi-class CE loss we use is as follows:

$$L_{CE} = -\frac{1}{N}\sum_{i=1}^{N}\sum_{c=1}^{C} y_{i,c} \cdot \hat{y}_{i,c} \tag{6}$$

where N is a batch of training examples, C is class of a multi-class classification problem ($C = 2$), $y_{i,c}$ is the label, $\hat{y}_{i,c}$ is the model output for the probability of the ith example belonging to the class c. The SCL function is:

$$L_{SCL} = \sum_{i=1}^{N} -\frac{1}{N_{y_i} - 1}\sum_{j=1}^{N} l_{i \neq j} l_{y_i = y_j} log \frac{\exp\left(\varphi(h_i) \cdot \frac{\varphi(h_j)}{\tau}\right)}{\sum_{k=1}^{N} l_{i \neq k} \exp\left(\varphi(h_i) \cdot \frac{\varphi(h_k)}{\tau}\right)} \tag{7}$$

where h_i is the concatenated fusion representations $h_i = [R_{ins}; R_{tar1}; R_{tar2}]$, N_{y_i} is the total number of examples in the batch that have the same label as y_i. $\varphi(\cdot) \in R^d$ denotes an encoder that outputs the final encoder hidden layer before the softmax projection. $\tau > 0$ denotes an adjustable scalar temperature parameter that controls the separation of classes. The final loss function consists of the CE loss function and SCL function. The formula is shown as follows:

$$L = \lambda L_{CE} + (1 - \lambda) L_{SCL} \tag{8}$$

where λ is a scalar weighting hyperparameter.

4 Experiments

4.1 Datasets

The experiment was conducted on BioCreative VI Precision Medicine Track Corpus [1]. In this paper, only the data set used for subtask 2 PPI extraction is discussed. As shown in Table 1, there were 1729 articles related to PPI in the training set and 704 articles in the test set. Annotate the protein entities in the training set and test set through GNormPlus toolkits1 [20], and normalize them to Entrez Gene ID.

Table 1. Statistics of the PPI datasets

Dataset	Documents	PPI
Training	1729	752
Testing	704	869
Total	2433	1621

For PPI extraction tasks, BC6PM organizers provide two evaluation indicators: 1) Exact Match: The prediction relationship counts only if the GeneID is the same as the

Table 2. Hyper-parameters

Group	Hyper-parameter	Value
Embedding layer	Learning rate	1×10^{-5}
	BERT embedding dimension	768
RC	Max sentence length	512
	Batch size	4
	Learning rate	5×10^{-5}
Gauss	epoch	10
	λ	0.6
	δ	4
SCL	μ	0
	τ	0.2

GeneID of the human annotation 2) HomoloGene Match: As long as the predicted gene identifiers are homologous with artificially annotated gene identifiers, this relationship is considered correct. Referring to other PPI extraction tasks, this article uses only the first evaluation method and uses an official evaluation script.

4.2 Pre-processing and Settings

For the model to identify the gene pairs that should be classified, a scheme of Tran et al. [21] was used for text substitution. When PPI was associated with only one gene ID, each reference to the involved gene was replaced with gene A and gene B or gene S, and the other references were replaced with gene N. We use PyTorch framework to realize the deep neural network model, with BioBERT [18] as the text embedding layer, and at the same time, we use right truncation for the text that exceeds the input limit of BioBERT. In addition, the Adam algorithm with different learning rates is used to train the BioBERT embedded layer and downstream layer of the model. The hyperparameter Settings are shown in Table 2.

4.3 Comparison with Other Works

The comparison of GRM with other baseline methods is as follows:

1) Rule-based method [8]: If more than N sentences refer to two genes in a given gene pair, the gene pair is considered to be a positive example. (if two genes are mentioned in more than N sentences in a given pair of sentences, then this pair of genes is considered a positive example.). This method is simple and effective and gets the best F score in PPI extraction task.
2) CNN-based method [22]: A standard CNN-based deep neural model for document-level binary relation classification of an entity pair.

3) Knowledge-aware attention network (KAN) [2]: A model that makes use of prior knowledge in the knowledge base and got a state-of-the-art F1 score in the PPI extraction task.
4) BioBERT-based Method [3]: This model obtained the most advanced F1 score of 43.14% in the PPI extraction task by using the auxiliary task strategy. In this part, we only discuss the method BioBERT-RC-only [3] that does not use the auxiliary task strategy under this method, and the comparison with the auxiliary task strategy method will be discussed in the next part.

All of these methods use the GNormPlus toolkit for gene identification and normalization. The calculation of indicators is based on accurate matching to ensure the fairness of comparison.

As can be seen from the Table 3, compared with BioBERT-RC-only [3] without auxiliary task, F1 score of GRM improved by 1.84%, and F1 score increased by 5.29% compared with model KAN [2], which proves the effectiveness of GRM.

Table 3. Performance comparison with baselines

Method	Precision (%)	Recall (%)	F1-score (%)
Rule-based [8]	38.90	30.10	33.94
CNN-based [22]	37.07	35.64	36.33
KAN [2]	41.62	35.44	38.28
BioBERT-RC-only [3]	44.16	39.59	41.75
GRM	50.90	37.84	43.57

In order to compare with the work of Wang et al. [3], we also added auxiliary tasks for training based on GRM. As shown in Table 4, when the named entity recognition task was added on the basis of GRM, the F1 value decreased slightly by 0.25% compared with BioBERT RC + NER [3]. After adding Triage task, F1 value increased by 0.62% compared with BioBERT RC + Triage [3]. It is proved that GRM is superior to Wang et al. [3]'s model with the addition of a multi-task strategy. However, the F1 score of GRM with the addition of multi-task strategy are not as good as those of our proposed model without multi-task strategy, which proves that multi-task strategy only yields gains for a limited number of models, but not all.

4.4 Ablation Study

In this section, we discuss the influence of each component of GRM, respectively. Firstly, we removed the Gaussian-enhanced module, and the BioBERT and SCL parts are retained in the model. It can be seen from Table 5 that the F1 score decreased by 1.16%. The experimental results verify the gain of the Gaussian-enhanced module for the GRM model. Secondly, to further discuss the influence of Gaussian probability distribution on the model, we replace the Gaussian probability distribution with mean pooling within

Table 4. Performance comparison with baselines (+Auxiliary Task)

Method	Precision (%)	Recall (%)	F1-score (%)
BioBERT RC + Triage [3]	45.78	39.36	42.33
BioBERT RC + NER [3]	48.42	38.90	43.14
GRM + NER	48.34	38.55	42.89
GRM + Triage	45.71	40.51	42.95
GRM	50.90	37.84	43.57

the target entity local window, and the F1 value decreases by 0.72%. The experimental results verify the necessity of Gaussian distribution. Thirdly, we removed the SCL part, and the BioBERT part and Gaussian-enhanced module are retained in the model. As can be seen from Table 5, F1 score decreased by 0.61%, and the experimental results proved the effectiveness of SCL part. These experimental results show that the Gaussian-enhanced module and SCL are helpful for PPI extraction, and they are complementary to each other.

Table 5. Ablation study

Method	Precision (%)	Recall (%)	F1-score (%)
w/o Gaussian	43.26	41.92	42.41
w/o SCL	44.61	41.43	42.96
GRM-mp	44.69	36.75	42.85
GRM	50.90	37.84	43.57

Notes. GRM-mp denotes mean pooling within the target entity local window.

4.5 Case Study

As shown in Fig. 3, we selected two articles from the PPI data set for detailed analysis. In analyzing the advantages and disadvantages of the GRM model, we compare the prediction results of BioBERT RC-only [3] and GRM, respectively. According to Case 1, when the ground truth is Negative, the result of the BioBERT RC-only [3] model is False-Positive, that is, the prediction is wrong, and GRM is True-Negative, that is, the forecast is correct. We believe that the GRM model effectively improves the extraction effect of PPI. In Case 2, if the ground truth is Positive, then BioBERT RC-only [3] is true-positive, GRM is False-Negative, and the GRM prediction is wrong. Since the target entity, in this case, is a single entity, we guess that the Gaussian probability distribution enhancement module in the GRM model plays a limited role at this time. In addition, the number of cases similar to Case 2 (n = 18) was much smaller than in Case 1 (n = 107).

In summary, although GRM has some limitations, overall, GRM can achieve incredible improvements in PPI extraction through significantly improved accuracy.

Case 1	Document instance	The retrieval function of the Gene_B requires PKA phosphorylation of its C - terminus . The Gene_B is a Golgi / intermediate compartment located integral membrane protein that carries out the retrieval of escaped ER proteins bearing a C - terminal KDEL sequence . This occurs throughout retrograde traffic mediated by COPI coated transport carriers . The role of the C - terminal cytoplasmic domain of the Gene_B in this process has been investigated . As revealed with a peptide - binding assay , this domain did not interact with both coatomer and ARF - GAP unless serine 209 was mutated to aspartic acid . In contrast ,alanine replacement of serine 209 inhibited coatomer / ARF - GAP recruitment , receptor redistribution into the ER , and intracellular retention of KDEL ligands . Serine 209 was phosphorylated by both cytosolic and recombinant Gene_A . Inhibition of endogenous PKA activity with H89 blocked Golgi ER transport of the native receptor but did not affect redistribution to the ER of a mutated form bearing aspartic acid at position 209 . We conclude that PKA phosphorylation of serine 209 is required for the retrograde transport of the Gene_B from the Golgi complex to the ER from which the retrieval of proteins bearing the KDEL signal depends .
	Gene A	ID: 5566;
	Gene B	ID: 10945;
	Label	Negative
	Prediction results	BioBERT RC-only: False-Positive GRM: True-Negative
Case 2	Document instance	Gene_S (Gene_S , Gene_S / Gene_N) , complement receptor 1 (Gene_N , Gene_N / Gene_N) , and p150 ,95 (Gene_N / Gene_N) . Binding of Gene_N to these receptors mediates leukocyte - adhesive functions in immune and inflammatory responses . In this report , we describe a cell – free assay using purified recombinant extracellular domains of Gene_S and a dimeric immunoadhesin of Gene_N . The binding of recombinant secreted Gene_S to Gene_N is divalent cation dependent (Mg2 + and Mn2 + promote binding) and sensitive to inhibition by antibodies that block Gene_S - mediated cell adhesion , indicating that its conformation mimics that of Gene_S on activated lymphocytes . These mutants , along with modeling studies , define the Gene_S binding site on Gene_N as residues E34 , Gene_N , M64 , Y66 , N68 , and Q73 , that are predicted to lie on the CDFG beta - sheet of the Ig fold . The mutant G32A also abrogates binding to Gene_S while retaining binding to all of the antibodies , possibly indicating a direct interaction of this residue with Gene_S .
	Gene S	ID: 3683
	Label	Positive
	Prediction results	BioBERT RC-only: True-Positive GRM: False-Negative

Fig. 3. Examples of the extracted results by different models

5 Conclusion and Future Work

In this paper, we propose a novel model GRM to improve the performance of PPI extraction by introducing the Gaussian probability distribution and supervised contrastive learning. Experiments on the BioCreative VI data set show that GRM leads to a new state-of-the-art performance (43.57% in F1-score). The ablation experiments proved the gain of Gaussian probability distribution and SCL for the pre-training model.

The study's main limitation was that only 56% of the gene pairs were identified using the GNormPlus toolkit. Since breaking through the bottleneck of BioNER and BioNEN is a necessary prerequisite for improving the performance of the PPI extraction task, we will explore the upstream study of relational classification in the future. In addition, we intend to give different weights not only at the token level but also at the sentence level to address the limitations of the current task.

Acknowledgment. This work is supported by grant from the Natural Science Foundation of China (No. 62072070).

References

1. Rezarta, I.D., et al.: Overview of the BioCreative VI precision medicine track: mining protein interactions and mutations for precision medicine. Database J. Biol. Databases Curation (2019)

2. Zhou, H., Zhuang, N.S., Lang, C., Du, L.: Knowledge-aware attention network for protein-protein interaction extraction (2020)
3. Wang, Y., Zhang, S., Zhang, Y., Wang, J., Lin, H.: Extracting protein-protein interactions affected by mutations via auxiliary task and domain pre-trained model. In: BIBM 2020 (2020)
4. Guo, M., Zhang, Y., Liu, T.: Gaussian transformer: a lightweight approach for natural language inference. In: EAAI 2019 (2019)
5. Sun, K., et al.: BioCreative V BioC track overview: collaborative biocurator assistant task for BioGRID. Database (2016)
6. Krallinger, M., Vazquez, M., Leitner, F., Salgado, D., Valencia, A.: The protein-protein Interaction tasks of BioCreative III: classification/ranking of articles and linking bio-ontology concepts to full text. BMC Bioinform. **12**(S8), S3 (2011)
7. Huang, M., Zhu, X., Payan, D.G., Qu, K., Ming, L.: Discovering patterns to extract protein-protein interactions from full biomedical texts, pp. 3604–3612 (2004)
8. Chen, Q., Chandrasekarasastry, N.P., Elangovan, A., Davis, M., Verspoor, K.M.: Document triage and relation extraction for protein-protein interactions affected by mutations, pp. 103–106 (2017)
9. Chowdhury, M.F.M., Lavelli, A.: Combining tree structures, flat features and patterns for biomedical relation extraction. In: Conference of the European Chapter of the Association for Computational Linguistics: 2012, pp. 420–429 (2012)
10. Phan, T.T.T., Ohkawa, T.: Protein-protein interaction extraction with feature selection by evaluating contribution levels of groups consisting of related features. BMC Bioinform. (2016)
11. Qian, L.H., Zhou, G.D.: Tree kernel-based protein-protein interaction extraction from biomedical literature. J Biomed. Inform. (2012)
12. Peng, Y., Lu, Z.: Deep learning for extracting protein-protein interactions from biomedical literature (2017)
13. Hadsell, R., Chopra, S., LeCun, Y.: Dimensionality reduction by learning an invariant mapping. In: 2006 IEEE Computer Society Conference on Computer Vision and Pattern Recognition (CVPR 2006), pp. 1735–1742. IEEE (2006)
14. Chen, T., Kornblith, S., Norouzi, M., Hinton, G.: A simple framework for contrastive learning of visual representations. In: ICML, pp. 1597–1607 (2020)
15. He, K., Fan, H., Wu, Y., Xie, S., Girshick, R.: Momentum contrast for unsupervised visual representation learning. In: CVPR, pp. 9726–9735 (2020)
16. Khosla, P., et al.: Supervised contrastive learning. In: NeurIPS, pp. 18661–18673 (2020)
17. Wang, Y., et al.: ClusterSCL: cluster-aware supervised contrastive learning on graphs, pp. 1611–1621. ACM (2022)
18. Lee, J., et al.: BioBERT: a pre-trained biomedical language representation model for biomedical text mining. Bioinformatics (2019)
19. Devlin, J., Chang, M.W., Lee, K., Toutanova, K.: BERT: pre-training of deep bidirectional transformers for language understanding (2018)
20. Wei, C.H., Kao, H.Y., Lu, Z.: GNormPlus: an integrative approach for tagging genes, gene families, and protein domains. Biomed. Res. Int. **2015**, 918710 (2015)
21. Tran, T., Kavuluru, R.: Exploring a deep learning pipeline for the BioCreative VI precision medicine task. In: BioCreative VI Workshop: 2017, pp. 107–110 (2017)
22. Tung, T., Ramakanth, K.: An end-to-end deep learning architecture for extracting protein-protein interactions affected by genetic mutations. Database J. Biol. Databases Curation (2018)

Distance Profiles of Optimal RNA Foldings

J. Liu[1], I. Duan[1], S. Santichaivekin[1], and R. Libeskind-Hadas[2](✉)

[1] Department of Computer Science, Harvey Mudd College, Claremont, USA
[2] Kravis Department of Integrated Sciences, Claremont McKenna College Claremont, Claremont, CA 91711, USA
rhadas@cmc.edu

Abstract. Predicting the secondary structure of RNA is an important problem in molecular biology, providing insights into the function of non-coding Rn As and with broad applications in understanding disease, the development of new drugs, among others. Combinatorial algorithms for predicting RNA foldings can generate an exponentially large number of equally optimal foldings with respect to a given optimization criterion, making it difficult to determine how well any single folding represents the entire space. We provide efficient new algorithms for providing insights into this large space of optimal RNA foldings and a research software tool, toRNAdo, that implements these algorithms.

Keywords: RNA folding · Nussinov Algorithm · Zuker Algorithm

1 Introduction

The secondary structure of RNA plays an important role in gene regulation and expression. For this reason, the problem of predicting RNA secondary structure, also called a *folding* of the RNA, has received considerable attention over more than four decades. Among the best-known algorithms for predicting RNA foldings are those due to Nussinov [10] and Zuker [15].

The Nussinov Algorithm finds a folding of the RNA sequence that maximizes the number of paired complementary bases (A-U, G-C, and, optionally, the weaker G-U "wobble" pairs) with no *pseudoknots.* A pseudoknot is a pair of "crossing" matched pairs; that is, matched pairs with indices (i, j) and (k, ℓ) where $i < k < j < \ell$. The Zuker Algorithm finds a folding that minimizes the free energy of the folding, again assuming no pseudoknots. While the structure of the Zuker and Nussinov dynamic programs are similar, the standard Nussinov Algorithm has time complexity $O(n^3)$ (which can be improved to $O(n^3/\log n)$

This work was funded by the U.S. National Science Foundation under Grant Number IIS-1419739 to Claremont McKenna College.

Supplementary Information The online version contains supplementary material available at https://doi.org/10.1007/978-3-031-23198-8_29.

M. S. Bansal et al. (Eds.): ISBRA 2022, LNBI 13760, pp. 315–329, 2022.
https://doi.org/10.1007/978-3-031-23198-8_29

[13]), while the Zuker Algorithm's more complex computation of free energies associated with different loop structures results in time complexity $O(n^4)$.

The restriction that foldings do not contain pseudoknots is required for the optimal substructure property exploited by the Nussinov and Zuker dynamic programs. In fact, finding minimum energy foldings with pseudoknots permitted is known to be computationally intractable (NP-complete) [7]. While comparatively rare [1], pseudoknots can arise and some efforts have been made to consider them, including complicated and slow combinatorial algorithms as well as machine learning approaches that have their own limitations [12]. As a result, the Zuker Algorithm remains a standard and has been implemented in a number of the most widely-used software tools for RNA folding [6,8,16].

However, in general, the number of equally optimal foldings (e.g., maximum number of matched base pairs in Nussinov's formulation or minimum free energy in Zuker's formulation) grows exponentially with the length of the RNA sequence. For example, Kiirala *et al.* performed experiments on 23S rRNA sequences with 117 sequences of average length 2726 and found that, on average, the number of Nussinov-optimal solutions was approximately 6×10^{130} [5].

An important and well-recognized challenge, therefore, is understanding the potentially large and diverse space of possible optimal foldings induced by a given optimization criterion. One approach is based on sampling a relatively small subset of optimal foldings and seeking to cluster them with respect to a given distance metric [2,9]. The medoids of those clusters can then be selected as good representatives of the space of optimal foldings [3]. An alternative approach to understanding the space of optimal foldings seeks to identify all of the base pair matches that are common to *all* foldings in a given set; for the sets of optimal foldings found by dynamic programming algorithms such as the Nussinov and Zuker algorithms, this can be done in polynomial-time [5].

Another approach for understanding the space of optimal foldings is based on computing statistics on the space with respect to a given distance metric. For example, the diameter of the space (the maximum distance between two foldings in the space) and the distribution of pairwise distances between all pairs of foldings are both potentially informative statistics and have been used successfully for other problems with large solution spaces [4,11].

In many studies, a single optimal folding, R, is selected and it is therefore desirable to indicate the extent to which all other optimal foldings differ from R. A relatively simple measure is the maximum distance between R and any other optimal folding with respect to a given distance metric. A more general and informative statistic is the distribution of the distances between all optimal foldings and R. In this paper we show that both of these statistics can be computed exactly (i.e., without using sampling methods) in polynomial time for a family of distance metrics, in spite of the fact that the space of optimal foldings is, in general, exponentially large. In particular, we show how this can be done for both the maximum base pairing (Nussinov Algorithm) and the minimum free energy (Zuker Algorithm) objectives. Specifically, the contributions of this paper are: (1) An efficient exact algorithm for finding the maximum distance from a given optimal folding to all optimal foldings found by the Nussinov and

Zuker Algorithms; (2) a generalization of this algorithm for computing the distribution of distances from a given optimal folding to all optimal foldings; and (3) a software tool, toRNAdo, that implements these algorithms.

2 Background

This section provides background required for the next section where our new algorithms are presented. Since there are a number of variations in the formulations of the Nussinov and Zuker Algorithms, and our new distance algorithms are formulation-dependent, we describe those algorithms here.

Throughout this paper, let S denote a given RNA string of length n, indexed from 1 to n. Let $S(i,j)$ denote the substring of S from indices i to j, inclusive. If $j < i$, we define $S(i,j)$ to be the empty string.

Let $\mathbf{comp}(i,j)$ be a Boolean-valued function that is True if the bases at indices i and j are complementary and False otherwise. Note that A-U and G-C are always assumed to be complementary and some formulations also allow G-U "wobble pairs" to be complementary.

A *folding* of string $S(i,j)$ is a set of ordered pairs $\{(i_1,j_1),\ldots,(i_m,j_m)\}$ such that $i \leq i_k < j_k \leq j$ for $1 \leq k \leq m$; each pair (i_k,j_k), called a *matched pair*, satisfies $\mathbf{comp}(i_k,j_k)$; and each index appears in at most one matched pair. We restrict our attention to foldings with no pseudoknots, that is, foldings where there exists no $1 \leq k, \ell \leq m$ such that $i_k < i_\ell < j_k < j_\ell$. Also note that when referring to an ordered pair of indices (i,j), we do not assume that they correspond to the indices of a matched pair in the RNA string unless explicitly stated.

2.1 The Nussinov Algorithm

The Nussinov Algorithm [10] seeks to maximize the total number of matched base pairs in a folding. Let $\mathbf{N}(i,j)$ denote the maximum number of matched base pairs in a folding of $S(i,j)$ for $1 \leq i,j \leq n$. We compute and store these values in a $n \times n$ dynamic programming table $\mathbf{N}$. When $j \leq i$, no matches are possible, so $N(i,j) = 0$. When $j \geq i+1$, we consider two cases: either i is unpaired in an optimal folding of $S(i,j)$, or i is paired with a complementary base at some index k, $i < k \leq j$. Therefore,

$$\mathbf{N}(i,j) = \max \begin{cases} \mathbf{N}(i+1,j) & i \text{ unpaired} \\ \max\limits_{\substack{i<k\leq j, \\ \mathbf{comp}(i,k)}} 1 + \mathbf{N}(i+1,k-1) + \mathbf{N}(k+1,j) & i \text{ paired with } k \end{cases}$$

Note that each entry in this recurrence relies on entries with a greater value of i. Thus, we compute the entries in the dynamic programming table by decreasing values of i. We can reconstruct the optimal solutions by recording the optimal choices at each step. Specifically, the traceback information is stored in a table $\mathbf{opt}$. Each entry $\mathbf{opt}(i,j)$ stores the set of all k such that $i < k \leq j$, and the base at index i is paired with the base at index k in some Nussinov-optimal folding for $S(i, j)$. Additionally, if i is unmatched in some Nussinov-optimal solution for $S(i, j)$, then i is included in $\mathbf{opt}(i, j)$ as well. The asymptotic time complexity of this algorithm, including the construction of the table **opt**, is $O(n^3)$, since there are $O(n^2)$ entries and computing each entry takes at most $O(n)$ steps.

2.2 The Zuker Algorithm

In contrast to the Nussinov Algorithm, which seeks to maximize the number of matched base pairs, the Zuker Algorithm seeks to minimize the free energy of a folding. There are a number of different formulations of the dynamic program for the Zuker Algorithm, all yielding the same set of optimal solutions. In this section we describe an adaptation of the formulation from [14] with the additional important property that when multiple cases are considered to compute the set of optimal foldings for $S(i, j)$, the sets of foldings for each of these cases are disjoint. This property is necessary when we compute the distribution of distances from one optimal folding to the set of all optimal foldings, in order to avoid overcounting solutions.

The Zuker Algorithm proceeds by identifying different types of RNA substructures, as each type of substructure may make a different contribution to the free energy. These structures are called *hairpin loops*, *stacking loops*, *internal loops*, and *multiloops*. Each of these loops, other than internal loops, is represented by a matched pair of indices (i, j), $i < j$, where the loop begins and ends; internal loops are parameterized by four indices (i, j, i', j'), $i < i' < j' < j$, to indicate that the loop has two ends, one at matched pair (i, j) and the other at matched pair (i', j'). The details of these loop topologies are not necessary for what follows, but the interested reader is referred to [14] as well as the Supplement (`shorturl.at/ELQX0`) which explicates the relationship between the original Zuker formulation and the equivalent one used here.

The energy functions of a hairpin loop (i, j), a stacking loop (i, j), and an internal loop (i, j, i', j') are given and denoted by $eH(i, j)$, $eS(i, j)$, and $eL(i, j, i', j')$, respectively [14]. Multiloops are computed as a function of three given positive constants a, b, and c; the free energy computation for multiloops is described below.

Now, we describe the formulation of the Zuker Algorithm. The equivalence of this formulation and the standard formulation [14] is easily verified. We calculate four $n \times n$ tables: **W**, **V**, **WM**, and **WM2**.[1] Let $\mathbf{W}(i, j)$ denote the minimum energy of a folding of $S(i, j)$. Note that the minimum energy for a folding of the entire RNA string is given by $\mathbf{W}(1, n)$. Let $\mathbf{V}(i, j)$ denote the minimum energy of a folding of $S(i, j)$ such that (i, j) is a matched pair. The terms $\mathbf{WM}(i, j)$ and $\mathbf{WM2}(i, j)$ are used to compute the minimum energy of foldings that are part of a multiloop.

We first compute $\mathbf{V}(i, j)$. A positive integer m is used to specify the minimum size of a permitted loop. In the base case where $j - i < m$ or $\mathbf{comp}(i, j)$ is False,

[1] **WM** is the multiloop DP table [14] and table **WM2** is introduced here as a "helper" table that allows us to avoid double-counting optimal solutions involving multiloops.

we have $\mathbf{V}(i,j) = \infty$ because no solution that matches the bases at indices i and j can exist. Otherwise, we take the minimum over the following four cases, corresponding to the four possible types of loops closed by (i,j) in an optimal folding:

$$\mathbf{V}(i,j) = \min \begin{cases} eH(i,j) & \text{Hairpin loop}(i,j) \\ eS(i,j) + \mathbf{V}(i+1,j-1) & \text{Stacking loop}(i,j) \\ \min\limits_{\substack{i<i'<j'<j \\ (i',j')\neq(i+1,j-1)}} eL(i,j,i',j') + \mathbf{V}(i',j') & \text{Internal loop } (i,j,i',j') \\ a + \mathbf{WM2}(i+1,j-1) & \text{Multiloop} \end{cases}$$

Next, we compute $\mathbf{W}(i,j)$. When $j \leq i$, the optimal folding of $S(i,j)$ contains no matched base pairs, so $\mathbf{W}(i,j) = 0$. Otherwise, we consider the two cases, where i either is unpaired, or paired with k for some $i < k \leq j$. Thus,

$$\mathbf{W}(i,j) = \min \begin{cases} \mathbf{W}(i+1,j) & i \text{ is unpaired} \\ \min\limits_{i<k\leq j} \mathbf{V}(i,k) + \mathbf{W}(k+1,j) & i \text{ is paired with } k \end{cases}$$

Next, we compute the minimum free energies due to multiloops. For $j \leq i$, we set $\mathbf{WM}(i,j) = \mathbf{WM2}(i,j) = \infty$ since no loop can exist on a substring of length 1 or 0. Otherwise,

$$\mathbf{WM}(i,j) = \min \begin{cases} \mathbf{WM}(i+1,j) + c & i \text{ is unpaired} \\ \min\limits_{i<k\leq j} \mathbf{V}(i,k) + b + c(j-k) & i \text{ is paired with } k\text{, no pairings in } S(k+1,j) \\ \min\limits_{i<k\leq j} \mathbf{V}(i,k) + \mathbf{WM}(k+1,j) + b & i \text{ is paired with} k\text{, more pairings in } S(k+1,j) \end{cases}$$

$$\mathbf{WM2}(i,j) = \min \begin{cases} \mathbf{WM2}(i+1,j) + c & i \text{ is unpaired} \\ \min\limits_{i<k<j} \mathbf{V}(i,k) + \mathbf{WM}(k+1,j) + b & i \text{ is paired with } k \end{cases}$$

As in the Nussinov Algorithm, it is easy to record all optimal choices at each step so that we can reconstruct all the energy-minimizing solutions. We store this traceback information in a $4 \times n \times n$ table **opt**, where each entry $\mathbf{opt}(\mathbf{T},i,j)$ records the set of optimal choices for the entry $\mathbf{T}(i,j)$ where $\mathbf{T} \in \{\mathbf{W}, \mathbf{V}, \mathbf{WM}, \text{and } \mathbf{WM2}\}$; each optimal choice is a list of recursive calls in the form of $(\mathbf{T}',i',j')$. For example, suppose when computing the entry $\mathbf{WM2}(i,j)$, we find that there are three optimal choices (all yielding the same minimum value for that entry), which are: (1) not pairing i, (2) pairing i with k_1, and (3) pairing i with k_2. Then, for this example,

$$\begin{aligned} \mathbf{opt}(\mathbf{WM2},i,j) = \Big\{ & \{(\mathbf{WM2}, i+1, j)\}, \\ & \{(\mathbf{V}, i, k_1), (\mathbf{WM}, k_1+1, j)\}, \\ & \{(\mathbf{V}, i, k_2), (\mathbf{WM}, k_2+1, j)\} \Big\}. \end{aligned}$$

An optimal folding for $\mathbf{WM2}(i,j)$ can then be obtained by choosing any one of those three choices and continuing the traceback process from those choices. The resulting optimal folding is the set of matched pairs (k,ℓ) that appear in each tuple $(\mathbf{V},k,\ell)$ encountered in this traceback, since $\mathbf{V}(k,\ell)$ is invoked iff k is matched to ℓ.

As noted earlier and easily verified, this formulation has the property that each optimal choice (i.e. a list of recursive calls) recorded in $\mathbf{opt}(\mathbf{T},i,j)$ corresponds to a disjoint set of optimal solutions for $\mathbf{T}(i,j)$.

The asymptotic time complexity of the Zuker Algorithm, including the construction of the traceback table **opt**, is $O(n^4)$, since there are $O(n^2)$ entries to fill and computing each entry takes at most $O(n^2)$ time. The most costly computation occurs in computing the internal loop case for $V(i,j)$, which takes $O(n^2)$, while computing all the other entries takes $O(n)$ time.[2]

2.3 Definitions and Notation

Recall that we use S to denote a given RNA string of length n indexed from 1 to n and $S(i,j)$ denotes the substring of S from indices i to j, inclusive. We use R to denote a given optimal folding (Nussinov- or Zuker-optimal). Let $R(i,j)$ denote the set of matched bases (i',j') in R where $i \leq i' < j' \leq j$.

For simplicity of exposition, we consider one distance metric on foldings, the *symmetric difference distance*, also known as the *BP distance* [2]: Given two foldings R_1 and R_2, the distance between the two foldings is equal to the number of matched pairs in R_1 that are not found in R_2 plus the number of matched pairs in R_2 that are not found in R_1. Henceforth, "distance" refers to this distance metric. We revisit distance metrics in Sect. 6.

The distribution of distances from a given optimal folding, R, to the set of all optimal foldings will be represented by a vector, where the ith element of the vector indicates the number of optimal foldings at distance exactly i from the folding R under consideration. Throughout this work, vectors are understood to be over the non-negative integers. For a vector v, let $v[i]$ denote the element at index i. Given two vectors u and v, the *sum* of u and v, denoted $u+v$ is defined by $(u+v)[i] = u[i] + v[i]$. The *convolution* of u and v, denoted $u * v$ is defined by:

$$(u * v)[i] = \sum_{j=0}^{i} u[j]v[i-j]$$

The convolution of a sequence of vectors $V = (v_1, \ldots, v_n)$ is denoted $\mathop{\ast}_{v \in V} v$. (Note that this extension is well-defined since convolution is associative.) Given a vector v, σ^j, the j-place *shift* of v is defined by:

$$\sigma^j(v)[i] = \begin{cases} v[i-j] : i \geq j \\ 0 \quad\quad\;\; : i < j \end{cases}$$

[2] A common heuristic bounds the interior loop size, which reduces the running time to $O(n^3)$.

Finally, we use the notation $\delta(X)$ to indicate the function that evaluates to 0 if the Boolean X is False and 1 otherwise.

3 Nussinov-Optimal Foldings

3.1 Nussinov Maximum Distance

We now describe a dynamic programming algorithm that, given an RNA sequence S and a Nussinov-optimal folding R of that sequence, calculates the maximum distance between R and all Nussinov-optimal foldings for S. A most distant Nussinov-optimal folding with respect to R can then be reconstructed through traceback.

The maximum distance algorithm begins by running the Nussinov Algorithm, which returns a traceback table **opt**. Recall from Sect. 2.1 that $\mathbf{opt}(i, j)$ stores the set of k, $i < k \leq j$, such that the base at index i is paired with the base at index k in some optimal folding for the substring $S(i, j)$ and, if i is unmatched in some Nussinov-optimal solution for that substring, i is also included in this set. Additionally, given the Nussinov-optimal folding R, we can efficiently construct a table **numpairs**, where $\mathbf{numpairs}(i, j)$ is the number of matched pairs in $R(i, j)$ for $1 \leq i, j \leq n$. In other words, $\mathbf{numpairs}(i, j)$ is the number of pairs $(i', j') \in R$ such that $i \leq i' < j' \leq j$. We construct **numpairs**, as follows.

For the base cases, when $j \leq i$, $\mathbf{numpairs}(i, j) = 0$ because any RNA substring of R of length 1 or less contains no matches. When $j \geq i + 1$, we compute the entries in increasing lengths of the substring, that is, by increasing the value of $j - i$. These entries are computed as follows:

$$\mathbf{numpairs}(i, j) = \mathbf{numpairs}(i + 1, j) + \begin{cases} 1 & \text{if } (i, k) \in R \text{ for some } i < k \leq j \\ 0 & \text{otherwise} \end{cases}$$

The **numpairs** table allows us to efficiently compute the number of matches (i', j') in $R(i, j)$ that either use or "cross" an index k, meaning that $i \leq i' \leq k$ and $k \leq j' \leq j$. Specifically, the number of such crossings, denoted by $\mathbf{numcrossings}(i, j, k)$, can be calculated as:

$$\mathbf{numcrossings}(i, j, k) = \mathbf{numpairs}(i, j) - \mathbf{numpairs}(i, k - 1) - \mathbf{numpairs}(k + 1, j)$$

Now, we can calculate the maximum distance between R and all Nussinov-optimal foldings of S by computing a $n \times n$ table **maxdistance**. Each entry $\mathbf{maxdistance}(i, j)$ represents the maximum distance between $R(i, j)$, the set of matched bases (i', j') in R where $i \leq i' < j' \leq j$, and any Nussinov-optimal folding of $S(i, j)$, for all $1 \leq i, j \leq n$.

Base Case: When $j \leq i$, the substring $S(i, j)$ has length 1 or 0, so both $R(i, j)$ and any Nussinov-optimal solution for $S(i, j)$ contain no matched bases. Therefore, for all $j \leq i$:

$$\mathbf{maxdistance}(i, j) = 0$$

Recursive Case: When $j \geq i+1$, we compute the entries $\mathbf{maxdistance}(i,j)$ in increasing values of $j-i$. For each entry, we consider two cases: (1) there exists k' such that $(i,k') \in R(i,j)$, that is, i is matched to k' in R for some $i < k' \leq j$, (2) there doesn't exist such k', which means i is either unmatched in R or matched with some ℓ where $\ell < i$ or $\ell > j$.

Case 1: i is matched to k' in $R(i,j)$ for some $i < k' \leq j$.

In this case, we must consider all the ways that i is matched or unmatched in the set of Nussinov-optimal solutions for $S(i,j)$. Recall that $\mathbf{opt}(i,j)$ stores the set of indices k, $i < k \leq j$, such that (i,k) is matched in some optimal solution for $S(i,j)$. Additionally, $i \in \mathbf{opt}(i,j)$ if i is unmatched in some optimal solution for $S(i,j)$. We partition the set of optimal solutions for $S(i,j)$ into two parts, depending on whether i is matched in the solution or not. We consider each part separately and then we calculate $\mathbf{maxdistance}(i,j)$ by taking the maximum of the two parts.

In **part1**, we compute the maximum distance between $R(i,j)$ and all Nussinov-optimal solutions for $S(i,j)$ where i is matched to some k, $i < k \leq j$:

$$\begin{aligned}\mathbf{part1} = \max_{\substack{k \in \mathbf{opt}(i,j)\\ k \neq i}} \{&\mathbf{maxdistance}(i+1,k-1) + \mathbf{maxdistance}(k+1,j)\\ &+ \mathbf{numcrossings}(i+1,j,k) + 2\delta(k \neq k')\}\end{aligned}$$

Note that for each k under consideration, the maximum distance for the substring $S(i,j)$ consists of the maximum distance for the substrings $S(i+1,k-1)$ and $S(k+1,j)$ (first line), the number of matched pairs in $R(i+1,j)$ that cross index k since these matched pairs are found in $R(i,j)$ but not in any optimal folding for $S(i,j)$ that matches i and k (second line), and the distances contributed by the matched pairs (i,k') and (i,k) (second line). If $k \neq k'$, then (i,k') and (i,k) are two different matched pairs and contribute 2 to the distance. If $k = k'$, then (i,k') and (i,k) are the same matched pair and make no contribution to the distance.

In **part2**, we compute the maximum distance between $R(i,j)$ and all Nussinov-optimal solutions for $S(i,j)$ where i is unmatched. If $i \notin \mathbf{opt}(i,j)$, such optimal solutions do not exist, so we set $\mathbf{part2} = -\infty$. If $i \in \mathbf{opt}(i,j)$, then $\mathbf{part2} = \mathbf{maxdistance}(i+1,j)+1$. Here, the matched pair $(i,k') \in R(i,j)$ contributes 1 to the distance because $R(i,j)$ matches i with k', where $i < k' \leq j$, but the Nussinov-optimal solutions under consideration do not match i. Thus,

$$\mathbf{part2} = \begin{cases} \mathbf{maxdistance}(i+1,j)+1 & \text{if unmatched } i \text{ is in } \mathbf{opt}(i,j)\\ -\infty & \text{otherwise.}\end{cases}$$

Case 2: i is unmatched in $R(i,j)$.

Similar to the previous case, we partition the set of Nussinov-optimal solutions into two parts. In **part1**, we compute the maximum distance between $R(i,j)$ and the set of Nussinov-optimal solutions for $S(i,j)$ where i is matched to some k, $i < k \leq j$:

$$\mathbf{part1} = \max_{\substack{k \in \mathbf{opt}(i,j) \\ k \neq i}} \{\mathbf{maxdistance}(i+1, k-1) + \mathbf{maxdistance}(k+1, j) + \mathbf{numcrossings}(i+1, j, k) + 1\}$$

The calculation is similar to the calculation for **part1** in Case 1 except for the last term: since i is unmatched in $R(i,j)$, any matching involving i in a Nussinov-optimal solution will contribute 1 to the distance. In **part2**, we again compute the maximum distance between $R(i,j)$ and all Nussinov-optimal solutions for $S(i,j)$ where i is unmatched:

$$\mathbf{part2} = \begin{cases} \mathbf{maxdistance}(i+1, j) & \text{if unmatched } i \text{ is in } \mathbf{opt}(i,j) \\ -\infty & \text{otherwise.} \end{cases}$$

This calculation differs from **part2** in Case 1 because neither $R(i,j)$ nor the Nussinov-optimal foldings under consideration match i.

Finally, in both Case 1 and Case 2, we find the maximum distance from all optimal solutions by

$$\mathbf{maxdistance}(i,j) = \max\{\mathbf{part1}, \mathbf{part2}\}$$

The asymptotic time complexity for the maximum distance algorithm is $O(n^3)$. The precomputation of **opt** takes time $O(n^3)$ and **numpairs** takes time $O(n^2)$. Filling the **maxdistance** DP table takes $O(n^3)$ as there are $O(n^2)$ cells to fill, and filling each cell takes $O(n)$ steps since we need to iterate through $O(n)$ possible values in $\mathbf{opt}(i,j)$.

3.2 Nussinov Distance Vector

The maximum distance algorithm can be extended to compute the *distance vector* $x = (x_0, x_1, \ldots, x_n)$ where x_i denotes the number of Nussinov-optimal foldings whose distance from R is exactly i. Note that $x[0] = 1$ since R is the only Nussinov-optimal folding with distance 0 from R, and that the maximum distance between any two Nussinov-optimal foldings is upper-bounded by n since a folding of string S of length n can have at most $\lfloor n/2 \rfloor$ matched pairs and thus two such foldings can differ in at most $2 \times \lfloor n/2 \rfloor \leq n$ matched pairs. Henceforth, we assume that all distance vectors have length $n+1$, with indices 0 through n, by appending zeros to the end if necessary. Recall that the notation $*$ denotes the convolution operator and $+$ denotes vector addition when its arguments are vectors.

We now describe the dynamic programming algorithm for computing $\mathbf{distanceVector}(i,j)$, the distance vector between $R(i,j)$ and the set of all Nussinov-optimal foldings for $S(i,j)$.

Base Case: When $j \leq i$, the substring $S(i,j)$ has length 1 or 0, so there is only one Nussinov-optimal solution for $S(i,j)$, which is the trivial solution. This trivial solution is distance 0 from $R(i,j)$ since $R(i,j)$ also contains no matched bases. Therefore, for all $1 \leq i \leq n, j \leq i$:

$$\mathbf{distanceVector}(i,j) = (1, 0, 0, \ldots, 0)$$

Recursive Case: When $j \geq i+1$, we compute the entries of **distanceVector** in increasing values of $j - i$, as we did for **maxdistance**. As before, we calculate the entries $\mathbf{distanceVector}(i,j)$ by cases.

Case 1: i is matched to k' in $R(i,j)$ for some $i < k' \leq j$.

In this case, $\mathbf{distanceVector}(i,j)$ is calculated by considering two parts in the partition of the set of Nussinov-optimal solutions. In **part1**, we consider the distribution of distances formed by $R(i,j)$ and the set of Nussinov-optimal solutions for $S(i,j)$ where i is matched to some $i < k \leq j$:

$$\mathbf{part1} = \sum_{\substack{k \in \mathbf{opt}(i,j) \\ k \neq i}} \sigma^{c_{i+1,j,k}+2\delta(k \neq k')}(\mathbf{distanceVector}(i+1, k-1) * \mathbf{distanceVector}(k+1, j))$$

where $c_{i+1,j,k} = \mathbf{numcrossings}(i+1, j, k)$. Here, each term in the sum corresponds to the distance vector between $R(i,j)$ and the set of Nussinov-optimal solutions for $S(i,j)$ that contain the match (i,k). Consider a Nussinov-optimal solution for $S(i,j)$, denoted $Q(i,j)$, that contains the match (i,k) and has a distance of exactly d from $R(i,j)$. That distance, d, is made up of the distance between $Q(i+1, k-1)$ and $R(i+1, k-1)$, the distance between $Q(k+1, j)$ and $R(k+1, j)$, plus the distance contributed by matches in $Q(i,j)$ and $R(i,j)$ not contained in either substrings $S(i+1, k-1)$ and $S(k+1, j)$. To calculate how many Nussinov-optimal solutions are of this type, we take the convolution of the $\mathbf{distanceVector}(i+1, k-1)$ and $\mathbf{distanceVector}(k+1, j)$ and shift the result by the number of matches in $R(i+1, j)$ that cross k and any distance contributed by the matches (i,k) and (i,k').

In **part2**, we compute the distribution of distances between $R(i,j)$ and all Nussinov-optimal solutions for $S(i,j)$ where i is unmatched. If $i \notin \mathbf{opt}(i,j)$, such optimal solutions do not exist, so we set $\mathbf{part2} = \mathbf{0}$, the zero vector. Therefore, we have:

$$\mathbf{part2} = \begin{cases} \sigma^1(\mathbf{distanceVector}(i-1, j)) & \text{if unmatched } i \text{ is in } \mathbf{opt}(i,j) \\ \mathbf{0} & \text{otherwise.} \end{cases}$$

The shift on $\mathbf{distanceVector}(i-1, j)$ accounts for the loss of the pairing $(i, k') \in R(i,j)$.

Case 2: i is unmatched in $R(i,j)$.

For the optimal solutions where i is matched to some k, we have:

$$\mathbf{part1} = \sum_{\substack{k \in \mathbf{opt}(i,j) \\ k \neq i}} \sigma^{c_{i+1,j,k}+1}(\mathbf{distanceVector}(i+1, k-1) * \mathbf{distanceVector}(k+1, j))$$

where $c_{i+1,j,k}$ is defined as in Case 1 above. This calculation is identical to the one in Case 1 except that we shift by 1 to account for the fact that i is unmatched in $R(i, j)$ but matched in the optimal solutions for $S(i, j)$ under consideration.

For the optimal solutions where i is unmatched, we have:

$$\mathbf{part2} = \begin{cases} \mathbf{distanceVector}(i+1, j) & \text{if unmatched } i \text{ is in } \mathbf{opt}(i, j) \\ \mathbf{0} & \text{otherwise.} \end{cases}$$

Finally, in both Case 1 and Case 2, we sum together **part1** and **part2** to account for all optimal solutions for $S(i, j)$:

$$\mathbf{distanceVector}(i, j) = \mathbf{part1} + \mathbf{part2}$$

The asymptotic time complexity of the distance vector algorithm is $O(n^4 \log n)$, since there are $O(n^2)$ cells to fill, filling each cell requires up to $O(n)$ convolutions, and each convolution takes $O(n \log n)$ time using Fast Fourier Transforms (FFTs).[3]

4 Zuker-Optimal Foldings

In this section, we describe a dynamic programming algorithm that computes the maximum distance from a given Zuker-optimal folding R to all other Zuker-optimal foldings for the given string S. In the interest of space, the extension to distance vectors can be found in the Supplementary materials (`shorturl.at/ELQX0`).

The distance algorithm begins by running the Zuker Algorithm which gives us the traceback table **opt**, where each entry $\mathbf{opt}(\mathbf{T}, i, j)$ records the set of optimal choices for obtaining the value at $\mathbf{T}(i, j)$, $\mathbf{T} \in \{\mathbf{W}, \mathbf{V}, \mathbf{WM}, \text{and } \mathbf{WM2}\}$. Recall that each optimal choice is a list of recursive calls in the form of $(\mathbf{T}', i', j')$ and the optimal choices in $\mathbf{opt}(\mathbf{T}, i, j)$ form a partition of the set of optimal solutions for $\mathbf{T}(i, j)$. We also precompute the table **numpairs** for R, defined and calculated in Sect. 3.1.

Now we can compute the maximum distance between R and all Zuker-optimal foldings of S by computing a $4 \times n \times n$ table **maxdistance**. Each entry **maxdistance**$(\mathbf{T}, i, j)$ computes the maximum distance between $R(i, j)$ and any folding that can be constructed through a traceback of the table entry $\mathbf{T}(i, j)$, i.e. any optimal solution for $\mathbf{T}(i, j)$. The final answer, the maximum distance between R, and all Zuker-optimal solutions is found in $\mathbf{maxdistance}(\mathbf{W}, 1, n)$ since the set of solutions constructed by tracing back the entry $\mathbf{W}(1, n)$ is the set of all Zuker-optimal solutions.

[3] Our implementation of the convolution operator in the accompanying toRNAdo software tool is not optimized and uses the naive $O(n^2)$ algorithm.

Base Case: When $j \leq i$, the substring $S(i,j)$ has length 1 or 0. A traceback of the entry $\mathbf{T}(i,j)$ will yield the trivial solution for $\mathbf{T} = \mathbf{W}$, and no solution for $\mathbf{T} = \mathbf{V}, \mathbf{WM}, \mathbf{WM2}$. Thus,

$$\mathbf{maxdistance}(\mathbf{T}, i, j) = \begin{cases} 0 & \text{if } \mathbf{T} = \mathbf{W} \\ -\infty & \text{otherwise} \end{cases}$$

Recursive Cases: We calculate the remaining entries of **maxdistance** in increasing order of $j - i$. For each (i,j) we calculate the cells in the order $T = \mathbf{V}, \mathbf{W}, \mathbf{WM}, \mathbf{WM2}$. We compute the entry $\mathbf{maxdistance}(\mathbf{T}, i, j)$ by considering the maximum distances between $R(i,j)$ and the different parts of the set of optimal solutions for $\mathbf{T}(i,j)$, partitioned by which optimal choice they correspond to in $\mathbf{opt}(\mathbf{T}, i, j)$. The maximum distance between $R(i,j)$ and the set of optimal solutions that make the choice e is given by:

$$\begin{aligned} \mathbf{choice}(e) = & \sum_{(\mathbf{T}',i',j') \in e} \mathbf{maxdistance}(\mathbf{T}', i', j') \\ & + \mathbf{numpairs}(i,j) - \sum_{(\mathbf{T}',i',j') \in e} \mathbf{numpairs}(i', j') \\ & + \lambda(\mathbf{T}, i, j) \end{aligned}$$

where

$$\lambda(\mathbf{T}, i, j) = \begin{cases} 0 & \text{if } \mathbf{T} \neq \mathbf{V} \\ -1 & \text{if } \mathbf{T} = \mathbf{V} \text{ and } (i,j) \in R(i,j) \\ 1 & \text{if } \mathbf{T} = \mathbf{V} \text{ and } (i,j) \notin R(i,j) \end{cases}$$

Consider a choice e that contains a set of recursive calls, each of the form $(\mathbf{T}', i', j')$. The optimal solutions for $\mathbf{T}(i,j)$ that use this choice e comprise optimal solutions for each of these entries $\mathbf{T}'(i', j')$. The first line sums up the maximum distances between $R(i', j')$ and optimal solutions for $\mathbf{T}'(i', j')$ for each recursive call $(\mathbf{T}', i', j') \in e$. The second line accounts for the matched pairs that are in $R(i,j)$ but not in the optimal solutions under consideration, analogous to the **numcrossings** term in the Nussinov distance algorithm. Finally, the term $\lambda(\mathbf{T}, i, j)$ adjusts the distance depending on whether the set of optimal solutions under consideration matches (i,j), which happens when $\mathbf{T} = \mathbf{V}$. If $\mathbf{T} = \mathbf{V}$ and (i,j) is also in $R(i,j)$, then we have overcounted the distance by 1 when we count (i,j) as a difference in the second line, so in this case $\lambda(e) = -1$. If $\mathbf{T} = \mathbf{V}$ but (i,j) is not in $R(i,j)$, then (i,j) is in all the optimal solutions under consideration but not in $R(i,j)$, and this difference has not been previously accounted for, so $\lambda(e) = 1$.

Finally, we maximize over all optimal choices:

$$\mathbf{maxdistance}(\mathbf{T}, i, j) = \max_{e \in \mathbf{opt}(\mathbf{T}, i, j)} \mathbf{choice}(e)$$

The asymptotic time complexity of this algorithm is $O(n^4)$, as there are $O(n^2)$ cells to fill, and each cell can take a maximum of $O(n^2)$ steps since there can be $O(n^2)$ optimal choices in $\mathbf{opt}(\mathbf{T}, i, j)$. (Recall that an interior loop (i, j, i', j') requires us to decide the optimal locations for both i' and j'. Therefore, there can be $O(n^2)$ optimal interior loops.)

5 Algorithm Implementation

The algorithms described in the previous section have been implemented in a Python command line software tool called toRNAdo (https://github.com/iisaduan/toRNAdo).

As an example, we used this tool to fold a 517bp Arabidopsis thaliana (thale cress) partial 5S ribosomal RNA sequence (Accession: AF198208.1:1..517: rRNA). Under the Nussinov Algorithm, there were more than 10^{36} optimal foldings, and under the Zuker Algorithm, there were more than 10^{16} optimal foldings. Figure 1 shows the histograms representing the distance vectors for the two algorithms relative to randomly selected optimal solutions. These runs took 417 s (Nussinov) and 4056 s (Zuker) on a 3.2 GHz Apple M1 processor with 16 GB RAM.

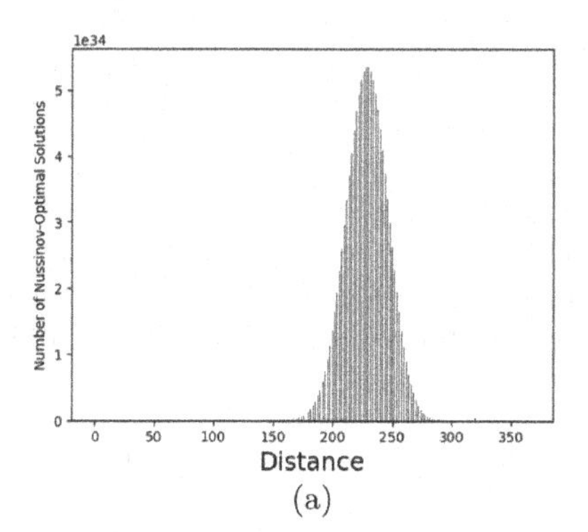

(a)

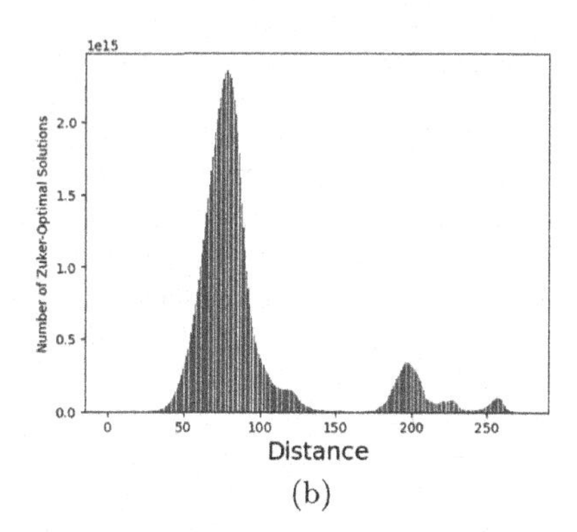

(b)

Fig. 1. Distance vector histograms for a 517bp Arabidopsis thaliana sequence relative to a randomly selected optimal solution for (a) Nussinov Algorithm (maximum distance: 366) and (b) Zuker Algorithm (maximum distance: 277). The horizontal axis indicates the distance from a particular randomly selected optimal folding R and the vertical axis indicates the number of optimal foldings that are at that distance from R, scaled by 10^{34} in (a) and 10^{15} in (b).

6 Conclusion

In this paper we have shown that the maximum distance and, more generally, the distribution of distances between a given optimal RNA folding and the set of all optimal RNA foldings can be found in polynomial time, in spite of the fact that the number of optimal foldings can be exponentially large in the length of the given RNA sequence. This provides new insights into the space of optimal RNA foldings that can be particularly important when presenting and interpreting a folding for an RNA sequence.

Our results generalize in a number of ways. First, the "base" folding R needs not be optimal; our algorithms are able to find the maximum distance and the distance vector for non-optimal foldings. Second, we can trace back the dynamic programming tables to find an actual folding that is of maximum distance from R. (This feature is implemented in toRNAdo.) Third, in this paper we used the symmetric difference distance metric. The algorithmic results presented here are generalizable to other distance metrics that assign a positive (but not necessarily unit) cost when a matched pair occurs in one folding but not in another. However, there are many other possible distance metrics on foldings [9] and determining which metrics can be used with our approach is a direction for future research.

Acknowledgements. The authors thank Harvey Mudd College for use of lab resources and the four anonymous reviewers for valuable feedback and suggestions.

References

1. Aalberts, D.P., Hodas, N.O.: Asymmetry in RNA pseudoknots: observation and theory. Nucleic Acids Res. **33**(7), 2210–2214 (2005). 10.1093/nar/gki508
2. Agius, P., Bennett, K.P., Zuker, M.: Comparing RNA secondary structures using a relaxed base-pair score. RNA **16**(5), 865–878 (2010)
3. Ding, Y., Chan, C.Y., Lawrence, C.E.: Clustering of RNA secondary structures with application to messenger Rn As. J. Mol. Biol. **359**(3), 554–571 (2006)
4. Haack, J., Zupke, E., Ramirez, A., Wu, Y.C., Libeskind-Hadas, R.: Computing the diameter of the space of maximum parsimony reconciliations in the duplication-transfer-loss model. IEEE/ACM Trans. Comput. Biol. Bioinf. **16**(1), 14–22 (2018)
5. Kiirala, N., Salmela, L., Tomescu, A.I.: Safe and complete algorithms for dynamic programming problems, with an application to RNA folding. In: 30th Annual Symposium on Combinatorial Pattern Matching (CPM 2019). Schloss Dagstuhl-Leibniz-Zentrum fuer Informatik (2019)
6. Lorenz, R., et al.: Vienna RNA package 2.0. Algorithms Molecular Biol. **6**(1), 1–14 (2011)
7. Lyngsø, R.B., Pedersen, C.N.: RNA pseudoknot prediction in energy-based models. J. Comput. Biol. **7**(3–4), 409–427 (2000)
8. Markham, N.R., Zuker, M.: Unafold. In: Bioinformatics, pp. 3–31. Springer (2008)
9. Moulton, V., Zuker, M., Steel, M., Pointon, R., Penny, D.: Metrics on RNA secondary structures. J. Comput. Biol. **7**(1–2), 277–292 (2000)
10. Nussinov, R., Pieczenik, G., Griggs, J.R., Kleitman, D.J.: Algorithms for loop matchings. SIAM J. Appl. Math. **35**(1), 68–82 (1978)
11. Santichaivekin, S., Mawhorter, R., Libeskind-Hadas, R.: An efficient exact algorithm for computing all pairwise distances between reconciliations in the duplication-transfer-loss model. BMC Bioinform. **20**(20), 1–11 (2019)
12. Singh, J., Hanson, J., Paliwal, K., Zhou, Y.: RNA secondary structure prediction using an ensemble of two-dimensional deep neural networks and transfer learning. Nat. Commun. **10**(1), 1–13 (2019)
13. Venkatachalam, B., Gusfield, D., Frid, Y.: Faster algorithms for rna-folding using the four-russians method. Algorithms for Molecular Biology **9**(1), 1–12 (2014)

14. Will, S.: Lecture notes from course 18.417, computational biology, MIT, fall 2011. http://www.math.mit.edu/classes/18.417/Slides/rna-prediction-zuker.pdf. Accessed 23 June 2022
15. Zuker, M.: On finding all suboptimal foldings of an RNA molecule. Science **244**(4900), 48–52 (1989)
16. Zuker, M.: Mfold web server for nucleic acid folding and hybridization prediction. Nucleic Acids Res. **31**(13), 3406–3415 (2003)

2D Photogrammetry Image of Adolescent Idiopathic Scoliosis Screening Using Deep Learning

Zhenda Xu[1(✉)], Jiazi Ouyang[3], Qiang Gao[2], Aiqian Gan[4], Qihua Zhou[1], Jiahao Hu[3], and Song Guo[1]

[1] Department of Computing, Hong Kong Polytechnic University, Hung Hom, Hong Kong
cszxu@comp.polyu.edu.hk
[2] Department of Rehabilitation Medicine, West China Hospital, Sichuan University, Chengdu, Sichuan, China
[3] Shenzhen ZeroDynamic Medical Technology Company Limited, Shenzhen, China
[4] Faculty of Science, The University of Sydney, Camperdown, Australia

Abstract. Adolescent idiopathic scoliosis (AIS) is a high incidence disease in adolescents, with a long treatment time and difficult to cure. As a consensus, the preliminary AIS screening is of crucial importance to detect the disease at an early stage and allows proactive interventions to prevent the disease from becoming worse and reduce future treatment. Currently, the conventional palpation or Adam forward leaning is the most widely used preliminary screening method considering the Axial Trunk Rotation (ATR) value calculated by scoliosis assessment equipment. However, this method relies heavily on the subject's standing posture and the doctor's experience. In this paper, we develop an efficient deep learning-based framework to enable a large-scale scoliosis screening by using only one unclothed two-dimensional (2D) human back image, without any X-radiation equipment. We classify the normal and abnormal scoliosis using ATR value as classification label which calculated from the human back three-dimensional (3D) point cloud. Our accuracy in the task of AIS classification reaches 81.3%, far exceeding the accuracy of visual observation by experienced doctor (65%), which can be used as a remote preliminary scoliosis screening method.

Keywords: Preliminary AIS screening · ATR · Deep learning · Human back · Classification

1 Introduction

AIS is the most common spinal disease in adolescents, with a global prevalence of 0.5–5.2% [19]. Without intervention, AIS will continue to develop before the adolescent bone matures, affecting the body appearance, cardiopulmonary function, and even paralysis [16]. However, as a chronic disease, it can usually be

M. S. Bansal et al. (Eds.): ISBRA 2022, LNBI 13760, pp. 330–342, 2022.
https://doi.org/10.1007/978-3-031-23198-8_30

found early in the disease, such as judging by abnormal posture and back muscle imbalance. Through timely regular review and health education, it can be effectively controlled and corrected. Therefore, it is considered that the prevention of such diseases is far more important than later treatment [17]. AIS is generally diagnosed by ATR angle or Cobb angle. Although Cobb angle is more authoritative in AIS diagnosis, it cannot be accurately calculated from human back shape, and additional invasive radiography is required to expose the characteristics of the whole spine. In the field of large-scale scoliosis screening, some research teams found through a large number of samples that judging whether adolescents have scoliosis by whether the ATR value exceeds 5° has achieved the highest sensitivity of 87% in the AIS classification task. [6], shows a good correlation with the radiographic measurements, which become a more popular and universal AIS screening standard in the world.

The routine AIS screening process includes preliminary screening, outpatient screening and instrument screening. As the first step of AIS diagnosis, preliminary screening is an important basis for disease awareness and treatment. In the initial screening, patients who are judged to be at risk of spinal abnormalities will receive follow-up outpatient screening and instrument screening according to the recommendations to further diagnose posture abnormalities or scoliosis.

At present, the common means of preliminary screening include general examination, Adam forward leaning test [1] and scoliosis meter examination [3]. General examination and forward leaning test require the subject to expose his back, stand naturally or make a standard flexion posture, and the examiner shall make a diagnosis through visual observation. Their accuracy is highly dependent on the clinician's diagnostic experience, leading to the subjectivity of screening. Scoliosis measurement instrument detects ATR value through scoliosis measurement instrument on the basis of forward bending posture or standing posture. ATR value is highly correlated with back information, and the current mainstream screening method is to calculate ATR value. Previously, a research team estimated the rotation angle of vertebrae by calculating the three-dimensional back surface ATR data of human body in standing position, and found that they had a high correlation [14]. Although it is cheap, easy to obtain and harmless, complex processes and screening instruments with different quality levels are difficult to ensure the standardization of the screening process.

The well-known diagnostic method X-ray examination, although it is professional and authoritative in diagnostic accuracy, the radiation injury of X-ray, high cost and environmental requirements limit its large-scale application in primary screening. In order to solve the problems of large-scale AIS preliminary screening. Many noninvasive AIS evaluation methods have been proposed [9], such as Moire topographic map or parallel light to display the back surface shape, but strict equipment use conditions hinder their popularization and universal application. Another non-radiation injury evaluation method [5] through ultrasound is introduced to image the shape of the spine, but it needs to apply media on the back of the human body, professional doctors to operate, and contact the back of the human body. This method has low efficiency and weak stability. In addition, the optical non-invasive surface measurement system [8] developed based on high-precision surface measurement equipment can realize three-dimensional reconstruction of the back or the whole torso, but it is usually very expensive.

Based on the above preliminary screening work for scoliosis, the main challenges are as follows: 1. The current mainstream non-radiation scoliosis assessment equipment are relatively large and not portable enough; 2. The screening equipment cost are high; 3. The screening equipment detection efficiency are low. Since there is a high correlation between the 2D human back image of AIS patients and ATR, and ATR is also the standard for the primary screening of scoliosis, we use deep learning technology to predict scoliosis based on the unclothed human back 2D image and ATR value. This system takes one single 2D image of the unclothed human back and uses ATR value as an indicator to classify whether adolescents have scoliosis. First, we collect the 2D back image of the same sample and its corresponding three-dimensional(3D) back point cloud data, and based on the point cloud data, we calculate the ATR value of the back of the human body, and mark whether there is scoliosis according to the size of the ATR value. At the same time, an experienced doctor will judge whether the sample has scoliosis through the naked eye. Then, we use the UNET model [12] and the YOLO mode [11] l to train the human back segmentation model to segment the human back region on the 2D image to achieve preprocessing. After that, we train EFFICIENTNET-B4 network [13] to classify images by using the 2D image of the human back and corresponding label calculated by using ATR values and 3D human back. The experimental results reveal that there is a high correlation between 2D human back image and ATR values. Using the result of classifying scoliosis by ATR value as a label can well realize the classification of scoliosis based on one singe 2D human back image. In summary, the contributions of this paper are outlined below:

1. We are the first to establish a 2D human back image and the corresponding 3D point cloud database for scoliosis screening. All data contain 2D human back image of the same sample and the corresponding 3D back point cloud.

2. We propose a fast, accurate and low-cost scoliosis screening algorithm. This method creatively uses ATR based classification results as labels. This system learns the correlation between the 2D human back image and the ATR related information calculated from the 3D point cloud of the unclothed human back. Finally, the system can classify scoliosis based on a single 2D image of the back and the screening effect is better than that observed by experienced doctors with the naked eye.

3. This method realizes the user's family independent screening without professional doctors and radiation.

2 Related Works

2.1 AIS Screening Using 3D Image

A research group proposed to use low-cost RGB-D camera to locate the anatomical landmarks on the back of the human body [17]. This method has high accuracy and stability, but the subjects need to take off their clothes in public, which is still limited to large-scale screening. Some other teams use the support vector machine model to classify the severity of scoliosis based on the 3D human back surface. However, the main classification ranges are the Cobb angle less than 30° and greater than 30°, which can not be used for preliminary screening [10].

2.2 AIS Screening Using 2D Image

With the image processing technology and deep learning technology rapidly developing, some researchers have applied image processing and deep learning algorithm to AIS screening by using one single 2D human unclothed back image. A research group proposed an edge-based contour model to reconstruct the contour of the back of the human body and the center of the spine. Although this method is very simple and has a certain degree of accuracy, the results are unstable and inaccurate due to the complex background and the uneven light on the back of the human body [7]. Another research team uses one single RGB image of the human back, the Cobb angle interval is predicted based on RESNET model to realize classification. The classification accuracy is only about 60%, which is similar to that of experienced doctors' visual diagnosis [19]. Deep learning methods are popular in many screening fields. However, there are less studies are proposed for AIS screening.

3 Proposed Method

3.1 Data Acquisition

1935 subjects participate in the experiment. They are healthy or have different degrees of scoliosis, and their ages are concentrated between 5 and 14 years old with ATR ranges from 1 to 15 degrees. All the images used in the experiment are taken when the subjects stood naturally facing the wall without upper clothes (above the sacrum and completely exposed the anatomical landmarks on the back). Kinect2 (Microsoft) sensor is located 1 m behind the subject, and the height of Kinect sensor is about the position of the fifth thoracic vertebra of the subject. The sensing plane of Kinect is still relatively parallel to the wall [17]. After the acquisition process lasting about 2 s, a group of RGB-D images of the back of the same person at the same time are obtained, which includes 2D human back image and depth map. The ATR value and classification label are obtained by 3D point cloud data, which calculated by depth map. We use 2D human back image as input and its corresponding ATR value as label for training and testing. In order to compare with the artificial naked eye diagnosis, an experienced doctor uses the naked eye to judge whether the sample has scoliosis. The doctor marks 1 for the samples with scoliosis, and 0 for those without scoliosis. We count the data observed by doctors with the naked eye for comparison. Our collection method and collected data example are shown in Fig. 1.

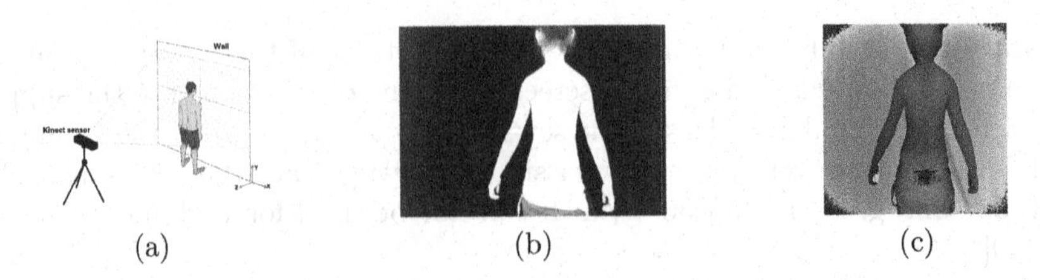

(a) (b) (c)

Fig. 1. The data acquisition method, 2D back image and depth map data are shown in (a), (b) and (c) respectively.

3.2 Scoliosis Risk Assessment Model

Figure 2 shows the framework of our model. Our scoliosis risk assessment model includes three modules: 2D image data processing, 3D human back ATR calculation and label classification, and scoliosis risk classification. The data processing module helps to remove interference features and retain the necessary image features for scoliosis risk assessment. The scoliosis risk classification module classifies the spinal abnormalities based on the label calculated by ATR value under international AIS criteria.

Image Preprocessing Module. Before assessing the risk of scoliosis, we process the original 2D images to obtain back images. As shown in Fig. 2, we first segment the people in the 2D image with UNET model. Using the model information our data pre-trained on the human matching dataset [4], we obtain an image mask in which the background area is black and the human back is white. Mask the image mask on the 2D image, and fill the background with black according to formula (1) to eliminate the interference caused by background information.

$$a_{ij} = min(c_{ij}, m_{ij}) \quad (1)$$

where a_{ij}, c_{ij} and m_{ij} respectively represent the pixel values with coordinates (i, j) in the segmented 2D image, original 2D image and human mask. Further, in order to extract the region of interest in the human back, we train a YOLO algorithm to identify the back region of interest from C7 to the sacrum. We use YOLO V5 released in June 2020 as the back region recognition model, which is faster, more flexible and lighter than the previously released version [2]. So as to obtain our training data.

ATR Calculation Module. We extract feature points from the 3D human back, as shown in Fig. 3(a). The following anatomical landmarks need to be located: Carinal point (C7), posterior superior iliac spine (PSIS), spinous process line (a series of lines from carinal bone to the midpoint of left and right PSIS).

In order to reduce the comparison error between different the reconstructions of the subjects and that of the same subject with different angle relative to the

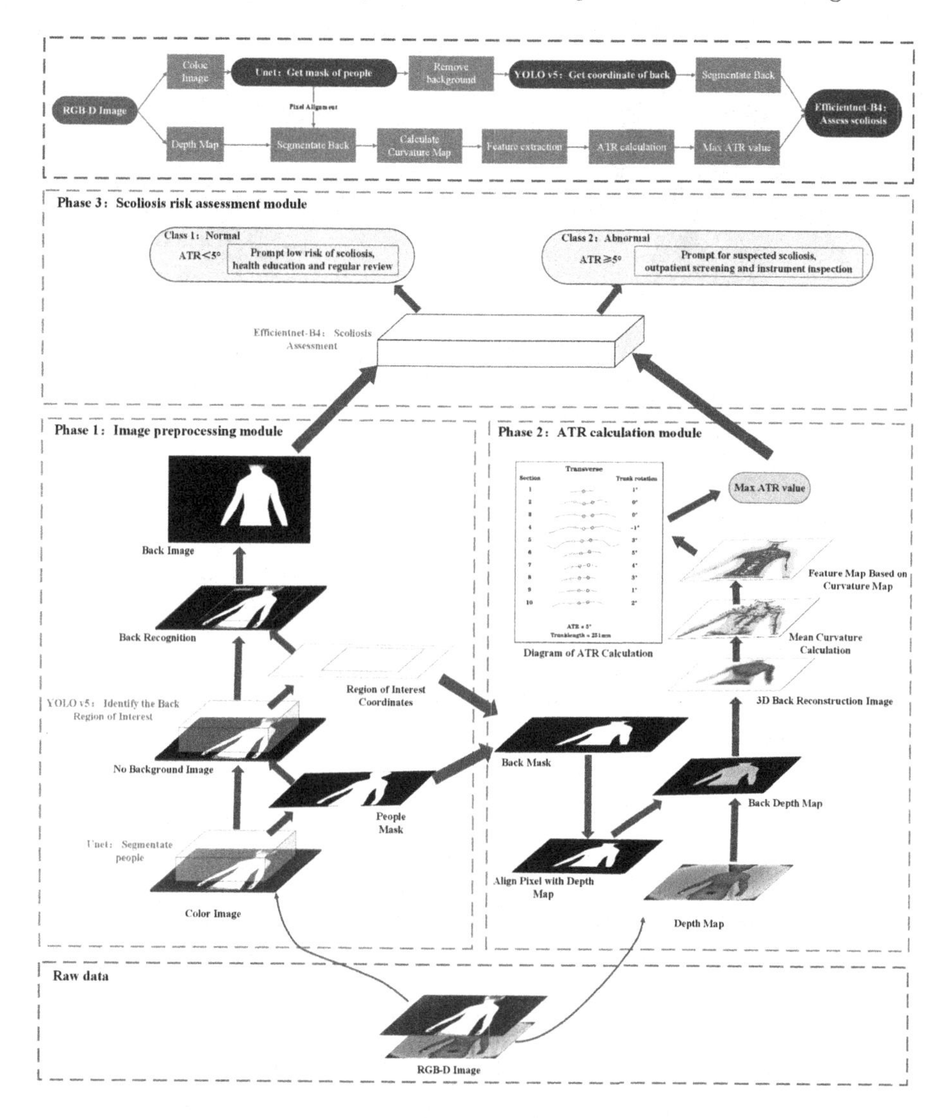

Fig. 2. The framework of proposal model.

main plane (XY plane)of the Kinect sensor. It is essential to rotate the trunk 3D point cloud into a body-fixed coordinates reference [17].

According to the method in [17], based on the Mean curvature and Gaussian curvature calculated by the 3D point cloud, we obtain the curvature map of the human back according to a certain curvature criterion, as shown in Fig. 3(b); and calculate the line of the spinous processes, as shown in the red line in Fig. 3(c). As shown in Fig. 3(d), 10 transverse sections ($y = 0$, plane, near horizontal) are divided by equal distance from C7 to the midpoint of left and right PSIS. In each section, the paramedian lines are 10% of the length of the back body to the

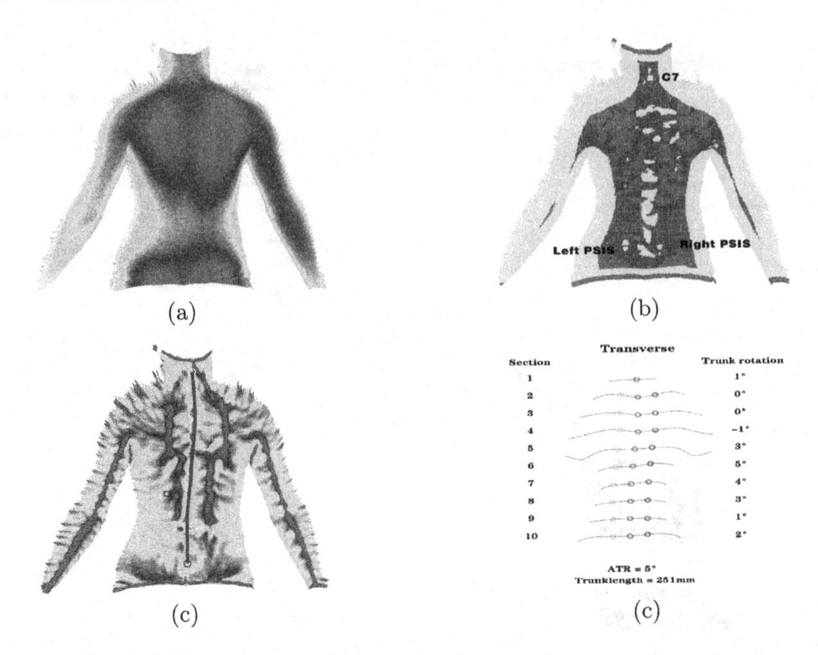

Fig. 3. Point cloud map of human back; Curvature anatomical feature point map; The spinous process line diagram and ATR value calculation diagram are represented by (a), (b), (c) and (d) respectively

left and right of the spinous process point, the red point represents the spinous process point, and the right and left paramedian line are represented by yellow and purple respectively. The trunk rotation angle is a positive number when the right side is higher; Then, we do polynomial fitting on the rotation angle calculated by 10 transverse sections, and calculate the value with the largest absolute value as the final ATR value in the results [14]. Finally, according to the international scoliosis classification standard, we label the data as 0 or 1 based on the ATR value as training and testing labels. 0 means that the sample has no scoliosis, and 1 means that the sample has scoliosis [18].

Scoliosis Risk Assessment Module. At present, the standard of medical treatment is to refer the subjects with ATR $\geq 5°$ to the hospital for X-ray evaluation, while the subjects with ATR $< 5°$ degrees should not be re screened [18]. The screening classification criteria based on ATR value are as follows:

1. If ATR $< 5°$, it indicates that the sample is normal with low scoliosis risk.
2. If ATR $\geq 5°$, It means that the sample is abnormal and is suspected of scoliosis.

We train EFFICIENTNET-B4 network to classify images and minimize the loss function (2) to obtain the best classifier f:

$$f = argmin \sum_{i=1}^{N} [y_i log(p_i) + (1 - y_i) log(1 - p_i)] \tag{2}$$

where n is the number of samples, y is the real label, and p is the prediction probability.

4 Experiment Results

4.1 Evaluation Metrics

Five evaluation indicators are used to evaluate our model, which are defined as follows:

$$Accuracy = \frac{TP + FN}{TP + FN + TN + FP} \tag{3}$$

$$F_1 = \frac{2 \times Precision \times Recall}{Precision + Recall} \tag{4}$$

$$Precision = \frac{TP}{TP + FP} \tag{5}$$

$$Recall/Sensitivity = \frac{TP}{TP + FN} \tag{6}$$

$$Specificity = \frac{TN}{TN + FP} \tag{7}$$

where TP and TN represent true positive and true negative respectively, which means that people with normal and suspected scoliosis are predicted to be the correct category; FP indicates false negative, indicating that subjects with suspected scoliosis are predicted to be normal; FN indicates false negative, indicating that the model wrongly judges normal people as suspected scoliosis. Their meaning is shown in Fig. 4.

In addition, the false positive rate (8) represents the misdiagnosis rate, that is, the percentage of actual normal samples judged as suspected scoliosis; The false negative rate (9) represents the missed diagnosis rate, that is, the percentage of

True / Predicate	Positive	Negative
Positive	True Positive (TP)	Fales Positive (FP)
Negative	Fales Negative (FN)	True Negative (TN)

Fig. 4. The meaning of TP, FP, TN, FN.

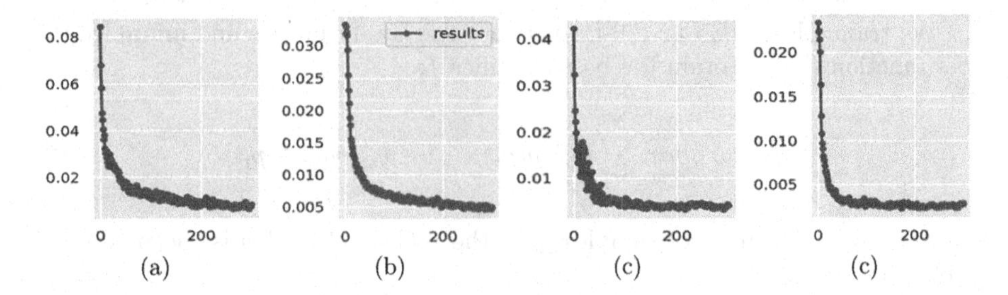

Fig. 5. The loss change of human back segmentation model in training and verification dataset: (a) is the GIoU (generalized intersection over union) loss function on the training dataset; (b) is the loss of target detection during training; (c) is the bounding box loss on the verification dataset; (d) is the loss of verification dataset target detection.

samples actually suspected of scoliosis but mistakenly recognized as normal. It is an important index to judge whether the model is available.

$$false\ positive\ rate = 1 - Specificity \tag{8}$$

$$false\ negative\ rate = 1 - Sensitivity \tag{9}$$

4.2 Model Training and Testing

The training and testing of all our deep learning models are performed on a Linux server with four NVIDIA Tesla V100.

Firstly, the YOLO V5 algorithm is trained on a private dataset containing 1935 images without background. They are labeled using Labelimg software [15], a rectangular frame is used to surround the back area, which is slightly larger than the area from the C7 to the sacrum. An experienced doctor help us to correct the labeling results. 80% of the data is divided into the training dataset and the remaining 20% is used in the testing process. The maximum number of iterations is 300 epochs and the batch size is 16. Other parameters are set use default.

The final model for human back recognition has an average map value of 0.999 when the IOU (Intersection Over Union) threshold is 0.5. When the step size is 0.05, the average map value is 0.945. Figure 5 shows the changes of loss, precision and recall on the training and verification dataset. Figure 6 shows part of the results of the human body segmentation and back segmentation process.

Subsequently, we train and test the scoliosis evaluation model. After ATR value calculation and labeling, 948 available images are marked as normal and the other 987 are marked as abnormal. We arbitrarily split the images for training and the rest for testing (75% of the images are used as training dataset and the rest as test dataset). When training the scoliosis risk assessment model, the dataset was enhanced by rotation and turnover, and it is expanded by 8 times. The input image size is adjusted to 380 × 380 to adapt to the network.

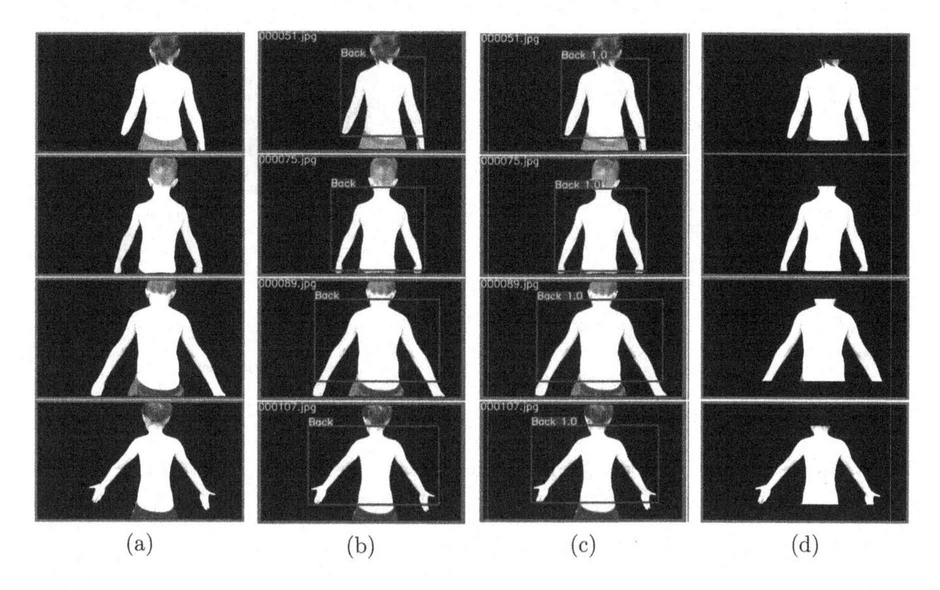

Fig. 6. Partial results of human segmentation, target detection label, human back recognition, and human back segmentation on test data. (a) is original 2D image; (b) is ground truth of human back detection; (c) is the segmentation result predicted by YOLO V5 model; (d) is segmented human back.

Table 1. Performance of our model scoliosis assessment task.

Metric	Normal	Abnormal
f1-score	0.832	0.788
Precision	0.754	0.907
Recall/Sensitivity	0.929	0.696
Specificity	0.696	0.929

The final scoliosis risk classification model accuracy is 0.813. Its ROC curve is shown in Fig. 7(a), and the AUC of the algorithm is 0.839. The confusion matrix on the test dataset is shown in Fig. 7(b). The results recorded in Table 1 show that the system has the precision and specificity of 0.907 and 0.929 to detect suspect scoliosis, indicating that this system can well detect the occurrence of abnormalities despite the possibility of missed detection. Most of the normal samples with wrong classification have ATR values close to 5 ° because these samples have small differences in 2D characteristics. However, for abnormal samples, our model has a high false negative rate (0.304), which means that it will predict some abnormal samples as normal. Most of the incorrectly classified abnormal samples have large ATR values, which means that they have severe scoliosis. The failure of this model is due to the lack of enough samples

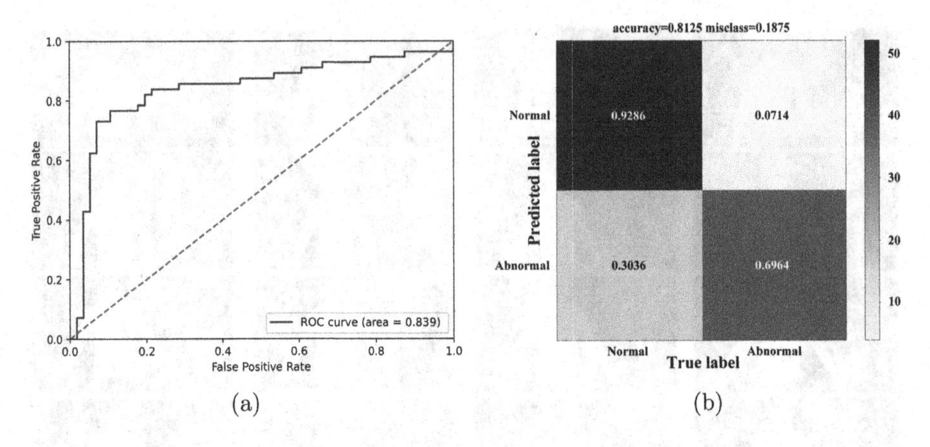

Fig. 7. Performance of the model on test dataset. (a) The ROC curve and AUC value of proposal method for discerning whether the ATR $> 5°$. (b) Confusion matrix on the test dataset

of severe scoliosis. In fact, these samples can be easily identified by naked eyes. An example of correctly classified samples is shown in Fig. 8.

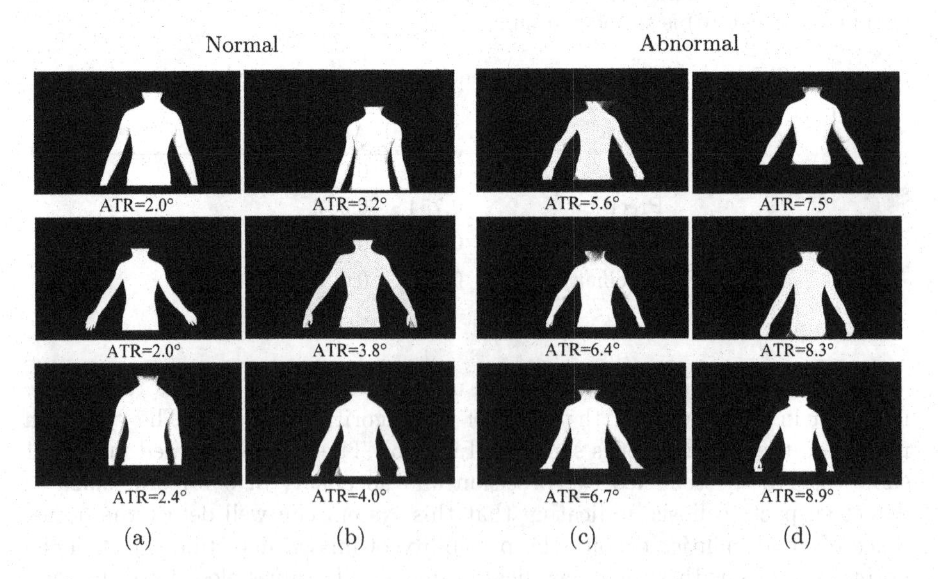

Fig. 8. Correctly classified samples. (a) For normal samples correctly classified, ATR$<$ $5°$ (label = 0); (b) For abnormal samples correctly classified, ATR $\geq 5°$ (label = 1)

5 Conclusion

In this paper, we propose a new preliminary AIS screening method, which puts the scoliosis screening power into user's hand. This method uses 2D human back image to classify scoliosis normal and abnormal based on deep learning algorithm, so as to realize the large-scale preliminary scoliosis screening. No need for professional doctors, high precision and no radiation are the main contributions of this paper. In addition, ATR is creatively classified as classification label and screening groundtruth, based on the high correlation between ATR value and scoliosis risk. Experimental results verify the effectiveness of the proposed model with state-of-the-art accuracy (81.3%) over naked eye test by professional doctors (65%).

References

1. Adams, W.: Lectures on the pathology and treatment of lateral and other forms of curvature of the spine. Churchill, J. & A (1882)
2. Bochkovskiy, A., Wang, C.Y., Liao, H.Y.M.: Yolov4: optimal speed and accuracy of object detection. arXiv preprint arXiv:2004.10934 (2020)
3. Bunnell, W.P.: An objective criterion for scoliosis screening. J. Bone Joint Surgery. Am. Vol. **66**(9), 1381–1387 (1984)
4. Chen, Q., Ge, T., Xu, Y., Zhang, Z., Yang, X., Gai, K.: Semantic human matting. In: Proceedings of the 26th ACM International Conference on Multimedia, pp. 618–626 (2018)
5. Church, B., Lasso, A., Schlenger, C., Borschneck, D.P., Mousavi, P., Fichtinger, G., Ungi, T.: Visualization of scoliotic spine using ultrasound-accessible skeletal landmarks. In: Medical Imaging 2017: Image-Guided Procedures, Robotic Interventions, and Modeling, vol. 10135, pp. 166–174. SPIE (2017)
6. Coelho, D.M., Bonagamba, G.H., Oliveira, A.S.: Scoliometer measurements of patients with idiopathic scoliosis. Braz. J. Phys. Ther. **17**, 179–184 (2013)
7. Pan, W., Hou, G., Zhang, C.: Automatic methods for screening and assessing scoliosis by 2-d digital images. In: International Conference on Intelligent Science and Big Data Engineering, pp. 392–400. Springer (2015)
8. Patias, P., Grivas, T.B., Kaspiris, A., Aggouris, C., Drakoutos, E.: A review of the trunk surface metrics used as scoliosis and other deformities evaluation indices. Scoliosis **5**(1), 1–20 (2010)
9. Pazos, V., Cheriet, F., Danserau, J., Ronsky, J., Zernicke, R.F., Labelle, H.: Reliability of trunk shape measurements based on 3-d surface reconstructions. Eur. Spine J. **16**(11), 1882–1891 (2007)
10. Ramirez, L., Durdle, N.G., Raso, V.J., Hill, D.L.: A support vector machines classifier to assess the severity of idiopathic scoliosis from surface topography. IEEE Trans. Inf Technol. Biomed. **10**(1), 84–91 (2006)
11. Redmon, J., Divvala, S., Girshick, R., Farhadi, A.: You only look once: unified, real-time object detection. In: Proceedings of the IEEE Conference on Computer Vision and Pattern Recognition, pp. 779–788 (2016)
12. Ronneberger, O., Fischer, P., Brox, T.: U-net: Convolutional networks for biomedical image segmentation. In: International Conference on Medical Image Computing and Computer-Assisted Intervention, pp. 234–241. Springer (2015)
13. Tan, M., Le, Q.: Efficientnet: Rethinking model scaling for convolutional neural networks. In: International Conference on Machine Learning, pp. 6105–6114. PMLR (2019)
14. Turner-Smith, A.R., Harris, J.D., Houghton, G.R., Jefferson, R.J.: A method for analysis of back shape in scoliosis. J. Biomech. **21**(6), 497–509 (1988)

15. Tzutalin, D.: Labelimg. GitHub Repository 6 (2015)
16. Weinstein, S.L., Dolan, L.A., Spratt, K.F., Peterson, K.K., Spoonamore, M.J., Ponseti, I.V.: Health and function of patients with untreated idiopathic scoliosis: a 50-year natural history study. JAMA **289**(5), 559–567 (2003)
17. Xu, Z., Zhang, Y., Fu, C., Liu, L., Chen, C., Xu, W., Guo, S.: Back shape measurement and three-dimensional reconstruction of spinal shape using one kinect sensor. In: 2020 IEEE 17th International Symposium on Biomedical Imaging (ISBI), pp. 1–5. IEEE (2020)
18. Yan, B., Lu, X., Nie, G., Huang, Y.: China urgently needs a nationwide scoliosis screening system. Acta Paediatr. **109**(11), 2416–2417 (2020)
19. Yang, J., Zhang, K., Fan, H., Huang, Z., Xiang, Y., Yang, J., He, L., Zhang, L., Yang, Y., Li, R., et al.: Development and validation of deep learning algorithms for scoliosis screening using back images. Commun. Biol. **2**(1), 1–8 (2019)

EMRShareChain: A Privacy-Preserving EMR Sharing System Model Based on the Consortium Blockchain

Xinglong Zhang, Peng Xi, Wenjuan Liu, and Shaoliang Peng(✉)

College of Computer Science and Electronic Engineering, HuNan University, Changsha 410082, China
{xinglongzhang,slpeng}@hnu.edu.cn

Abstract. Electronic medical record (EMR) Sharing can reduce duplicate medical tests, reduce treatment costs and improve the quality of medical services. However, most medical organizations currently use the internal network to track patients but don't implement data sharing with other medical organizations. Meanwhile, the EMR is also highly vulnerable to theft and tampering by malicious attackers.

To solve the above problems, we design a privacy-preserving EMR sharing system model based on consortium blockchain, called EMRShareChain. In EMRShareChain, multiple medical organizations firstly spontaneously build up a medical consortium. The EMR generated by a patient's treatment at a medical organization is encrypted and stored securely in the cloud, and the index of the EMR is stored in the blockchain. With the above storage method, not only the risk of EMR leakage is reduced, but also the EMR cannot be tampered with arbitrarily. Then, other organizations in the medical consortium can find the patient's relevant EMR in the blockchain by keywords in the index and keyword searchable encryption, and obtain the EMR key and EMR storage address generated by the conditional proxy re-encryption after the patient's authorization to achieve access to the EMR and data sharing. In addition, an improved PBFT consensus algorithm (I-PBFT) based on comprehensive scoring mechanism is proposed to improve the blockchain block-out efficiency and stability, making it more suitable for medical scenarios. Finally, a prototype implementation of the above model is carried out based on the Hyperledger Fabric framework. The Hyperledger fabric chaincode is accessible at https://github.com/xl13455/EMRShareChain and code can be downloaded.

Keywords: EMR sharing · Consortium blockchain · Searchable encryption · Conditional proxy re-encryption

1 Introduction

With the advancement of information technology, the medical industry has shown the development trend of informatization. The emergence of EMR solves

M. S. Bansal et al. (Eds.): ISBRA 2022, LNBI 13760, pp. 343–355, 2022.
https://doi.org/10.1007/978-3-031-23198-8_31

the problems of easy loss and damage of paper medical records. However, most medical organizations currently use the internal network to keep track of their patients but don't implement data sharing with other medical organizations and increase the cost and difficulty of medical services, which brings about the phenomenon of data silos [1]. And EMR is also highly vulnerable to theft and tampering by malicious attackers, jeopardizing patient privacy and interests [2].

The rise of cloud computing has provided a solution for data sharing [3]. Medical organization uploads patients' EMRs to the cloud sharing platform in real-time, and other medical organizations in the consortium can access EMR, enabling data sharing. However, the above data sharing scheme has many problems, as the cloud is semi-trustworthy, the cloud may peek into the patient's EMR and conspire with doctors to modify the patient's EMR for profit [4]. Later, many research scholars [5,6] proposed a secure EMR sharing scheme based on cloud and attribute-based encryption, where EMR is encrypted by a set of attributes, and only users who meet the attribute permissions can access the data. However, the EMR stored in the cloud is still subject to tampering.

Fortunately, the rise of blockchain and cryptography offers the possibility of privacy protection and data sharing. Blockchain is a peer-to-peer distributed database with excellent features such as decentralization, group maintenance, tamper-evident and transparent data on the chain, making it ideal as an underlying technology for data sharing. And cryptography can protect the privacy of data, making it accessible only to authorized users.

A lot of decentralized EMR privacy protection and sharing schemes have emerged. Dubovitskaya et al. [7] used blockchain, cloud, and symmetric encryption for EMR sharing. In the scheme, EMR data is encrypted using symmetric key to ensure data privacy, and blockchain ensures that the data is difficult to tamper with. The Patient can authorize other organizations in the medical consortium symmetric key to enable data sharing. However, the scheme imposes patient to a heavy key management burden and communication overhead. Dubovitskaya et al. [8] proposed an EMR sharing scheme based on public key cryptosystem and blockchain. The scheme uses a symmetric key to encrypt the EMR and then uses the patient's public key to encrypt the symmetric key to protect data privacy. The scheme then implements data sharing by the patient encrypting the symmetric key using the doctor's public key. However, the frequent encryption and decryption work imposes a computational burden on the patient. The EMR sharing scheme based on blockchain and attribute-based encryption [9–11] stores EMR in the cloud after encryption by patient-set access policies, and stores storage address in the blockchain. Each medical organization in the medical consortium finds the relevant EMR through the blockchain, then downloads the encrypted EMR from the storage address and decrypts it using its own attribute private key. However, attribute-based encryption is deeply affected by encryption efficiency, and the change of the access policy has been a challenge.

In this paper, a privacy-preserving EMR sharing system model based on consortium blockchain is proposed. The specific contributions are as follows:

(1) A collaborative cloud-chain secure storage approach is proposed to protect data privacy and integrity.
(2) A joint-design of conjunctive-keyword searchable encryption and conditional proxy re-encryption is proposed to achieve data search and secure sharing.
(3) An improved PBFT consensus algorithm based on a comprehensive scoring mechanism is proposed to improve the performance and stability of the blockchain outgoing blocks.

The remainder of this paper is organized as follows. In Sect. 2 we briefly introduce the preliminary. In Sect. 3, we describe the architecture and workflow of the EMRShareChain. In Sect. 4, we test the performance and analyze security of EMR. Finally, we conclude the full paper in Sect. 5.

2 Preliminary

2.1 Blockchain

Blockchain is a technical solution for storing, transferring and exchanging network data through its own distributed nodes without relying on third parties. Decentralization, group maintenance, non-tamperability and transparency are the features of blockchain [12]. The current blockchains can be divided into three categories: public blockchain, private blockchain and consortium blockchain.

The openness of the consortium blockchain is between public blockchain and private blockchain and is jointly managed by multiple organizations. Only authorized users in authorized organizations can initiate and view blockchain transactions, such as Hyperledger Fabric [13]. In this paper, we build the Hyperledger Fabric consortium blockchain by different medical organizations spontaneously providing nodes and sharing data by storing the EMR index in the blockchain.

2.2 The Conditional Proxy Re-encryption

Conditional proxy re-encryption(CPRE) [14] is able to re-encrypt the eligible ciphertext under Alice's public key into ciphertext under Bob's public key with the help of a semi-trusted proxy. And proxy has no access to information about the original plaintext. A CPRE algorithm consists of the following six function.

$CPRE.KeyGen(i)$: This function generates the key pair (pk_i, sk_i) for user i.

$CPRE.ReKeyGen(sk_i, pk_i, pk_j, c)$: This function takes the private key sk_i of user i, the public keys pk_i of user i and pk_j of user j, a condition value c as input, and generates a re-encryption key $rk_{i,j}$.

$CPRE.Encrypt(pk_i, m, c)$: This function takes the public key pk_i of user i, the plaintext m and a conditional value c as input and generates a ciphertext c_i as output.

$CPRE.reEncrypt(rk_{i,j}, c_i)$: This function takes the re-encryption key $rk_{i,j}$, the ciphertext c_i as input, and generates a re-encrypted ciphertext d_j as output.

$CPRE.Decrypt(sk_i, c_i)$: This function takes the private key sk_i of a user i, a ciphertext c_i as input, and generates a plaintext m as output.

$CPRE.ReDecrypt(sk_j, d_i)$: This function takes the private key sk_j of a user j, a re-encrypted ciphertext d_i as input, and generates a plaintext m as output.

2.3 The Public-Key Encryption with Conjunctive Keyword Search

The public key encryption with conjunctive keyword search (PKSE) [15] enables data users to search files containing several keywords over a public key encryption setting. The scheme consists of the following four functions.

$PKSE.KeyGen(i)$: This function generates key pair (pk_i, sk_i) for user i.

$PKSE.keywordIndex(pk_i, W)$: This function takes the public key pk_i of user i and a keyword set W as input, and generates a encrypted keyword CW.

$PKSE.Trapdoor(sk_i, Q)$: This function takes the private key sk_i of user i and a keyword query set Q as input, and generates a search trapdoor CT.

$PKSE.Search(CW, CT)$: This function takes the encrypted keyword set CW and a keyword search trapdoor CT as input, and generates a value. If CT is included in CW, the server outputs 1, otherwise 0.

3 EMRShareChain: The Proposed System Model

3.1 System Architecture

The EMRShareChain adopts a three-layer architecture, including data generation layer, data storage layer, and data sharing layer, as shown in Fig. 1. The functional description of each layer is specified as follows.

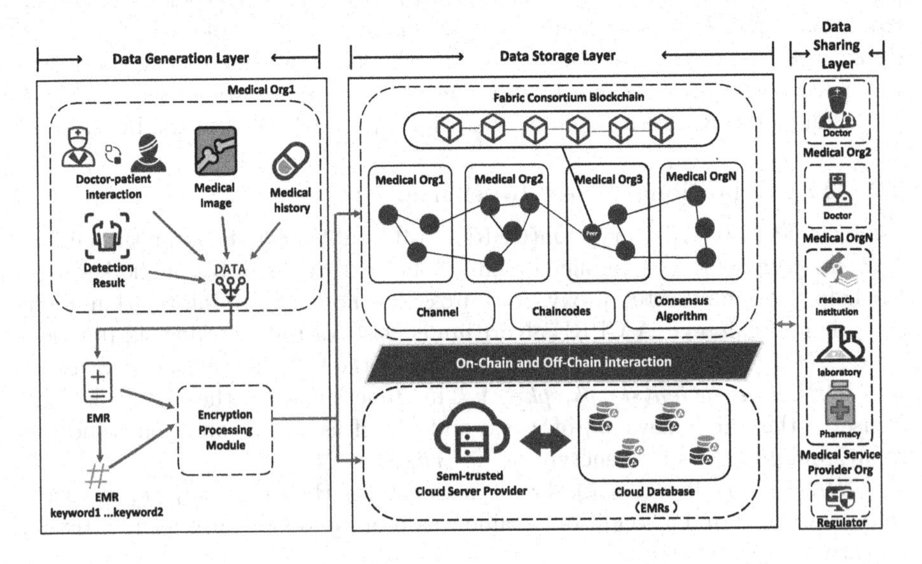

Fig. 1. The Architecture Of EMRShareChain system model

(1) **Data generation layer.** In this layer, an original EMR is generated when a patient is treated at one of the medical organizations in the medical consortium. Since the EMR contains much sensitive information, the patient has the right to entrust the doctor to encrypt the EMR using a symmetric key.
(2) **Data storage layer.** In this layer, we use cloud service provider (CSP) and blockchain to store EMR and the EMR index, respectively.

- CSP: CSP is responsible for storing the encrypted EMR and returning the corresponding storage address after successful storage.
- Consortium blockchain: The blockchain is responsible for storing index to realize data sharing, where the index contains id, digital digest, digital signature, encrypted keywords, encrypted symmetric key, and encrypted storage address.

(3) **Data sharing layer.** At this layer, any organization or user in the medical consortium authorized by the patient can search the EMR on the blockchain and obtain the re-encrypted symmetric key and the re-encrypted storage address. The organization or user can use the private key to decrypt the ciphertext and obtain the EMR to provide better care to patients or conduct scientific research.

3.2 System Workflow

In the following, we will describe the workflow of EMRShareChain. The whole workflow is divided into four phases: system initialization phase, data generation phase, data processing and storage phase, and data search and sharing phase.

System Initialization Phase. During this phase, the purpose is to build the blockchain network and complete user registration. (i) Regarding the blockchain construction, consortium blockchain includes three types of nodes: CA node, peer nodes, and orderer nodes. (ii) Regarding user registration, each user of the organization needs to provide real identity information for registration and will get the key pairs and digital certificates for joining the blockchain network.

CA Join Phase. The CA is an important part of the consortium blockchain and is managed by government personnel. Any organization and user joining the blockchain must have key pairs and digital certificates issued by the CA.

Orderer Join Phase. The orderer service is responsible for sorting and packaging blockchain transactions. Each node of orderer service is provided by the medical organization M_j. The M_j provides the real identity information RID to the CA for registration, and then CA generates the identity key pair $(sk_{iden(M_j)}, pk_{iden(M_j)})$, TLS key pair $(sk_{tls(M_j)}, pk_{tls(M_j)})$, and identity digital certificate $x509Cert_{iden(M_j)}$ and TLS digital certificate $x509Cert_{tls(M_j)}$ needed to build the orderer node for it. Among them, the identity key is used to indicate the identity of the organization or user, and tls key is used to ensure the security of the communication. Then, M_j sets up the orderer service.

Peer Join Phase. All medical organizations need to provide server as peer node to maintain the blockchain ledger. Similarly, the medical organization M_j needs to obtain key pairs and digital certificates required to build the peer node. Then M_j sets up the peer node service. Further, the peer node needs to join the corresponding blockchain channel and install the chaincode (cc).

User Join Phase. The user of all organization needs to go through verification when they join the blockchain network. Taking the patient user of the medical organization as an example, the patient P_i provides real identity information RID to CA. Then, CA generates the key pair and digital certificate for P_i.

In addition to this, the patient P_i also needs to generate two key pairs locally. Specifically, encryption key pair $(sk_{enc(P_i)}, pk_{enc(P_i)})$ are used to implement encryption and decryption of the data, and search key pair $(sk_{search(P_i)}, pk_{search(P_i)})$ is used to implement encryption and search of keywords.

Data Generation Phase. During this phase, the patient P_i is treated by doctor D_i at one medical organization M_j in the medical consortium, and the interaction data, test results and medical images is collected and formed into an EMR.

Data Processing and Storage Phase. During this phase, the EMR generated by patient P_i will be encrypted and uploaded to CSP. The doctor D_i first encrypts the EMR using a symmetric key Key to generate the encrypted data EMR_e, then uploads data EMR_e to the CSP and obtains the corresponding storage address Url.

The doctor D_i then encrypts the keyword $Keyword$ of the EMR, the symmetric key Key used to encrypt the EMR, and the storage address Url using the public key of patient P_i and the conditional value c to generate the encrypted keyword $Keyword_e$, the encrypted symmetric key Key_e and the encrypted storage address Url_e.

After encryption operations are completed, the doctor D_i further calculates the digital digest $digest_{EMR_e}$ of the encrypted EMR and generates digital signature $sign_{EMR_e}$ using the own identity private key $sk_{iden(D_i)}$.

Finally, The doctor D_i then feedbacks the information generated during the above operations to the patient P_i. After receiving the feedback information, the patient P_i extracts the necessary information to build the EMR index $EMRIndexEntity$ as shown in Fig. 2 and uploads it to the blockchain.

Data Search and Sharing Phase. During this phase, data users who wish to share access the patient's EMR to provide better medical service or conduct research. Here we take a doctor as an example, when the patient P_i is referred to another medical institution M_{j+1}, the doctor D_i' needs to obtain the patient's previous EMR to assist in diagnosis.

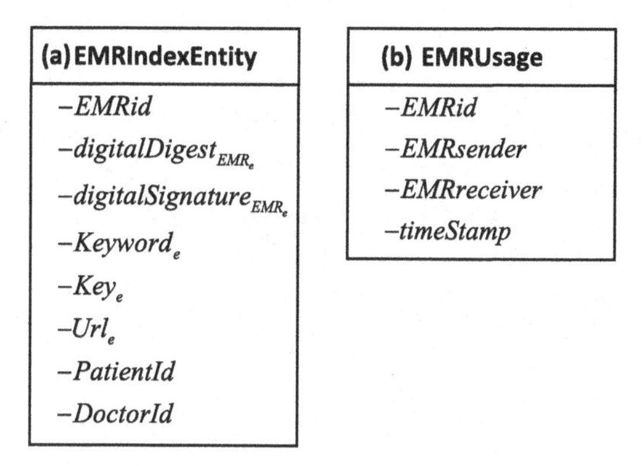

Fig. 2. The EMRIndexEntity and EMRUsage data structure

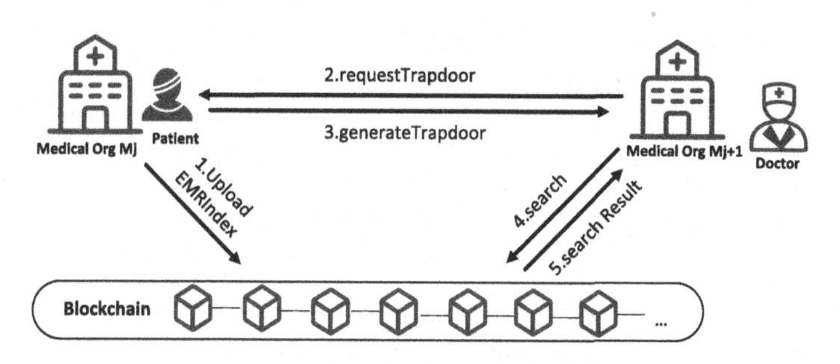

Fig. 3. The data search phase

The doctor D_i' first needs to obtain a search trapdoor *trapdoor* generated by the patient P_i based on search keywords Sw. The doctor then invokes the search chaincode with the search trapdoor *trapdoor* as an input parameter, and the search chaincode will return the matching EMR ID to the doctor D_i'. And the flow of the search phase is shown in Fig. 3.

Because EMRs are private, any user who wants to actually access the EMR plaintext must be authorized by the patient P_i.

The doctor first needs to generate a $dataAccessRequest$ based on information such as the EMRid and the public encryption key of the doctor, and sends it to the patient. If the patient agrees to the access request, the patient will generates a re-encryption key $reEncKey$ using his encryption private key sk_{encP_i} and the doctor's encryption public key $pk_{enc_{(D_i')}}$ along with the conditional values c. Then, the patient sends the EMRid, re-encryption key to the master node of the consortium blockchain. The master node performs the re-encryption work, and sends the re-encrypted key Key_{reEnc} and re-encrypted storage address Url_{reEnc} to the doctor D_i'.

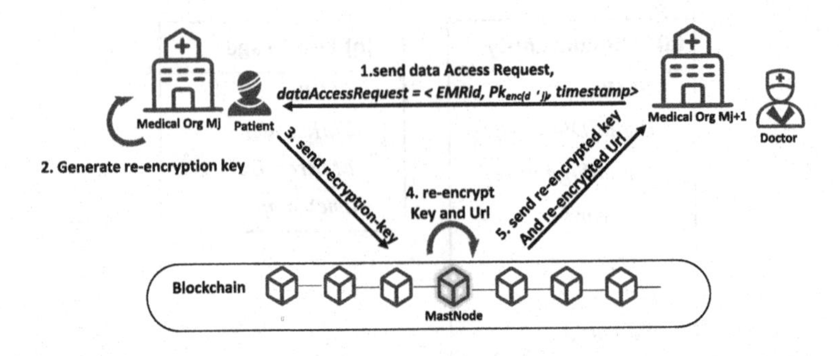

Fig. 4. The data sharing phase

The doctor D_i' can decrypt the re-encrypted storage address and then download the EMR from the CSP. At this point, the data sharing phase is completed. And the flow of the sharing phase is shown in Fig. 4.

At the same time, the patient needs to construct the data usage record $EMRUsage$ as shown in the Fig. 2 to record the details of each data usage, and then the patient uploads it $EMRUsage$ to the blockchain network. The patient can then query the data usage through the blockchain.

I-PBFT (Improved PBFT Consensus Mechanism). PBFT algorithm, as a state machine copy replication algorithm, can provide $(n-1)/3$ fault tolerance (n is the total number of nodes in the blockchain network). The mechanism can not only start and run on fewer nodes but also does not require a lot of computing power to maintain. Considering that the node number of medical organizations is small, the PBFT is more suitable for medical scenarios.

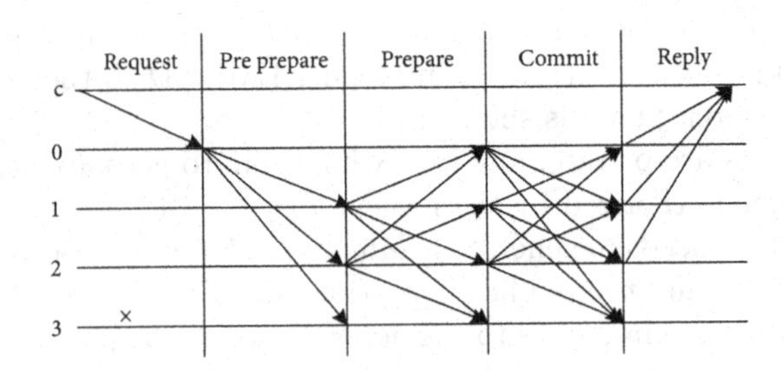

Fig. 5. The principle of PBFT consensus mechanism

This model adopts the Fabric consortium blockchain and uses the PBFT consensus algorithm, in which the nodes in PBFT are equivalent to the orderer nodes in the Fabric blockchain, and the principle of PBFT mechanism is shown in Fig. 5. In PBFT, the nodes are divided into master and slave nodes. The master node is responsible for sorting the transactions and distributing them to the slave nodes, and generating blocks after the nodes reach consensus. However, in the original PBFT, the master node is randomly selected, which will cause frequent replacement of master node when computationally inadequate or potentially malicious medical organization act as the master node, affecting the blockchain's block-out performance and stability. Therefore, we improve on the original PBFT by changing the selection method of the master node from randomly selecting medical organization to serving as the one with the high overall score.

The score of the medical organization nodes is shown in Eq. 1, and the final score ($finalScore$) of each node is calculated based on the hardware level score ($hardwareScore$) and the consensus performance score ($consensusScore$).

$$finalScore = 0.2 * hardwareScore + 0.8 * consensusScore \tag{1}$$

The hardware level score is calculated by combining the CPU processing frequency score ($h1$), CPU core count score ($h2$), memory score ($h3$), and disk score ($h4$), and bandwidth score ($h5$), as shown in Eq. 2. The k_i of Eq. 2 indicates the different weighting factors. And it should be noted that the hardware level score is normalized.

$$hardwareScore = k1 * h1 + k2 * h2 + k3 * h3 + k4 * h4 + k5 * h5 \tag{2}$$

The node consensus performance score is calculated from the number of consensus successes (cs), consensus failures (cf), consensus successes as master node (mcs), and consensus failures as master node (mcf) in the past n consensus times by the node, as shown in the Eq. 3. It is worth noting that the node consensus level scores are normalized.

$$consensusScore = e^{\frac{(cs+mcs)-(\alpha \times cf+\beta \times mcf)}{n}} \tag{3}$$

4 Model Comparison and Performance Test

4.1 Model Comparison

We compare EMRShareChain system model with other sharing schemes in six aspects: whether it is based on blockchain, whether it requires tokens, power requirement, consensus algorithm, storage method, whether it achieves data privacy protection and fine-grained access control. For the convenience of representation, the six aspects are represented by F1, F2, F3, F4, F5, and F6. The comparison of different schemes is shown in Table 1.

Table 1. Comparison of EMRShareChain system with other systems or schemes.

Items	[5]	[6]	[7]	[8]	[9]	[10]	[11]	Ours
F1	F	F	T	T	T	T	T	T
F2	–	–	F	F	F	F	T	F
F3	–	–	Low	Low	Low	Low	High	Low
F4	–	–	PBFT	PBFT	PBFT	-	EOS	I-PBFT
F5	Cloud	Cloud	Cloud	Cloud	IPFS	Cloud	IPFS	Cloud
F6	T	T	T	T	T	T	T	T

Our scheme uses blockchain and cloud collaboration for reliable data storage. However, Zhang and Wang's schemes [5,6] use the cloud alone for storage and does not guarantee data integrity. Dubovitskaya's schemes [7,8] use the original PBFT consensus algorithm, and Liu [9] also uses the default PBFT consensus algorithm. Gao's scheme [11] uses the EOS consensus algorithm. Compared with the above schemes, our scheme uses an improved PBFT consensus algorithm (I-PBFT) to improve the efficiency and stability of blockchain out blocks. Moreover, our scheme uses consortium blockchain, which not only does not require tokens but also does not require strong power support. All schemes use encryption algorithms for data privacy protection and access control to ensure that only authorized users and institutions can share access to EMR data, but in our scheme, patients only need to generate a re-encryption key to complete access authorization operations, greatly reducing key management and communication overhead.

4.2 Performance Evaluation

The performance test is carried out to evaluate the efficiency of data processing and sharing. In order to simulate the real environment, the blockchain network CA node, orderer nodes and peer nodes are simulated by deploying docker containers in the host. The host computer is Intel(R) Xeon(R) CPU E78890 v3 and the operating system is Ubuntu20.

In the data processing and storage phase, we perform the data encryption operation (T_{de}), the digital digest calculation operation of encrypted data (T_{ddc}), the digital signature calculation operation of encrypted data (T_{dsc}), the encrypted data upload operation (T_{edu}), the keyword encryption operation (T_{kwe}), the storage address encryption operation (T_{sde}), key encryption operation (T_{ke}), EMR index upload operation (T_{eiu}) and EMR index query operation (T_{eiq}) are tested for their time cost overhead (ms). And the test results are shown in Table 2.

Table 2. The time cost overhead of operations in data processing and storage phase

File size	T_{de}	T_{ddc}	T_{dsc}	T_{edu}	T_{kwe}	T_{sde}	T_{ke}	T_{eic}	T_{eiu}	T_{eiq}
64KB	15	1	5	5	150	135	115	1	51	44
256KB	18	2	5	7	150	135	115	1	51	44
1024KB	28	6	5	12	150	135	115	1	51	44

In the data search and sharing phase, we test the time cost overhead of the search trapdoor generation operation (T_{stg}), data search operation (T_{ds}), the re-encryption key generation operation (T_{rkg}), encryption key and storage address re-encryption operation (T_{ksr}), encryption key and storage address decryption operation (T_{ksd}), encrypted data download operation (T_{edd}), encrypted data decryption operation (T_{edd2}), EMR usage record construction operation (T_{euc}), EMR usage record upload operation (T_{euu}), and EMR usage record query operation (T_{euq}). And the test results are shown in Table 3.

Table 3. The time cost overhead of operations in the data search and sharing phase

File size	T_{stg}	T_{ds}	T_{rkg}	T_{ksr}	T_{ksd}	T_{edd}	T_{edd2}	T_{euc}	T_{euu}	T_{euq}
64 KB	12	72	16	285	64	1114	8	1	35	34
256 KB	12	72	16	285	64	1333	11	1	35	34
1024 KB	12	72	16	285	64	2745	17	1	35	34

4.3 Security Analysis

Data Confidentiality. EMR is encrypted with symmetric key and stored in the cloud server, and the symmetric key and storage address are stored on the blockchain after encryption. Without the patient's private key, the attacker cannot get the plaintext information of EMR, which effectively ensures the confidentiality of EMR data.

Data Integrity. The digital digest of the encrypted EMR is calculated and stored in the blockchain. Because the data on the blockchain is difficult to tamper with, the digital digest stored on the blockchain can be used to verify the integrity of the data.

Data Authenticity. The encrypted EMR is signed by the doctor and stored permanently in the blockchain, which binds the authenticity of the EMR to the doctor's reputation and ensures data authenticity.

5 Conclusion

In this paper, a privacy-preserving EMR sharing system model based on consortium blockchain is proposed, called EMRShareChain. In EMRShareChain, the EMR generated by a patient after a visit to a medical organization is stored encrypted in the cloud, and the EMR index is stored in the tamper-proof consortium blockchain. With the above joint storage method of cloud and blockchain, the problem of EMR data being easily leaked and maliciously tampered with is solved. Meanwhile, the joint-design of conjunctive-keyword searchable encryption and conditional proxy re-encryption enables organizations in a medical consortium to quickly retrieve the relevant EMR on the blockchain and achieve secure shared access to EMR with patient authorization. Further, the improved PBFT blockchain consensus algorithm proposed in conjunction with the actual medical scenario effectively improves the model performance and stability. The implementation of the proposed EMRShareChain will enable secure sharing of EMR data between different medical organizations, thus greatly facilitating cross-institution treatment and effectively providing easy access to data for scientific research.

Acknowledgements. This work was supported by National Key R&D Program of China 2022YFC3400404; NSFC Grants U19A2067; Science Foundation for Distinguished Young Scholars of Hunan Province (2020JJ2009); Science Foundation of Changsha Z202069420652, kq2004010; JZ20195242029, JH20199142034; The Funds of State Key Laboratory of Chemo/Biosensing and Chemometrics, the National Supercomputing Center in Changsha (http://nscc.hnu.edu.cn/), and Peng Cheng Lab.

References

1. Liu, J., Li, X., Ye, L., Zhang, H., Du, X., Guizani, M.: BPDS: a blockchain based privacy-preserving data sharing for electronic medical records. In: 2018 IEEE Global Communications Conference (GLOBECOM), pp. 1–6. IEEE (2018)
2. Zhao, Y., et al.: Research on electronic medical record access control based on blockchain. Int. J. Distrib. Sens. Netw. **15**(11), 1550147719889330 (2019)
3. Agyekum, K.O.B.O., Xia, Q., Sifah, E.B., Cobblah, C.N.A., Xia, H., Gao, J.: A proxy re-encryption approach to secure data sharing in the internet of things based on blockchain. IEEE Syst. J. **16**(1), 1685–1696 (2021)
4. Kelbert, F., et al.: Securecloud: secure big data processing in untrusted clouds. In: Design, Automation & Test in Europe Conference & Exhibition (DATE), pp. 282–285. IEEE (2017)
5. Zhang, A., Wang, X., Ye, X., Xie, X.: Lightweight and fine-grained access control for cloud-fog-based electronic medical record sharing systems. Int. J. Commun. Syst. **34**(13), e4909 (2021)
6. Wang, T., Zhou, Y., Ma, H., Zhang, R.: Enhanced dual-policy attribute-based encryption for secure data sharing in the cloud. Secur. Commun. Netw. 2022 (2022)
7. Dubovitskaya, A., Xu, Z., Ryu, S., Schumacher, M., Wang, F.: Secure and trustable electronic medical records sharing using blockchain. In: AMIA Annual Symposium Proceedings, vol. 2017, p. 650. American Medical Informatics Association (2017)

8. Dubovitskaya, A., Baig, F., Xu, Z., Shukla, R., Zambani, P.S., Swaminathan, A., Jahangir, M.M., Chowdhry, K., Lachhani, R., Idnani, N., et al.: ACTION-EHR: patient-centric blockchain-based electronic health record data management for cancer care. J. Med. Internet Res. **22**(8), e13598 (2020)
9. Liu, J., Wu, M., Sun, R., Du, X., Guizani, M.: BMDS: a blockchain-based medical data sharing scheme with attribute-based searchable encryption. In: ICC 2021-IEEE International Conference on Communications, pp. 1–6. IEEE (2021)
10. Zhang, L., Peng, M., Wang, W., Su, Y., Cui, S., Kim, S.: Secure and efficient data storage and sharing scheme based on double blockchain (2021)
11. Gao, H., Ma, Z., Luo, S., Xu, Y., Wu, Z.: BSSPD: a blockchain-based security sharing scheme for personal data with fine-grained access control. Wirel. Commun. Mob. Comput. 2021 (2021)
12. Zheng, Z., Xie, S., Dai, H.N., Chen, X., Wang, H.: Blockchain challenges and opportunities: a survey. Int. J. Web Grid Serv. **14**(4), 352–375 (2018)
13. Androulaki, E., et al.: Hyperledger fabric: a distributed operating system for permissioned blockchains. In: Proceedings of the Thirteenth EuroSys Conference, pp. 1–15 (2018)
14. Paul, A., Selvi, S.S.D., Rangan, C.P.: A provably secure conditional proxy re-encryption scheme without pairing. Cryptology ePrint Archive (2019)
15. Farràs, O., Ribes-González, J.: Provably secure public-key encryption with conjunctive and subset keyword search. Int. J. Inf. Secur. **18**(5), 533–548 (2019). https://doi.org/10.1007/s10207-018-00426-7

Simulating Spiking Neural Networks Based on SW26010pro

Zhichao Wang[1,2], Xuelei Li[2(✉)], Jintao Meng[2], Yi Pan[2], and Yanjie Wei[2(✉)]

[1] Southern University of Science and Technology, Shenzhen 518055, China
[2] Shenzhen Institute of Advanced Technology, Chinese Academy of Sciences, Shenzhen 518055, China
{xl.li,yj.wei}@siat.ac.cn

Abstract. The spiking neural network (SNN) simulators play a significant role in modeling neural systems and the study of brain function. Currently, many simulators using CPU or GPU have been developed. However, these simulators usually show low efficiency, resulting from the random synaptic connections, random spiking events, and random synaptic delay and plastic properties in the SNN models. To overcome the problem of random memory access etc., a new simulator named SWsnn is developed based on a new Chinese processor, SW26010pro. SW26010pro consists of six core groups (CGs), and each CG has 16 MB of local direct memory (LDM) (similar to L1/L2 cache), which is enough to store neuron data for a long time. By rearranging the synaptic and neuron data, SWsnn ensures that most of the random memory access occurs in the neuron data, and the reusability of the LDM is improved obviously. The results illustrate that the proposed SWsnn runs faster than other GPU-based simulators.

Keywords: Spiking neural network simulation · Computer simulation · SW26010pro

1 Introduction

Spiking neural network (SNN) models can help study brain function and validate the hypotheses of neuroscience [1]. Furthermore, the SNN simulators are tools used to help neuroscientists model and study the required SNN models on the computer. In the past decades, some general CPU-based SNN simulators have been developed, such as NEST [2], NEURON [3], Brian [4], GENESIS [5]. These simulators have undergone several iterations of versions, providing a relatively friendly interface and supporting rich functional modules to simulate neuronal behavior. Brian2 [6] is a continuation of the Brian software, which uses code generation techniques to make the simulator more flexible and easy to use. Brian has complete functions, but it does not support distributed simulation and its simulated scale and running speed are limited. NEST and NEURON support distributed simulations on computer clusters, and thus can support larger scale

The original version of this chapter was revised: An incorrect grant number in the acknowledgement section has been corrected. The correction to this chapter is available at https://doi.org/10.1007/978-3-031-23198-8_35

M. S. Bansal et al. (Eds.): ISBRA 2022, LNBI 13760, pp. 356–368, 2022.
https://doi.org/10.1007/978-3-031-23198-8_32

network simulations. NEST focuses on the simulation of the entire neural network system, with a core written in C++ and external interfaces written in Python. NESTML [7] supports the specification of neuronal models in NEST using a precise and concise syntax based on Python. However, these SNN simulators face the challenges of low simulation efficiency, and the wall clock time for brain simulation is usually 2–3 orders of magnitude higher than that for real-time brain activity.

With the development of GPU technology, the newly designed SNN simulators achieve better performance with the help of GPU. GeNN [8] used a single GPU to simulate the visual cortex of rhesus monkeys and achieved good performance, and it also allows users to implement other custom models. The new generation of GeNN [9] has also been greatly improved over previous versions, and performance has been greatly enhanced. GeNN proposes a simulation approach called 'procedural connectivity', where connectivity and synaptic weights are generated 'on the fly' rather than stored and retrieved from memory, and memory access time is thus eliminated. However it is difficult to generate synaptic weights and connectivity when simulating plastic synaptic models. PyGeNN [10] is a python package that provides a friendly interface to all GeNN functions, allowing scientists to use GeNN in a python-based way. Brian2GeNN [11] is a Python-based package that connects GeNN and Brian. It does not require users to know much about GPU, C++ or GeNN.

CARLsim [12,13] provides abundant functional modules that support the simultaneous operation of multiple GPUs and CPUs. It is also more suitable for large-scale neural network simulation, but its simulation speed is much slower than GeNN. BindsNET [14] is a Python package for simulating and building spiking neural networks. Better suited for machine learning and reinforcement learning, BindsNET is built on the PyTorch machine learning library and can run on CPUs and GPUs.

Despite the apparent acceleration of these GPU-based simulators, serious problems still affect the performance of SNN simulation. Due to the randomness and sparsity of the synaptic connections between neurons and the uncertainty and randomness of the frequency of spiking events in neurons, the memory access during the spiking transmission is random and global, which makes the memory access cost expensive. This is a common problem with current GPU- and CPU-based simulators. Therefore, it is necessary to consider a new processor architecture to implement SNN simulation.

In order to reduce the overhead of random memory access, a new simulator SWsnn is designed on the SW26010pro processor. The SW26010pro processor has 16MB of local direct memory (LDM) in each core group (CG), which is much larger than the shared (or cached) memory in GPUs and can hold enough data. The read and write speed in LDM is similar to normal L1/L2 cache. In the process of SNN simulation, most random access only occurs in LDM by rearranging data structures, so the cost of random access is reduced.

SWsnn takes advantage of SW26010pro processor to overcome the problem of random memory access in GPU- and CPU-based simulators and obtains much better performance. Besides, SWsnn also has a simple operation. Neuroscientists can quickly model a spiking neural network using the program's interfaces without understanding the simulator in detail. In the second section, we will introduce the essential elements of the simulator, neuron, and synaptic units. In the third section, the structure of SW26010pro and the design of SWsnn will be introduced. In the fourth section, the performance of SWsnn will be displayed and analyzed. In the fifth section, we give the conclusion and discussion.

2 Simulator Model

2.1 Neuron Types

Scientists have obtained many computational models of neurons by studying complex neurons in the past decades. Single compartment neuron is widely used because it ignores the morphological features of the neuron, abstracts the neuron as a point, and considers only the effects of the current and spiking action of the neuron. The Hodgkin-Huxley (H-H) neuron model, the Izhikevich neuron model, and the Leaky-Integrate-and-Fire (LIF) neuron model are all single compartment neurons. SWsnn supports LIF model and Izhikevich model.

The Hodgkin-Huxley model [15] is biologically well explained. However, the difficulty in obtaining parameters for all aspects of the nerve cell often prevents us from applying this model. LIF neuron model [16] is widely used because it is very simple in design but still remains some key biological properties. It is described by the following equation.

$$\tau_m \frac{\mathrm{d}u}{\mathrm{d}t} = -\left[u - u_{rest}\right] + RI \tag{1}$$

Here u is the membrane potential, τ_m is the membrane time constant, I is the current, u_{rest} is the resting membrane potential, and R is membrane resistance.

The Izhikevich neuron model [17] is relatively complex to compute and better simulates biological properties. It is described by a simple set of equations as follows.

$$\frac{dv}{dt} = 0.04v^2 + 5v + 140 - u + I \tag{2}$$

$$\frac{du}{dt} = a(bv - u) \tag{3}$$

Here v is the membrane potential voltage, u is the recovery variable, and I is the sum of the external incoming currents transmitted from the connected synapses. a, b, c and d are changeable parameters to represent different types of neurons. a is the time scale of the recovery variable, and the smaller the value of a is, the slower the variable u recovers. b is the sensitivity of the recovery variable fluctuations. c is the reset value of v after sending a spike, and d is the reset value of u after sending a spike.

$$\text{if } v \geq 30mV, \text{ then } \begin{cases} v \leftarrow c \\ u \leftarrow u + d \end{cases} \tag{4}$$

When the voltage u exceeds 30 mv , the recovery variable u and the voltage v are reset to simulate the changes inside the neuron after the spike occurs.

2.2 Synapse Types

Two main synaptic models are used in computational neuroscience: the current-based (CUBA) synaptic model and the conductance-based (COBA) synaptic model. The CUBA synapses ignore voltage-dependent properties and calculate currents by weight accumulation. The COBA synaptic model is more complex. In a real neural system, the synaptic current depends on the voltage of the post-neuron, which is considered in the conductance-based synaptic model. The following is a model of CUBA synapses.

$$I_j^{syn} = \sum_{i=1}^{N} s_{ij} w_{ij} \tag{5}$$

where w_{ij} is the weight of the synaptic connection from the i-th pre-neuron to the j-th post-neuron, and I_j^{syn} is the total synaptic current at the j-th post-neuron. s_{ij} is 1 if the i-th pre-neuron is spiking and 0 otherwise.

2.3 Simulation Strategy

The general simulation strategies for neural networks are divided into clock-driven and event-driven simulation algorithms. The clock-driven strategy divides the simulated time into time steps, and at each time step, all neurons do synchronized updates, which applies to most neuronal models and has been widely used. Numerical computations of the spiking neuron model generally use Euler and Runge-Kutta methods. To improve the accuracy, smaller time steps are usually used. SWsnn supports both Euler and Runge-Kutta's fourth-order methods.

3 Simulator Design

The SWsnn simulator can be used to simulate spiking neural networks. It is written in C/C++ language and runs on Linux systems using a compiler designed for the SW26010pro processor. SWsnn simulator supports more general neuron models, supports multiple neuron population connections, does not require much compilation time, and can be used more efficiently. SWsnn supports setting its time step within 1 ms as needed and allows for more accurate simulation by setting its own delay time for the spiking transmission process. Next, the spiking neural network simulator's essential components will be described.

3.1 SW26010pro

SW26010pro is a continuation of the SW26010 processor and performs better than before. SW26010pro consists of six core groups (CGs), and each core group consists of one manager processing element (MPE, main core) and 64 computing processing elements (CPEs, slave cores). About 97% of the computing power of the SW26010pro processor is on the slave cores (Fig. 1). Compared with GPU,

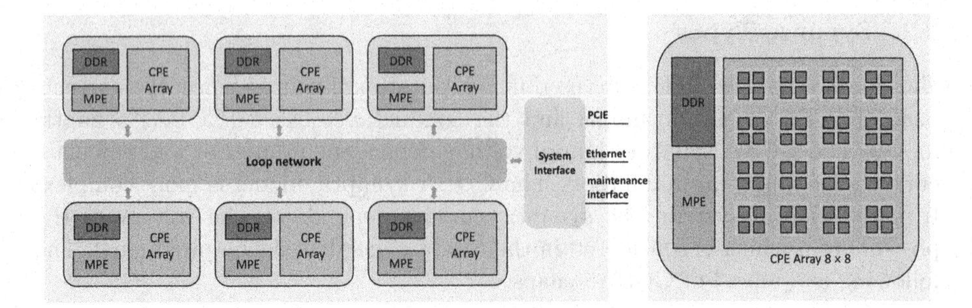

Fig. 1. Schematic diagram of SW26010pro processor.

SW26010pro has one significant advantage: each slave core has 256 KB of local data memory (LDM) space, and each CG has 16 MB LDM in total. LDM is similar to L2 cache, so it can avoid too much time consuming due to the memory access. The characteristics of LDM make SW26010pro more appropriate for the SNN algorithm than GPUs. SW26010pro has also achieved good results in other areas. SW_Qsim [18] is a quantum simulator based on tensor networks, implemented on SW26010pro, and the paper won the Gordon bell prize in 2021.

3.2 SWsnn Model Design

In SNN simulation, two types of data need to be stored and processed. Synaptic data occupies most memory because each neuron connects nearly a thousand or more synapses. In SW26010pro, each CG has 16 GB of the main memory and 16 MB LDM. In order to make full use of memory space, reduce memory access costs and improve the reusability of the LDM, SWsnn put neuron data in LDM and synaptic data in main memory. In SW26010pro, each slave core has its number, and the memory of each slave core is independent. Therefore, neuron data is placed evenly in different slave cores according to their number. Each neuron connects to thousands of synapses, and each synapse corresponds to a pre-neuron and a post-neuron. The memory of the synapse is set up as an array of two-dimensional structures, with rows arranged according to the post-neuron number and columns arranged according to the pre-neuron number.

For the simulation of a simple neuron, it is only necessary to perform iterative calculations according to the mathematical formula of the neuron model to determine whether to fire and deliver spiking events. However, factors such as spiking transmission time and memory reading time should be considered in the simulation of larger neural networks. SWsnn divided the simulation process into neuron update part and network update part (Fig. 2).

Neuron update part refers to the differential equation iterative calculation of all neuron voltages and currents. The network update part refers to read the weight and delay information from the synapse and add the current to the neuron according to the collected excitation neuron information. The neuron update part runs on 64 slave cores, and their calculations are independent. The

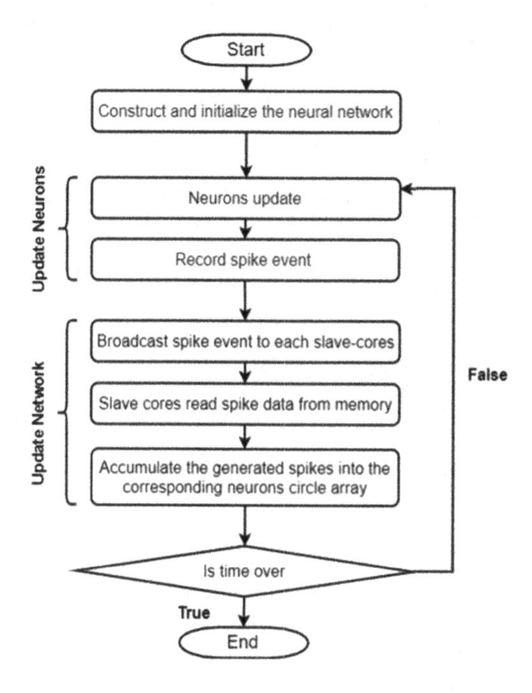

Fig. 2. Spiking neural network update flow chart.

information of each neuron is read from the LDM on the slave core and calculated according to the iterative formula of the neuron model used. The neuron number is recorded when the neuron membrane voltage reaches the threshold. Because iterative computation is computationally intensive, vectorization is added to the compiler layer to optimize the program.

In the network update part, each slave core needs to traverse the synaptic information, read the synaptic information of its core group of neurons, and add the calculated current to the neuron. Reading messages directly from the main memory is time-consuming, so direct memory access (DMA) is used to read the main memory into the LDM. DMA is a fast way to read information from the main memory to LDM. Since the number of neurons connected to each synapse is random, reading data from an LDM can significantly reduce the time consumed by random access. Since spiking transmission from pre-neurons to post-neurons requires delay time, a circular array is set behind each neuron, and the current corresponding to the delay time is placed in different positions of the circular array, and the size of the circular array is set to correspond to the time step and the spiking transmission delay. After a spike occurs, the current is put into different locations in the circle array depending on the delay. Discontinuous addressing also takes time because the delay is randomly distributed. However, the circular array is in the LDM, thus reducing read and write time consumption. As shown in Fig. 3, the neuron reads and calculates the current at the following positions in the circular array after each time step.

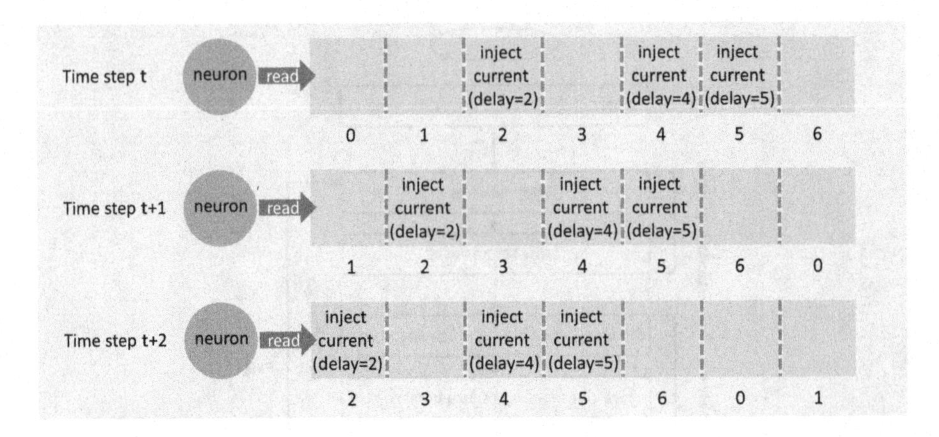

Fig. 3. Each time step, each neuron reads the current value for the corresponding time of the circle array.

As mentioned above, SWsnn splits the simulation process into two parts, and each time step needs to update neurons and read synapses from the main memory to continue the computation. The delay of the spike in the transmission process is within a range, so there is no need to read the synaptic information at each time step, and reading before the shortest delay will not alter the experimental results. At the same time, in the simulation process, tasks are allocated to the slave cores through the main core. However, each task distribution process will cause the overhead of allocating memory and scheduling threads, which is very time-consuming. It also reduces program time by reading synapses before the minimum delay, eliminating the need to assign tasks to each time step.

4 Simulator Result

In this section, the spiking balance network and cortical microcircuit balance network are constructed and experimented. The performance of SWsnn and GeNN running the same model is compared, and the neural network simulations of different scales are tested on SWsnn.

4.1 Spiking Balancing Network

Izhikevich designed the Izhikevich-type neuron model, and different neuron types correspond to different parameters of the neuron model [17]. Meanwhile, Izhikevich constructed a spiking neural network that describes the cerebral cortex of mammals, which can display similar collective dynamics and rhythms. SWsnn also used this model to simulate and realize an 80–20 random spiking neural network with 80% excitatory and 20% inhibitory neurons. The network consists of 1,000 neurons, each neuron connecting 100 neurons. This network shows sleep-like oscillations observed in the mammalian cortex.

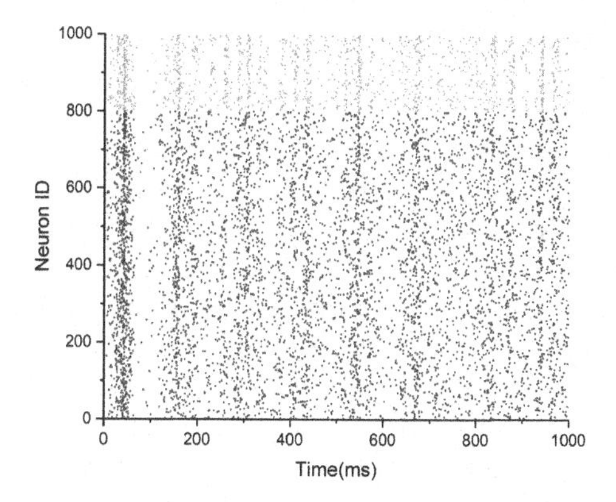

Fig. 4. Raster plot shows the spike times of neurons in the cortical microcirculation model. Each point represents the corresponding numbered neuron where the spiking occurred.

The model consists of two groups of neurons, inhibitory and excitatory, which are directly connected to each other and self-connected. The parameters a, b, c and d of inhibitory and excitatory neurons are different so that the membrane voltage will react differently under the same current stimulation. SWsnn utilizes the same model and experimental parameters as Izhikevich, so the results in the raster plot are similar, and the error is due to the random number setting. As shown in Fig. 4, the 800th to 1000th neurons are inhibitory neurons, and the rest are excitatory neurons.

Neural networks are randomly connected, but still exhibit regular behavior. This could represent the regular behavior of the mammalian cerebral cortex during wakefulness. Different states of neuronal activity in other brain regions can be simulated by changing the strength of synaptic connections.

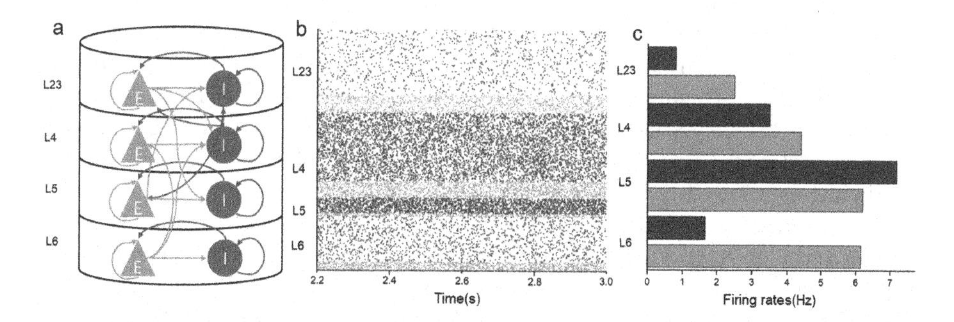

Fig. 5. Figure a is a brief description of the cortical microcircuit model and connections. Red triangles represent excitatory neurons, blue circles represent inhibitory neurons, and arrows represent corresponding connections. Figure b describes the excitation grids of excitatory and inhibitory neurons at different model layers. Figure c records the excitation frequencies of different layers of the microcircuit model.

4.2 Cortical Microcircuit Model

Potjans and Diesmann constructed a network model of the local cortical microcircuit model, in which spontaneous activity is asynchronous and irregular, and cell type specific discharge rates are consistent with those recorded in vivo in awake animals [19]. The cortical microcircuit model describes a square centimeter of epidermal cells, with 77,169 neurons and 3×10^8 synaptic connections. The model is simulated using LIF neurons, divided into four layers: L2/3, L4, L5, and L6. Each layer of neurons is divided into excitatory neurons and inhibitory neurons. The model is simulated on SW26010pro, using the same connection parameters and a direct current power supply as the stimulus source. Since the complete model takes up much memory, we scale the model by one-tenth, and the result is shown in Fig. 5.

In Fig. 5.b, the raster plot shows the spiking time of neurons in the cortical microcircuit model. Each point represents the neuron spiking at this moment. Different currents stimulate different layers, and the stimulated frequencies are also different. The experimental raster diagram is similar to that of Potjans and Diesmann's paper. The excitation frequencies of excitatory and inhibitory neurons in different layers are counted, and the histogram of excitation frequencies of different types of neurons in different layers is shown in Fig. 5.c. Blue represents inhibitory neurons, and red represents excitatory neurons.

4.3 Performance Analysis

Simulations of spiking neural network simulators are time-consuming when simulating large network models, and in 2018, the 1 mm^2 macaque monkey visual cortex was simulated on NEST on an IBM Blue Gene/Q supercomputer, which took about 12 min to simulate 1s biological time. SWsnn is developed on SW26010pro processor. It fully uses the architecture of SW26010pro to optimize the program to balance ease of use and computing performance. And GeNN is one of the fastest SNN simulators at present. The version of GeNN we use is 4.0.0.

The performance of SWsnn and GeNN was compared in the test. SWsnn runs in a Linux environment on a single core group of the SW26010pro processor. SW26010pro single core group has a single-precision peak performance of 2.2TPLOPS. GeNN uses the NVIDIA A100 PCIe processor in a Linux environment and has a single-precision peak performance of 19.5TPLOPS, and the bandwidth is about 1.6 TB/s.

The simulation times of neural networks of different sizes were tested on SWsnn and GeNN, and the probability of each neuron connecting was set to 0.1. The time step was set to 0.1 ms, and the neuron spike frequency was set 10 Hz. Both experiments use the Izhikevich model of neurons and the iterative method of the Runge-Kutta method. While a core group is used on the processor to run SWsnn programs, GeNN runs primarily on the GPU, and the two processors are not similar. While not perfect, the comparison using peak performance still reflects the performance advantage of SWsnn.

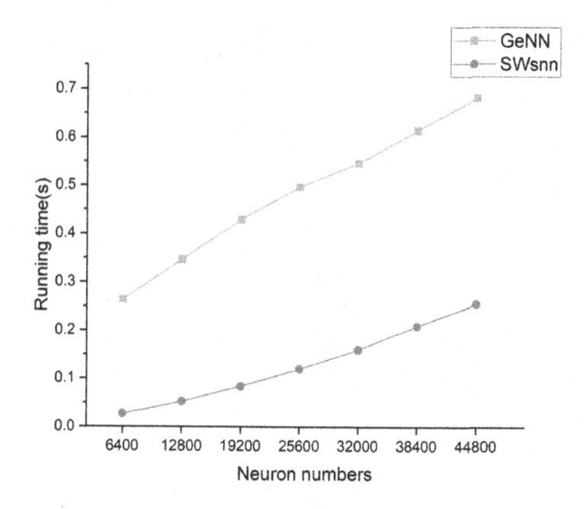

Fig. 6. Running time of different neural network sizes for SWsnn using SW26010pro and GeNN simulations using NVIDIA A100.

The computational performance results of SWsnn and GeNN are shown in Fig. 6. The simulation time of SWsnn is converted from the peak performance of SW26010pro and NVIDIA A100 graphics cards. It can be observed from the picture that SWsnn has a faster simulation speed under the scale of neural network experiments. This is also due to the full use of the SW26010pro processor structure to achieve performance optimization. However, due to the low bandwidth of SWsnn, the slope of simulation time of SWsnn increases with the size of the neural network, while the change of GeNN is small.

Figure 7 shows the SWsnn performance at different frequencies with different neural network scales. A simple model was used to test SWsnn performance using Izhikevich neurons with the same probability of connection between neurons as 0.1, and each neuron was simulated with the same current. As shown in the figure above, as the scale of the neural network increases, the program's running time becomes more protracted, while the time of updating the neural network in the program increases nonlinearly. This is because the increase in the neural network size will cause an exponential increase in the size of synapses to read from the main memory, while the growth of neurons is linear, and the calculation time required is also linear. With the increase of neuron excitation frequencies, the update time of the neural network increases linearly, while the update part of neurons remains unchanged. This is because the size of neurons does not change, while the synapses read from the main memory increase linearly.

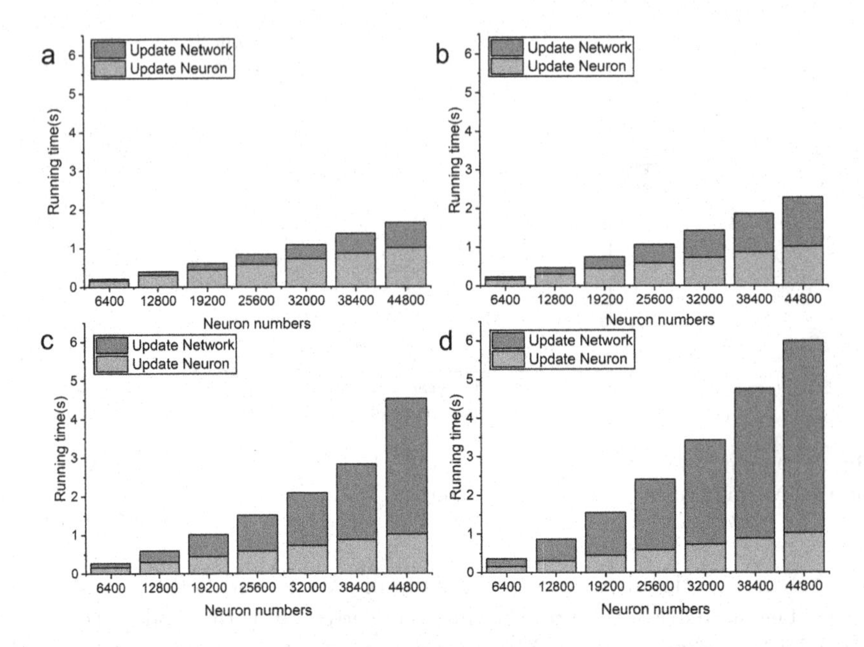

Fig. 7. Running time comparison between SWsnn neuron update and neural network update. Figure a, b, c, and d respectively show the time-consuming diagrams at excitation frequencies 5 Hz, 10 Hz, 20 Hz, 40 Hz.

5 Conclusion

This paper proposed an efficient parallel SNN simulator, SWsnn, which uses the new Chinese SW26010pro processor to solve the random memory access problem in the SNN algorithm. The simulator takes full advantage of the processor features of SW26010pro, such as sufficient LDM and independent slave cores, and speeds up the simulation process by optimizing synaptic and neuron data. Experimental results show that SWsnn runs faster than other simulation software while maintaining the same peak performance. In addition, SWsnn is user-friendly and neuroscientists can easily use SWsnn to simulate the activity of populations of neurons.

Acknowledgement. This work was partly supported by the Key Research and Development Project of Guangdong Province under grant no. 2021B0101310002; the National Key Research and Development Program of China under grant no. 2018YFB0204403; the Shenzhen Basic Research Fund under grant no. JCYJ20210324102007021, RCYX20200714114734194 and JSGG20201102163800001. We would also like to thank the funding support by the Youth Innovation Promotion Association (Y2021101), CAS to Yanjie Wei. And we would like to thank the SW26010pro processor provided by sunway supercomputer.

References

1. Eliasmith, C., et al.: A large-scale model of the functioning brain. Science **338**(6111), 1202–1205 (2012)
2. Gewaltig, M.O., Diesmann, M.: NEST (neural simulation tool). Scholarpedia **2**(4), 1430 (2007)
3. Migliore, M., Cannia, C., Lytton, W.W., et al.: Parallel network simulations with NEURON. J Comput. Neurosci. **21**, 119–12. (2016). https://doi.org/10.10007/1234567890
4. Goodman, D.F.M., Brette, R.: The brian simulator. Front. Neurosci. **3**(2), 192–197 (2009)
5. Bower, J.M., Beeman, D.: The Book of GENESIS: Exploring Realistic Neural Models with the GEneral NEural Simulation System, 2nd edn. Springer, New York (1998). https://doi.org/10.1007/978-1-4612-1634-6
6. Stimberg, M., Brette, R., Goodman, D.F.M.: Brian 2, an intuitive and efficient neural simulator. Elife **8**, e47314 (2019)
7. Plotnikov, D., et al. (eds.): NESTML: A Modeling Language for Spiking Neurons. Gesellschaft für Informatik, Bonn (2016)
8. Yavuz, E., Turner, J., Nowotny, T.: GeNN: a code generation framework for accelerated brain simulations. Sci. Rep. **6**, 18854 (2016)
9. Knight, J.C., Nowotny, T.: Larger GPU-accelerated brain simulations with procedural connectivity. Nat. Comput. Sci. **1**(2), 136–142 (2021)
10. Knight, J.C., Komissarov, A., Nowotny, T.: PyGeNN: a Python library for GPU-enhanced neural networks. Front. Neuroinform., 15 (2021)
11. Stimberg, M., Goodman, D.F.M., Nowotny, T.: Brian2GeNN: accelerating spiking neural network simulations with graphics hardware. Sci. Rep. **10**(1), 1–12 (2020)
12. Beyeler, M., Carlson, K.D., Chou, T.S., et al.: CARLsim 3: a user-friendly and highly optimized library for the creation of neurobiologically detailed spiking neural networks. In: 2015 International Joint Conference on Neural Networks (IJCNN), p. 1–8. IEEE (2015)
13. Chou, T.S., Kashyap, H.J., Xing, J., et al.: CARLsim 4: an open source library for large scale, biologically detailed spiking neural network simulation using heterogeneous clusters. In: 2018 International joint conference on neural networks (IJCNN), pp. 1–8. IEEE (2018)
14. Hazan, H., Saunders, D.J., Khan, H., et al.: BindsNET: a machine learning-oriented spiking neural networks library in Python. Front. Neuroinform., 89 (2018)
15. Hodgkin, A.L., Huxley, A.F.: A quantitative description of membrane current and its application to conduction and excitation in nerve. J. Physiol. **117**(4), 500 (1952)
16. Kistler, W.M., Gerstner, W.: Stable propagation of activity pulses in populations of spiking neurons. Neural Comput. **14**(5), 987–997 (2002)
17. Izhikevich, E.M.: Simple model of spiking neurons. IEEE Trans. Neural Netw., 14(6), 1569–1572 (2003)
18. Li, F., Liu, X., Liu, Y., et al.: SW_Qsim: a minimize-memory quantum simulator with high-performance on a new sunway supercomputer. In: Proceedings of the International Conference for High Performance Computing, Networking, Storage and Analysis, pp. 1–13 (2021)

19. Potjans, T.C., Diesmann, M.: The cell-type specific cortical microcircuit: relating structure and activity in a full-scale spiking network model. Cereb. Cortex **24**(3), 785–806 (2014)
20. Schmidt, M., et al.: A multi-scale layer-resolved spiking network model of resting-state dynamics in macaque visual cortical areas. PLoS Comput. Biol. **14**, e1006359 (2018)

Entropy Based Clustering of Viral Sequences

Akshay Juyal[1], Roya Hosseini[1], Daniel Novikov[1], Mark Grinshpon[2(✉)], and Alex Zelikovsky[1]

[1] Department of Computer Science, Georgia State University, Atlanta, USA
{ajuyal1,ahosseini3,dnovikov1}@student.gsu.edu, alexz@gsu.edu
[2] Department of Mathematics and Statistics, Georgia State University, Atlanta, USA
mgrinshpon@gsu.edu

Abstract. Clustering viral sequences allows us to characterize the composition and structure of intrahost and interhost viral populations, which play a crucial role in disease progression and epidemic spread. In this paper we propose and validate a new entropy based method for clustering aligned viral sequences considered as categorical data. The method finds a homogeneous clustering by minimizing information entropy rather than distance between sequences in the same cluster. We have applied our entropy based clustering method to SARS-CoV-2 viral sequencing data. We report the information content extracted from the sequences by entropy based clustering. Our method converges to similar minimum-entropy clusterings across different runs and limited permutations of data. We also show that a parallelized version of our tool is scalable to very large SARS-CoV-2 datasets.

Keywords: Categorical data · Clustering · Entropy · Monte Carlo algorithm · Viral genomic sequences

1 Introduction

Clustering viral sequences allows us to characterize the composition and structure of intrahost and interhost viral populations, which play a crucial role in disease progression and epidemic spread. For intrahost populations, clustering allows us to detect the distinct viral variants present in the patient, including minor low-frequency variants, which can cause immune escape, drug resistance, and an increase of virulence and infectivity [1,3,5,7,9,15,16]. Furthermore, such minor variants are often responsible for transmissions and establishment of infection in new hosts [4,8,17].

In this paper we propose a Monte Carlo entropy minimization method for clustering of viral sequences considered as categorical data. The method finds

A. Juyal and R. Hosseini—Joint first authors.

M. S. Bansal et al. (Eds.): ISBRA 2022, LNBI 13760, pp. 369–380, 2022.
https://doi.org/10.1007/978-3-031-23198-8_33

a homogeneous clustering by minimizing information entropy rather than distance between sequences in the same cluster. We discuss advantages and disadvantages of both entropy and distance based approaches, and further validate the meaningful information content extracted by entropy based clustering. We demonstrate that the proposed method is stable, moving towards the same minimal entropy configuration across multiple runs. We also show that it is fast and scalable to hundreds of thousands of sequences.

By clustering viral populations across different hosts, we determine major strains of closely related viral samples, which is helpful for tracking transmissions and informing public health strategies [2]. For transmission tracking, clustering can identify the source of an outbreak and whether the source is present in the sampled population. It can also determine whether two viral samples belong to the same outbreak, and whether one infected the other [13]. Therefore, using clustering to obtain an accurate characterization of viral mutation profiles from infected individuals is essential for viral research, therapeutics, and epidemiological investigations.

Viral sequences are strings from a fixed nucleotide alphabet, and hence they can be viewed as vectors of categorical data. In the best possible clustering, sequences in each cluster will be as homogeneous as possible in each site. Typically, this is achieved by minimizing the Hamming distances between sequences within the same cluster or the distances to the cluster's consensus [18]. However, Hamming distance carries the implicit assumption that all mutations at all sites are of equal cost, and also it does not consider the distribution of values in a given category, counting each mismatch equally.

In this paper we propose to use entropy based clustering for viral sequences. Entropy considers the distributions of nucleotides in each site, allowing us to capture different kinds of mismatches. Minimizing entropy instead of distance also avoids the need to introduce the abstraction of equal transition costs, implicit in Hamming distance. Thus, entropy as an objective for clustering makes fewer assumptions on the data and is more informative for clustering categorical data.

We have applied our entropy based clustering method to the SARS-CoV-2 viral sequencing data. The unprecedented effort in sequencing its genome has created vast databases of SARS-CoV-2 sequences, such as GISAID [10]. Clustering techniques can provide new insights into the evolution of the virus, assist with phylogenetic and phylodynamic analyses, and offer new tools for constructing transmission networks to help with understanding the spread of the pandemic [2].

We validate effectiveness of the entropy based clustering of viral sequences on real datasets. We measure the information content extracted from the sequences by the resulting clustering. We also demonstrate that our method converges to the same minimum-entropy clustering across different runs, thus marking stability of the method. Finally, we describe a tag selection procedure, which selects the highest entropy sites to represent sequences leading to a significant decrease in runtime without major loss of information.

2 Methods

In this section, first we define the entropy and Hamming distance of clustering of a set of aligned viral sequences. We use each of these two measures as an objective function to be minimized for clustering. Then we describe a Monte Carlo clustering algorithm, which is a modification of an algorithm proposed in [12]. Finally, we describe a tag selection preprocessing step, which significantly reduces the runtime of the algorithm.

2.1 Entropy Based Clustering of Viral Sequences

Entropy of a category across a set of categorical vectors quantifies the heterogeneity of values in the category. Entropy is low when a single value is highly frequent, and it is at its highest when all values are equally frequent in the category. Since we are treating viral sequences as vectors of categorical data, the categories here are sites along the sequence, and their values are from the nucleotide alphabet $\{\mathtt{A}, \mathtt{C}, \mathtt{G}, \mathtt{T}\}$. A clustering with minimal entropy will have the highest possible homogeneity of nucleotides in each site for sequences in the same cluster.

Formally, we have a set S of aligned nucleotide sequences on a set X of genomic sites. Since the sequences $s \in S$ are aligned, they can be viewed as rows of a matrix, and the sites $x \in X$, can be viewed as columns of this matrix. Let the alphabet $\mathcal{A} = \{\mathtt{A}, \mathtt{C}, \mathtt{G}, \mathtt{T}\}$ be the four nucleotides, not counting the gap (-) character. Following [12], the entropy $H(C_x)$ of a site $x \in X$ in cluster C is defined as

$$H(C_x) = -\sum_{s \in C} \sum_{a \in \mathcal{A}} p(s_x = a) \cdot \log p(s_x = a). \qquad (1)$$

Note that $p(s_x = a)$, the probability that a sequence $s \in C$ has nucleotide $a \in \mathcal{A}$ at site x, essentially amounts to the *relative frequency* of the nucleotide a in cluster C at site x (ignoring gap characters).

The entropy $H(C)$ of a cluster C of viral sequences on a set X of sites is then defined as

$$H(C) = \sum_{x \in X} H(C_x), \qquad (2)$$

that is, we simply sum up the entropies at the individual sites.

Finally, given a clustering $\mathcal{C}$ of the set S, the *entropy* of $\mathcal{C}$ is defined as follows:

$$H(\mathcal{C}) = \sum_{C \in \mathcal{C}} \frac{|C|}{|S|} \cdot H(C) = \frac{1}{|S|} \sum_{C \in \mathcal{C}} |C| \cdot H(C). \qquad (3)$$

In other words, the entropy of clustering $\mathcal{C}$ is the sum of cluster entropies weighted by their relative sizes.

In [12], the authors prove that the entropy defined in Eq. (3) is a convex function, allowing any optimization procedure to reach a global minimum. It is because of this property that we can use techniques aimed directly at minimizing clustering entropy as the objective.

2.2 Hamming Distance Based Clustering of Viral Sequences

Similarly, we define a different clustering objective as Hamming distance (HD) instead of entropy. This objective is the sum of the Hamming distances from each sequence s to the consensus of the cluster containing s.

Formally, for a cluster C and for a site $x \in X$, the Hamming distance from the consensus letter in this cluster at this site is

$$HD(C_x) = \sum_{a \in \mathcal{A}} C_x(a) - \max_{a \in \mathcal{A}} \{C_x(a)\} \tag{4}$$

where $C_x(a)$ is the number of occurrences of the letter $a \in \mathcal{A}$ in site x in cluster C.

Then the Hamming distance $HD(C)$ of a cluster C of viral sequences on a set X of sites is defined as

$$HD(C) = \sum_{x \in X} HD(C_x), \tag{5}$$

and the *Hamming distance* of the clustering $\mathcal{C}$ is defined as

$$HD(\mathcal{C}) = \sum_{C \in \mathcal{C}} HD(C). \tag{6}$$

2.3 Algorithm Description

In general, Monte Carlo methods optimize an objective by attempting random changes and accepting a change only if it improves the objective. In our case, the objective is to minimize either the clustering entropy or the clustering Hamming distance, defined in Subsects. 2.1 and 2.2 above. A trial step consists of moving a randomly selected sequence to a different randomly selected cluster, and accepting the move only if the objective function is reduced.

Algorithm 1, Monte Carlo based clustering, implements this approach, with several modifications intended to improve its runtime and the quality of its outputs.

The algorithm takes an existing clustering as its starting point. In our experiments, we use clusterings generated by the CliqueSNV tool [11] from the datasets described in Subsect. 3.1. We supply such clustering as an input to the algorithm in order to generate a new clustering with reduced clustering entropy $H(\mathcal{C})$ or reduced Hamming distance $HD(\mathcal{C})$. Additional inputs to the algorithm are two parameters I and K, where I defines the number of consecutive rejected trials before stopping, and K defines the relative objective function reduction threshold for accepting a move.

To achieve the goal of finding a new clustering with reduced entropy or Hamming distance as the objective, the algorithm applies Monte Carlo optimization by repeatedly trying to move a randomly selected sequence from its current cluster to another randomly selected cluster; any such move is accepted only if the relative improvement to the objective function is higher than the threshold value K.

Algorithm 1: Monte Carlo based clustering

Input:
Initial clustering (by default, from CliqueSNV)
Number of rejected moves: I (by default, $I = 800$)
Relative difference: K (by default, $K = 0.00001$)
Output:
Clustering with reduced entropy or Hamming distance
Initializations:
1 Compute nucleotide counts for each column in each cluster
2 Compute entropy (resp., Hamming disatnce) for each cluster and H, clustering entropy (resp., clustering Hamming distance)
3 Initialize number of rejected moves $T = 0$
Iteration:
4 **while** $T \leqslant I$ **do**
5 Pick a random sequence s
6 Move s from its cluster A to a randomly selected cluster B, $B \neq A$
7 Update the nucleotide counts for A and B
8 Compute Δ, overall entropy (resp., Hamming distance) reduction after moving s from A to B
9 **if** $\Delta/H \geqslant K$ **then**
10 Accept the move
11 $H = H - \Delta$
12 $T = 0$
13 **else**
14 $T = T + 1$
15 Move s from B back to A
16 **end**
17 **end**

In the initialization phase, lines 1–3 of the algorithm, it starts by computing nucleotide counts for each column in each cluster, which are then used to compute the values of the entropy and the Hamming distance for each cluster, as well as the overall clustering entropy or Hamming distance, H.

A Monte Carlo optimization procedure is then implemented in the loop in lines 4–17 as follows:

- Line 5: Randomly pick a sequence s.
- Line 6: Remove sequence s from its cluster A and place into another randomly selected cluster B.
- Line 7: Recalculate entropy (resp., Hamming distance) for both clusters A and B.
- Line 8: Compute the entropy (resp., Hamming distance) reduction between the new and the previous clusterings.
- Lines 9–16: If entropy (resp., Hamming distance) has reduced by at least the relative difference parameter K, keep the new clustering; otherwise, revert to the previous clustering. By default, we set $K = 0.00001$.

- Repeat lines 4–17 until the clustering converges. Specifically, the algorithm will stop if we do not accept any moves for sufficiently long time, i.e., if the clustering does not change for I consecutive iterations. By default, we set $I = 800$.

2.4 Tag Selection

To improve runtime, we apply a preprocessing tag selection step that allows us to represent sequences by a smaller subset of sites. Preferring tags with highest variability, the procedure chooses the n sites with highest entropy, where n is some predefined value. Then clustering proceeds with each sequence now of length n corresponding to the selected sites.

3 Settings for Validation of Clustering Methods

We validate entropy based clustering by estimating improvement over an existing clustering technique. To that end, we apply this Monte Carlo based algorithm to the clustering obtained by the CliqueSNV tool [11].

3.1 Datasets

For validation, we use two of the datasets of SARS-CoV-2 sequences that were also used by Melnyk et al. in [14]. For both datasets, an initial clustering was obtained by the CliqueSNV-based method.

D1: This dataset includes all sequences submitted to the global GISAID viral database [10] from the beginning of the pandemic up until the beginning of March 2020. It consists of 3688 aligned SARS-CoV-2 sequences, all sequences 29891 nucleotides long. CliqueSNV produced an initial clustering of this dataset consisting of 28 clusters.

D2: This dataset includes all sequences submitted to the UK-based EMBL-EBI database from the end of January 2020 to the end of December 2020 [6]. It consists of 148000 aligned SARS-CoV-2 sequences, all sequences 29903 nucleotides long. CliqueSNV produced an initial clustering of this dataset consisting of 15 clusters.

3.2 Tag Selection Effects on Runtime

To measure the effects of tag selection on runtime, the proposed method was run on the D1 dataset of 3688 sequences, using the initial clustering generated by CliqueSNV as a starting point. The tag selection procedure was employed to produce four subdatasets consisting of the same sequences, but of reduced lengths of 100, 1000, 3000, and 5100 tags. We expect that the Monte Carlo method, when applied to a dataset consisting of shorter sequences, will be able to take more trial steps in the same amount of time, and thus reduce its clustering entropy quicker. The total number of SNPs in the input data was exactly 5100; all other positions did not mutate. Thus, this largest number of tags contains information equivalent to the full length sequences for the purposes of clustering.

The program was run on each length of sequences. Every hour, the current clusterings of each run were evaluated by their entropy on the full-length sequences. These hourly entropy values are shown on an entropy-over-time graph, Fig. 1, which compares the speed of entropy reduction for different sequence lengths.

3.3 Discerning Signal from Noise with Monte Carlo Based Clustering Optimization

We estimate the amount of meaningful information extracted by the clusterings obtained by our entropy minimization method. To distinguish between sample-specific noise and meaningfully extracted information, we run our method on a perturbed version of the input with the same starting entropy. For this experiment we use the D1 dataset with 3688 sequences, alongside the initial clustering from CliqueSNV of 28 clusters.

The permutation procedure is as follows. Within each cluster, every site is shuffled into a random permutation. Importantly, by respecting clusters during permutation, the initial nucleotide frequencies within each site in each cluster stay the same. Thus, the permuted input has the same starting entropy as the original input. What changed is the haplotypes being clustered.

We run the program on both of these inputs for exactly 100000 Monte Carlo trials each, accepting all moves that reduce entropy. We compare the resulting entropy reductions between the two runs. Any entropy reduction present in the permuted data is sample-specific noise extracted by our method, while the difference in resulting entropies between the original and permuted inputs corresponds to the amount of meaningful information extracted by our method.

3.4 Stability of Optimized Clustering

Now we evaluate the robustness of our method against slight permutations of the input data as well as changes in random seed. Rather than completely shuffling each site as in Sect. 3.3, we only shuffle a small percentage p of nucleotides at each site. We still respect clusters when permuting the data, to ensure that nucleotide frequencies in each site in each cluster remain unchanged.

We chose two values of p to create slightly permuted data sets for validation, $p = 1\%$ and $p = 5\%$, to be compared with the original data with 0% permutation. For each of the three datasets (two permuted and one original), we run our method three times, on two different objectives: first minimizing entropy, and second minimizing Hamming distance between sequences and their cluster consensus. As a result, for each degree of permutation and for each Monte Carlo objective, we obtain three minimum entropy clusterings.

The Rand index, measuring the degree of agreement between two clusterings, is measured between the initial clustering and all resulting clusterings, to get a sense of how far away the resulting clusterings have moved from the initial one under varying degrees of permutation. Further, we also measure the Rand index between the resulting clusterings, to determine whether the proposed method converges to similar clusterings across multiple runs.

4 Validation Results

We ran the proposed method on the cluster hardware consisting of 128 cores Intel® Xeon® CPU E7-4850 v4 CPU @ 2.10 GHz, with 3 TB of RAM, running Ubuntu 16.04.7 LTS.

We ran our entropy based Monte Carlo clustering algorithm on different selections of tags, selected based on highest entropy contribution. Figure 1 shows the results for different numbers of tags. The maximum number of tags, 5100, corresponds to all SNP positions present in the input data; all other sites were homogeneous. Since homogeneous sites have zero entropy, they can be ignored from entropy calculations. Therefore, the yellow line representing 5100 tags corresponds to using all sites.

When the number of tags is 1000, using tag selection yields better results of reducing entropy for some time durations (up to 3 h). But in general, the effect of using tags on reducing entropy is not significant.

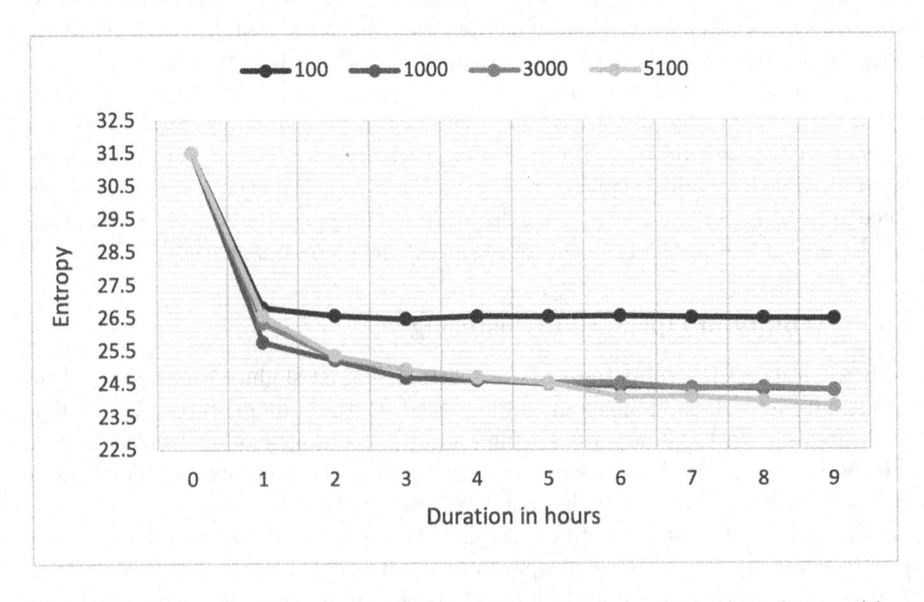

Fig. 1. Entropy reduction over runtime for different numbers of selected tags. After 1 h, the 1000 tags representation was able to reach the lowest entropy.

Compared to previous work [14], we were able to reduce the runtime by 95.83%. For accomplishing this, we initially stored the counts of each nucleotide across a given tag in a cluster of sequences.

After further improving our entropy based Monte Carlo clustering algorithm and implementing parallel computing, we made it scalable for running on large data sets. This version performs 12K–13K iterations on average per hour, which is approximately **10** times faster than all our previous implementations.

Interestingly, after running the algorithm for **83K** iterations on the D2 dataset (which contains 148000 aligned sequences each 29903 nucleotides long) originally distributed across 15 clusters, we came to the conclusion that in instances where sequence data is clustered into a smaller number of clusters there are some clusters where sequences are more dense than in others. For instance, after this run, in the output clustering the biggest cluster consisted of 34995 sequences, while the smallest had only 3571 sequences.

Therefore, moving one sequence at a time is not always beneficial for overall entropy reduction. We tried to tackle this problem by updating our move acceptance threshold to accept even smaller positive changes to entropy, from our previous relative difference threshold of $K = 10^{-5}$ to $K = 10^{-7}$. Although this change made our algorithm run for many more iterations before stopping (recall that the algorithm stops when it reaches the stopping threshold of $I = 800$ unsuccessful moves), the reduction to the overall entropy was still very small.

We believe that moving similar sequences together within clusters rather than moving just one could be a possible way of overcoming the problem we face here.

4.1 Picking Signal Over Noise in Clustering

By minimizing entropy on the permuted data, we find that the method reduces entropy to 29.6, while on the original, unshuffled data the method reaches a much lower entropy of 24 (see Table 1). This difference in entropies, $29.6 - 24 = 5.6$, of resulting clusterings accounts for the amount of meaningful information, which is not noise, that our method was able to extract from the real data.

Table 1. Results after running Monte Carlo for 1000 tags selected in decreasing order of entropy across 100 datasets obtained by applying the random permutation procedure described in Sect. 3.3 to the D1 dataset 100 times. *Average iterations in Monte Carlo*: 53804.39. *Average successful moves*: 615.31.

Entropy MC reduced					Hamming distance MC reduced				
Initial	Original		Permuted		Initial	Original		Permuted	
31.524	Avg	Min	Avg	Min	1008.41	Avg	Min	Avg	Min
	24.77	24.7	29.62	28.65		373.14	369.61	770.39	689.97

4.2 Stability of Monte Carlo Output

Table 2 shows the results of stability validation, in which we compare clustering similarity for varying degrees of permutation of the input data (see Subsect. 3.4).

The first column compares resulting clusterings to the initial clustering. Without any permutations, the resultant clustering moves significantly further away from the initial one, giving a Rand index of 0.93. As the permutation degree increases, we observed that the clusterings produced by the Monte Carlo algorithm do not move as far away from the initial clustering; in other words, even after Monte Carlo was applied, the resulting clusterings had high degree of agreement with the initial clustering.

The second column in Table 2 gives the average Rand index between multiple runs of Monte Carlo for a given permutation. We see that for all degrees of permutation the method stably converges towards similar clusterings, with Rand index scores of 0.97–0.98. The same trends can be observed when using Hamming distance to cluster consensus as the objective.

Table 2. Clustering similarity (Rand index) across three choices of degree of permutation. The proposed method was run three times for each permuted instance, each run consisting of 100000 Monte Carlo trials. Reported are average Rand index similarity of the resulting clusterings to the initial clustering, as well as between resulting clusterings.

% permutation	Cluster similarity (Rand index)			
	Entropy		Hamming	
	With original	With runs	With original	With runs
0	0.936476	0.970195	0.936114	0.970135
1	0.978898	0.979435	0.936126	0.970374
5	0.980458	0.980688	0.936180	0.970616

4.3 Results for Large Datasets

Running our entropy based Monte Carlo method on the large dataset D2, which consist of 143000 aligned sequences, we get the initial entropy for our initial clustering from CliqueSNV of 80.2750171. After 82786 iterations, the final entropy for the resultant clustering was 79.509444, and the total runtime for this was 9 h 15 min.

5 Conclusions

We have developed a scalable method to find minimum-entropy clusterings of datasets viral genomic sequences. The method is scalable to hundreds of thousands of sequences, and is made even faster without significant loss of accuracy by picking a subset of tags with maximum entropy to represent the sequences. We estimate the amount of meaningful information extracted by the method. We also show that our method converges toward similar minimum-entropy clusterings across multiple runs, demonstrating its stability.

For future directions, we believe the Monte Carlo entropy minimization approach can be improved by using simulated annealing, whose tolerance of suboptimal moves can allow us to escape local minima. We are also going to add Monte Carlo entropy minimization method to CliqueSNV's clustering of intrahost populations.

References

1. Beerenwinkel, N., et al.: Computational methods for the design of effective therapies against drug resistant HIV strains. Bioinformatics **21**(21), 3943–3950 (2005). https://doi.org/10.1093/bioinformatics/bti654
2. Bousali, M., et al.: SARS-CoV-2 molecular transmission clusters and containment measures in ten European regions during the first pandemic wave. Life **11**(3) (2021). https://doi.org/10.3390/life11030219
3. Campo, D.S., et al.: Drug resistance of a viral population and its individual intrahost variants during the first 48 hours of therapy. Clin. Pharmacol. Ther. **95**(6), 627–635 (2014). https://doi.org/10.1038/clpt.2014.20
4. Campo, D.S., et al.: Accurate genetic detection of hepatitis C virus transmissions in outbreak settings. J. Infect. Dis. **213**(6), 957–965 (2016). https://doi.org/10.1093/infdis/jiv542
5. Douek, D.C., Kwong, P.D., Nabel, G.J.: The rational design of an AIDS vaccine. Cell **124**(4), 677–681 (2006). https://doi.org/10.1016/j.cell.2006.02.005
6. EMBL-EBI: EMBL's European Bioinformatics Institute. https://www.ebi.ac.uk/
7. Gaschen, B., et al.: Diversity considerations in HIV-1 vaccine selection. Science **296**(5577), 2354–2360 (2002). https://doi.org/10.1126/science.1070441
8. Glebova, O., et al.: Inference of genetic relatedness between viral quasispecies from sequencing data. BMC Genomics (2017). https://doi.org/10.1186/s12864-017-4274-5
9. Holland, J., De La Torre, J., Steinhauer, D.: RNA virus populations as quasispecies. Curr. Topics Microbiol. Immunol., 1–20 (1992)
10. Khare, S., et al.: GISAID's role in pandemic response. China CDC Weekly **3**(49), 1049–1051 (2021). https://doi.org/10.46234/ccdcw2021.255
11. Knyazev, S., et al.: Accurate assembly of minority viral haplotypes from next-generation sequencing through efficient noise reduction. Nucleic Acids Res. **49**(17), e102–e102 (2021). https://doi.org/10.1093/nar/gkab576
12. Li, T., Ma, S., Ogihara, M.: Entropy-based criterion in categorical clustering. In: Proceedings, Twenty-First International Conference on Machine Learning, ICML 2004, vol. 3, pp. 536–543 (2004). https://doi.org/10.1145/1015330.1015404
13. Melnyk, A., Knyazev, S., Vannberg, F., Bunimovich, L., Skums, P., Zelikovsky, A.: Using earth mover's distance for viral outbreak investigations. BMC Genomics **21**(582) (2020). https://doi.org/10.1186/s12864-020-06982-4
14. Melnyk, A., et al.: From Alpha to Zeta: identifying variants and subtypes of SARS-CoV-2 via clustering. J. Comput. Biol. J. Comput. Mol. Cell Biol. **28**(11), 1113–1129 (2021). https://doi.org/10.1089/cmb.2021.0302
15. Rhee, S.Y., Liu, T.F., Holmes, S.P., Shafer, R.W.: HIV-1 subtype B protease and reverse transcriptase amino acid covariation. PLOS Comput. Biol. **3**(5), 1–8 (2007). https://doi.org/10.1371/journal.pcbi.0030087

16. Skums, P., Bunimovich, L., Khudyakov, Y.: Antigenic cooperation among intra-host HCV variants organized into a complex network of cross-immunoreactivity. Proc. Natl. Acad. Sci. **112**(21), 6653–6658 (2015). https://doi.org/10.1073/pnas.1422942112
17. Skums, P., et al.: QUENTIN: reconstruction of disease transmissions from viral quasispecies genomic data. Bioinformatics **34**(1), 163–170 (2017). https://doi.org/10.1093/bioinformatics/btx402
18. de la Vega, W.F., Karpinski, M., Kenyon, C., Rabani, Y.: Approximation schemes for clustering problems. In: Proceedings of the Thirty-Fifth Annual ACM Symposium on Theory of Computing, STOC 2003, pp. 50–58. Association for Computing Machinery (2003). https://doi.org/10.1145/780542.780550

A Tensor Robust Model Based on Enhanced Tensor Nuclear Norm and Low-Rank Constraint for Multi-view Cancer Genomics Data

Qian Qiao, Sha-Sha Yuan, Junliang Shang, and Jin-Xing Liu(✉)

School of Computer Science, Qufu Normal University, Rizhao 276826, China
sdcavell@126.com

Abstract. Cancer genomics data often contain multi-view resources of different data types, which provide rich complementary information. Analyzing multi-view cancer genomics data can effectively advance cancer research. The main process of analyzing multi-view cancer genomics data in this paper is to discover the relationship between cancers and genes by clustering cancer samples and identification of differentially expressed genes. To make full use of the consistency and complementarity between genomics data, we propose a new tensor robust model based on enhanced tensor nuclear norm and low-rank constraint (EPTR-TV). First, we define the concept of the enhanced partial sum of tensor nuclear norm (EPSTNN). It dramatically improves the flexibility of the tensor nuclear norm (TNN), effectively avoiding some errors brought by TNN when approximating tensor rank. Then, the anisotropic spatial-temporal total variation (TV) regularization is introduced, which enables the model to exploit the relationship between the structures of tensor data while focusing on the details of the tensor data features. In addition, EPSTNN and TV regularization are unified into the low-rank tensor framework for in-depth analysis of cancer genomics data. Finally, the iterative optimization problem of EPTR-TV is solved by alternating direction method of multipliers (ADMM). Experimental results from clustering and feature selection experiments performed on three multi-view cancer genomics datasets show that EPTR-TV outperforms other comparative models. These suggest that EPTR-TV plays an important role in identifying cancer subtypes and finding new carcinogenic information, thus providing important insights into cancer mechanisms.

Keywords: Cluster analysis · Enhanced tensor nuclear norm · Feature selection · Cancer genomics data · Multi-view model

1 Introduction

Cancer is one of the most life-threatening diseases in nowadays society. It is critical to develop models for analyzing multi-view cancer genomics data. At present, many models have been applied, among which robust principal component analysis (RPCA) is the most common [1]. Although the matrix models have been widely used in various fields, they are not accurate in dealing with multi-view data.

M. S. Bansal et al. (Eds.): ISBRA 2022, LNBI 13760, pp. 381–388, 2022.
https://doi.org/10.1007/978-3-031-23198-8_34

Recently, Lu *et al.* proposed tensor robust principal component analysis (TRPCA) [2]. Its purpose is to explore the relationship between the internal structures of multi-view data. Tensor models have been widely used in biological information research. Hu *et al.* first proposed the use of TRPCA to select common differentially expressed genes (DEG) from data of multiple genomics [3]. A model called HTRPCA was proposed to cluster cancer samples, and its effectiveness is proved [4].

Furthermore, a model called outlier-robust tensor principle component analysis (OR-TPCA) has also been proposed to solve high-dimensional tensor problems [5]. Both TRPCA and OR-TPCA directly depend on tensor singular value decomposition (T-SVD) [6, 7]. But T-SVD may produce some inevitable deviations. Therefore, Liu *et al.* offered improved robust tensor principal component analysis (IRTPCA) [8]. The partial sum of the tensor nuclear norm (PSTNN) is used in place of TNN [9]. PSTNN does not operate on large singular values, but shrinks small singular values, which will not show the case of rank deficiency.

Based on the above theoretical knowledge, a new tensor nuclear norm is defined in this paper, named the enhanced partial sum of the tensor nuclear norm (EPSTNN). It adds weight to the top N singular values, highlighting important data information and avoiding the situation of rank inadequacy. Furthermore, to further focus on the characteristic details of the correlation between genes and cancers in the multi-view cancer genomics data, we introduce total variation (TV) regularization [10] to constrain the low-rank component. Our contributions are:

Firstly, a new framework called tensor robust model based on EPSTNN and TV (EPTR-TV) is proposed, which can be employed for feature gene selection and sample clustering experiments in multi-view cancer genomics datasets.

Secondly, in this paper, EPSTNN is defined. EPSTNN allows EPTR-TV to increase the difference of the importance degree between the internal information of the data. It can effectively extract low-rank structures from multi-view cancer genomics data and capture the correlation between the spatial structures of multi-view cancer genomics data.

Thirdly, TV is introduced into EPTR-TV as a regularization constraint for the low-rank component. After TV is submitted, the model avoids the situation of rank deficiency and focuses more on the characteristic details of the internal data structures.

2 Materials and Model

2.1 Tensor Preliminaries

In the paper, we use Euler letters to represent tensors, e.g., $\mathcal{M} \in \mathbb{R}^{n_1 \times n_2 \times n_3}$. $\overline{\mathcal{M}}$ is the Fourier transformed tensor of the tensor $\mathcal{M}$. $\mathcal{M}^{(i)}$ is the i-th frontal slice of $\mathcal{M}$.

Definition 2.1. (Tensor average rank) [11] The tensor average rank of $\mathcal{M}$ can be denoted as $\text{rank}_a(\mathcal{M})$, which is expressed as follows:

$$\text{rank}_a(\mathcal{M}) = (1/n_3)\text{rank}(\text{b}circ(\mathcal{M})), \tag{1}$$

Definition 2.2. (TNN) [11] The TNN of $\mathcal{M}$ is defined as the sum of the tensor singular values. Recently, it has been proved in [11] that TNN can be defined as (2):

$$\|\mathcal{M}\|_* = (1/n_3)\sum_{i=1}^{n_3}\left\|\overline{\mathcal{M}}^{(i)}\right\|_* \tag{2}$$

2.2 EPSTNN

We propose a new non-convex proxy, approximating the rank of tensor, which is called EPSTNN, as follows. It arranges the singular values of the first N in order of size and imposes weight on them.

$$\|\mathcal{M}\|_{\mathrm{EPSTNN}} = \sum_{i=1}^{n_3}\left(\left\|\overline{\mathcal{M}}^{(i)}\right\|_{w=topN} + \left\|\overline{\mathcal{M}}^{(i)}\right\|_{p=N}\right). \tag{3}$$

$\|\cdot\|_{w=topN}$ represents the weighting operation on the first N large singular values. In addition, $\|\cdot\|_{p=N} = \sum_{i=N+1}^{\min(m,n)}\sigma_i(\cdot)$ [9]. To solve the problem based on EPSTNN, we propose the enhanced partial sum of the singular value threshold operator:

$$\mathbb{E}_{\mathrm{N},w,\tau}(Y) = U_{Y1}T_w[S_{Y1}]V_{Y1}^H + U_{Y2}T_\tau[S_{Y2}]V_{Y2}^H. \tag{4}$$

2.3 TV Regularization

The definition of TV is as follows:

$$\|\mathcal{M}\|_{\mathrm{TV}} = \sum_{i_1,i_2,i_3}\begin{pmatrix}|\mathcal{M}(i_1,i_2,i_3) - \mathcal{M}(i_1-1,i_2,i_3)| \\ +|\mathcal{M}(i_1,i_2,i_3) - \mathcal{M}(i_1,i_2-1,i_3)| \\ +|\mathcal{M}(i_1,i_2,i_3) - \mathcal{M}(i_1,i_2,i_3-1)|\end{pmatrix}, \tag{5}$$

Let $\mathbf{D}(\mathcal{M}) = [\mathbf{D}_1(\mathcal{M}), \mathbf{D}_2(\mathcal{M}), \mathbf{D}_3(\mathcal{M})]$ denote the three-dimensional difference operator. Then TV is defined as the l_1-norm of the difference results, as follows:

$$\|\mathcal{M}\|_{TV} = \|\mathbf{D}(\mathcal{M})\|_1. \tag{6}$$

2.4 EPTR-TV Model and Its Solution

This paper defines a biological data processing model named EPTR-TV to explore the correlation between cancers and genes.

$$\min_{\mathcal{Z},\mathcal{L},\mathcal{D},\mathcal{H}} \|\mathcal{Z}\|_{\mathrm{EPSTNN}} + \lambda_1\|\mathcal{L}\|_1| + \lambda_2\|\mathcal{H}\|_1, s.t. \mathcal{Z} = \mathcal{D}, \mathcal{L} = \mathbf{D}(\mathcal{D}), \mathcal{G} = \mathcal{Z} + \mathcal{H}. \tag{7}$$

Through the alternating direction method of multipliers (ADMM), the above problem can be transformed into several subproblems for iterative solution. The solution process of EPTR-TV is summarized in Table 1.

Update $\mathcal{Z}$:

$$\mathcal{Z}^{k+1} = \|\mathcal{Z}\|_{\mathrm{EPSTNN}} + \frac{\mu^k}{2}\left\|\mathcal{Z} - (\mathcal{G} - \mathcal{H} + \mathcal{D} + (\mathcal{Y}_3 - \mathcal{Y}_1)/\mu^k)\right\|_F^2. \tag{8}$$

Update $\mathcal{D}$:

$$\begin{cases} \mathrm{H}_{\mathcal{D}} = \mathcal{Y}_1^k + \mathbf{D}^*(\mathcal{Y}_2^k) + \mu^k \mathcal{Z}^{k+1} + \mu^k \mathbf{D}^*(\mathcal{L}^k), \\ \mathcal{D}^{k+1} = \mathrm{ifft}(\mathrm{fft}(\mathrm{H}_{\mathcal{D}})/(\mu^k 1 + \mu^k(|\mathrm{fft}(\mathbf{D}_1)|^2 + |\mathrm{fft}(\mathbf{D}_2)|^2 + |\mathrm{fft}(\mathbf{D}_3)|^2))). \end{cases} \tag{9}$$

Update $\mathcal{L}$:

$$\mathcal{L}^{k+1} = \arg\min_{\mathcal{L}} \lambda_1 \|\mathcal{L}\|_1 + < \mathcal{Y}_2^k, \mathcal{L} - \mathbf{D}(\mathcal{D}^{k+1}) > + \frac{\mu^k}{2}\left\|\mathcal{L} - \mathbf{D}(\mathcal{D}^{k+1})\right\|_F^2. \tag{10}$$

Update $\mathcal{H}$:

$$\mathcal{H}^{k+1} = \arg\min_{\mathcal{H}} \lambda_2 \|\mathcal{H}\|_1 + \frac{\mu^k}{2}\left\|\mathcal{H} - (\mathcal{G} - \mathcal{Z}^{k+1} + \mathcal{Y}_3^k/\mu^k)\right\|_F^2. \tag{11}$$

Table 1. Solving EPTR-TV by ADMM.

Input: $\mathcal{G} \in \mathbb{R}^{n_1 \times n_2 \times n_3}$; **Output:** $\mathcal{D}$, $\mathcal{H}$

While not converged do

1. Update $\mathcal{Z}^{k+1}$ by (8)
2. Update $\mathcal{D}^{k+1}$ by (9)
3. Update $\mathcal{L}^{k+1}$ by (10)
4. Update $\mathcal{H}^{k+1}$ by (11)
5. Update $\mathcal{Y}_1^{k+1}$ by $\mathcal{Y}_1^{k+1} = \mathcal{Y}_1^k + \mu^k(\mathcal{Z}^{k+1} - \mathcal{D}^{k+1})$
6. Update $\mathcal{Y}_2^{k+1}$ by $\mathcal{Y}_2^{k+1} = \mathcal{Y}_2^k + \mu^k(\mathcal{L}^{k+1} - \mathbf{D}(\mathcal{D}^{k+1}))$
7. Update $\mathcal{Y}_3^{k+1}$ by $\mathcal{Y}_3^{k+1} = \mathcal{Y}_3^k + \mu^k(\mathcal{G} - \mathcal{Z}^{k+1} - \mathcal{H}^{k+1})$

End while

3 Results and Discussion

To test the performance of EPTR-TV, feature selection experiments and clustering experiments are designed on tensor datasets. In addition, we choose RPCA [12], TRPCA [2], T-TRPCA [11], OR-TPCA [5], IRTPCA [8], PSTNN [9] and EPSTNN (in this paper) as comparison algorithms.

3.1 Datasets

The original data are downloaded from the TCGA website [13]. Using twelve cancer data, pancreatic adenocarcinoma (PAAD), cholangiocarcinoma (CHOL), colon adenocarcinoma (COAD), esophageal carcinoma (ESCA), head and neck squamous cell carcinoma (HNSC), stomach adenocarcinoma (STAD), rectum adenocarcinoma (READ), liver hepatocellular carcinoma (LIHC), lung adenocarcinoma (LUAD), uterine corpus endometrial carcinoma (UCEC), cervical squamous cell carcinoma and endocervical adenocarcinoma (CESC), and uterine carcinosarcoma (UCS), we obtained three multi-cancer composite datasets, SRLL (the combination of STAD, READ, LIHC, and LUAD), PCCEH (the combination of PAAD, CHOL, COAD, ESCA, and HNSC), and UCU (the combination of UCEC, CESC, and UCS). Table 2 lists the detailed information of the three datasets. The three dimensions in the three-dimensional tensor dataset are genes, data types, samples.

Table 2. Descriptions of the datasets.

Datasets	Samples	Genes	Data types
SRLL	1276	17462	Gene expression, copy number variation, methylation
PCCEH	1055	16951	Gene expression, copy number variation, methylation
UCU	737	17243	Gene expression, copy number variation, methylation

3.2 Feature Selection and Analysis

Feature selection can reveal pathogenic gene associated with a variety of cancers and promote cancer research at the gene level. Each frontal slice is extracted from the differentially expressed tensor $\mathcal{H}$. Then we calculate the absolute value of frontal slice and sum the columns to get:

$$h_j = \sum_{i=1}^{n_3} \left|H_{ij}\right|, \hat{h} = \left(h_1, h_2 \cdots h_{n_1}\right), \tag{12}$$

We sort the DEG according to the level of gene relevance score (GRS). Then the top 500 genes are selected and searched on the GeneCards website [14] for gene details. Table 3 lists the genes with the highest GRS selected by the models in three datasets and the average values of the selected genes. HRN, HRS, and ARS represent the gene name with the highest GRS, the highest GRS, and the average GRS, respectively. On the three real datasets, EPTR-TV can select the most effective key gene, which shows that it has better data mining performance.

Table 4 lists the two genes with the highest GRS screened out by the EPTR-TV model in the datasets. Next, we conduct the case analysis of gene SMAD4 with the highest relevance score on the SRLL dataset and analyze its relationship with cancers in detail. SMAD4 is a tumor suppressor gene, and its germline mutation is an autosomal dominant disorder characterized by susceptibility to gastrointestinal cancers [15]. Recently, Jiang

Table 3. The results of extraction of DEG.

	SRLL			PCCEH			UCU		
	HRN	HRS	ARS	HRN	HRS	ARS	HRN	HRS	ARS
RPCA	KRT20	108.49	33.57	MAP2K2	61.3	23.65	TP63	138.08	92.99
TRPCA	KRT7	141.05	34.98	CXCR4	91.71	33.97	PGR	178.50	113.74
T-TRPCA	MSH2	178.88	36.13	CXCR4	91.71	37.85	BRAF	268.86	153.31
OR-TPCA	MSH2	178.88	34.75	NOTCH1	109.02	34.83	EP300	173.54	118.37
IRTPCA	HRAS	197.24	36.09	CDKN1A	119.73	38.70	CDKN2A	329.87	145.49
PSTNN	NRAS	192.27	37.89	CDKN1A	119.73	34.01	ERBB2	321.95	174.73
EPSTNN	CASP8	201.86	40.86	CDKN1B	125.95	**41.59**	KRAS	333.80	143.24
EPTR-TV	SMAD4	**211.86**	**42.39**	KRAS	**172.62**	36.00	EGFR	**378.98**	**183.20**

Table 4. The two genes with the highest GRS are selected in each dataset.

Dataset	Gene abbreviation	Gene official name	Related diseases
SRLL	SMAD4	SMAD Family Member 4	Endometrial cancer and Gastrulation
	MSH2	MutS Homolog 2	Lynch Syndrome I and Mismatch Repair Cancer Syndrome 2
PCCEH	KRAS	KRAS Proto-Oncogene, GTPase	colorectal cancer, non-small cell lung cancer, and others
	MET	MET Proto-Oncogene, Receptor Tyrosine Kinase	Renal Cell Carcinoma, Papillary, 1 and Deafness, Autosomal Recessive 97
UCU	EGFR	Epidermal Growth Factor Receptor	Inflammatory Skin And Bowel Disease, Neonatal, 2 and Lung Cancer
	MUC1	Mucin 1, Cell Surface Associated	Tubulointerstitial Kidney Disease, Autosomal Dominant, 2 and Syringoma

et al. conducted studies aimed at identifying new molecular markers to predict which locally advanced rectal cancers might be resistant to neoadjuvant chemoradiotherapy [16]. The results showed that SMAD4 mutations were associated with neoadjuvant chemoradiotherapy resistance in locally advanced rectal cancer. Yu *et al.* mentioned that SMAD4 could inhibit cell invasion in liver hepatocellular carcinoma [17]. Bian *et al.* explored the correlation between the protein changes and clinical results. The results showed that the immunohistochemical status of SMAD4 could predict the prognosis of patients with lung adenocarcinoma [18].

3.3 Clustering Results and Analysis

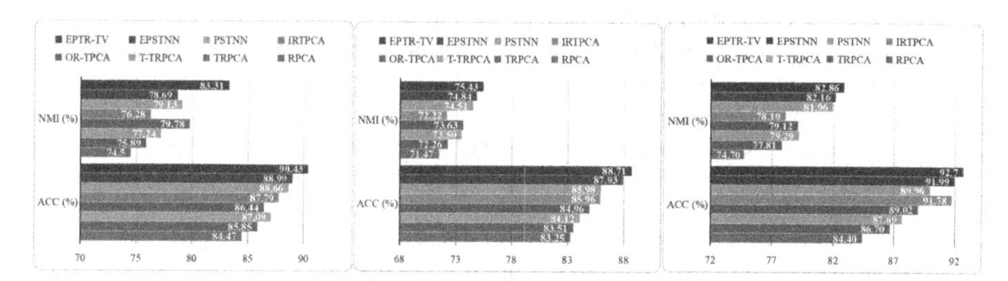

Fig. 1. Clustering results on SRLL dataset, PCCEH dataset and UCU dataset respectively.

Various evaluation indexes are used to judge the clustering performance of the models, including accuracy (ACC), and normalized mutual information (NMI). Figure 1 shows the results of the evaluation indexes of the above methods on the three datasets.

In Fig. 1, the matrix model RPCA is at least 1% lower than the tensor data analysis models. Overall, the experimental results of EPSTNN and method based on EPSTNN (EPTR-TV) are better than those of PSTNN and models based on TNN. Compared with other comparative models, the EPTR-TV model with TV can prevent the spatial structure of multi-view data from being destroyed. EPTR-TV can well present the important complementary information among cancers.

4 Conclusion

A new model called EPTR-TV is proposed to analyze multi-view cancer genomics data. In this model, we define EPSTNN. EPSTNN provides a greater degree of differentiation in the information carried by the internal structure of the data. It allows vital information to get more attention and makes the extraction of the low-rank component more accurate. Furthermore, we combine TV as the constraint term of the low-rank tensor decomposed from the original tensor. TV uses the correlation information within and between data structures, which can retain the local features of data. It can also play a role in increasing the accuracy of the extraction of the sparse component. The EPTR-TV model is compared with the comparison methods, and the experimental results show that the EPTR-TV model has great potential in the diagnosis and treatment of cancers.

It is worth noting that EPSTNN has a higher calculation cost. In the future, we will further explore the rapid algorithms based on tensor decomposition and apply them to the study of multi-view cancer genomics data. In addition, low-rank structure extraction is still the focus of our research.

Acknowledgments. This work was supported in part by the National Natural Science Foundation of China under Grant Nos. 61872220 and 62172254.

References

1. Liu, J.X., Xu, Y., Zheng, C.H., Kong, H., Lai, Z.H.: RPCA-based tumor classification using gene expression data. IEEE/ACM Trans. Comput. Biol. Bioinform. **12**(4), 1 (2014)
2. Lu, C., Feng, J., Chen, Y., Liu, W., Lin, Z., Yan, S.: Tensor robust principal component analysis: exact recovery of corrupted low-rank tensors via convex optimization. Comput. Vis. Pattern Recognit., 5249–5257 (2016)
3. Hu, Y., Liu, J.-X., Gao, Y.-L., Li, S.-J., Wang, J.: Differentially expressed genes extracted by the tensor robust principal component analysis (TRPCA) method. Complexity **2019**, 1–13 (2019). https://doi.org/10.1155/2019/6136245
4. Zhao, Y.Y., Jiao, C.N., Wang, M.L., Liu, J.X., Zheng, C.H.: HTRPCA: hypergraph regularized tensor robust principal component analysis for sample clustering in tumor omics data. Interdiscip. Sci. Comput. Life Sci. (6) (2021)
5. Zhou, P., Feng J.: Outlier-robust tensor PCA. Comput. Vis. Pattern Recognit., 3938–3946 (2017)
6. Kilmer, M.E., Martin, C.D.: Factorization strategies for third-order tensors. Linear Algebra Appl. **435**(3), 641–658 (2011). https://doi.org/10.1016/J.LAA.2010.09.020
7. Braman, K.: Third-order tensors as linear operators on a space of matrices. Linear Algebra Appl. **433**(7), 1241–1253 (2010). https://doi.org/10.1016/J.LAA.2010.05.025
8. Liu, Y., Chen, L., Zhu, C.: Improved robust tensor principal component analysis via low-rank core matrix. IEEE J. Sel. Top. Signal Process. **12**(6), 1378–1389 (2018). https://doi.org/10.1109/JSTSP.2018.2873142
9. Zhang, L., Peng, Z.: Infrared small target detection based on partial sum of the tensor nuclear norm. Remote Sens. **11**(4) (2019). https://doi.org/10.3390/RS11040382
10. He, W., Zhang, H., Zhang, L., Shen, H.: Total-variation-regularized low-rank matrix factorization for hyperspectral image restoration. IEEE Trans. Geosci. Remote Sens. **54**(1), 178–188 (2016). https://doi.org/10.1109/TGRS.2015.2452812
11. Lu, C., Feng, J., Chen, Y., Liu, W., Lin, Z., Yan, S.: Tensor robust principal component analysis with a new tensor nuclear norm. IEEE Trans. Pattern Anal. Mach. Intell. **42**(4), 925–938 (2020). https://doi.org/10.1109/TPAMI.2019.2891760
12. Oh, T.-H., Tai, Y.-W., Bazin, J.-C., Kim, H., Kweon, I.S.: Partial sum minimization of singular values in robust PCA: algorithm and applications. IEEE Trans. Pattern Anal. Mach. Intell. **38**(4), 744–758 (2016). https://doi.org/10.1109/TPAMI.2015.2465956
13. Tomczak, K., Czerwińska, P., Wiznerowicz, M.: Review The Cancer Genome Atlas (TCGA): an immeasurable source of knowledge. Contemp. Oncol. Współczesna Onkologia **2015**(1), 68–77 (2015)
14. Safran, M., Dalah I., Alexander, J., Rosen, N., Iny Stein, T., Shmoish, M., et al.: GeneCards version 3: the human gene integrator. Database **2010** (2010)
15. Howe, J.R., Shellnut, J., Wagner, B., Ringold, J.C., Sayed, M.G., Ahmed, A.F., et al.: Common deletion of SMAD4 in juvenile polyposis is a mutational hotspot. Am. J. Hum. Genet. **70**(5), 1357–1362 (2002)
16. Jiang, D., Wang, X., Wang, Y., Philips, D., Meng, W., Xiong, M., et al.: Mutation in BRAF and SMAD4 associated with resistance to neoadjuvant chemoradiation therapy in locally advanced rectal cancer. Virchows Arch. **475**(1), 39–47 (2019)
17. Yu, M., Lin, Y., Zhou, Y., Jin, H., Hou, B., Wu, Z., et al.: MiR-144 suppresses cell proliferation, migration, and invasion in hepatocellular carcinoma by targeting SMAD4. OncoTargets Ther. **9**, 4705 (2016)
18. Bian, C., Li, Z., Xu, Y., Wang, J., Xu, L., Shen, H.: Clinical outcome and expression of mutant P53, P16, and Smad4 in lung adenocarcinoma: a prospective study. World J. Surg. Oncol. **13**(1), 1–8 (2015)

Correction to: Simulating Spiking Neural Networks Based on SW26010pro

Zhichao Wang, Xuelei Li, Jintao Meng, Yi Pan, and Yanjie Wei

Correction to: Chapter "Simulating Spiking Neural Networks Based on SW26010pro" in: M. S. Bansal et al. (Eds.): *Bioinformatics Research and Applications*, LNBI 13760, https://doi.org/10.1007/978-3-031-23198-8_32

In the originally published version of chapter 32, one of the grant numbers in the acknowledgment section was stated incorrectly. This has been corrected.

The updated original version of this chapter can be found at
https://doi.org/10.1007/978-3-031-23198-8_32

M. S. Bansal et al. (Eds.): ISBRA 2022, LNBI 13760, p. C1, 2023.
https://doi.org/10.1007/978-3-031-23198-8_35

Author Index

Printed in the United States
by Baker & Taylor Publisher Services